The Impact of Probiotics/Prebiotics and Functional Foods in Human Health

The Impact of Probiotics/Prebiotics and Functional Foods in Human Health

Guest Editor

Nelson P. Guerra

MDPI

Basel • Beijing • Wuhan • Barcelona • Belgrade • Novi Sad • Cluj • Manchester

Guest Editor
Nelson P. Guerra
Department of Analytical and
Food Chemistry
University of Vigo
Ourense
Spain

Editorial Office
MDPI AG
Grosspeteranlage 5
4052 Basel, Switzerland

This is a reprint of the Special Issue, published open access by the journal *Nutrients* (ISSN 2072-6643), freely accessible at: https://www.mdpi.com/journal/nutrients/special_issues/44ODO4M2WJ.

For citation purposes, cite each article independently as indicated on the article page online and as indicated below:

Lastname, A.A.; Lastname, B.B. Article Title. *Journal Name* **Year**, *Volume Number*, Page Range.

ISBN 978-3-7258-6097-5 (Hbk)
ISBN 978-3-7258-6098-2 (PDF)
https://doi.org/10.3390/books978-3-7258-6098-2

Contents

About the Editor

Nelson P. Guerra

Nelson P. Guerra is a Full Professor of Biochemistry at the University of Vigo, Spain. He earned his degree in Chemical Engineering from the University of Havana (Cuba) in 1989 and obtained his PhD in Food and Technology Science from the University of Vigo in 1999. His research focuses on the valorization of food industry by-products and low-commercial-value fruits for the development of high-value-added products such as functional beverages, bacteriocins, and probiotics.

Dr. Pérez Guerra has participated in 38 research projects—5 of them as Principal Investigator—and 13 industry contracts. He has supervised eight doctoral theses and numerous master's and undergraduate projects. He is the author or co-author of more than 90 scientific publications, including 76 articles indexed in the Journal Citation Reports, with an h-index of 26 and over 1800 citations (Scopus).

He has been awarded four research-assessment periods (sexenios, 2000–2005, 2006–2011, 2012–2017, and 2018–2023) and one knowledge-transfer period (2009–2016), as well as four teaching-assessment periods (quinquenios, 2000–2006, 2006–2011, 2011–2016, and 2016–2021). In addition, he has received five excellence awards for research and teaching granted by the Xunta de Galicia, Spain.

Currently, Dr. Pérez Guerra is a member of the EQ11–BiotecnIA research group at the University of Vigo.

Preface

This Reprint brings together a tailored selection of recent contributions that explore the influence of probiotics, prebiotics and functional foods on human health. Its scope encompasses mechanistic investigations of microbiota–host interactions, translational studies of bioactive food components, and clinical data addressing emerging applications. The aim of this volume is to provide readers with a coherent and up-to-date overview of how dietary interventions leveraging microbial and functional food strategies can support immune, metabolic, cognitive and gastrointestinal systems. The purpose is twofold: first, to serve as a reference resource summarizing current findings in the field; and second, to stimulate new avenues of research by highlighting unresolved questions and promising trends. The motivation for compiling this Reprint arises from the accelerating pace of discovery in microbiome science, the growing interest in personalized nutrition approaches, and the need for an accessible synthesis of evidence bridging food science, nutrition and clinical practice. It is addressed to researchers, clinicians, food technologists and nutrition professionals seeking to deepen their understanding of how functional foods and microbial-based interventions may contribute to health promotion and disease prevention. We trust that this Reprint will facilitate interdisciplinary dialogue and foster the translation of emerging knowledge into actionable strategies.

Nelson P. Guerra
Guest Editor

nutrients

MDPI

Editorial

The Impact of Probiotics, Prebiotics, and Functional Foods on Human Health

Nelson Pérez Guerra

Department of Analytical and Food Chemistry, Faculty of Sciences, University of Vigo, 32004 Ourense, Spain; nelsonpg@uvigo.gal

The impact of probiotics, prebiotics, and functional foods on human health continues to be a dynamic and expanding area of research. Numerous studies have explored their potential to modulate gut microbiota, enhance immune function, and address various chronic conditions, highlighting the significant implications of nutritional science in health maintenance and disease prevention. Reflecting this growing interest, this Special Issue presents 15 papers contributed by researchers from diverse fields (including nutrition, biotechnology, biomedicine, clinical pharmacology, metabolic diseases, food science, microbiology, medicine, pediatrics, and neonatology) examining the potential applications of probiotics, prebiotics, and functional foods in promoting human health.

Furthermore, the articles featured in this Special Issue provide compelling evidence that combinations of prebiotics and probiotics, as well as metabolites produced by probiotic strains, can play a valuable role in the management and treatment of various human diseases. For example, the combination of inulin with resistant dextrin (another prebiotic compound) significantly reduced gas production, making it a more tolerable option, particularly for older adults, who are more susceptible to gut microbiota imbalances. Notably, this blend not only enhanced microbial diversity, but also tended to lower the pH in the fermentation medium, indicating a more favorable gut environment (contribution 1).

Thua-Nao, a traditional fermented soybean product from Thailand, demonstrated a capacity to suppress preneoplastic lesions in the liver and colon of rats exposed to carcinogens. The protective effects were associated with both modulation of the gut microbiota and the presence of bioactive compounds such as isoflavones, including genistein and daidzein. These findings suggest the potential of Thua-Nao as a dietary intervention for cancer prevention (contribution 2).

Probiotic supplementation also continues to show promise. A randomized, double-blind trial on Lactobacillus paragasseri OLL2716 revealed that daily consumption of yogurt containing this probiotic significantly alleviated epigastric pain and improved gastrointestinal symptoms. These results support the therapeutic role of probiotics in managing common digestive complaints, such as postprandial fullness and early satiety (contribution 3).

Fruit by-products, such as fiber-rich fractions from persimmon peels and pomace, are also emerging as valuable functional food sources. Rich in phenolic compounds, these fractions promoted the growth of beneficial bacteria like Faecalibacterium prausnitzii, and exhibited anti-inflammatory properties. This study highlights both the potential for food waste valorization, and the role of dietary fibers in supporting both gut health and systemic inflammation reduction (contribution 4).

A combination of Bifidobacterium animalis subsp. lactis GCL2505 with inulin significantly increased resting energy expenditure in overweight individuals. This suggests that modifying the gut microbiota through targeted probiotic and prebiotic intake could enhance metabolic efficiency and support weight management, marking a notable development in the understanding of gut microbiota's influence on human metabolism (contribution 5).

Sleep deprivation is known to negatively impact gut health, yet pre-colonization with Faecalibacterium prausnitzii was found to protect against intestinal barrier dysfunction in sleep-deprived mice. The treatment reduced inflammation, increased butyrate production, and restored gut microbiota balance. This provides promising evidence for the use of probiotics to mitigate the gastrointestinal effects of poor sleep (contribution 6).

Another compelling study examined the influence of Lacticaseibacillus paracasei strain Shirota on mental performance. Office workers who consumed this probiotic via fermented milk experienced improvements in attention and stress-related physiological markers, such as heart rate variability and theta power on electroencephalogram. These findings reinforce the emerging role of the gut–brain axis in cognitive health and workplace performance (contribution 7).

Further evidence from studies involving GCL2505 and inulin points to their efficacy in reducing visceral and total fat area, as well as in increasing bifidobacteria counts in overweight participants. This supports the hypothesis that gut microbiota modulation can influence fat metabolism and body composition (contribution 8). Additionally, this same combination improved cognitive function, particularly in attention and executive function domains, likely through gut microbiota-mediated reductions in inflammation, offering a novel approach to combat cognitive decline (contribution 9).

Brassica juncea, a glucosinolate-rich plant, has shown potential in addressing metabolic conditions such as obesity and liver disease. Treatments with whole-plant extracts or isolated glucosinolates ameliorated hepatic steatosis and liver injury in high-fat-diet-fed rats. These results suggest that bioactive plant compounds may serve as natural therapeutic agents for non-alcoholic steatohepatitis and related metabolic disorders (contribution 10).

Data from the U.S. National Health and Nutrition Examination Survey (NHANES, 2011–2014) revealed that nonfood probiotics and prebiotics may protect against cognitive decline, particularly in older men. Supplement users exhibited significantly better cognitive scores, suggesting that even non-dietary supplementation may contribute meaningfully to brain health in aging populations (contribution 11).

A preclinical study using a chronic kidney disease (CKD) model demonstrated that resistant maltodextrin and chitosan oligosaccharide supplementation supported gut barrier integrity and promoted beneficial bacteria such as Lactobacillus, Bifidobacteria, Akkermansia, and Roseburia in rats with chronic kidney disease. This intervention may offer a novel strategy to mitigate gut-derived inflammation and complications in CKD (contribution 12).

Comprehensive reviews further emphasize the diversity and specificity of prebiotic and probiotic actions. For instance, inulin and fructo-oligosaccharides (FOS) were shown to prevent microbial dysbiosis, strengthen intestinal barrier function, and modulate immune responses, highlighting the importance of prebiotic variety in maintaining health (contribution 13). In pediatric populations, bioactive compounds derived from probiotics (such as short-chain fatty acids, bacteriocins, exopolysaccharides, vitamins, and gamma-aminobutyric acid) have demonstrated positive effects on immune function, neurodevelopment, and protection against gastrointestinal disorders. These findings reinforce the value of integrating probiotics and postbiotics into child health strategies (contribution 14) [1]. Finally, early-life probiotic intake has shown promise in modulating immune tolerance to food antigens, suggesting a role in both the prevention and treatment of food allergies in infants and children. While mechanisms require further investigation, these studies point to the potential of probiotics in managing allergic diseases (contribution 15) [2].

The ingestion of probiotics, prebiotics, or synbiotics in older adults has shown promising results, particularly in improving gut microbiota composition and promoting healthy aging (contributions 1, 3, 5, 9) [3,4]. However, there is conflicting evidence regarding their effects on humoral immunity markers, immune cell population levels and activity, incidence and duration of infectious diseases [3], as well as cognition (contribution 11), physical function, frailty, mood, length of hospitalization, and mortality [5]. Therefore, more rigorous and high-quality research is needed to clarify their broader role in healthy aging and to identify the most effective probiotic combinations [3–5].

In summary, this collection of research offers diverse insights into how prebiotics, probiotics, and functional foods may positively influence health outcomes across a wide range of populations and conditions. From metabolic and gastrointestinal health to cognitive function and cancer prevention, the evidence underscores the crucial role of gut microbiota modulation in contemporary nutritional and medical science.

However, as some studies have been conducted using animal models (contributions 2, 6, 10, 12), extrapolating these findings to human health should be approached with great caution, as the results have not yet been validated in humans.

Twenty-eight manuscripts were submitted for consideration for the Special Issue, and all of them were subject to a rigorous review process. In total, fifteen papers were finally accepted for publication and inclusion in this Special Issue (twelve articles and three reviews). The contributions are listed below:

1. Yoshida, K.; Kokubo, E.; Morita, S.; Sonoki, H.; Miyaji, K. Combination of Inulin and Resistant Dextrin Has Superior Prebiotic Effects and Reduces Gas Production During In Vitro Fermentation of Fecal Samples from Older People. *Nutrients* **2024**, *16*, 4262.
2. Taya, S.; Dissook, S.; Ruangsuriya, J.; Yodkeeree, S.; Boonyapranai, K.; Chewonarin, T.; Wongpoomchai, R. Thai Fermented Soybean (Thua-Nao) Prevents Early Stages of Colorectal Carcinogenesis Induced by Diethylnitrosamine and 1,2-Dimethylhydrazine Through Modulations of Cell Proliferation and Gut Microbiota in Rats. *Nutrients* **2024**, *16*, 3506
3. Yamada, N.; Kobayashi, K.; Nagira, A.; Toshimitsu, T.; Sato, A.; Kano, H.; Hojo, K. The Beneficial Effects of Regular Intake of *Lactobacillus paragasseri* OLL2716 on Gastric Discomfort in Healthy Adults: A Randomized, Double Blind, Placebo Controlled Study. *Nutrients* **2024**, *16*, 3188.
4. López-Bermudo, L.; Moreno-Chamba, B.; Salazar-Bermeo, J.; Hayward, N.; Morris, A.; Duncan, G.; Russell, W.; Cárdenas, A.; Ortega, Á.; Escudero-López, B.; Berná, G.; Martí Bruña, N.; Duncan, S.; Neacsu, M.; Martin, F. Persimmon Fiber-Rich Ingredients Promote Anti-Inflammatory Responses and the Growth of Beneficial Anti-Inflammatory Firmicutes Species from the Human Colon. *Nutrients* **2024**, *16*, 2518.
5. Baba, Y.; Azuma, N.; Saito, Y.; Takahashi, K.; Matsui, R.; Takara, T. Effect of Intake of Bifidobacteria and Dietary Fiber on Resting Energy Expenditure: A Randomized, Placebo-Controlled, Double-Blind, Parallel-Group Comparison Study. *Nutrients* **2024**, *16*, 2345.
6. Wang, X.; Li, Y.; Wang, X.; Wang, R.; Hao, Y.; Ren, F.; Wang, P.; Fang, B. Faecalibacterium prausnitzii Supplementation Prevents Intestinal Barrier Injury and Gut Microflora Dysbiosis Induced by Sleep Deprivation. *Nutrients* **2024**, *16*, 1100.
7. Kikuchi-Hayakawa, H.; Ishikawa, H.; Suda, K.; Gondo, Y.; Hirasawa, G.; Nakamura, H.; Takada, M.; Kawai, M.; Matsuda, K. Effects of *Lacticaseibacillus paracasei* Strain Shirota on Daytime Performance in Healthy Office Workers: A Double-Blind, Randomized, Crossover, Placebo-Controlled Trial. *Nutrients* **2023**, *15*, 5119.

8. Baba, Y.; Saito, Y.; Kadowaki, M.; Azuma, N.; Tsuge, D. Effect of Continuous Ingestion of Bifidobacteria and Inulin on Reducing Body Fat: A Randomized, Double-Blind, Placebo-Controlled, Parallel-Group Comparison Study. *Nutrients* **2023**, *15*, 5025.
9. Azuma, N.; Mawatari, T.; Saito, Y.; Tsukamoto, M.; Sampei, M.; Iwama, Y. Effect of Continuous Ingestion of Bifidobacteria and Dietary Fiber on Improvement in Cognitive Function: A Randomized, Double-Blind, Placebo-Controlled Trial. *Nutrients* **2023**, *15*, 4175.
10. Sheu, M.; Yeh, M.; Tsai, M.; Wang, C.; Chang, Y.; Wang, C.; Huang, H. Glucosinolates Extracts from *Brassica juncea* Ameliorate HFD-Induced Non-Alcoholic Steatohepatitis. *Nutrients* **2023**, *15*, 3497.
11. Chen, J.; Yang, N.; Peng, Y.; Zhou, H.; Li, Q. Association between Nonfood Pre- or Probiotic Use and Cognitive Function: Results from NHANES 2011–2014. *Nutrients* **2023**, *15*, 3408.
12. Anegkamol, W.; Kamkang, P.; Hunthai, S.; Kaewwongse, M.; Taweevisit, M.; Chuaypen, N.; Rattanachaisit, P.; Dissayabutra, T. The Usefulness of Resistant Maltodextrin and Chitosan Oligosaccharide in Management of Gut Leakage and Microbiota in Chronic Kidney Disease. *Nutrients* **2023**, *15*, 3363.
13. Pedrosa, L.; de Vos, P.; Fabi, J. From Structure to Function: How Prebiotic Diversity Shapes Gut Integrity and Immune Balance. *Nutrients* **2024**, *16*, 4286.
14. Guamán, L.; Carrera-Pacheco, S.; Zúñiga-Miranda, J.; Teran, E.; Erazo, C.; Barba-Ostria, C. The Impact of Bioactive Molecules from Probiotics on Child Health: A Comprehensive Review. *Nutrients* **2024**, *16*, 3706.
15. Di Costanzo, M.; Vella, A.; Infantino, C.; Morini, R.; Bruni, S.; Esposito, S.; Biasucci, G. Probiotics in Infancy and Childhood for Food Allergy Prevention and Treatment. *Nutrients* **2024**, *16*, 297.

Funding: This research received no external funding.

Acknowledgments: The editor would like to express his sincere gratitude to all the authors of the manuscripts published in this Special Issue for their valuable contributions and professionalism, which were essential in shaping this engaging topic. Their work highlights the effectiveness of probiotics, prebiotics, and functional foods in promoting health and well-being.

Conflicts of Interest: The author declares no conflicts of interest.

References

1. Guamán, L.; Carrera-Pacheco, S.; Zúñiga-Miranda, J.; Teran, E.; Erazo, C.; Barba-Ostria, C. The Impact of Bioactive Molecules from Probiotics on Child Health: A Comprehensive Review. *Nutrients* **2024**, *16*, 3706. [CrossRef] [PubMed]
2. Di Costanzo, M.; Vella, A.; Infantino, C.; Morini, R.; Bruni, S.; Esposito, S.; Biasucci, G. Probiotics in Infancy and Childhood for Food Allergy Prevention and Treatment. *Nutrients* **2024**, *16*, 297. [CrossRef] [PubMed]
3. Hutchinson, A.N.; Bergh, C.; Kruger, K.; Sũsserová, M.; Allen, J.; Améen, S.; Tingö, L. The Effect of Probiotics on Health Outcomes in the Elderly: A Systematic Review of Randomized, Placebo-Controlled Studies. *Microorganisms* **2021**, *9*, 1344. [CrossRef] [PubMed]
4. Chenhuichen, C.; Cabello-Olmo, M.; Barajas, M.; Izquierdo, M.; Ramírez-Vélez, R.; Zambom-Ferraresi, F.; Martínez-Velilla, N. Impact of Probiotics and Prebiotics in the Modulation of the Major Events of the Aging Process: A Systematic Review of Randomized Controlled Trials. *Exp. Gerontol.* **2022**, *164*, 111809. [CrossRef] [PubMed]
5. Coutts, L.; Ibrahim, K.; Tan, Q.Y.; Lim, S.E.R.; Cox, N.J.; Roberts, H.C. Can Probiotics, Prebiotics and Synbiotics improve Functional Outcomes for Older People: A Systematic Review. *Eur. Geriatr. Med.* **2020**, *11*, 975–993. [CrossRef] [PubMed]

nutrients

MDPI

Article

Combination of Inulin and Resistant Dextrin Has Superior Prebiotic Effects and Reduces Gas Production During In Vitro Fermentation of Fecal Samples from Older People

Kazuma Yoshida *, Eri Kokubo, Shunsuke Morita, Hirofumi Sonoki and Kazuhiro Miyaji

Health Care & Nutritional Science Institute, R&D Division, Morinaga Milk Industry Co., Ltd., 5-1-83, Higashihara, Zama 252-8583, Kanagawa, Japan; e-karita@morinagamilk.co.jp (E.K.); s-morita@morinagamilk.co.jp (S.M.); h-sonoki@morinagamilk.co.jp (H.S.); k_miyazi@morinagamilk.co.jp (K.M.)

* Correspondence: kazuma-yoshida693@morinagamilk.co.jp; Tel.: +81-46-252-3057

Abstract: Background: Older people are more susceptible to deterioration of the gut microbiota. Prebiotics help improve the gut microbiota. Inulin, a major prebiotic, stimulates the growth of *Bifidobacterium*; however, it produces a large amount of gas, which leads to abdominal symptoms. Methods: In this study, in vitro fecal fermentation was performed using fecal samples from seven older people (mean subject age, 73.4 years; five men and two women) to examine whether combining inulin with another prebiotic material, resistant dextrin, could lead to decreased gas production and show prebiotic effects. Results: The *Bifidobacterium* counts and short-chain fatty acid production did not differ significantly between the inulin 0.5% group and the inulin 0.25% plus resistant dextrin 0.25% combination group. However, the inulin 0.25% plus resistant dextrin 0.25% combination group had lower gas production than the inulin 0.5% group ($p < 0.10$). Furthermore, compared with the inulin 0.5% group, the 0.25% combination group showed significantly greater gut microbiota diversity and tended toward a lower pH in the fermentation medium at the end of fermentation ($p = 0.09$). These effects are believed to be due to the combination of inulin, which is highly selective for *Bifidobacterium* and rapidly utilized by the gut microbiota, and resistant dextrin, which is slowly utilized by various bacterial genera. Conclusions: These findings suggest that the inulin plus resistant dextrin combination has superior prebiotic effects in older people and causes less gas production than inulin alone.

Keywords: gut microbiota; prebiotics; inulin; resistant dextrin; gas production; older people; fecal fermentation

1. Introduction

The human gut microbiota deteriorates with aging, leading to a decrease in the abundance of beneficial bacteria, including the *Bifidobacterium* genus [1,2]. Gut microbiota deterioration has been found to impair intestinal function and the immune system [3,4]; additionally, it is associated with diseases such as diabetes and kidney disease [5]. Therefore, the improvement of the gut microbiota is particularly crucial for older individuals.

A prebiotic is defined as "a substrate that is selectively utilized by host microorganisms conferring a health benefit"; prebiotics include nondigestible carbohydrates, such as fibers and oligosaccharides, that help improve the gut microbiota [6]. The ability of the gut microbiota to access non-digestible carbohydrates varies depending on several factors, including their chain length, molecular weight, and chemical structure [7]. In general, oligosaccharides are rapidly utilized by the gut microbiota owing to their short chain lengths and low molecular weights. In contrast, soluble dietary fiber, which has a long chain length and high molecular weight, is slowly utilized by the gut microbiota.

Inulin, one of the major prebiotics, is a type of fructan composed of β-(2, 1) fructosyl-fructose bonds. It has a linear structure. Although inulin is a soluble dietary fiber, its structure is similar to that of oligosaccharides, and it can be easily utilized by the gut microbiota [8]. Inulin is highly selective for *Bifidobacterium* and reportedly exhibits prebiotic effects, including promoting *Bifidobacterium* growth (bifidogenic effect) and short-chain fatty acid (SCFA) production [9,10]. However, inulin is rapidly utilized in large quantities owing to its easy utilization by the gut microbiota, leading to significant gas production in the intestine, which can cause abdominal symptoms such as bloating [11,12]. In addition, inulin cannot exert a prebiotic effect in the distal colon as it is primarily utilized in the proximal colon and does not reach the distal colon [13].

High gas production from inulin can be mitigated by combining it with other prebiotics. For example, studies on in vitro human fecal fermentation have demonstrated that combining inulin with acacia gum, which is slowly utilized by the gut microbiota, leads to bifidogenic effects and decreased gas production [13]. Combining other slowly utilized prebiotic materials with inulin may yield similar effects. However, the utilization of prebiotic materials by the gut microbiota depends on their composition and structure. Therefore, it is necessary to determine if combining inulin with these materials can produce similar effects.

Resistant dextrin is a soluble fiber synthesized from starch found in wheat and maize. Similar to normal dextrin, resistant dextrin is characterized by its 1→4 glucoside linkages. In addition to its 1→6 linkages, resistant dextrin also has several branched structures, such as 1→2 and 1→3 linkages [14]. Resistant dextrin slows the rise in blood glucose and triglyceride levels after meals, and it is widely used in health-food products in Japan [15,16]. Resistant dextrin is utilized more slowly and produces less gas than inulin [17]. Therefore, a resistant dextrin plus inulin combination may exert prebiotic effects and cause lesser gas production than inulin alone. This could be used to develop health-food products for older people. However, this aspect has not yet been investigated in detail.

In vitro fecal fermentation assays are conducted to study the effects of prebiotics on the stimulation of the growth of beneficial bacteria and the production of useful metabolites [7]. This approach helps verify the prebiotic effects by measuring both the microbiota and metabolite levels in the fermentation medium. Additionally, when prebiotics are utilized, useful metabolites, such as SCFAs, are produced and the pH of the medium decreases, thus mimicking the conditions in the human intestinal tract [18]. Therefore, the prebiotic utilization rate can be determined by measuring the fermentation medium pH [13]. Furthermore, the amount of gas produced by the gut microbiota can be measured by examining the gas production in the fermentation vessel [17,19]. Although examining these effects in human studies is challenging, they can be easily verified using in vitro fecal fermentation.

The selection of individuals is crucial in fecal fermentation studies because gut microbiota characteristics can vary among individuals. The response to prebiotics has been found to differ between older and younger individuals because of age-related changes in the gut microbiota [20,21]. Studies comparing the effects of prebiotics on older people mainly focus on single prebiotic materials [22–24]. To the best of our knowledge, no studies have compared the gas production and prebiotic effects of prebiotic combinations with those of each material alone in older people. Therefore, in this study, we specifically used fecal samples from older individuals to investigate the effects of prebiotics in this age group.

The objective of this study was to examine whether using a combination of inulin and resistant dextrin would provide prebiotic effects, such as bifidogenic effects and increased SCFA production, and whether this combination would cause lesser gas production than inulin alone. Therefore, we performed an in vitro fecal fermentation study with fecal samples from older people.

2. Materials and Methods

2.1. Sample Collection

We collected fecal samples from healthy older participants in Japan. Participants who were suspected of having an infectious bowel disease were excluded to prevent infection during fermentation. Participants were recruited through a clinical trial recruitment site https://www.seikatsu-kojo.jp/ (accessed on 8 December 2024). To recruit participants, the exam details were posted on the recruiting site by the agency operating this site. Thirteen participants were recruited from the pool of healthy older Japanese individuals who were registered on the recruiting site. The participants collected the fecal samples in small stool tubes and stored them under anaerobic conditions using an AnaeroPouch (Mitsubishi Gas Chemical, Tokyo, Japan), which was then kept in the household freezer. To limit the burden on the participants, we requested that each participant provide only one fecal sample. Within 48 h of collection, the frozen fecal samples were shipped with dry ice to the Morinaga Milk Industry laboratory and stored at −80 °C.

Five participants were excluded from the study due to improper sample collection methods, such as the failure to maintain anaerobic conditions. Additionally, one sample did not show *Bifidobacterium* and was therefore not used in this study, which focused on *Bifidobacterium* growth. Finally, seven samples were used for fermentation (mean subject age, 73.4 years; age range, 72–75 years; five men and two women). Written informed consents were obtained from the participants before the start of the study. The study protocol was approved by the Ethics Committee of the Japan Conference of Clinical Research (Tokyo, Japan; protocol code: NFSFA-01; date of approval: 18 October 2019).

2.2. In Vitro Fecal Fermentation

Fecal fermentation was performed using a pH-controlled multichannel jar fermenter (Bio Jr. 8; ABLE, Tokyo, Japan), following the methods described in a previous study [22]. A carbon source was added to yeast extract, casitone, and fatty acid (YCFA) medium (Supplementary Table S1), and 100 mL of medium was filled in each vessel. The fermentation process was performed under anaerobic conditions (100% CO_2) at a temperature of 37 °C, with constant stirring at 200 rpm and the pH maintained above 6.0. Inulin (Fuji FF; Fuji Nihon Seito Co., Tokyo, Japan), resistant dextrin (Fibersol-2; Matsutani Chemical Industry Co., Ltd., Hyogo, Japan), and a mixture of both were used as carbon sources. In the single carbon source groups, inulin (Inu0.5% group) or resistant dextrin (RD0.5% group) was added at a concentration of 0.5%. In the combination groups, a mixture containing 0.25% of each carbon source (Inu0.25%RD0.25% group) or 0.5% of each carbon source (Inu0.5%RD0.5% group) was used.

The frozen stored fecal samples were thawed and diluted with saline to a concentration of 10%. A volume of 100 μL of diluted fecal samples was added to each vessel. A 100 μL sample of the fecal dilution was used for bacterial community analysis. After 4 h of fermentation, the CO_2 supply was stopped, and a gas collection bag (Laboratory for Expiration Biochemistry Nourishment Metabolism Co., Ltd., Nara, Japan) was connected to each vessel. The exhaust port was sealed with a clip, and the vessel was sealed for gas measurement. The dissolved oxygen concentration was monitored during fermentation to ensure that anaerobic conditions were maintained. If the dissolved oxygen concentration exceeded 0.5%, indicating a failure to maintain anaerobic conditions, fecal fermentation was repeated. The medium's pH was measured every 30 min for 24 h; the pH value at 23.5 h was used as the 24 h value for data analysis. After 24 h of fermentation, the gas collection bags were removed, and the medium was collected for SCFA and gut microbiota analyses. The amount of gas accumulated in the bag was measured using a syringe. Centrifugation of the diluted fecal sample remainder and the medium after 24 h of fermentation was performed at 8000× g for 3 min at 4 °C. The sediment was collected for bacterial analysis, and the supernatant was used for SCFA measurements and thin-layer chromatography (TLC). Fecal fermentation was performed once for each fecal sample.

2.3. Microbiota Analysis

The bead-beating method was used to extract the total DNA. Following extraction, the V3–V4 region of the bacterial 16S rRNA gene was amplified and sequenced according to a previously described method [25].

2.4. RT-PCR

RT-PCR was performed using a QuantStudio3® RT-PCR system (Thermo Fisher Scientific, Waltham, MA, USA) and TB Green® Premix Ex Taq™ Tli RNaseH Plus (TaKaRa Bio Inc., Shiga, Japan). The primers F: 5′-CTCCTGGAAACGGGTGG-3′ and R: 5′-GGTGTTCTTCCCGATATCTACA-3′ were used to determine the total *Bifidobacterium* count, as previously described [26]. To determine the total bacterial count, the primers F: 5′-CCTACGGGRSGCAGCAG-3′ and R: 5′-ATTACCGCGGCTGCT-3′ were used, as previously described [27]. The amplification process was conducted as follows: it commenced with an initial hold at 95 °C for 10 s, followed by 40 cycles each at 95 °C for 3 s, 55 °C for 20 s, and 72 °C for 30 s. Fluorescent products were detected at the end of each cycle, and melting curves were generated from 60 °C to 95 °C, with increments of 0.1 °C/s. All the samples were evaluated in duplicate. The detection limit was set at 10^5 copies/mL.

2.5. SCFA Analysis

SCFAs (acetic acid, propionic acid, and butyric acid) were analyzed via liquid chromatography performed using a Shimadzu Organic Acid Analysis System (Shimadzu Co., Kyoto, Japan). The culture supernatants were diluted 10 times and filtered through a 0.22 μm membrane filter (TORAST disc nylon membrane; Shimadzu Co.) before analysis. The HPLC system consisted of three Shim-pack Fast-OA columns (size, 100 mm × 7.8 mm internal diameter [ID]) connected in series, along with a Shim-pack Fast-OA guard column (size, 10 mm × 4.0 mm ID). The eluent used was 5 mmol/L *p*-toluenesulfonic acid, and the reaction solution comprised 5 mmol/L *p*-toluenesulfonic acid, 100 μmol/L ethylenediaminetetraacetic acid, and 20 mmol/L Bis-Tris (Shimadzu Co.). The flow rate and oven temperature were set at 0.8 mL/min and 50 °C, respectively. The conductivity detector CDD-10AVP (Shimadzu) was used for the measurements.

2.6. TLC Analysis

TLC was performed to measure the resistant dextrin and inulin residues remaining in the medium's supernatant at the start of incubation and end of fermentation, following a previously described method [25,28]. The samples were spotted onto precoated silica gel 60 TLC aluminum plates (Merck, Darmstadt, Germany) and developed using a solvent composed of ethyl acetate, acetic acid, 2-propanol, formic acid, and water in a ratio of 25:10:5:1:15. The plates were then treated with a reagent comprising diphenylamine, aniline, and phosphoric acid. This reagent was prepared by blending 100 mL acetone (FUJIFILM Wako Pure Chemical Corp., Osaka, Japan), 1 g diphenylamine (Nacalai Tesque Inc., Kyoto, Japan), 1 mL aniline (FUJIFILM Wako Pure Chemical Corp.), and 10 mL phosphoric acid (Hayashi Pure Chemical Industry, Ltd., Osaka, Japan). After development, the samples were visualized by heating in an oven.

2.7. Statistical Analysis

Statistical analysis was performed using JMP version 13.2.1 (SAS Institute, Cary, NC, USA). The gas production and the Shannon index were compared using the Steel–Dwass test. The bacterial counts, SCFAs, and pH were compared using the Tukey–Kramer multiple comparison test. The correlation between the gas production and the bacterial population was calculated using Spearman's rank correlation coefficient. Statistical significance was set at $p < 0.05$.

3. Results

3.1. Gas Production

The gas production in the Inu0.5% group and Inu0.5%RD0.5% combination group was significantly higher than that in the RD0.5% group ($p < 0.05$, Figure 1). Additionally, the gas production in the Inu0.25%RD0.25% combination group was lower than that in the Inu0.5% group ($p < 0.10$).

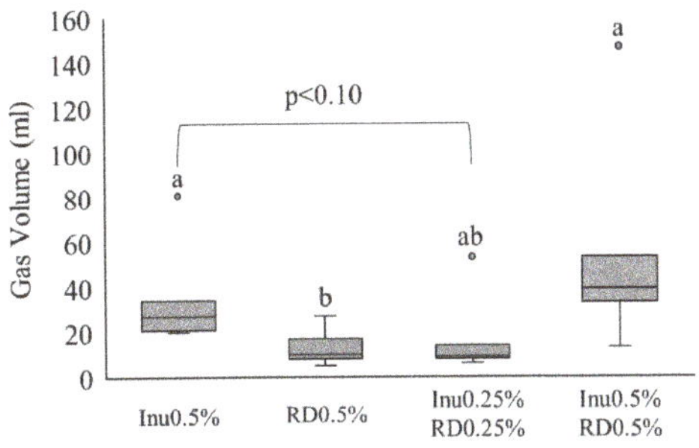

Figure 1. Gas production during 4–24 h of fermentation. Different letters indicate significant differences ($p < 0.05$, Steel–Dwass test).

3.2. Microbiota Composition

The bacterial composition differed among the groups (Figure 2A). The proportion of *Bifidobacterium* was highest in the Inu0.5% group (26.1%), followed by that in the Inu0.5%RD0.5% combination group (17.7%), the Inu 0.25%RD0.25% combination group (15.1%), and the RD0.5% group (5.8%). Additionally, the *Prevotella* genus proportion was higher in the combination groups than in the prebiotic alone groups. The microbiota diversity (α-diversity: Shannon index) was the lowest in the Inu0.5% group; the diversity in the Inu0.25%RD0.25% combination group was significantly higher than that in the Inu0.5% group ($p < 0.05$, Figure 2B).

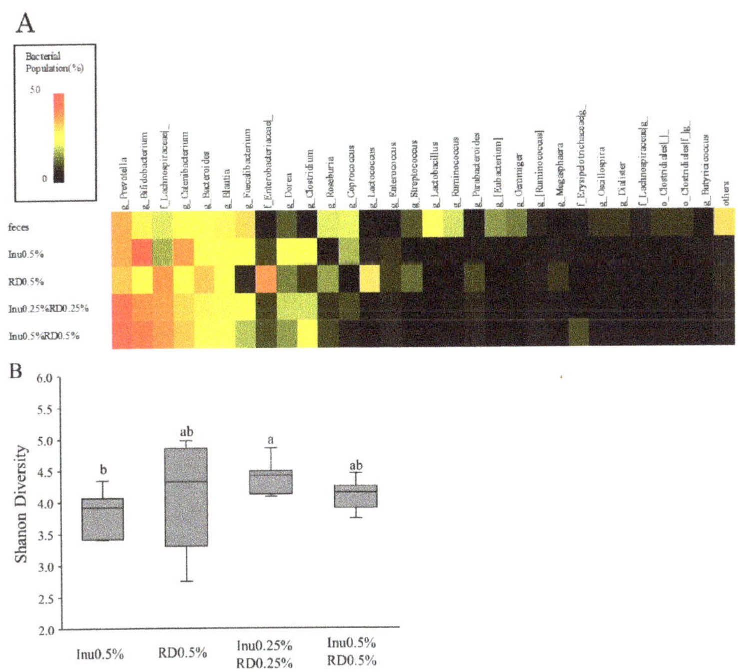

Figure 2. Bacterial population and microbiota diversity after 24 h of fermentation. (**A**) Top 30 predominant bacterial genera in the feces and fecal fermentation samples. Colors indicate the percentage of each bacterial population. (**B**) Box plot showing the α-diversity measured using the Shannon index after 24 h of fermentation. Different letters indicate significant differences ($p < 0.05$, Steel–Dwass test).

The correlation between the bacterial genera and the gas production was examined to investigate the bacteria involved in gas production. A significant correlation was observed only for the *Clostridium* genus (R = 0.58). However, this genus was detected in only one sample, and no significant correlation was observed with the other bacterial genera (Supplementary Table S2A,B).

3.3. *Bifidobacterium and Total Bacterial Counts*

The *Bifidobacterium* and total bacterial counts were determined using RT-PCR. In all the inulin-containing groups (Inu0.5%, Inu0.25%RD0.25%, and Inu0.5%RD0.5%), the *Bifidobacterium* count was significantly higher than that in the RD0.5% group (Table 1). The total bacterial count was the highest in the Inu0.5%RD0.5% combination group, followed by the Inu0.5% group and the Inu0.25%RD0.25% combination group, which had almost equal counts, and the RD0.5% group. Notably, the total bacterial count was significantly higher in the Inu0.5%RD0.5% group than in the RD0.5% group.

Table 1. *Bifidobacterium* and total bacterial cell counts after 24 h in the fecal fermentation sample, determined using RT-PCR.

	Total_Bifidbacterium		Total_Bacteria	
	Cell Number (/mL)		Cell Number (/mL)	
	Mean	SD	Mean	SD
Inu0.5%	9.64 [a]	0.26	10.06 [ab]	0.16
RD0.5%	8.44 [b]	0.33	9.86 [b]	0.41
Inu0.25%RD0.25%	9.45 [a]	0.22	10.05 [ab]	0.17
Inu0.5%RD0.5%	9.69 [a]	0.27	10.26 [a]	0.17

Different letters indicate significant differences ($p < 0.05$, Tukey–Kramer multiple comparison test).

3.4. *SCFA Production*

The acetate level in all the inulin-containing groups was significantly higher than that in the RD0.5% group (Table 2). The propionate and butyrate levels did not significantly differ between the groups, but all the inulin-containing groups had higher levels than the RD0.5% group. Additionally, the total acetate, propionate, and butyrate levels in all the inulin groups were significantly higher than that in the RD0.5% group; furthermore, the Inu0.5%RD0.5% combination group showed a significantly higher total level than the Inu0.25%RD0.25% combination group.

Table 2. Short-chain fatty acids in the medium after 24 h of fecal fermentation, measured using HPLC.

	Acetic Acid (mM)		Propionic Acid (mM)		Butyric Acid (mM)		Acetic + Propionic + Butyric Acid (mM)	
	Mean	SD	Mean	SD	Mean	SD	Mean	SD
Inu0.5%	87.3 [ab]	7.5	16.1	6.6	6.0	4.5	109.3 [ab]	15.2
RD0.5%	58.6 [c]	14.4	12.4	3.9	2.9	3.6	73.9 [c]	18.6
Inu0.25% RD0.25%	77.7 [b]	5.5	14.7	6.0	5.5	2.9	97.9 [b]	10.8
Inu0.5% RD0.5%	100.2 [a]	6.8	14.3	8.6	5.6	4.8	120.0 [a]	12.0

Different letters indicate significant differences ($p < 0.05$, Tukey–Kramer multiple comparison test).

3.5. *pH*

The medium's pH in the Inu0.5% group increased after 14 h of fermentation (Figure 3). After 24 h of fermentation, it was higher ($p = 0.09$) than that in the Inu0.25%RD0.25% combination group and significantly higher than that in the Inu0.5%RD0.5% combination group (Table 3). The RD0.5% group showed a slower decrease than all the inulin groups, with the pH being significantly higher than that of the other groups at 12 h after fermentation.

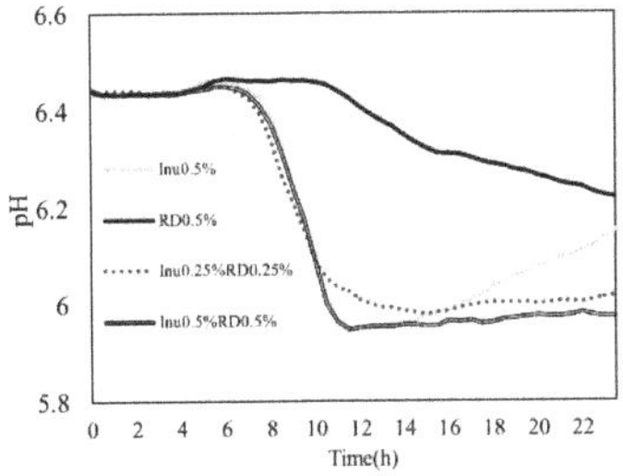

Figure 3. Changes in the pH of the fermentation medium during the 24 h fermentation process. Values are presented in terms of the mean (n = 7). Inu0.5%, RD0.5%, RD0.25%Inu0.25%, and RD0.5%Inu0.5% are depicted using a light gray line, a black line, a dotted line, and double lines, respectively.

Table 3. pH after 0, 12, and 24 h of fermentation.

	pH_0 h		pH_12 h		pH_24 h	
	Mean	**SD**	**Mean**	**SD**	**Mean**	**SD**
Inu0.5%	6.44	0.02	5.95 ^{b}	0.01	6.15 ^{ab}	0.09
RD0.5%	6.44	0.01	6.41 ^{a}	0.06	6.22 ^{a}	0.17
Inu0.25%RD0.25%	6.44	0.01	6.01 ^{b}	0.06	6.02 ^{bc}	0.05
Inu0.5%RD0.5%	6.44	0.01	5.95 ^{b}	0.01	5.97 ^{c}	0.05

Different letters indicate significant differences ($p < 0.05$, Tukey–Kramer multiple comparison test).

3.6. Inulin and Resistant Dextrin Residues in the Culture Supernatant

The inulin and resistant dextrin residues in the culture supernatant after fermentation were analyzed using TLC (Figure 4). Inulin spots were not detected for the supernatant after fermentation, indicating that all of the inulin had been used by the fecal microbiota. In contrast, resistant dextrin spots remained after fermentation, confirming that the resistant dextrin was retained after 24 h of fermentation.

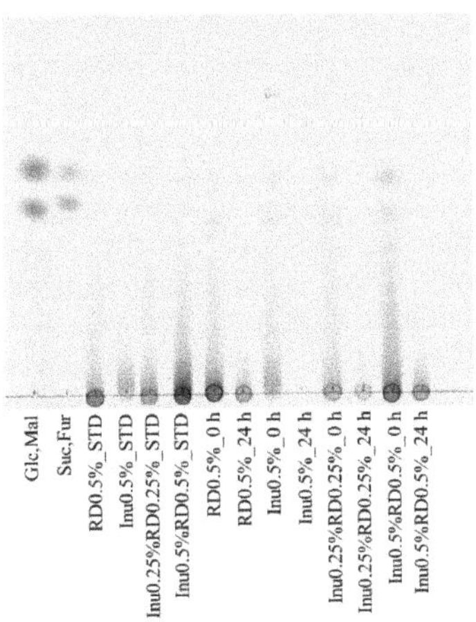

Figure 4. Thin-layer chromatography (TLC) showing the resistant dextrin and inulin degradation in the fecal fermentation supernatants at 0 (0 h) and 24 (24 h) h after fermentation. A representative sample image is shown. Glucose (Glc), maltose (Mal), sucrose (Suc), and fructose (Fur) spots are size markers. The standard (STD) was prepared by dissolving each prebiotic in distilled water.

4. Discussion

In this study, in vitro fermentation was performed using fecal samples obtained from older individuals. The results suggest a combination of inulin and resistant dextrin can elicit prebiotic effects, including a bifidogenic response and SCFA production. Moreover, gas production was also reduced when compared to using inulin alone. Additionally, we found that the inulin plus resistant dextrin combination increased the diversity of the gut microbiota and helped to maintain a low pH in the culture medium. These results suggest that the combination of inulin and resistant dextrin may have superior prebiotic effects than each prebiotic material alone. To the best of our knowledge, this study not only examines the prebiotic effects of inulin and resistant dextrin combinations but also provides valuable insights as it utilizes fecal samples from older people.

The Inu0.25%RD0.25% combination group showed lesser gas production than the Inu0.5% group. This finding suggests that reducing the total inulin level leads to a decrease in gas production, consistent with the findings of a previous study that showed that replacing some of the inulin with other prebiotics reduced gas production [13]. Human studies have also shown that replacing some of the fructo-oligosaccharide (a high gas-producing material) with acacia gum (a slowly fermented material) can lead to reduced gas production and alleviate abdominal symptoms, such as belching [29]. These findings suggest that, if the total intake of prebiotics is the same, an inulin and resistant dextrin combination would be more effective in reducing abdominal symptoms than inulin alone.

In this study, we examined the Inu0.5%RD0.5% combination group to assess whether gas production can be additively increased when inulin is combined with resistant dextrin. As we expected, the Inu0.5%RD0.5% combination group showed more gas production than the Inu0.5% group, suggesting an additive increase in gas production due to the combination with resistant dextrin. However, the Inu0.25%RD0.25% combination group, which had half the amount of inulin and resistant dextrin compared to the Inu0.5%RD0.5% combination group, produced less than half the amount of gas as that group. Thus, the findings suggest that gas production is not necessarily related to the amount of prebiotics added, and there might be a certain threshold for the prebiotic quantity and blending ratio that leads to gas production. Previous human studies have reported that inulin and fructo-oligosaccharide do not cause abdominal symptoms when administered at low doses; however, they can cause abdominal symptoms once the intake exceeds a certain threshold [11,30]. An in vitro study that assessed the prebiotic effects of an inulin and polydextrose mixture in two different ratio combinations (inulin:polydextrose = 1:1 or 1:2), found that the amount of gas production was the same regardless of the combination ratios [31]. Therefore, examining different amounts and blending ratios of inulin and resistant dextrin combinations could clarify the amounts and blending ratios that do not cause gas production.

Bifidobacterium is a beneficial taxon that maintains intestinal and immune function [32]. *Bifidobacterium* counts decline with age and are also reduced in sarcopenia [1,33]. Therefore, increasing the *Bifidobacterium* counts in older people is important. Inulin is highly selective for *Bifidobacterium*; some studies have suggested that the combination of inulin with other prebiotics, which are less selective for *Bifidobacterium*, can potentially enhance or maintain a bifidogenic effect [19,31]. In the current study, although the Inu0.25%RD0.25% combination group showed a slight decrease in the *Bifidobacterium* count and proportion compared to the Inu0.5% group, these values were significantly higher than those in the RD0.5% group. A recent human study has also demonstrated an increased *Bifidobacterium* count when an inulin plus resistant dextrin combination was used [34]. Therefore, replacing some of the inulin with resistant dextrin could help increase or maintain the bifidogenic effects, with less gas production than when administering inulin alone.

The SCFAs, particularly acetate, propionate, and butyrate, produced by the gut microbiota from prebiotics are important for host energy regulation and immunomodulation; however, their production is reduced in older people [35,36]. In the current study, the Inu0.25%RD0.25% combination group showed a slightly lower acetate level than the Inu0.5% group but a significantly higher level than the RD0.5% group. Acetate is a beneficial metabolite that promotes glucagon-like peptide 1 secretion, thereby increasing energy expenditure and fat oxidation [37]. *Bifidobacterium* is an acetate-producing bacterial genus [38]. The Inu0.25%RD0.25% combination group also demonstrated an increase in the *Bifidobacterium* counts, explaining the increased acetate levels compared to those of the RD0.5% group. Additionally, the Inu0.5%RD0.5% combination group had a higher acetate level than the Inu0.5% group. These findings suggest that *Bifidobacterium* rapidly grows by utilizing inulin and that the growing *Bifidobacterium* utilizes resistant dextrin, resulting in increased acetic acid production. Propionate and butyrate were also elevated in the Inu0.5% group and combination groups compared to the RD0.5% group, although the differences were not statistically significant. These results suggest that using an inulin plus resistant dextrin combination leads to an increase in SCFA production.

This study showed that the inulin plus resistant dextrin combination had superior prebiotic effects than inulin alone, that is, it increased the microbiota diversity and decreased the fermentation medium's pH. A decrease in the gut microbiota diversity is associated with various diseases, such as obesity [39,40]. It is suggested that prebiotic intake is effective in maintaining this diversity [41]. In the current study, the Inu0.25%RD0.25% combination group had greater microbiota diversity than the Inu0.5% group. The utilization of prebiotics by different bacteria depends on the prebiotic structure; prebiotics with more complex structures may be utilized by a greater variety of bacteria, leading to an increase in the microbiota diversity [42]. Increased diversity was also observed in another study that investigated the effects of inulin combined with pectin, which has a more complex structure than inulin [43]. Furthermore, an in vivo study showed that a combination of galacto-oligosaccharides and inulin, which are highly selective for *Bifidobacterium*, with dietary fiber, increased various bacterial species owing to their low selectivity, resulting in increased microbiota diversity [44]. Resistant dextrin is utilized by various bacterial species and has been reported to particularly increase the abundance of the *Fusicatenibacter* genus [45]. In this study, the resistant dextrin plus inulin combination may also increase the diversity, suggesting that combinations of inulin with other soluble fibers with low selectivity may increase the microbiota diversity compared to using inulin alone.

An increased intestinal pH inhibits the growth of beneficial bacteria, such as butyrate-producing bacteria, and increases the number of bacteria producing putrefactive products harmful to the host [18,46]. The decrease in the intestinal pH is induced by SCFAs derived from prebiotics [18]. Consistent with the results of a previous study, the pH decreased faster in the inulin-only group than in the resistant-dextrin-only group [17]. However, during fermentation, the pH of the inulin-only group gradually increased and was higher than that of the resistant-dextrin-only group at the end of fermentation. Inulin residue was not detected in the supernatant after fermentation, suggesting that all the inulin had been utilized by the gut microbiota during fermentation. Therefore, although the amount of SCFAs at 24 h was higher in the inulin-only group than in the resistant-dextrin-only group, the production of SCFAs from inulin decreased later in the fermentation process and the lower pH could not be maintained. Similar results were observed in a fecal fermentation study comparing the prebiotic effects of several arabinoxylan materials. Some materials increased the pH at the end of fermentation, suggesting that the materials may have been utilized completely [47]. Even in vivo, in the distal colon, where prebiotics are less accessible, the production of SCFAs by the gut microbiota is limited, resulting in a higher pH than that of the proximal colon [48]. In contrast, when inulin was combined with resistant dextrin, a lower pH was maintained even after 24 h of fermentation. The presence of resistant dextrin residue in the culture supernatant after fermentation suggested that SCFA production continued during fermentation. Additionally, no increase in the pH was observed in the combination groups, which may be due to the utilization of resistant dextrin after inulin consumption by the gut microbiota. When prebiotics with different fermentation speeds are combined, the faster material is utilized first, then the slower material is utilized, suggesting the prebiotic effect is sustained for longer [49,50]. These results suggest that the inulin plus resistant dextrin combination was effective in lowering the intestinal pH and maintaining the acidic condition.

The main limitation of this study is its small sample size. Therefore, these findings may not be generalizable to other older populations. However, our study provides valuable insights, despite the small sample size, owing to its analysis of fecal samples from older people. A previous in vitro fecal fermentation study had reported that middle-aged individuals produced more gas from prebiotics than young adults; this finding suggests that abdominal symptoms are more likely to occur with an increase in age [21]. Although the current study did not directly compare fecal samples from older people with those from younger people, it showed that the inulin plus resistant dextrin combination may have prebiotic effects in older people, with lesser gas production than that with inulin alone. In addition, this study provides limited mechanistic insights into how the combination of inulin and resistant dextrin leads to the observed effects. Future studies should compare samples from people of different age groups and health statuses, and examine the prebiotic dose-dependence, to provide further insights into the mechanisms. Finally, this in vitro study administrated prebiotics only once, so it is unclear whether similar effects would be observed with long-term intake in humans. Further clinical trials are required to confirm whether this composition exhibits prebiotic effects while reducing abdominal symptoms in older people.

5. Conclusions

The inulin plus resistant dextrin combination showed prebiotic effects, such as increased *Bifidobacterium* counts and SCFA production, and lesser gas production, than inulin alone. Additionally, it showed potential benefits, such as an increase in the microbiota diversity and maintenance of a lower pH. These effects are believed to be attributable to the combination of prebiotic materials that are highly selective for *Bifidobacterium* and are rapidly utilized by the gut microbiota along with prebiotic materials that are slowly utilized by various bacterial genera. These findings can be applied to the development of food products combining inulin with resistant dextrin for older people, which would provide a prebiotic that also addresses the issue of abdominal symptoms.

Supplementary Materials: The following supporting information can be downloaded at https://www.mdpi.com/article/10.3390/nu16244262/s1, Supplementary Table S1. YCFA medium composition; Supplementary Table S2A. Spearman correlation between gas production and top 30 bacterial genera; Supplementary Table S2B. Percentage of g_*Clostridium* population for each subject after 24 h of fermentation.

Author Contributions: Conceptualization, K.Y. and E.K.; methodology, K.Y. and E.K.; formal analysis, K.Y.; investigation, K.Y. and E.K.; resources, K.M.; data curation, K.Y.; writing—original draft preparation, K.Y.; writing—review and editing, E.K, S.M. and H.S.; visualization, K.Y.; supervision, H.S.; project administration, K.M. All authors have read and agreed to the published version of the manuscript.

Funding: This study was funded by Morinaga Milk Industry Co., Ltd., Tokyo, Japan. The funder provided support in the form of salaries for all authors.

Institutional Review Board Statement: This study was conducted in accordance with the Declaration of Helsinki, and it was approved by the Ethics Committee of the Japan Conference of Clinical Research (Tokyo, Japan; protocol code NFSFA-01 and date of approval 18 October 2019).

Informed Consent Statement: Written informed consent was obtained from all the subjects involved in this study.

Data Availability Statement: The data presented in this study are available on request from the corresponding author.

Acknowledgments: The authors would like to thank Odamaki, Mizuno, and Mitsuyama of Morinaga Milk Industry for their technical support in the analysis of the gut microbiota.

Conflicts of Interest: K.Y., E.K., S.M., H.S. and K.M. are employees of and receive a salary from Morinaga Milk Industry Co., Ltd.

References

1. Odamaki, T.; Kato, K.; Sugahara, H.; Hashikura, N.; Takahashi, S.; Xiao, J.Z.; Abe, F.; Osawa, R. Age-related changes in gut microbiota composition from newborn to centenarian: A cross-sectional study. *BMC Microbiol.* **2016**, *16*, 90. [CrossRef] [PubMed]
2. Ghosh, T.S.; Shanahan, F.; O'Toole, P.W. The gut microbiome as a modulator of healthy ageing. *Nat. Rev. Gastroenterol. Hepatol.* **2022**, *19*, 565–584. [CrossRef]
3. Peterson, C.T.; Sharma, V.; Elmén, L.; Peterson, S.N. Immune homeostasis, dysbiosis and therapeutic modulation of the gut microbiota. *Clin. Exp. Immunol.* **2015**, *179*, 363–377. [CrossRef] [PubMed]
4. Zhang, S.; Wang, R.; Li, D.; Zhao, L.; Zhu, L. Role of gut microbiota in functional constipation. *Gastroenterol. Rep.* **2021**, *9*, 392–401. [CrossRef]
5. Chen, Y.; Zhou, J.; Wang, L. Role and mechanism of gut microbiota in human disease. *Front. Cell. Infect. Microbiol.* **2021**, *11*, 625913. [CrossRef] [PubMed]
6. Gibson, G.R.; Hutkins, R.; Sanders, M.E.; Prescott, S.L.; Reimer, R.A.; Salminen, S.J.; Scott, K.; Stanton, C.; Swanson, K.S.; Cani, P.D.; et al. Expert consensus document: The international scientific association for probiotics and prebiotics (ISAPP) consensus statement on the definition and scope of prebiotics. *Nat. Rev. Gastroenterol. Hepatol.* **2017**, *14*, 491–502. [CrossRef] [PubMed]
7. Ashaolu, T.J.; Ashaolu, J.O.; Adeyeye, S.A.O. Fermentation of prebiotics by human colonic microbiota in vitro and short-chain fatty acids production: A critical review. *J. Appl. Microbiol.* **2021**, *130*, 677–687. [CrossRef]
8. Stribling, P.; Ibrahim, F. Dietary fibre definition revisited—The case of low molecular weight carbohydrates. *Clin. Nutr. ESPEN* **2023**, *55*, 340–356. [CrossRef]
9. Kolida, S.; Gibson, G.R. Prebiotic capacity of inulin-type fructans. *J. Nutr.* **2007**, *137*, 2503S–2506S. [CrossRef]
10. van der Beek, C.M.; Canfora, E.E.; Kip, A.M.; Gorissen, S.H.M.; Olde Damink, S.W.M.; van Eijk, H.M.; Holst, J.J.; Blaak, E.E.; Dejong, C.H.C.; Lenaerts, K. The prebiotic inulin improves substrate metabolism and promotes short-chain fatty acid production in overweight to obese men. *Metabolism* **2018**, *87*, 25–35. [CrossRef]
11. Bonnema, A.L.; Kolberg, L.W.; Thomas, W.; Slavin, J.L. Gastrointestinal tolerance of chicory inulin products. *J. Am. Diet. Assoc.* **2010**, *110*, 865–868. [CrossRef]
12. Bruhwyler, J.; Carreer, F.; Demanet, E.; Jacobs, H. Digestive tolerance of inulin-type fructans: A double-blind, placebo-controlled, cross-over, dose-ranging, randomized study in healthy volunteers. *Int. J. Food Sci. Nutr.* **2009**, *60*, 165–175. [CrossRef] [PubMed]
13. Marzorati, M.; Qin, B.; Hildebrand, F.; Klosterbuer, A.; Roughead, Z.; Roessle, C.; Rochat, F.; Raes, J.; Possemiers, S. Addition of acacia gum to a FOS/inulin blend improves its fermentation profile in the simulator of the human intestinal microbial ecosystem (SHIME®). *J. Funct. Foods* **2015**, *16*, 211–222. [CrossRef]
14. Slizewska, K.; Kapusniak, J.; Barczynska, R.; Jochym, K. Resistant dextrins as prebiotic. In *Carbohydrates—Comprehensive Studies on Glycobiology and Glycotechnology*; Chang, C.F., Ed.; Intechopen: London, UK, 2012. [CrossRef]
15. Iwatani, S.; Yamamoto, N. Functional food products in Japan: A review. *Food Science and Human Wellness.* **2019**, *8*, 96–101. [CrossRef]
16. Shintani, T. Processing and Application of Starch in Food in Japan. *J. Food Sci. Nutr. Res.* **2020**, *3*, 140–152. [CrossRef]
17. Noack, J.; Timm, D.; Hospattankar, A.; Slavin, J. Fermentation profiles of wheat dextrin, inulin and partially hydrolyzed guar gum using an in vitro digestion pretreatment and in vitro batch fermentation system model. *Nutrients* **2013**, *5*, 1500–1510. [CrossRef]
18. Davani-Davari, D.; Negahdaripour, M.; Karimzadeh, I.; Seifan, M.; Mohkam, M.; Masoumi, S.J.; Berenjian, A.; Ghasemi, Y. Prebiotics: Definition, types, sources, mechanisms, and clinical applications. *Foods* **2019**, *8*, 92. [CrossRef]
19. Ghoddusi, H.B.; Grandison, M.A.; Grandison, A.S.; Tuohy, K.M. In vitro study on gas generation and prebiotic effects of some carbohydrates and their mixtures. *Anaerobe* **2007**, *13*, 193–199. [CrossRef] [PubMed]
20. Arnold, J.W.; Roach, J.; Fabela, S.; Moorfield, E.; Ding, S.; Blue, E.; Dagher, S.; Magness, S.; Tamayo, R.; Bruno-Barcena, J.M.; et al. The pleiotropic effects of prebiotic galacto-oligosaccharides on the aging gut. *Microbiome* **2021**, *9*, 31. [CrossRef] [PubMed]
21. Chen, J.; Pi, X.; Liu, W.; Ding, Q.; Wang, X.; Jia, W.; Zhu, L. Age-related changes of microbiota in midlife associated with reduced saccharolytic potential: An in vitro study. *BMC Microbiol.* **2021**, *21*, 47. [CrossRef]
22. Kono, G.; Yoshida, K.; Kokubo, E.; Ikeda, M.; Matsubara, T.; Koyama, T.; Iwamoto, H.; Miyaji, K. Fermentation supernatant of elderly feces with inulin and partially hydrolyzed guar gum maintains the barrier of inflammation-induced Caco-2/HT29-MTX-E12 co-cultured cells. *J. Agric. Food Chem.* **2023**, *71*, 1510–1517. [CrossRef] [PubMed]
23. Liu, Y.; Gibson, G.R.; Walton, G.E. An in vitro approach to study effects of prebiotics and probiotics on the faecal microbiota and selected immune parameters relevant to the elderly. *PLoS ONE* **2016**, *9*, e0162604. [CrossRef] [PubMed]
24. Miguez, B.; Vila, C.; Venema, K.; Parajo, J.C.; Alonso, J.L. Prebiotic effects of pectooligosaccharides obtained from lemon peel on the microbiota from elderly donors using an in vitro continuous colon model (TIM-2). *Food Funct.* **2020**, *11*, 9984–9999. [CrossRef] [PubMed]
25. Murakami, R.; Hashikura, N.; Yoshida, K.; Xiao, J. Z.; Odamaki, T. Growth promoting effect of alginate on Faecalibacterium prausnitzii through cross-feeding with Bacteroides. *Food Res. Int.* **2021**, *144*, 110326. [CrossRef] [PubMed]
26. Matsuki, T.; Watanabe, K.; Fujimoto, J.; Kado, Y.; Takada, T.; Matsumoto, K.; Tanaka, R. Quantitative PCR with 16S rRNA-gene-targeted species-specific primers for analysis of human intestinal bifidobacteria. *Appl. Environ. Microbiol.* **2004**, *70*, 167–173. [CrossRef]

27. Milani, C.; Hevia, A.; Foroni, E.; Duranti, S.; Turroni, F.; Lugli, G.A.; Sanchez, B.; Martín, R.; Gueimonde, M.; van Sinderen, D.; et al. Assessing the fecal microbiota: An optimized ion Torrent 16S RRNA gene-based analysis protocol. *PLoS ONE* **2013**, *8*, e68739. [CrossRef]
28. Naidoo, K.; Ayyachamy, M.; Permaul, K.; Singh, S. Enhanced fructooligosaccharides and inulinase production by a Xanthomonas campestris Pv. Phaseoli KM 24 mutant. *Bioprocess Biosyst. Eng.* **2009**, *32*, 689–695. [CrossRef]
29. Goetze, O.; Fruehauf, H.; Pohl, D.; Giarrè, M.; Rochat, F.; Ornstein, K.; Menne, D.; Fried, M.; Thumshirn, M. Effect of a prebiotic mixture on intestinal comfort and general wellbeing in health. *Br. J. Nutr.* **2008**, *100*, 1077–1085. [CrossRef]
30. Bouhnik, Y.; Vahedi, K.; Achour, L.; Attar, A.; Salfati, J.; Pochart, P.; Marteau, P.; Flourié, B.; Bornet, F.; Rambaud, J.C. Short-chain fructo-oligosaccharide administration dose-dependently increases fecal bifidobacteria in healthy humans. *J. Nutr.* **1999**, *129*, 113–116. [CrossRef]
31. Zhu, L.; Guo, F.; Guo, Z.; Chen, X.; Qian, X.; Li, X.; Li, X.; Li, J.; Wang, X.; Jia, W. Potential health benefits of lowering gas production and bifidogenic effect of the blends of polydextrose with inulin in a human gut model. *Front. Nutr.* **2022**, *9*, 934621. [CrossRef]
32. Gavzy, S.J.; Kensiski, A.; Lee, Z.L.; Mongodin, E.F.; Ma, B.; Bromberg, J.S. *Bifidobacterium* mechanisms of immune modulation and tolerance. *Gut Microbes.* **2023**, *15*, 2291164. [CrossRef] [PubMed]
33. Ticinesi, A.; Nouvenne, A.; Cerundolo, N.; Catania, P.; Prati, B.; Tana, C.; Meschi, T. Gut microbiota, muscle mass and function in aging: A focus on physical frailty and sarcopenia. *Nutrients* **2019**, *11*, 1633. [CrossRef]
34. Hess, A.L.; Benítez-Páez, A.; Blædel, T.; Larsen, L.H.; Iglesias, J.R.; Madera, C.; Sanz, Y.; Larsen, T.M. MyNewGut Consortium. *The effect of inulin and resistant maltodextrin on weight loss during energy restriction: A randomised, placebo-controlled, double-blinded intervention. Eur. J. Nutr.* **2020**, *59*, 2507–2524. [CrossRef]
35. Portincasa, P.; Bonfrate, L.; Vacca, M.; De Angelis, M.; Farella, I.; Lanza, E.; Khalil, M.; Wang, D.Q.H.; Sperandio, M.; Di Ciaula, A. Gut microbiota and short chain fatty acids: Implications in glucose homeostasis. *Int. J. Mol. Sci.* **2022**, *23*, 1105. [CrossRef]
36. Rampelli, S.; Candela, M.; Turroni, S.; Biagi, E.; Collino, S.; Franceschi, C.; O'Toole, P.W.; Brigidi, P. Functional metagenomic profiling of intestinal microbiome in extreme ageing. Aging. *Albany* **2013**, *5*, 902–912. [CrossRef]
37. Hernández, M.A.G.; Canfora, E.E.; Jocken, J.W.E.; Blaak, E.E. The short-chain fatty acid acetate in body weight control and insulin sensitivity. *Nutrients* **2019**, *11*, 1943. [CrossRef] [PubMed]
38. Wolin, M.J.; Zhang, Y.; Bank, S.; Yerry, S.; Miller, T.L. NMR detection of $^{13}CH_3{}^{13}COOH$ from 3-^{13}C-glucose: A signature for *Bifidobacterium* fermentation in the intestinal tract. *J. Nutr.* **1998**, *128*, 91–96. [CrossRef] [PubMed]
39. Agus, A.; Clément, K.; Sokol, H. Gut microbiota-derived metabolites as central regulators in metabolic disorders. *Gut* **2021**, *70*, 1174–1182. [CrossRef]
40. Avalos-Fernandez, M.; Alin, T.; Métayer, C.; Thiébaut, R.; Enaud, R.; Delhaes, L. The respiratory Microbiota alpha-diversity in chronic lung diseases: First systematic review and meta-analysis. *Respir. Res.* **2022**, *23*, 214. [CrossRef]
41. Hills, R.D.; Pontefract, B.A.; Mishcon, H.R.; Black, C.A.; Sutton, S.C.; Theberge, C.R. Gut microbiome: Profound implications for diet and disease. *Nutrients* **2019**, *11*, 1613. [CrossRef]
42. Chung, W.S.F.; Walker, A.W.; Louis, P.; Parkhill, J.; Vermeiren, J.; Bosscher, D.; Duncan, S.H.; Flint, H.J. Modulation of the human gut microbiota by dietary fibres occurs at the species level. *BMC Biol.* **2016**, *14*, 3. [CrossRef] [PubMed]
43. Chung, W.S.F.; Walker, A.W.; Vermeiren, J.; Sheridan, P.O.; Bosscher, D.; Garcia-Campayo, V.; Parkhill, J.; Flint, H.J.; Duncan, S.H. Impact of carbohydrate substrate complexity on the diversity of the human colonic microbiota. *FEMS Microbiol. Ecol.* **2019**, *95*, fiy201. [CrossRef] [PubMed]
44. Cheng, W.; Lu, J.; Li, B.; Lin, W.; Zhang, Z.; Wei, X.; Sun, C.; Chi, M.; Bi, W.; Yang, B.; et al. Effect of functional oligosaccharides and ordinary dietary fiber on intestinal microbiota diversity. *Front. Microbiol.* **2017**, *8*, 1750. [CrossRef] [PubMed]
45. Nishimoto, Y.; Mizuguchi, Y.; Mori, Y.; Ito, M.; Miyazato, S.; Kishimoto, Y.; Yamada, T.; Fukuda, S. Resistant maltodextrin intake reduces virulent metabolites in the gut environment: A randomized control study in a Japanese cohort. *Front. Microbiol.* **2022**, *13*, 644146. [CrossRef] [PubMed]
46. Peled, S.; Livney, Y.D. The role of dietary proteins and carbohydrates in gut microbiome composition and activity: A review. *Food Hydrocoll.* **2021**, *120*, 106911. [CrossRef]
47. Demuth, T.; Edwards, V.; Bircher, L.; Lacroix, C.; Nyström, L.; Geirnaert, A. In vitro colon fermentation of soluble arabinoxylan is modified through milling and extrusion. *Front. Nutr.* **2021**, *8*, 707763. [CrossRef]
48. Cummings, J.H.; Macfarlane, G.T. The control and consequences of bacterial fermentation in the human colon. *J. Appl. Bacteriol.* **1991**, *70*, 443–459. [CrossRef]
49. Rogers, T.E.; Pudlo, N.A.; Koropatkin, N.M.; Bell, J.S.K.; Moya, B.M.; Jasker, K.; Martens, E.C. Dynamic responses of *Bacteroides thetaiotaomicron* during growth on glycan mixtures. *Mol. Microbiol.* **2013**, *88*, 876–890. [CrossRef]
50. Tuncil, Y.E.; Nakatsu, C.H.; Kazem, A.E.; Arioglu-Tuncil, S.; Reuhs, B.; Martens, E.C.; Hamaker, B.R. Delayed utilization of some fast-fermenting soluble dietary fibers by human gut microbiota when presented in a mixture. *J. Funct. Foods* **2017**, *32*, 347–357. [CrossRef]

nutrients MDPI

Article

Thai Fermented Soybean (Thua-Nao) Prevents Early Stages of Colorectal Carcinogenesis Induced by Diethylnitrosamine and 1,2-Dimethylhydrazine Through Modulations of Cell Proliferation and Gut Microbiota in Rats

Sirinya Taya [1,*], Sivamoke Dissook [2], Jetsada Ruangsuriya [1,2], Supachai Yodkeeree [2], Kongsak Boonyapranai [3], Teera Chewonarin [2] and Rawiwan Wongpoomchai [2]

[1] Functional Food Research Unit, Multidisciplinary Research Institute, Chiang Mai University, Chiang Mai 50200, Thailand
[2] Department of Biochemistry, Faculty of Medicine, Chiang Mai University, Chiang Mai 50200, Thailand
[3] Research Institute for Health Sciences, Chiang Mai University, Chiang Mai 50200, Thailand
* Correspondence: sirinya.t@cmu.ac.th

Abstract: Background: Thua-nao is a traditional fermented soybean product widely consumed in the northern areas of Thailand. There has been little research on the biological activity of Thua-nao, particularly its anticancer properties. Objectives: The objective of this study was to examine the cancer chemopreventive effects of dried Thua-nao on liver and colorectal carcinogenesis induced by carcinogens in rats. Methods: Rats were injected with diethylnitrosamine (DEN) and 1,2-dimethylhydrazine (DMH) to induce preneoplastic lesions. Rats orally received dried Thua-nao for 13 weeks. The preneoplastic lesions, including glutathione *S*-transferase placental form (GST-P)-positive foci and aberrant crypt foci (ACF), were evaluated in the liver and colon, respectively. The cancer chemopreventive mechanisms of dried Thua-nao on liver and colorectal carcinogenesis were examined. Results: Dried Thua-nao administration suppressed colorectal aberrant crypt foci. Moreover, dried Thua-nao reduced proliferation cell nuclear antigen (PCNA)-positive cells in the colon. Interestingly, dried Thua-nao modulated the gut microbiota in DEN- and DMH-induced rats. Isoflavones, including genistein and daidzein, represent promising chemopreventive agents in dried Thua-nao. Conclusions: In conclusion, these results highlight the cancer chemopreventive effect of dried Thua-nao in DEN and DMH-induced colorectal carcinogenesis through cell proliferation reduction and gut microbiota modulation.

Keywords: Thai fermented soybean; Thua-nao; aberrant crypt foci; cell proliferation; gut microbiota

1. Introduction

Cancer is a serious global public health issue, with colon and liver cancers being common and destructive [1]. Colon cancer, or colorectal cancer (CRC), ranks as the third most prevalent cancer globally and is the second leading cause of cancer-related mortality worldwide [1]. A sedentary lifestyle, obesity, smoking, and excessive alcohol consumption are among the lifestyle factors that can influence colorectal cancer [2]. Primary prevention strategies, including the implementation of early detection through screening, the avoidance of risk factors, and the maintenance of a healthy lifestyle, can significantly reduce the incidence and impact of colorectal cancer [2]. Liver cancer, or hepatocellular carcinoma (HCC), ranks as the sixth most prevalent cancer globally and is the third primary cause of cancer-related mortality worldwide [1]. Common risk factors for liver cancer are frequently associated with chronic liver disease, including hepatitis B or C infection, cirrhosis, or environmental factors such as heavy alcohol intake, smoking, and exposure to aflatoxin [3]. Both colorectal cancer and liver cancer have common risk factors, including lifestyle decisions such as diet, obesity, and alcohol intake. Particularly, diet is one of the environmental factors that is associated with an increased risk of cancer. However, dietary patterns that are appropriate may mitigate the likelihood of developing cancer [4,5]. The American Cancer Society (ACS) provides dietary and physical activity recommendations for preventing the progression of cancer [6]. These recommendations include the consumption of foods that are rich in nutrients, vegetables, legumes, fruits, and whole grains [6]. Furthermore, several studies have indicated that consumption of vegetables, unrefined grains, and fermented foods is associated with a reduced risk of cancer; naturally occurring chemicals in these foods act as chemopreventive agents against cancer [7–9].

Currently, researchers are investigating the influence of microbial community alterations on cancer. These microbial communities play important roles in several types of cancers, including breast, liver, and colorectal cancer. Tens of thousands of microorganisms live in the human digestive system, including bacteria, archaea, fungi, protozoa, and viruses, with bacteria making up the vast majority [10]. Gut dysbiosis is a condition characterized by changes in the composition and function of the gut microbiota due to an imbalance between beneficial and harmful microorganisms. Dysbiosis can be classified into three categories: depletion of helpful microorganisms, proliferation of harmful microorganisms, and reduction in microbial variety [11]. Dysbiosis is a contributing factor to various clinical illnesses, especially cancers [11]. Nutrition profoundly affects the composition of the human gut microbiota. For instance, a diet abundant in high-fermented foods enhances the diversity of bacteria in the body while concurrently reducing inflammatory indicators [12].

Thua-nao is a traditional fermented soybean product that is commonly consumed in the northern regions of Thailand, including Chiang Mai, Chiang Rai, Lampang, and Mae Hong Son. In addition, there are similar fermented soybean products in many countries, including natto in Japan, kinema in India, and chongkukjang in Korea [13]. The traditional preparation of Thua-nao involves immersing the soybeans overnight and boiling them for 4 h until they become completely soft. Subsequently, the cooked soybeans are naturally fermented in bamboo baskets lined with banana leaves for a few days. Although fresh Thua-nao can be consumed after fermentation, most people prefer steaming or grilling before consumption. Cooked Thua-nao can be shaped into a flat disk and sun-dried to create a variety form known as Thua-nao kab. The dried product can be mashed into a fine powder and stored for several months [13]. The previous study found that the proximate analysis of commercial Thua-nao collected from six local markets in Chiang Mai comprised 38.94–42.06% crude protein, 20.37–25.22% fat, 12.92–28.06% crude fiber, 4.70–5.44% ash, and 57.22–64.78% moisture [14]. Thua-nao is naturally fermented by indigenous microorganisms, especially the *Bacillus* spp. group [15,16]. Pakwan and colleagues reported that *Bacillus* was the most common bacteria in Thua-nao, while Lactic acid bacteria (LAB) of the family *Leuconostocaceae* were also present in substantial quantities [15]. Consequently, Thua-nao is a nutrient-rich source of beneficial bacteria that has the potential to be a probiotic food [15]. Furthermore, Thua-nao is abundant in bioactive compounds, including isoflavones and short-chain fatty acids; it also possesses biological activities, such as antioxidant activity, which are beneficial for health [14,16]. Nonetheless, the exact health benefits of Thua-nao consumption are still unclear, and the chemopreventive efficacy of Thua-nao against cancer in animal models remains inadequately explored. Diethylnitrosamine (DEN) and 1,2-dimethylhydrazine (DMH) are recognized as carcinogens for the liver and colon, respectively, and both undergo metabolism by cytochrome P450 in the dual organ carcinogenicity assay [17]. In line with the three Rs of animal research, these carcinogens were given to the same rats in this investigation to minimize the number of animals involved in the experimental process [17–19]. Therefore, to elucidate the cancer chemopreventive properties of Thua-nao, a model of liver and colon carcinogenesis induced by DEN and DMH in rats was employed to examine its effects on the early stages of liver and colorectal carcinogenesis in rats, as well as its potential inhibitory mechanisms.

2. Materials and Methods

2.1. Preparation of Thua-Nao

Soybeans were purchased from a commercial brand in Thailand. After being soaked overnight, soybeans were boiled for three hours and naturally fermented in an incubator at 37 °C for 3 days. The fresh Thua-nao was dried in a hot air oven at 50 °C for 48 h, subsequently being milled with a powder grinder. The dried powder was stored at −20 °C until use. The animal equivalent dose (AED) of dried Thua-nao was calculated based on a 75 kg Thai adult's average daily Thua-nao intake, which is roughly 12 g (160 mg/kg bw/day). A conversion factor of 6.2 is utilized to multiply by the human dose in order to calculate AED based on the difference in body surface area between rats and humans [20]. Therefore, the equivalent human dose of dried Thua-nao (160 mg/kg bw/day) is equivalent to a dose of 992 mg/kg bw/day in rats. In the cancer chemopreventive study, the daily animal doses of dried Thua-nao were 100 and 1000 mg/kg bw. Additionally, the BAM method was used to count the *Escherichia coli* and coliform bacteria in the dried powder of Thua-nao in order to check the food safety in Thua-nao for pathogenic microorganism contamination [21].

2.2. Chemical Constituents of Dried Thua-Nao

Dried Thua-nao powder was soaked in 70% ethanol for two days and then extracted again. The filtrate was evaporated and freeze-dried to obtain an ethanolic extract. This extract was used to measure the contents of phenolic compounds and isoflavones.

The content of total phenolic compounds in the ethanolic extract was measured spectrophotometrically using the Folin–Ciocalteu method [22].

The isoflavone contents of the ethanolic extract were examined by high-performance liquid chromatography (HPLC). The analysis was performed using a reverse phase C18 column (Agilent 4.6 mm × 250 mm, 5 μm) and analyzed on an Agilent HPLC 1260 (Agilent Technologies, Santa Clara, CA, USA). Gradient elution was carried out using a mixture of 1% trifluoroacetic acid (TFA) in water and methanol. The flow rate was 1 mL/min, and the injected volume was 10 μL. The absorbances at 254 and 260 nm were monitored. The isoflavone contents were determined and quantified using calibration curves for daidzin, daidzein, genistin, genistein, glycitin, and glycitein.

2.3. Animal and Experimental Protocol

The Institutional Animal Care and Use Committee of the Faculty of Medicine, Chiang Mai University, approved the animal study protocol (no. 26/2565). The animal study protocol was performed in accordance with the relevant guidelines and regulations. This protocol follows the recommendations in the ARRIVE guidelines. Male Wistar rats (three-week-old rats) were acquired from Nomura Siam International Co., Ltd., Bangkok, Thailand. All rats were maintained under standard conditions at 25 °C with a 12 h light/dark cycle and were provided with a commercial basal diet (C.P. mice feed 082G, Samut Prakan, Thailand, Table S1) and water.

The animal experimental design is illustrated in Figure 1. All rats were randomly divided into five groups. Rats in groups 1 through 3 were injected with 100 mg/kg bw of DEN (i.p.) on days 0, 4, and 11 and 40 mg/kg bw of DMH (s.c.) on days 0 and 7, whereas rats in groups 4 and 5 were injected with 0.9% normal saline solution. Dried Thua-nao at 100 and 1000 mg/kg bw were oral gavage fed 5 days a week for 13 weeks, whereas distilled water was used as a vehicle. Throughout the experiment, body weight and intake of diet and water were recorded. At the end of the experiment, rats were euthanized with overdosed isoflurane. Blood was collected for aspartate aminotransferase (AST) and alanine aminotransferase (ALT) determinations at the Small Animal Hospital, Faculty of Veterinary Medicine, Chiang Mai University. Internal organs, including the liver, spleen, and kidneys, were collected and weighed. The liver was sectioned into three pieces and fixed in 10% buffered neutral formalin (BNF) for determinations of glutathione S-transferase placental form (GST-P)-positive foci and proliferation cell nuclear antigen (PCNA)-positive cells by immunohistochemistry. The colon was expanded with 10% BNF and placed on ice for

30 min. Then, the colon was divided into three segments, namely proximal, distal, and rectum segments, and flattened in 10% BNF for evaluation of aberrant crypt foci (ACF); cross-section paraffin-embedded colonic tissues were used for determining PCNA-positive cells by immunohistochemistry. Furthermore, feces were collected from the rat anus and kept at −80 °C for analysis of the composition of the fecal intestinal microbiota.

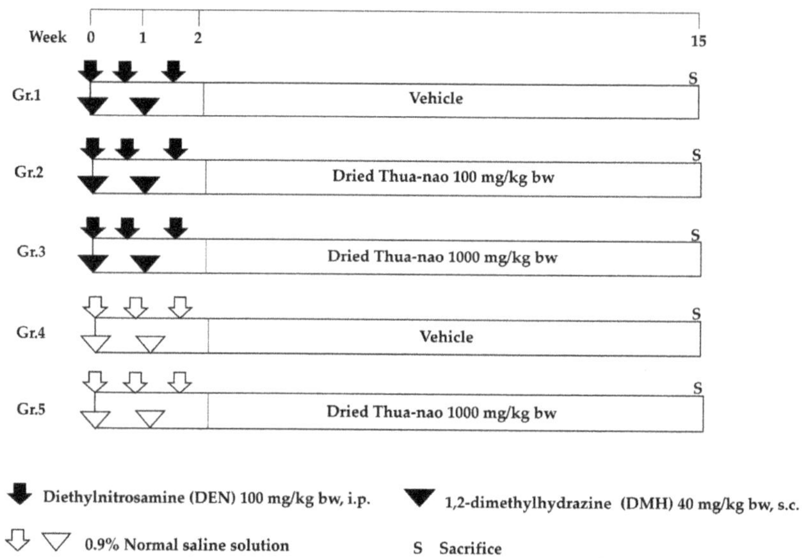

Figure 1. Animal experimental protocol for a cancer chemopreventive study of dried Thua-nao in DEN- and DMH-treated rats.

2.3.1. Determination of Preneoplastic Lesions in DEN- and DMH-Treated Rats

The preneoplastic lesions in the liver were investigated by immunohistochemical staining with the anti-GST-P antibody (MBL, Nagoya, Japan), as previously described [23]. Briefly, liver sections of 4 μm thickness were deparaffinized and dehydrated. The liver sections were soaked in 3% hydrogen peroxide (H_2O_2) and 1% skimmed milk to prevent the activity of pseudoperoxidase and non-specific binding of proteins, respectively. Following incubation with a rabbit polyclonal anti-rat GST-P antibody, the liver sections were incubated with an anti-rabbit IgG biotinylated antibody. Subsequently, the liver sections were soaked with diaminobenzidine (DAB), while hematoxylin was used as a counterstain. The area and number of GST-P-positive foci exceeding 0.02 mm^2 were analyzed. The results were represented as the area and number of GST-P-positive foci per liver area (cm^2).

The ACF in the colon was evaluated using methylene blue staining. Methylene blue staining was applied to each segment's flattened colon. The number and size of ACF were counted under the light microscope following Bird's criteria [24,25]. The results were expressed as the size and number of ACF per rat.

2.3.2. Determination of PCNA-Positive Cells by Immunohistochemistry

The immunohistochemistry of PCNA-positive cells was analyzed in the liver and colon as detailed by Chariyakornkul et al. [18]. The paraffin-embedded liver sections and colon cross-sections were deparaffinized. The sections were first incubated with a monoclonal mouse anti-rat PCNA antibody (BioLegend, San Diego, CA, USA), followed by biotinylated antibodies and an Elite avidin-biotin complex kit (Vector Laboratories, Inc., Burlingame, CA, USA). The PCNA-positive cells were counted using a light microscope.

2.3.3. Investigation of the Fecal Gut Microbiota Composition in Rats

The gut bacterial profiles were analyzed using 16s rRNA amplicon sequencing. First, bacterial DNA was extracted from frozen feces using a QIAamp DNA microbiome kit according to the instructions provided in the user manual. The hypervariable V4 region of the 16S rRNA gene from gut microbiota was sequenced using primers 515F (5′-GTGCCAGCMGCCGCGGTAA-3′) and 806R (5′-GGACTACHVGGGTWTCTAAT-3′). Following PCR amplification, paired-end sequencing (2 × 250 bp) was performed on the Illumina NovaSeq 6000 platform, according to the manufacturer's 16S genomic sequencing library preparation protocol. The raw sequencing data were processed to remove low-quality reads, merge overlapping reads, and identify and remove chimeric sequences. High-quality clean sequences were then annotated at various taxonomic levels (phylum to genus) using the Quantitative Insights Into Microbial Ecology 2 (QIIME2; version 2021.8.0). Taxonomy classification was performed against the SILVA 16S rRNA gene reference database release 138. The alpha diversity metrics, including species richness, were reflected by the Shannon index. Beta diversity was evaluated using the Unifrac phylogenetic distance. Data visualization was performed using R software version 4.3.2.

2.4. Statistical Analysis

All data are presented as mean ± SD values. The significance of differences between groups was assessed by one-way ANOVA, followed by Tukey's multiple comparison test, utilizing GraphPad Prism 9.0 software (GraphPad Software, Boston, MA, USA). Statistical significance was considered as p-value < 0.05.

The Kruskal-Wallis test was applied to assess differences in alpha-diversity among the groups. For pairwise comparisons between groups, the Wilcoxon rank-sum test was used. Additionally, PERMANOVA was employed to evaluate statistical differences in beta-diversity across groups.

3. Results

3.1. Chemical Components and Microbial Contamination of Dried Thua-Nao

The chemical constituents of dried Thua-nao are shown in Table 1. One hundred grams of dried Thua-nao contained 1180.21 ± 26.63 mg of phenolic compounds, as measured by spectrophotometry. The major isoflavones in dried Thua-nao were daidzein and genistein. To consider food safety in terms of pathogenic microorganism contamination in Thua-nao, total coliform bacteria were determined primarily, and it was found that there were >1100 most probable number (MPN)/g of total coliform bacteria. However, fecal coliform bacteria and *E. coli* were 3.6 MPN/g and <3 MPN/g, respectively. It is concluded that the Thua-nao used in this study is safe from bacterial contamination based on the presence of fecal coliform bacteria and *E. coli* being less than the limits (fecal coliform bacteria and *E. coli* < 500 and 3 MPN/g, respectively).

Table 1. The contents of total phenolic compounds and isoflavones of dried Thua-nao.

Compounds	Dried Thua-Nao
Spectrophotometry (per 100 g dried Thua-nao)	
Total phenolic compounds (mg GAE)	1180.21 ± 26.63
HPLC (mg per 100 g dried Thua-nao)	
Daidzin	1.68 ± 0.01
Glycitin	ND
Genistin	4.51 ± 0.34
Daidzein	43.45 ± 1.51
Glycitein	6.61 ± 0.02
Genistein	40.71 ± 0.51

Values are expressed as mean ± SD. Gallic acid equivalent (GAE); not detected (ND).

3.2. *Effect of Dried Thua-Nao on Preneoplastic Lesion Formation in the Liver and Colon of DEN- and DMH-Treated Rats*

According to overall findings, the administration of dried Thua-nao to rats receiving NSS led to a notable rise in their general parameters, including final body weight and diet consumption, compared to rats treated with only NSS. The injections of DEN and DMH tended to reduce the final body weight of the rats, while administration of dried Thua-nao at a low dose tended to increase the final body weight and significantly increase diet intake when compared to DEN- and DMH-treated rats (Table 2). Furthermore, administering dried Thua-nao did not affect the absolute and relative kidney and spleen across the groups. Nevertheless, the administration of dried Thua-nao at a high dose to rats treated with NSS resulted in a significant increase in the absolute weight of the liver, although no significant difference was observed in the relative weight of the liver (Table 3). The administration of DEN and DMH induced liver damage in rats, as detected by increasing serum AST and ALT levels; however, feeding dried Thua-nao did not reduce serum AST and ALT when compared to DEN- and DMH-treated rats (Figure 2). Interestingly, feeding dried Thua-nao did not change levels of liver function enzyme when compared to NSS-treated rats, suggesting that dried Thua-nao is non-hepatotoxic to rats.

Table 2. Effects of dried Thua-nao on general observations.

Group	Treatments	Body Weight (g)		Consumption (Per Rat Per Day)	
		Initial	Final	Diet (g)	Water (mL)
1	DEN + DMH	113 ± 8	453 ± 52	21.3 ± 1.8	29.3 ± 3.1
2	DEN + DMH + TN 100 mg/kg bw	115 ± 10	482 ± 45	23.6 ± 1.7 **	28.7 ± 3.9
3	DEN + DMH + TN 1000 mg/kg bw	114 ± 13	469 ± 21	21.3 ± 1.3	26.5 ± 3.9
4	NSS	115 ± 6	478 ± 35	20.3 ± 1.3	29.7 ± 5.0
5	NSS + TN 1000 mg/kg bw	116 ± 5	543 ± 30 *	24.4 ± 1.6 *	31.9 ± 2.2

The data are expressed as mean ± SD. Diethylnitrosamine (DEN); 1,2-dimethylhydrazine (DMH); normal saline solution (NSS); Thua-nao (TN). * Significantly different compared to the NSS-treated group ($p < 0.05$). ** Significantly different compared to the DEN- and DMH-treated group ($p < 0.05$).

Table 3. Effects of dried Thua-nao on some vital organ weights in rats.

Group	Treatments	Liver		Kidney		Spleen	
		Absolute (g)	Relative (%)	Absolute (g)	Relative (%)	Absolute (g)	Relative (%)
1	DEN + DMH	13.80 ± 2.07	3.09 ± 0.69	3.04 ± 1.08	0.70 ± 0.37	0.89 ± 0.27	0.20 ± 0.09
2	DEN + DMH + TN 100 mg/kg bw	14.58 ± 1.76	3.03 ± 0.24	3.07 ± 0.25	0.64 ± 0.04	0.89 ± 0.11	0.18 ± 0.02
3	DEN + DMH + TN 1000 mg/kg bw	13.66 ± 1.42	2.92 ± 0.30	2.99 ± 0.16	0.64 ± 0.03	0.86 ± 0.08	0.18 ± 0.02
4	NSS	12.78 ± 1.32	2.67 ± 0.19	2.70 ± 0.27	0.57 ± 0.04	0.77 ± 0.06	0.16 ± 0.02
5	NSS + TN 1000 mg/kg bw	16.36 ± 1.54 *	3.01 ± 0.22	3.07 ± 0.29	0.56 ± 0.03	0.70 ± 0.10	0.13 ± 0.02

The data are presented as mean ± SD. Diethylnitrosamine (DEN); 1,2-dimethylhydrazine (DMH); normal saline solution (NSS); Thua-nao (TN). * Significantly different compared to the NSS-treated group ($p < 0.05$).

Dried Thua-nao neither induced GST-P-positive foci nor ACF in NSS-treated rats (Table 4), suggesting that dried Thua-nao is non-carcinogenic to rat liver and colon. Rats treated with DEN and DMH exhibited a marked elevation in GST-P-positive foci and ACF formation compared to NSS-treated rats. The administration of dried Thua-nao at a dose of 1000 mg/kg bw tended to reduce the number and size of GST-P-positive foci in DEN- and DMH-treated rats, but the results were not significantly different when compared with carcinogen-induced rats. Notably, the dried Thua-nao at a dose of 1000 mg/kg bw significantly decreased the number of ACF in the colon of DEN- and DMH-treated rats, while the size of the aberrant crypt did not change. The results suggest that dried Thua-nao could reduce the development of colorectal carcinogenesis in rats. Additionally, administering dried Thua-nao to rats treated with NSS did not cause the formation of GST-P-positive foci in the liver or ACF in the colon, indicating that dried Thua-nao does not have carcinogenic effects in the rat liver and colon.

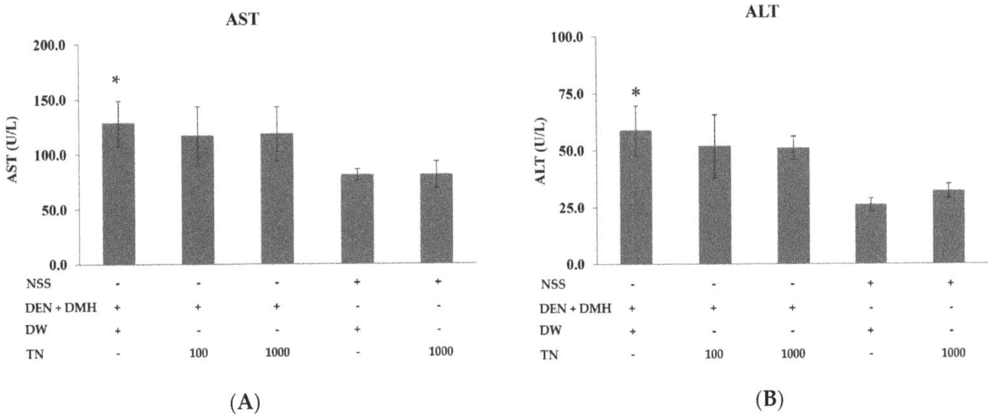

Figure 2. Effects of dried Thua-nao on serum liver function enzyme levels in rats. The levels of (**A**) aspartate aminotransferase (AST) and (**B**) alanine aminotransferase (ALT) in rats. Diethylnitrosamine (DEN); 1,2-dimethylhydrazine (DMH); normal saline solution (NSS); Thua-nao (TN); distilled water (DW). * Significantly different compared to the NSS-treated group ($p < 0.05$).

Table 4. Effects of dried Thua-nao on preneoplastic lesion formation in DEN- and DMH-treated rats.

Group	Treatments	Liver		Colon	
		Number of GST-P-Positive Foci/Liver Area (cm^2)	Area of GST-P-Positive Foci (mm^2)/Liver Area (cm^2)	Total Aberrant Crypt Foci/Rat	Aberrant Crypt/Focus
1	DEN + DMH	20.8 ± 7.4 *	1.12 ± 0.51 *	148.9 ± 52.2 *	4.90 ± 0.44 *
2	DEN + DMH + TN 100 mg/kg bw	19.1 ± 5.7	0.97 ± 0.20	126.1 ± 57.7	5.11 ± 0.43
3	DEN + DMH + TN 1000 mg/kg bw	14.9 ± 4.8	0.69 ± 0.60	87.8 ± 43.5 **	4.51 ± 0.44
4	NSS	0.0 ± 0.0	0.00 ± 0.00	0.0 ± 0.0	0.00 ± 0.00
5	NSS + TN 1000 mg/kg bw	0.0 ± 0.0	0.00 ± 0.00	0.0 ± 0.0	0.00 ± 0.00

The data are shown as mean ± SD. Diethylnitrosamine (DEN); 1,2-dimethylhydrazine (DMH); normal saline solution (NSS); Thua-nao (TN). * Significantly different compared to the NSS-treated group ($p < 0.05$). ** Significantly different compared to the DEN- and DMH-treated group ($p < 0.05$).

3.3. *Effect of Dried Thua-Nao on Cell Proliferation in Liver and Colon of DEN- and DMH-Treated Rats*

Following the administration of dried Thua-nao at a high dose, which demonstrated a decrease in preneoplastic lesions of carcinogen-treated rats, the mechanisms of cell proliferation were investigated. The number of PCNA-positive cells in the liver and colon of rats that were treated with DEN and DMH significantly increased when compared to rats that were treated with NSS. Administration of the dried Thua-nao at a high dose significantly reduced the number of PCNA-positive cells in both the liver and the colon when compared to DEN- and DMH-treated rats (Figure 3). Based on these findings, dried Thua-nao may prevent the formation of preneoplastic lesions by modulating cell proliferation in the liver and colon carcinogenesis.

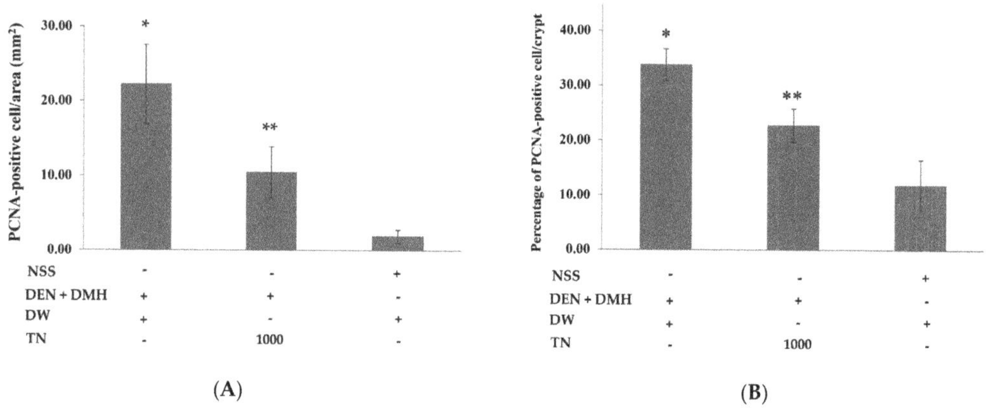

Figure 3. Effect of dried Thua-nao on cell proliferation in rat liver and colon tissues. PCNA-positive cells in the (**A**) liver and (**B**) colon were quantified by immunohistochemical staining. Diethylnitrosamine (DEN); 1,2-dimethylhydrazine (DMH); normal saline solution (NSS); Thua-nao (TN); distilled water (DW). * Significantly different compared to the NSS-treated group ($p < 0.05$). ** Significantly different compared to the DEN- and DMH-treated group ($p < 0.05$).

3.4. Effect of Dried Thua-Nao on Bacterial Profile in Rats

In this study, the impact of Thua-nao on the gut microbiota of rats treated with DEN and DMH was investigated. Our in-depth 16S rRNA amplicon sequencing analysis showed that the gut microbiome's composition and diversity changed significantly after Thua-nao administration.

This study found that the alpha diversity metrics, particularly the Shannon index, showed a significant change in the Thua-nao administered to the NSS-treated rats (NSS_TN) compared to the NSS-treated rats (NSS_DW). These changes indicate a notable increase in microbial diversity, suggesting that consuming Thua-nao contributes to a more robust gut microbiota. This finding was depicted in Figure 4A, which shows a boxplot comparison of the alpha diversity between groups. Furthermore, beta-diversity analysis, utilizing Unifrac phylogenetic distance, demonstrated a clear separation between the groups (Figure 4B), signifying distinct gut bacterial communities in each group. Such differentiation underpins Thua-nao's influential role in modulating the gut microbiome landscape, emphasizing its dietary significance.

The characterized bacterial composition at various taxonomic levels, from phyla to genera, is presented in Figure 5, revealing six bacterial phyla with differences in their relative abundance across the four groups (Figure 5A). Firmicutes were found to have a higher-level abundance in all of the groups, suggesting a foundational presence in the gut microbiome. At the genus level, a total of 155 microbial genera were annotated (Table S2), with the top 11 in abundance shown in Figure 5B, providing a clear visualization of the microbial distribution in each group.

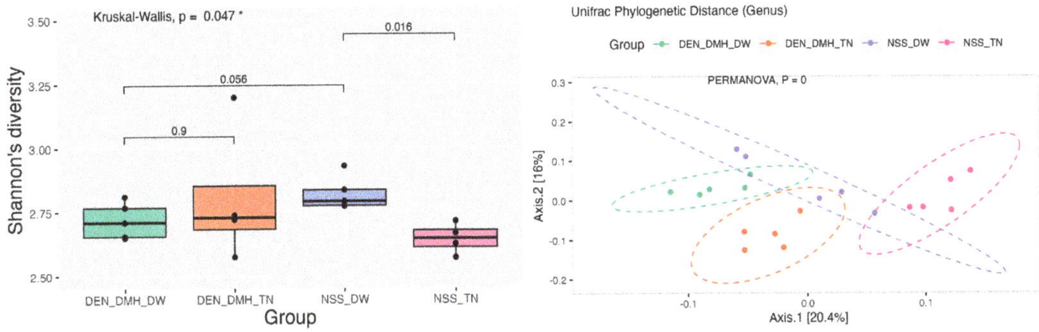

(A) **(B)**

Figure 4. Alternation in bacterial diversity within each group. (**A**) Boxplot of gut microbial alpha-diversity based on Shannon index. (**B**) Beta-diversity of gut microbiota composition structure based on Unifrac phylogenetic distance. Diethylnitrosamine (DEN); 1,2-dimethylhydrazine (DMH); normal saline solution (NSS); Thua-nao (TN); distilled water (DW); n = 5 rats in each group. Statistical significance was determined using the Kruskal-Wallis test to assess differences among groups, with Wilcoxon rank-sum tests used to determine pairwise differences between groups. The level of statistical significance is denoted as * $p < 0.05$.

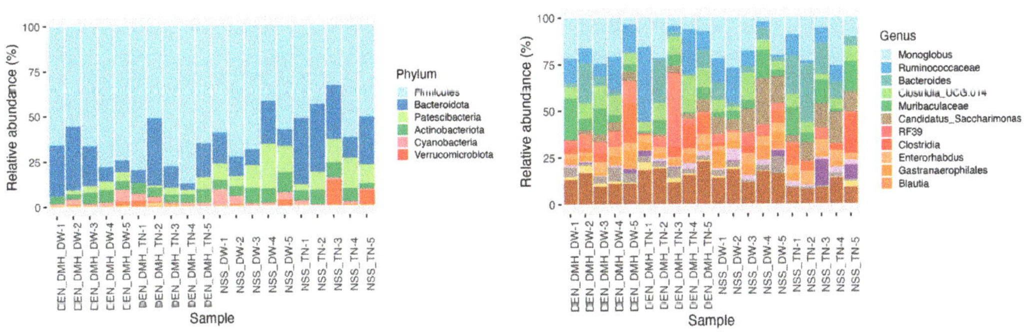

(A) **(B)**

Figure 5. Percent relative abundances of the taxonomic compositions in each group at (**A**) the phylum and (**B**) genus levels. Diethylnitrosamine (DEN); 1,2-dimethylhydrazine (DMH); normal saline solution (NSS); Thua-nao (TN); distilled water (DW); n = 5 rats in each group.

Furthermore, the relative abundance of the differential microbiota in each group is presented in Figure 6. Rats injected with DEN and DMH (DEN_DMH_DW) showed a marked increase in the abundance of *Alloprevotella* and *Oscillibacter* as compared to the NSS-treated group (NSS_DW). Administering dried Thua-nao to DEN- and DMH-treated rats (DEN_DMH_TN) significantly reduced the abundance of *Alloprevotella* and *Defluviitaleaceae_UCG-011*, along with an increase in the abundance of *Enterococcus*, *Lactococcus*, *Leuconostoc*, *Lactococcus lactis*, and *Staphylococcus sciuri* when compared to DEN- and DMH-treated rats (DEN_DMH_DW).

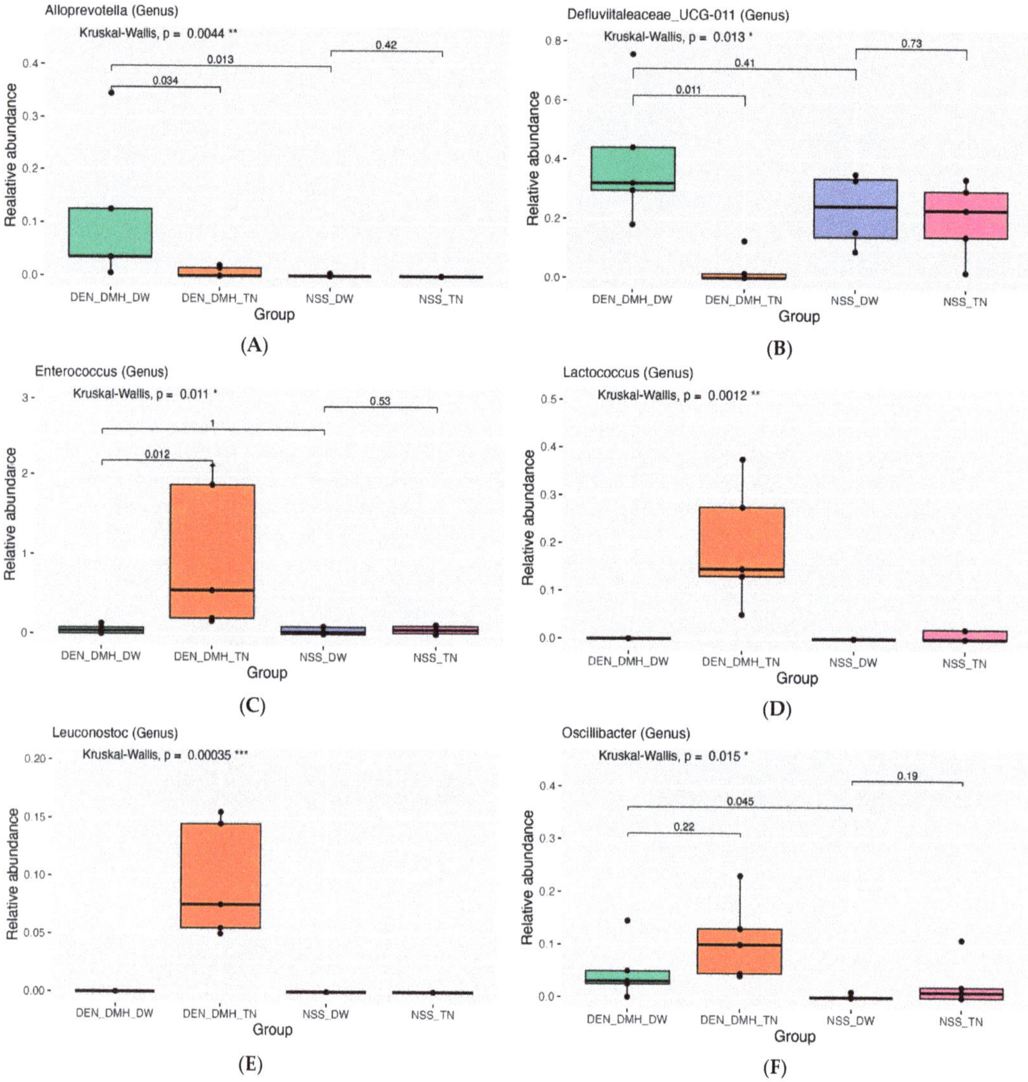

Figure 6. *Cont.*

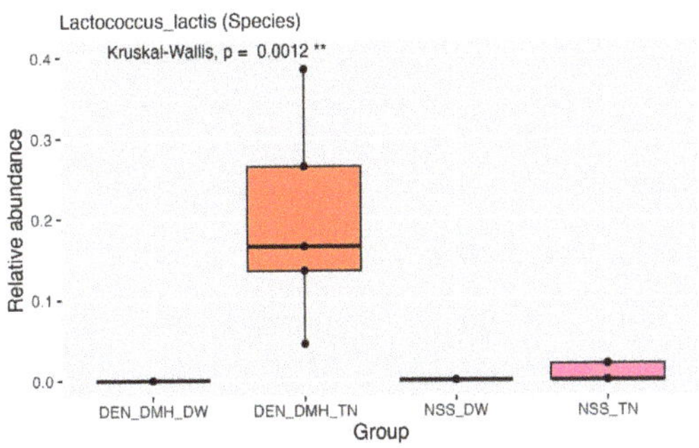

(G)

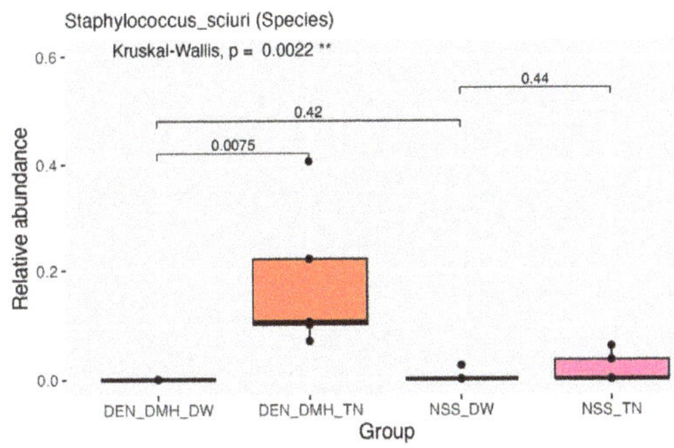

(H)

Figure 6. Relative abundance of the differential microbiota in each group. Relative abundance of (**A**) *Alloprevotella*, (**B**) *Defluviitaleaceae_UCG-011*, (**C**) *Enterococcus*, (**D**) *Lactococcus*, (**E**) *Leuconostoc*, (**F**) *Oscillibacter*, (**G**) *Lactococcus lactis*, and (**H**) *Staphylococcus sciuri*. Diethylnitrosamine (DEN); 1,2-dimethylhydrazine (DMH); normal saline solution (NSS); Thua-nao (TN); distilled water (DW); $n = 5$ rats in each group. Statistical significance was determined using the Kruskal-Wallis test to assess differences among groups, with Wilcoxon rank-sum tests used to determine pairwise differences between groups. Levels of statistical significance are denoted as * $p < 0.05$, ** $p < 0.01$, *** $p < 0.001$.

4. Discussion

Fermented soybean products are frequently employed as seasoning agents or condiments for enhancing the taste of food. Various fermented soybean products originating from diverse regions of Asia include tempeh, natto, miso, soy sauce, doenjang, and Thua-nao [13,26,27]. Fermented soybean products exhibit various bioactive compounds and potential health benefits, including antidiabetic, antioxidant, anti-inflammatory, antihyperlipidemic, and anticancer activities [27,28]. Thua-nao is a traditional fermented soybean product that is widely consumed in northern Thailand. However, there have been few studies on Thua-nao's biological activities, particularly anticancer activity. The present study reveals the cancer chemopreventive effects of Thua-nao in a rat model of liver and colon carcinogenesis induced by DEN and DMH. DEN is commonly employed as a hepatocarcinogen in the process of chemically induced hepatocarcinogenesis, whereas DMH is frequently utilized as a colon carcinogen in animal models of chemically induced colon carcinogenesis. The dual organ carcinogenicity model, employing DEN and DMH as initiators, was utilized to induce the early stages of liver and colon carcinogenesis in rats [17]. GST-P is highly expressed during hepatocarcinogenesis in rats and has been used as a reliable marker for experimental hepatocarcinogenesis in rats [29]. In addition, ACF can be used as a reliable marker for chemically induced colorectal carcinogenesis in rats [30]. This model served as a valuable tool for investigating potential cancer chemopreventive agents [18,19].

Dried Thua-nao consumption at a high dose significantly increased both the final body and the liver weights, but the relative organ weight showed no statistical difference. The normalization of organ weights to the body is one of the procedures to reduce variations among groups in which their body weights are varied and finally presented in organ-to-body weight ratios [31]. In addition, those ratios are considered critical parameters for drug and substance toxicity determinations. Consequently, when the relative liver weight shows no statistical difference, it can be inferred that administering dried Thua-nao had no effect on the liver.

Co-administration of DEN and DMH induced the formation of GST-P-positive foci in the liver and ACF in the colon. Thua-nao, however, did not inhibit GST-P-positive foci in DEN- and DMH-treated rats. Interestingly, Thua-nao reduced the total number of ACF but had no effect on the size of ACF in the colons of carcinogen-treated rats. Furthermore, the inhibitory effect of dried Thua-nao was observed only when administered in high doses, which were equal to human daily consumption. This underlines the role of Thua-nao in regulating the formation of preneoplastic lesions in the colon following carcinogen exposure in rats. In DMH-induced ACF development in rats, liver CYP2E1 catalyzes the biotransformation of DMH to methylazoxymethanol (MAM), which interacts with DNA, causing adduction at O^6-methylguanine (O^6-MeG), ultimately resulting in mutations and tumors [32]. In colonic mucosa cells, MAM, an electrophilic methylating chemical, can methylate DNA at guanine bases [33]. The number of ACF is thought to be related to O^6-MeG. Furthermore, data have been published that indicate a correlation between the decrease in O^6-MeG in the colonic mucosa and the decrease in ACF, or adenocarcinoma, in the rat colon [34]. Reducing phase I enzyme activity is, therefore, an effective approach for preventing reactive intermediates from generating DNA adducts [35,36]. Moreover, the conjugated MAM is hydrolyzed by intestine bacterial β-glucuronidase, which releases free MAM [37]. After free MAM is reabsorbed into the colonic epithelial cell, DNA adduct and mutation occur [38]. As a result, one potential approach to preventing colorectal carcinogenesis is to reduce the activity of bacterial β-glucuronidase. Numerous studies indicated that *Ficus dubia* latex extract and purple rice extract suppressed DMH-induced colorectal carcinogenesis by modifying xenobiotic metabolism in the liver and colon, resulting in decreased DNA adduct levels and a reduction in ACF numbers during the initiation stage [36,39].

One of the most important factors in cancer development is an imbalance between cell growth and apoptosis [40]. The proliferating cellular nuclear antigen (PCNA) is a reliable biomarker for cell proliferation [41]. This finding reveals that the administration of Thua-nao at a high dose significantly reduced the number of PCNA-positive cells in the colon in carcinogen-treated rats. Furthermore, this study reveals that genistein and daidzein are the most abundant isoflavones found in dried Thua-nao. These isoflavone aglycones are converted from their isoflavone glycosides by β-glycosidase, which is produced by microorganisms during the fermentation process [27]. Furthermore, recent studies indicate that many traditional Asian fermented soy products, such as Japanese natto, Indian kinema, and Thai Thua-nao, have a much higher content of isoflavone aglycone compared to unfermented soybeans [42,43]. A systematic review and meta-analysis reported that soy isoflavone consumption was significantly associated with a reduced risk of colorectal cancer [44]. Several studies have discovered that genistein effectively inhibited the development of colon cancer caused by carcinogens in in vitro studies. Zhang et al. reported that genistein suppresses azoxymethane-induced WNT/β-catenin signaling, avoiding early colon neoplasia [45]. Moreover, Sekar et al. found that genistein reduced ACF in DMH-induced rats by activating nuclear factor-erythroid 2 related factor 2 (Nrf-2) and modulating the expression of proliferative markers, including PCNA [46]. In human colon cancer cells, genistein promotes G2/M cell cycle arrest and apoptosis through an ATM/p53-dependent mechanism [47]. Hence, the probable cancer chemopreventive compounds in Thua-nao would likely comprise isoflavones, including genistein and daidzein.

The gut microbiota has a crucial role in the initiation and promotion of various types of cancer, especially malignancies in the gastrointestinal tract. Bacteria can indeed initiate chronic inflammation of the gastric mucosa, leading to permanent alterations in intestinal epithelial cells and increasing the susceptibility of individuals to cancer [48,49]. Notably, the gut microbiota in colorectal cancer is linked to certain strains of *Bacteroides fragilis*, *Escherichia coli*, *Clostridium* spp., *Streptococcus gallolyticus*, *Fusobacterium nucleatum*, *Streptococcus bovis*, and *Enterococcus faecalis* [49–51]. However, the role of *E. faecalis* remains controversial [52]. *E. faecalis* is a member of the *Firmicutes* phylum and is occasionally employed as a probiotic product [53]. Nevertheless, under certain circumstances, *E. faecalis* can have pathogenic effects and can be detrimental to the development of colorectal cancer due to its capacity to disrupt the DNA of colonic epithelial cells [54]. In this study, no significant difference in the abundance of *Enterococcus* was observed between carcinogen-treated rats and those treated with NSS. However, the administration of dried Thua-nao resulted in an increased abundance of this genus in rats treated with DEN and DMH. Furthermore, de Almeida et al. conducted an analysis of the correlation between *E. faecalis* and the development of cancer.

They observed that the data only indicated an increase in the concentration of the bacteria but did not provide any information on its specific functional roles in the development of colorectal cancer [52]. In this study, *Alloprevotella* abundance was increased in DEN- and DMH-treated rats. Previous research revealed that *Alloprevotella* was enriched in tumor tissues or feces of individuals with colorectal cancer or adenomas and involved in colorectal carcinogenesis [55,56]. However, administering dried Thua-nao to DEN- and DMH-treated rats resulted in a decrease in the abundance of *Alloprevotella*, which is inconsistent with previous studies [57]. According to Yu et al., a microecological preparation increased the amount of *Alloprevotella* compared to rats that were treated with azoxymethane/dextran sodium sulfate [57]. Moreover, administering dried Thua-nao to DEN- and DMH-treated rats revealed an increase in the amount of *Lactococcus* and *Leuconostoc* in the phylum *Firmicutes*, especially *Lactococcus lactis*, which are known as probiotics. Hosseini et al. found that nisin, a low molecular weight antibacterial peptide that is produced by *L. lactis*, has inhibitory effects on cancer cell proliferation, which are linked to the reduced expression of cyclin D1 in the SW480 cancer cell line [58]. This study proposes that dried Thua-nao has the capacity to modulate pathogenic bacteria and beneficial bacteria, hence potentially mitigating the formation of preneoplastic lesions. Nonetheless, clinical investigations about the cancer-preventive efficacy of Thua-nao should be further undertaken.

5. Conclusions

These findings indicate the cancer chemopreventive potential of dried Thua-nao in the early stages of colorectal carcinogenesis. Moreover, the inhibitory mechanism of Thua-nao may contribute to the reduction in cell growth and the modulation of gut microbiota. The probable cancer chemopreventive compounds in Thua-nao are isoflavones, including genistein and daidzein. In conclusion, Thua-nao may be a promising food source of beneficial chemopreventive substances for the prevention of cancer or the promotion of health. Moreover, further research is necessary to obtain additional evidence of its effects on chemopreventive potential via clinical studies.

Supplementary Materials: The following supporting information can be downloaded at: https://www.mdpi.com/article/10.3390/nu16203506/s1, Table S1: The nutritional composition of commercial basal diets; Table S2: Differential abundance analysis results (Genus levels).

Author Contributions: Conceptualization, S.T. and R.W.; methodology, S.T., S.D. and S.Y.; software, S.D.; formal analysis, S.T. and S.D.; investigation, S.T., S.D., J.R., S.Y. and K.B.; resources, S.T. and J.R.; writing—original draft, S.T. and S.D.; writing—review and editing, S.T. and S.D.; visualization, S.T. and S.D.; supervision, S.T.; project administration, S.T.; funding acquisition, T.C. All authors have read and agreed to the published version of the manuscript.

Funding: This research was funded by Fundamental Fund 2023, Chiang Mai University, grant number FF66/023.

Institutional Review Board Statement: The Institutional Animal Care and Use Committee of the Faculty of Medicine, Chiang Mai University, authorized the animal study protocol (protocol no. 26/2565, approved on 10 January 2023).

Informed Consent Statement: Not applicable.

Data Availability Statement: The original contributions presented in the study are included in the article/Supplementary Material; further inquiries can be directed to the corresponding author.

Acknowledgments: This research project was supported by Fundamental Fund 2023, Chiang Mai University. The authors would like to acknowledge John Tucker, Faculty of Humanities, Chiang Mai University, for his assistance in proofreading the English language in this article.

Conflicts of Interest: The authors declare no conflicts of interest.

References

1. Bray, F.; Laversanne, M.; Sung, H.; Ferlay, J.; Siegel, R.L.; Soerjomataram, I.; Jemal, A. Global cancer statistics 2022: GLOBOCAN estimates of incidence and mortality worldwide for 36 cancers in 185 countries. *CA Cancer J. Clin.* **2024**, *74*, 229–263. [CrossRef]
2. WHO. Colorectal Cancer. Available online: https://www.who.int/news-room/fact-sheets/detail/colorectal-cancer?gad_source=1&gclid=CjwKCAjwvKi4BhABEiwAH2gcwxWZyGFHvohtmCVWhmAYG1t5O_XncXLWcUehzAPalpq0OqIdbxhXGBoCR8YQAvD_BwE (accessed on 6 August 2024).
3. NCI. Liver Cancer Causes, Risk Factors, and Prevention. Available online: https://www.cancer.gov/types/liver/what-is-liver-cancer/causes-risk-factors (accessed on 7 August 2024).
4. Marino, P.; Mininni, M.; Deiana, G.; Marino, G.; Divella, R.; Bochicchio, I.; Giuliano, A.; Lapadula, S.; Lettini, A.R.; Sanseverino, F. Healthy Lifestyle and Cancer Risk: Modifiable Risk Factors to Prevent Cancer. *Nutrients* **2024**, *16*, 800. [CrossRef] [PubMed]
5. Kerschbaum, E.; Nussler, V. Cancer Prevention with Nutrition and Lifestyle. *Visc. Med.* **2019**, *35*, 204–209. [CrossRef]
6. Rock, C.L.; Thomson, C.; Gansler, T.; Gapstur, S.M.; McCullough, M.L.; Patel, A.V.; Andrews, K.S.; Bandera, E.V.; Spees, C.K.; Robien, K.; et al. American Cancer Society guideline for diet and physical activity for cancer prevention. *CA A. Cancer J. Clin.* **2020**, *70*, 245–271. [CrossRef]
7. Lim, H.J.; Park, I.S.; Jeong, S.J.; Ha, G.S.; Yang, H.J.; Jeong, D.Y.; Kim, S.Y.; Jung, C.H. Effects of Cheonggukjang (Fermented Soybean) on the Development of Colitis-Associated Colorectal Cancer in Mice. *Foods* **2023**, *12*, 383. [CrossRef]
8. Aghajanpour, M.; Nazer, M.R.; Obeidavi, Z.; Akbari, M.; Ezati, P.; Kor, N.M. Functional foods and their role in cancer prevention and health promotion: A comprehensive review. *Am. J. Cancer Res.* **2017**, *7*, 740–769.
9. Khuanphram, N.; Taya, S.; Kongtawelert, P.; Wongpoomchai, R. Sesame Extract Promotes Chemopreventive Effect of Hesperidin on Early Phase of Diethylnitrosamine-Initiated Hepatocarcinogenesis in Rats. *Pharmaceutics* **2021**, *13*, 1687. [CrossRef]
10. Gill, S.R.; Pop, M.; Deboy, R.T.; Eckburg, P.B.; Turnbaugh, P.J.; Samuel, B.S.; Gordon, J.I.; Relman, D.A.; Fraser-Liggett, C.M.; Nelson, K.E. Metagenomic analysis of the human distal gut microbiome. *Science* **2006**, *312*, 1355–1359. [CrossRef]
11. Pandey, H.; Tang, D.W.T.; Wong, S.H.; Lal, D. Gut Microbiota in Colorectal Cancer: Biological Role and Therapeutic Opportunities. *Cancers* **2023**, *15*, 866. [CrossRef]
12. Wastyk, H.C.; Fragiadakis, G.K.; Perelman, D.; Dahan, D.; Merrill, B.D.; Yu, F.B.; Topf, M.; Gonzalez, C.G.; Van Treuren, W.; Han, S.; et al. Gut-microbiota-targeted diets modulate human immune status. *Cell* **2021**, *184*, 4137–4153 e4114. [CrossRef]
13. Chukeatirote, E. Thua nao: Thai fermented soybean. *J. Ethn. Foods* **2015**, *2*, 115–118. [CrossRef]
14. Dajanta, K.; Chukeatirote, E.; Apichartsrangkoon, A. Nutritional and Physicochemical Qualities of Thua Nao (Thai Traditional Fermented Soybean). *Chiang Mai J. Sci.* **2012**, *39*, 562–574.
15. Pakwan, C.; Chitov, T.; Chantawannakul, P.; Manasam, M.; Bovonsombut, S.; Disayathanoowat, T. Bacterial compositions of indigenous Lanna (Northern Thai) fermented foods and their potential functional properties. *PLoS ONE* **2020**, *15*, e0242560. [CrossRef]
16. Wongsurawat, T.; Sutheeworapong, S.; Jenjaroenpun, P.; Charoensiddhi, S.; Khoiri, A.N.; Topanurak, S.; Sutthikornchai, C.; Jintaridth, P. Microbiome analysis of thai traditional fermented soybeans reveals short-chain fatty acid-associated bacterial taxa. *Sci. Rep.* **2023**, *13*, 7573. [CrossRef]
17. Punvittayagul, C.; Chariyakornkul, A.; Chewonarin, T.; Jarukamjorn, K.; Wongpoomchai, R. Augmentation of diethylnitrosamine-induced early stages of rat hepatocarcinogenesis by 1,2-dimethylhydrazine. *Drug Chem. Toxicol.* **2019**, *42*, 641–648. [CrossRef]
18. Chariyakornkul, A.; Inboot, N.; Taya, S.; Wongpoomchai, R. Low-polar extract from seed of Cleistocalyx nervosum var. paniala modulates initiation and promotion stages of chemically-induced carcinogenesis in rats. *Biomed. Pharmacother.* **2021**, *133*, 110963. [CrossRef]
19. Phannasorn, W.; Pharapirom, A.; Thiennimitr, P.; Guo, H.; Ketnawa, S.; Wongpoomchai, R. Enriched Riceberry Bran Oil Exerts Chemopreventive Properties through Anti-Inflammation and Alteration of Gut Microbiota in Carcinogen-Induced Liver and Colon Carcinogenesis in Rats. *Cancers* **2022**, *14*, 4358. [CrossRef]
20. Nair, A.B.; Jacob, S. A simple practice guide for dose conversion between animals and human. *J. Basic Clin. Pharm.* **2016**, *7*, 27–31. [CrossRef] [PubMed]
21. Feng, P.; Weagant, S.D.; Grant, M.A.; Burkhardt, W. Enumeration of Escherichia coli and the Coliform Bacteria. In *Food and Drug Administration (FDA)*, 8th ed.; Bacteriological Analytical Manual Online: Silver Spring, MD, USA; Berlin, Germany, 1998.
22. Inboot, W.; Taya, S.; Chailungka, A.; Meepowpan, P.; Wongpoomchai, R. Genotoxicity and antigenotoxicity of the methanol extract of Cleistocalyx nervosum var. paniala seed using a Salmonella mutation assay and rat liver micronucleus tests. *Mol. Cell. Toxicol.* **2012**, *8*, 19–24. [CrossRef]
23. Thumvijit, T.; Taya, S.; Punvittayagul, C.; Peerapornpisal, Y.; Wongpoomchai, R. Cancer chemopreventive effect of Spirogyra neglecta (Hassall) Kutzing on diethylnitrosamine-induced hepatocarcinogenesis in rats. *Asian Pac. J. Cancer Prev.* **2014**, *15*, 1611–1616. [CrossRef]
24. Bird, R.P. Observation and quantification of aberrant crypts in the murine colon treated with a colon carcinogen: Preliminary findings. *Cancer Lett.* **1987**, *37*, 147–151. [CrossRef] [PubMed]
25. Bird, R.P.; Good, C.K. The significance of aberrant crypt foci in understanding the pathogenesis of colon cancer. *Toxicol. Lett.* **2000**, *112–113*, 395–402. [CrossRef] [PubMed]
26. Dimidi, E.; Cox, S.R.; Rossi, M.; Whelan, K. Fermented Foods: Definitions and Characteristics, Impact on the Gut Microbiota and Effects on Gastrointestinal Health and Disease. *Nutrients* **2019**, *11*, 1806. [CrossRef]

27. do Prado, F.G.; Pagnoncelli, M.G.B.; de Melo Pereira, G.V.; Karp, S.G.; Soccol, C.R. Fermented Soy Products and Their Potential Health Benefits: A Review. *Microorganisms* **2022**, *10*, 1606. [CrossRef]
28. Qiao, Y.; Zhang, K.; Zhang, Z.; Zhang, C.; Sun, Y.; Feng, Z. Fermented soybean foods: A review of their functional components, mechanism of action and factors influencing their health benefits. *Food Res. Int.* **2022**, *158*, 111575. [CrossRef] [PubMed]
29. Ito, N.; Tamano, S.; Shirai, T. A medium-term rat liver bioassay for rapid in vivo detection of carcinogenic potential of chemicals. *Cancer Sci.* **2003**, *94*, 3–8. [CrossRef]
30. Wargovich, M.J.; Brown, V.R.; Morris, J. Aberrant crypt foci: The case for inclusion as a biomarker for colon cancer. *Cancers* **2010**, *2*, 1705–1716. [CrossRef]
31. Sellers, R.S.; Mortan, D.; Michael, B.; Roome, N.; Johnson, J.K.; Yano, B.L.; Perry, R.; Schafer, K. Society of Toxicologic Pathology Position Paper: Organ Weight Recommendations for Toxicology Studies. *Toxicol. Pathol.* **2007**, *35*, 751–755. [CrossRef] [PubMed]
32. Megaraj, V.; Ding, X.; Fang, C.; Kovalchuk, N.; Zhu, Y.; Zhang, Q.Y. Role of hepatic and intestinal p450 enzymes in the metabolic activation of the colon carcinogen azoxymethane in mice. *Chem. Res. Toxicol.* **2014**, *27*, 656–662. [CrossRef]
33. Samanta, S.; Swamy, V.; Suresh, D.; Rajkumar, M.; Rana, B.; Rana, A.; Chatterjee, M. Protective effects of vanadium against DMH-induced genotoxicity and carcinogenesis in rat colon: Removal of O(6)-methylguanine DNA adducts, p53 expression, inducible nitric oxide synthase downregulation and apoptotic induction. *Mutat. Res.* **2008**, *650*, 123–131. [CrossRef]
34. Suaeyun, R.; Kinouchi, T.; Arimochi, H.; Vinitketkumnuen, U.; Ohnishi, Y. Inhibitory effects of lemon grass (*Cymbopogon citratus Stapf*) on formation of azoxymethane-induced DNA adducts and aberrant crypt foci in the rat colon. *Carcinogenesis* **1997**, *18*, 949–955. [CrossRef] [PubMed]
35. Giftson Senapathy, J.; Jayanthi, S.; Viswanathan, P.; Umadevi, P.; Nalini, N. Effect of gallic acid on xenobiotic metabolizing enzymes in 1,2-dimethyl hydrazine induced colon carcinogenesis in Wistar rats—A chemopreventive approach. *Food Chem. Toxicol.* **2011**, *49*, 887–892. [CrossRef] [PubMed]
36. Hu, R.; Chantana, W.; Pitchakarn, P.; Subhawa, S.; Chantarasuwan, B.; Temviriyanukul, P.; Chewonarin, T. Ficus dubia latex extract prevent DMH-induced rat early colorectal carcinogenesis through the regulation of xenobiotic metabolism, inflammation, cell proliferation and apoptosis. *Sci. Rep.* **2022**, *12*, 15472. [CrossRef] [PubMed]
37. Weisburger, J.H. Colon carcinogens: Their metabolism and mode of action. *Cancer* **1971**, *28*, 60–70. [CrossRef]
38. Manoj, G.; Thampi, B.S.; Leelamma, S.; Menon, P.V. Effect of dietary fiber on the activity of intestinal and fecal beta-glucuronidase activity during 1,2-dimethylhydrazine induced colon carcinogenesis. *Plant Foods Hum. Nutr.* **2001**, *56*, 13–21. [CrossRef]
39. Summart, R.; Chewonarin, T. Purple rice extract supplemented diet reduces DMH- induced aberrant crypt foci in the rat colon by inhibition of bacterial beta-glucuronidase. *Asian Pac. J. Cancer Prev.* **2014**, *15*, 749–755. [CrossRef]
40. Mason, E.F.; Rathmell, J.C. Cell metabolism: An essential link between cell growth and apoptosis. *Biochim. Biophys. Acta.* **2011**, *1813*, 645–654. [CrossRef]
41. Bologna-Molina, R.; Mosqueda-Taylor, A.; Molina-Frechero, N.; Mori-Estevez, A.D.; Sanchez-Acuna, G. Comparison of the value of PCNA and Ki-67 as markers of cell proliferation in ameloblastic tumors. *Med. Oral Patol. Oral Cir. Bucal* **2013**, *18*, e174–e179. [CrossRef]
42. Li-Jun, Y.; Li-Te, L.; Zai-Gui, L.; Tatsumi, E.; Saito, M. Changes in isoflavone contents and composition of sufu (fermented tofu) during manufacturing. *Food Chem.* **2004**, *87*, 587–592. [CrossRef]
43. Yamabe, S.; Kobayashi-Hattori, K.; Kaneko, K.; Endo, H.; Takita, T. Effect of soybean varieties on the content and composition of isoflavone in rice-koji miso. *Food Chem.* **2007**, *100*, 369–374. [CrossRef]
44. Yu, Y.; Jing, X.; Li, H.; Zhao, X.; Wang, D. Soy isoflavone consumption and colorectal cancer risk: A systematic review and meta-analysis. *Sci. Rep.* **2016**, *6*, 25939. [CrossRef] [PubMed]
45. Zhang, Y.; Li, Q.; Zhou, D.; Chen, H. Genistein, a soya isoflavone, prevents azoxymethane-induced up-regulation of WNT/beta-catenin signalling and reduces colon pre-neoplasia in rats. *Br. J. Nutr.* **2013**, *109*, 33–42. [CrossRef]
46. Sekar, V.; Anandasadagopan, S.K.; Ganapasam, S. Genistein regulates tumor microenvironment and exhibits anticancer effect in dimethyl hydrazine-induced experimental colon carcinogenesis. *BioFactors* **2016**, *42*, 623–637. [CrossRef]
47. Zhang, Z.; Wang, C.-Z.; Du, G.-J.; Qi, L.-W.; Calway, T.; He, T.-C.; Du, W.; Yuan, C.-S. Genistein induces G2/M cell cycle arrest and apoptosis via ATM/p53-dependent pathway in human colon cancer cells. *Int. J. Oncol.* **2013**, *43*, 289–296. [CrossRef] [PubMed]
48. Zou, S.; Fang, L.; Lee, M.H. Dysbiosis of gut microbiota in promoting the development of colorectal cancer. *Gastroenterol. Rep.* **2018**, *6*, 1–12. [CrossRef]
49. Gagniere, J.; Raisch, J.; Veziant, J.; Barnich, N.; Bonnet, R.; Buc, E.; Bringer, M.A.; Pezet, D.; Bonnet, M. Gut microbiota imbalance and colorectal cancer. *World J. Gastroenterol.* **2016**, *22*, 501–518. [CrossRef] [PubMed]
50. Kostic, A.D.; Chun, E.; Robertson, L.; Glickman, J.N.; Gallini, C.A.; Michaud, M.; Clancy, T.E.; Chung, D.C.; Lochhead, P.; Hold, G.L.; et al. Fusobacterium nucleatum potentiates intestinal tumorigenesis and modulates the tumor-immune microenvironment. *Cell Host Microbe* **2013**, *14*, 207–215. [CrossRef]
51. Biarc, J.; Nguyen, I.S.; Pini, A.; Gosse, F.; Richert, S.; Thierse, D.; Van Dorsselaer, A.; Leize-Wagner, E.; Raul, F.; Klein, J.P.; et al. Carcinogenic properties of proteins with pro-inflammatory activity from Streptococcus infantarius (formerly S.bovis). *Carcinogenesis* **2004**, *25*, 1477–1484. [CrossRef]
52. de Almeida, C.V.; Taddei, A.; Amedei, A. The controversial role of Enterococcus faecalis in colorectal cancer. *Therap. Adv. Gastroenterol.* **2018**, *11*, 1756284818783606. [CrossRef]

53. Gong, J.; Bai, T.; Zhang, L.; Qian, W.; Song, J.; Hou, X. Inhibition effect of Bifidobacterium longum, Lactobacillus acidophilus, Streptococcus thermophilus and Enterococcus faecalis and their related products on human colonic smooth muscle in vitro. *PLoS ONE* **2017**, *12*, e0189257. [CrossRef]
54. Huycke, M.M.; Abrams, V.; Moore, D.R. Enterococcus faecalis produces extracellular superoxide and hydrogen peroxide that damages colonic epithelial cell DNA. *Carcinogenesis* **2002**, *23*, 529–536. [CrossRef] [PubMed]
55. Zhang, Y.; Yu, X.; Yu, E.; Wang, N.; Cai, Q.; Shuai, Q.; Yan, F.; Jiang, L.; Wang, H.; Liu, J.; et al. Changes in gut microbiota and plasma inflammatory factors across the stages of colorectal tumorigenesis: A case-control study. *BMC Microbiol.* **2018**, *18*, 92. [CrossRef] [PubMed]
56. Hua, H.; Sun, Y.; He, X.; Chen, Y.; Teng, L.; Lu, C. Intestinal Microbiota in Colorectal Adenoma-Carcinoma Sequence. *Front. Med.* **2022**, *9*, 888340. [CrossRef] [PubMed]
57. Yu, W.; Zhang, J.; Chen, Z.; Wang, S.; Ruan, C.; Zhou, W.; Miao, M.; Shi, H. Inhibitory Effect of a Microecological Preparation on Azoxymethane/Dextran Sodium Sulfate-Induced Inflammatory Colorectal Cancer in Mice. *Front Oncol.* **2020**, *10*, 562189. [CrossRef]
58. Hosseini, S.S.; Goudarzi, H.; Ghalavand, Z.; Hajikhani, B.; Rafeieiatani, Z.; Hakemi-Vala, M. Anti-proliferative effects of cell wall, cytoplasmic extract of Lactococcus lactis and nisin through down-regulation of cyclin D1 on SW480 colorectal cancer cell line. *Iran. J. Microbiol.* **2020**, *12*, 424–430. [CrossRef]

nutrients

MDPI

Article

The Beneficial Effects of Regular Intake of *Lactobacillus paragasseri* OLL2716 on Gastric Discomfort in Healthy Adults: A Randomized, Double-Blind, Placebo-Controlled Study

Naruomi Yamada [1], Kyosuke Kobayashi [2], Akika Nagira [1], Takayuki Toshimitsu [1], Asako Sato [1], Hiroshi Kano [2] and Kenichi Hojo [1,*]

1 Health Science Research Unit, Division of Research and Development, Meiji Co., Ltd., Tokyo 192-0919, Japan; naruomi.yamada@meiji.com (N.Y.); akika.nagira.aa@meiji.com (A.N.); takayuki.toshimitsu@meiji.com (T.T.); asako.satou@meiji.com (A.S.)

2 Wellness Science Labs, Meiji Holdings Co., Ltd., Tokyo 192-0919, Japan; kyousuke.kobayashi@meiji.com (K.K.); hiroshi.kano@meiji.com (H.K.)

* Correspondence: kenichi.houjou@meiji.com; Tel.: +81-42-632-5847

Abstract: We investigated the effects of *Lactobacillus paragasseri* OLL2716 on gastrointestinal symptoms in healthy adults with gastric complaints. In this randomized, double-blind, placebo-controlled trial, 174 healthy Japanese adults were randomly assigned to an OLL2716 or placebo group, and each group consumed 85 g of yogurt containing *L. paragasseri* OLL2716 or placebo yogurt daily for 12 weeks. The primary endpoint was the change in gastric symptoms from baseline as per the participants' questionnaires at 6 and 12 weeks. The secondary endpoints were changes from baseline in the short-form Nepean Dyspepsia Index (SF-NDI), the Gastrointestinal Symptom Rating Scale (GSRS), and the Council on Nutrition Appetite Questionnaire-Japanese (CNAQ-J) scores at 6 and 12 weeks. The primary endpoint data showed that the changes in "epigastric pain" at 6 and 12 weeks were significantly decreased in the OLL2716 group compared with those in the placebo group. Additionally, the changes in "epigastric pain syndrome-like symptoms" were significantly decreased in the OLL2716 group compared with those in the placebo group at 6 weeks. The SF-NDI items that improved at 6 weeks were "irritable, tense, or frustrated", "enjoyment of eating or drinking", and "tension", which are sub-scales related to mental stress. The items "Over-all" in the GSRS and "feeling hungry" in the CNAQ-J significantly improved in the OLL2716 group compared with the placebo group at 12 weeks. The results suggest that regular intake of *L. paragasseri* OLL2716 may improve both gastric discomfort and mental stress in healthy adults with gastric complaints, such as postprandial fullness or early satiety.

Keywords: probiotics; *Lactobacillus paragasseri* OLL2716; gastrointestinal discomfort; autonomic nervous system; mental stress; appetite

1. Introduction

Mental stress in daily life is thought to be related to gastrointestinal dysfunction and subsequent harmful symptoms because it disrupts the autonomic nervous system. Several studies have suggested that disturbances in the autonomic nervous system can cause delayed gastric motility [1–3]. A recent systematic review and meta-analysis supported the association between optimally measured delayed gastric emptying and upper gastrointestinal symptoms [4]. Furthermore, several studies concerning the stomach–brain axis have been performed, and they have reported the existence of a mutual communication between the brain and stomach via the autonomic nervous system [5–7].

Mental stress affects gastric motility, visceral perception, and the secretion of gastric hormone [8]. In contrast, stomach dysfunction, such as stomach pain and overloaded stomach acid, causes mental stress because it directly impairs the quality of life [9,10].

According to epidemiological studies, 50% of functional gastroenteropathies are caused by psychological stress, whereas the remaining 50% are caused by gastrointestinal tract dysfunction [11–13]. Although the cause-and-effect relationship remains unclear [14], maintaining both mental health and normal stomach function is essential.

Recently, several probiotic strains have been reported to exert beneficial effects on gastrointestinal function and discomfort [3,15,16]. *Lactobacillus paragasseri* OLL2716 (formerly known as *Lactobacillus gasseri*) has been reported to have beneficial effects, such as suppressive effects on *Helicobacter pylori* infection and the improvement of functional dyspepsia (FD), which is a condition with persistent abdominal symptoms mainly affecting the epigastrium [17,18]. Otomi et al. reported that *L. paragasseri* OLL2716 may ameliorate autonomic nervous system disorders [15]. Ohtsu et al. reported that *L. paragasseri* OLL2716 improved delayed gastric emptying and salivary amylase concentrations in healthy adults with stomach dysfunction [3]. In this study, we investigated the effects of *L. paragasseri* OLL2716 on gastric discomfort and mental stress in healthy adults with gastric complaints.

2. Materials and Methods

2.1. Study Design

A randomized, double-blind, placebo-controlled, parallel-group trial was conducted by assigning participants to either the OLL2716 or the placebo group (Figure 1). This study was conducted between June 2021 and April 2022 at eight hospitals and clinics (Ageo Central Second Hospital, Saitama, Japan; Kanauchi Medical Clinic, Tokyo, Japan; Kanazawabunko Hospital, Kanagawa, Japan; Fuefuki Central Hospital, Yamanashi, Japan; MY Medical Clinic Shibuya, Tokyo, Japan; Musashisakai Clinic, Tokyo, Japan; Medicaltopia Soka Hospital, Saitama, Japan; and Nihonbashi Sakura Clinic, Tokyo, Japan). The study was approved by the Ethics Committees of Meiji Co., Ltd. (Tokyo, Japan) and Ageo Central Second Hospital (Saitama, Japan). In compliance with the ethical principles of the Declaration of Helsinki and the ethical guidelines for epidemiological studies, the participants were fully informed of the purpose and content of the study, and written informed consent was obtained from all participants before participating in the study. The study protocol was registered on 12 November 2020, using the University Hospital Medical Information Network (UMIN) Clinical Trial Registration System (UMIN000042422). CONSORT 2010 checklist can be found in Table S1.

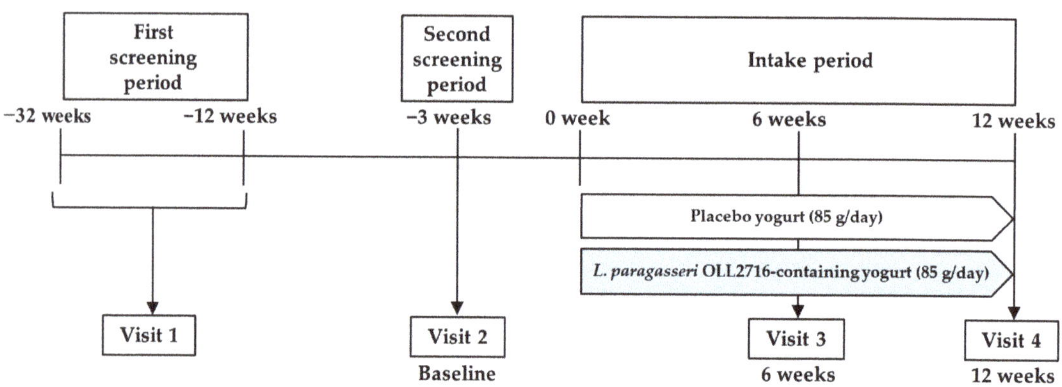

Figure 1. Study design and schedule to investigate the effects of *L. paragasseri* OLL2716 on gastric discomfort and mental stress in healthy adults with gastric complaints.

2.2. Participants

Healthy Japanese adults aged 20–64 years were recruited for this study. Eligibility was determined based on whether participants met the inclusion and exclusion criteria. The inclusion criteria were as follows: (1) healthy Japanese men and women aged 20–64 years at the time of consent; (2) participants who felt "mild" to "slightly severe" postprandial fullness or early satiety on the 'Individual Gastric Symptom Scores' at the first screening test and baseline (second screening).

The exclusion criteria were as follows: (1) receiving medical care or treatment for diabetes or gastrointestinal-, eating-, or stress-related psychiatric disorders in the past 6 months; (2) treatment with medication for dyspepsia symptoms in the past 6 months; (3) taking low-dose aspirin or nonsteroidal anti-inflammatory drugs (NSAIDs) in the past 6 months or more; (4) meeting at least one of the first screening or baseline functional dyspepsia Rome IV criteria; (5) serum anti-*H. pylori* antibody titer of $\geq$10 U/mL at the first *H. pylori* screening test and having received *H. pylori* eradication treatment; (6) severe heartburn or acid reflux at either the first screening or baseline; (7) suspected diabetes mellitus, dyslipidemia, gastrointestinal disease, or severe renal impairment at the first screening or baseline; (8) undergoing any treatment other than those listed previously (excluding transient treatments such as those for the common cold); (9) judged to have a good appetite on the Council on Nutrition Appetite Questionnaire-Japanese (CNAQ-J) at either the first screening or baseline; (10) m-FSSG scores for "postprandial fullness" and "early satiety" were both "0" at baseline; (11) answer was the "best (score 1)" on Q1 of the SF-36v2 at baseline; (12) a total score of 2 on the short-form Nepean Dyspepsia Index (SF-NDI) "irritable, tense, or frustrated because of your stomach problems" at baseline; (13) taking medicines that may affect gastric symptoms for at least 3 days per week for the past one month or more; (14) regular use of foods containing lactic acid bacteria such as yogurt, health foods, or supplements for at least 1 month; (15) taking medicines (such as antibiotics) that affect lactic acid bacteria; (16) smokers; (17) normal alcohol consumption exceeding 40 g per day; (18) food allergies; (19) pregnant, planning or hoping to become pregnant during the study period, or breastfeeding; (20) dental or oral problems that cause bleeding during saliva test; (21) marked lifestyle changes; (22) unable to perform the various testing procedures; (23) would or might travel for more than 1 week during the study period; (24) participated in other clinical trials in the past month or plan to participate in other clinical trials during the study period; and (25) judged by the investigator to be unsuitable for participation. The m-FSSG and SF-36v2 were used only as exclusion criteria during volunteer screening and were not used for the efficacy evaluation. Ultimately, the clinicians made a comprehensive judgment to ensure that no such patients were included in this study.

2.3. Study Protocol

The allocation manager randomly assigned participants to two groups, one receiving the test food and the other receiving the control food, using a block randomization method, with age and sex as adjustment factors. The placebo group was asked to ingest 85 g of yogurt per serving, which consisted of raw milk, dairy products, sugar, a sweetener (stevia), and water fermented with *Lactobacillus delbrueckii* subsp. *bulgaricus* and *Streptococcus thermophilus*. The OLL2716 group was asked to ingest yogurt containing *L. paragasseri* OLL2716 (with $\geq 10^9$ colony-forming units of *L. paragasseri* OLL2716 per serving), added to the same yogurt described for the placebo group. The nutritional values per serving (85 g) of placebo yogurt and yogurt containing *L. paragasseri* OLL2716 were 68 kcal energy, 2.9 g protein, 2.6 g fat, 8.3 g carbohydrate, and 102 mg calcium. After confirming that they were indistinguishable in flavor and appearance, the participants ingested one (85 g) of the assigned test foods per day for 12 weeks. The timing of consumption was not limited. In addition to the person in charge of allocation who assigned the participants to two groups and determined the group of participants, the person who decided the identification number of the test food products and the person who assigned the identification number of the test food products to the group number were separated. As a result, the participants and all study staff were blinded to the test food by using a method in which the three pieces of information necessary for key opening were not collected.

2.4. Helicobacter Pylori Screening Test

Serum anti-*H. pylori* antibody titers were measured during the first screening conducted by SRL (Tokyo, Japan), and participants with a titer of 10 U/mL or higher were excluded from the study because they were *H. pylori*-infected.

2.5. FD Rome IV Diagnostic Criteria

Participants who answered at least one question regarding postprandial fullness, early satiety, epigastric pain, or epigastric burning at baseline were suspected of having FD using a questionnaire based on the international diagnostic criteria (Rome IV) at first screening and baseline [19]. Two FD subtypes have been defined: (1) postprandial fullness and early satiety, classified as postprandial distress syndrome (PDS); and (2) epigastric pain and burning, classified as epigastric pain syndrome (EPS) [20]. Therefore, participants suspected of having FD symptoms, such as PDS and EPS, based on the Rome IV questionnaire were excluded according to the exclusion criteria, and participants with chronic stomach discomfort were also excluded from this study.

2.6. Individual Gastric Symptom Scores (Questionnaire for Gastric Symptoms of the Participants)

A questionnaire on the severity of individual FD and accompanying symptoms was completed during the baseline period and after 6 and 12 weeks of test food intake, as described in a previous study [18]. Participants rated the severity of symptoms (postprandial fullness, early satiety, epigastric bloating, epigastric pain, epigastric burning, heartburn, reflux feeling of gastric acid, nausea, belching, and abdominal bloating) occurring in the prior week on a seven-point Likert scale [18,21,22]: (1) none (absence of symptoms); (2) extremely mild (symptoms could be entirely ignored); (3) mild (symptoms easily tolerated); (4) moderate (symptoms noticed by the patient, but did not affect daily activities); (5) moderate-to-severe (symptoms occasionally limited daily activities); (6) severe (symptoms often limited daily activities); and (7) extremely severe (considerable interference with daily activities, often requiring rest). Furthermore, symptoms including "postprandial fullness" and "early satiety" were categorized as PDS-like symptoms; those including "epigastric pain" and "epigastric burning" as EPS-like symptoms; and symptoms including both PDS- and EPS-like symptoms as FD-like symptoms. This questionnaire has been widely used not only in previous clinical studies on *L. paragasseri* OLL2716 [18] but also as a tool for evaluating upper gastrointestinal symptoms [21,22].

2.7. Short-Form Nepean Dyspepsia Index (SF-NDI)

The SF-NDI consists of several items that assess health-related quality of life [23,24]. The questionnaire was administered at the first screening, baseline, and 6 and 12 weeks after consumption of the test foods. Five sub-scales, namely tension (two items, "general emotional well-being" and "irritable, tense, or frustrated"), interference with daily activities [two items, "fun (ability)" and "fun (enjoyment)"], eating/drinking [two items, "eat or drink (ability)" and "eating or drinking (enjoyment)"], knowledge/control [two items, "wondered (always)" and "thought (very serious illness)"], and work/study [two items, "work or study (ability)" and "work or study (enjoyment)"], were derived.

2.8. Gastrointestinal Symptom Rating Scale (GSRS)

The questionnaire was administered at baseline and at 6 and 12 weeks after consumption of the test foods. The GSRS is a 15-item, self-administered questionnaire that assesses the severity of a wide range of gastrointestinal symptoms [25,26]; Each of the 15 questions on a Likert scale could be scored on a scale of 1–7, with a total score ranging from 15 to 105 points; however, they were divided by the number of questions to evaluate the score. Five sub-scales, namely reflux syndrome (RS, two items, "heartburn" and "acid regurgitation"), abdominal pain syndrome (AP, three items, "abdominal pain", "sucking sensations in the epigastrium", and "nausea and vomiting"), indigestion syndrome (IS, four items, "borborygmus", "abdominal distension", "eructation", and "increased flatus"), constipation syndrome (CS, three items, "decreased passage of stools", "hard stools", and "feeling of incomplete evacuation"), and diarrhea syndrome (DS, three items, "increased passage of stools", "loose stools", and "urgent need for defecation"), were derived. Furthermore, symptoms including RS, AP, and IS were categorized as upper gastrointestinal (GI)-related, whereas symptoms including CS and DS were categorized as lower GI-related.

2.9. Council on Nutrition Appetite Questionnaire-Japanese (CNAQ-J)

The CNAQ-J consists of eight categories: (1) appetite, (2) feeling full, (3) feeling hungry, (4) food tastes, (5) food tastes compared with those when younger, (6) meal frequency per day, (7) feeling sick or nauseated when eating, and (8) usual mood. Each item is answered on a five-point scale ranging from 1 to 5, with the total score ranging from 8 to 40 [27,28]. Additionally, we defined the total score of feeling full and feeling hungry as "pre- and post-meal satisfaction" and the total score of appetite, feeling full, and feeling hungry as "eating satisfaction" for evaluation. The CNAQ-J was administered at the first screening, baseline, and 6 and 12 weeks after consumption of the test foods.

2.10. Statistical Analyses

The effect size of *L. paragasseri* OLL2716 in the present study was predicted based on a previous study [18] that examined its effectiveness in improving functional dyspepsia, because no studies have examined the effectiveness of *L. paragasseri* OLL2716 in healthy adults on the improvement of upper gastrointestinal symptoms. In a previous study [18], the disappearance rate of PDS symptoms in individuals with PDS symptoms was 37.5% in the OLL2716 group (n = 48) and 17.8% in the placebo group (n = 45), with a difference in disappearance rates of 19.7%. In the current study targeting healthy adults with PDS-like symptoms, the difference in disappearance rates between the OLL2716 and placebo groups was 19.7%, similar to that in the previous study [18]. At a two-sided significance level of 5% and power of 80%, the required sample size for each group was calculated to be 78. Considering an approximately 10% dropout rate, the target number of participants in the study was set at 90 for each group.

Intergroup comparisons at baseline were performed using the chi-square test for categorical variables and unpaired Student's t-test for continuous variables. Gastrointestinal symptoms were compared between the two groups using the Wilcoxon signed rank and Fisher's exact tests. As this study included healthy adults with low-severity upper gastrointestinal symptoms, the improvement rate instead of the disappearance rate was used for the evaluation. To compare the ratio of improvement in gastrointestinal symptoms between the groups, logistic regression analysis with group factors was performed to evaluate the odds ratio, 95% confidence interval, and p-value for the placebo group. Data are expressed as means ± standard deviation. All statistical analyses were two-sided, and the significance level was set at 5% with no adjustment for multiple comparisons. Data analysis was conducted using SAS version 9.4 (SAS Institute Inc., Cary, NC, USA) and Bell Curve for Excel 2016 version 3.20 (Social Survey Research Information Co., Ltd., Tokyo, Japan).

3. Results

3.1. Participant Selection and Baseline Characteristics

A total of 487 Japanese individuals (21–62 years old) who provided written, informed consent were recruited and selected using two screening tests: (1) first screening and (2) second screening (baseline). According to the protocol, after the screening, 174 participants were enrolled in the present study and randomly assigned to the OLL2716 and placebo groups, with 87 participants in each group (Figure 2).

No significant differences were observed in the participants' background characteristics (Table 1). All 174 participants completed the intake period of 12 weeks and were included in the analysis. No adverse effects or serious adverse events were observed in either group. Additionally, the average intake rate was 99.9% in both the OLL2716 and placebo groups, with the lowest intake rate among the participants being 97.5%. Compliance was confirmed by having the participants record their daily intake in a lifestyle diary.

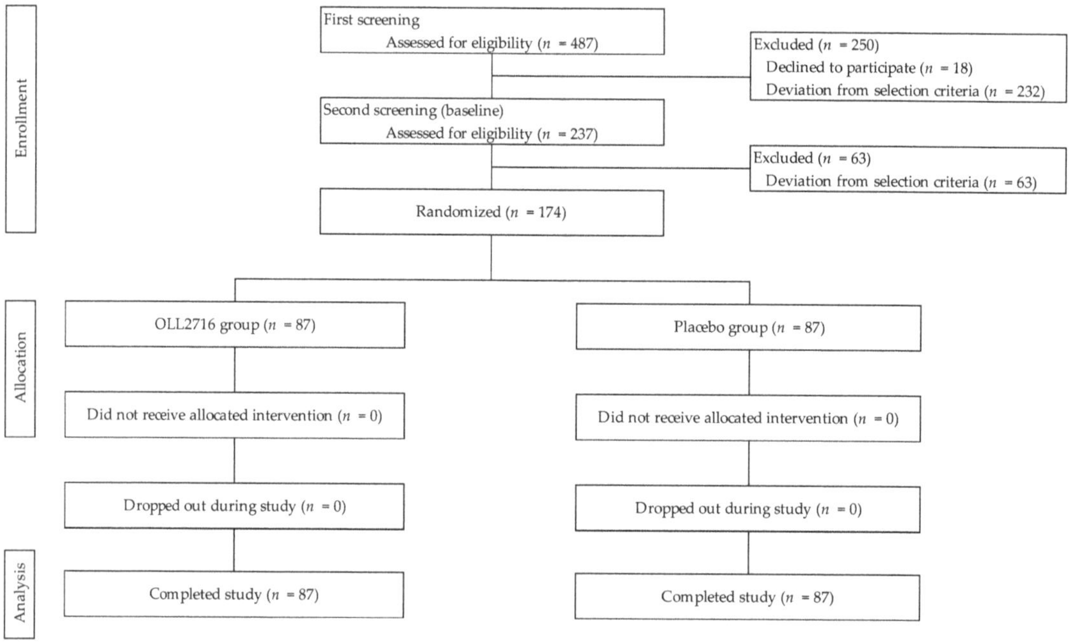

Figure 2. Flow chart of participants in this study.

Table 1. Baseline characteristics of the participants.

Characteristics	OLL2716 (n = 87)	Placebo (n = 87)	p-Value
Female	55	56	0.875 [1]
Male	32	31	
Age (year)	37.4 ± 11.9	37.0 ± 11.1	0.792 [2]
Postprandial fullness score	2.5 ± 0.7	2.5 ± 0.8	0.839 [2]
Early satiety score	2.1 ± 0.9	2.1 ± 1.0	0.814 [2]
BMI [kg(m^2)$^{-1}$]	22.2 ± 3.7	22.4 ± 3.5	0.777 [2]

Data are shown as mean ± standard deviation. Intergroup comparison: [1] Chi-square test, [2] *t*-test; BMI, body mass index.

3.2. Primary Endpoint

In the present study, the primary endpoint was the change in gastric symptoms in the participants' questionnaires, similar to a previous study [18]. Table 2 shows the changes in scores and the number of participants with improved scores. The number of participants with improved scores is further presented as the proportion of the number of participants with improved scores to the number of evaluated participants. The changes in epigastric pain at 6 and 12 weeks were −0.5 ± 1.3 and −0.7 ± 1.2 in the OLL2716 group and −0.1 ± 1.3 and −0.2 ± 1.4 in the placebo group, respectively. The data showed that changes in epigastric pain in the OLL2716 group were significantly improved compared with those in the placebo group at 6 and 12 weeks. The changes in EPS-like symptoms in the OLL2716 group (−0.9 ± 2.1) were significantly improved compared with those in the placebo group (−0.3 ± 2.3) at 6 weeks. Similarly, the number of participants with improved scores for epigastric pain and EPS-like symptoms in the OLL2716 group was 45 and 52, respectively, which was a significant improvement compared with that in the placebo group (30 and 34, respectively) at 6 weeks. No significant differences in other symptoms were noted between the two groups.

3.3. Secondary Endpoints

3.3.1. SF-NDI

In the present study, several items concerning the stress caused by gastric symptoms were improved by the intake of the yogurt containing *L. paragasseri* OLL2716 (Table 3). Specifically, the changes in "irritable, tense, or frustrated", "enjoyment of eating or drinking", and tension (summary of two items), which are the sub-scales related to mental stress, in the OLL2716 group at 6 weeks were improved compared with those in the placebo group. The number of participants with improved scores for "enjoyment of eating or drinking" in the OLL2716 group was significantly more than that in the placebo group at 6 weeks (Table 3).

3.3.2. GSRS

The changes in the GSRS scores were similar between the two groups (Table 4). The number of participants with improved scores for "lower GI symptoms" and "Over-all" in the OLL2716 group at 12 weeks was significantly more than that in the placebo group. No significant differences in other symptoms were noted between the two groups.

3.3.3. CNAQ-J

Changes in the CNAQ-J scores for feeling hungry at 12 weeks improved with the intake of yogurt containing *L. paragasseri* OLL2716 (Table 5). In addition, changes at 12 weeks in the CNAQ-J scores for "pre- and post-meal satisfaction", including feeling full and feeling hungry and for "eating satisfaction", including appetite, feeling full, and feeling hungry, were significantly improved in the OLL2716 group, compared with those in the placebo group. No significant differences in other symptoms were noted between the two groups.

Table 2. Score change before ingestion and the number of participants with improved scores in the Individual Gastric Symptom Scores.

Items	Group	0 Week	6 Weeks	12 Weeks	Score Change before Ingestion (⊿)				Number of Participants with Improved Scores (%)			
					6 Weeks–0 Week	*p* Value [1]	12 Weeks–0 Week	*p* Value [1]	6 Weeks	*p* Value [2]	12 Weeks	*p* Value [2]
1. Postprandial fullness	OLL2716	2.5 ± 0.7	2.0 ± 0.8	1.7 ± 0.9	−0.5 ± 0.8	0.564	−0.8 ± 1.0	0.745	37 (42.5%)	0.879	50 (57.5%)	1.000
	Placebo	2.5 ± 0.8	1.9 ± 0.9	1.6 ± 1.0	−0.6 ± 1.1		−0.9 ± 1.2		39 (44.8%)		49 (56.3%)	
2. Early satiety	OLL2716	2.1 ± 0.9	1.9 ± 1.0	1.6 ± 1.1	−0.2 ± 1.0	0.483	−0.5 ± 1.0	0.841	29 (33.3%)	0.528	37 (42.5%)	0.760
	Placebo	2.1 ± 1.0	1.8 ± 1.0	1.6 ± 1.1	−0.3 ± 1.0		−0.5 ± 1.1		34 (39.1%)		40 (46.0%)	
3. Epigastric bloating	OLL2716	2.3 ± 1.1	1.9 ± 1.0	1.6 ± 1.1	−0.5 ± 1.1	0.362	−0.7 ± 1.1	0.861	46 (52.9%)	0.544	48 (55.2%)	1.000
	Placebo	2.2 ± 1.2	1.9 ± 1.3	1.5 ± 1.1	−0.4 ± 1.2		−0.7 ± 1.4		41 (47.1%)		48 (55.2%)	
4. Epigastric pain	OLL2716	1.6 ± 1.1	1.1 ± 1.1	1.0 ± 1.0	−0.5 ± 1.3	0.020	−0.7 ± 1.2	0.028	45 (51.7%)	0.032	47 (54.0%)	0.095
	Placebo	1.3 ± 1.2	1.2 ± 1.1	1.1 ± 1.0	−0.1 ± 1.3		−0.2 ± 1.4		30 (34.5%)		35 (40.2%)	
5. Epigastric burning	OLL2716	1.3 ± 1.2	0.9 ± 1.0	1.0 ± 1.0	−0.4 ± 1.0	0.256	−0.4 ± 1.0	0.289	33 (37.9%)	0.753	34 (39.1%)	0.754
	Placebo	1.1 ± 1.2	0.9 ± 0.9	0.9 ± 1.0	−0.3 ± 1.3		−0.2 ± 1.4		30 (34.5%)		31 (35.6%)	
6. Heartburn	OLL2716	1.5 ± 1.2	1.1 ± 1.0	1.0 ± 1.0	−0.5 ± 1.1	0.550	−0.6 ± 1.1	0.322	37 (42.5%)	1.000	50 (57.5%)	0.225
	Placebo	1.4 ± 1.2	1.1 ± 1.1	1.0 ± 0.9	−0.3 ± 1.2		−0.4 ± 1.3		36 (41.4%)		41 (47.1%)	
7. Reflex feeling of gastric acid	OLL2716	1.5 ± 1.1	1.0 ± 1.0	1.0 ± 1.0	−0.5 ± 1.0	0.258	−0.5 ± 1.1	0.094	41 (47.1%)	0.166	39 (44.8%)	0.442
	Placebo	1.2 ± 1.2	0.9 ± 1.0	1.0 ± 1.0	−0.3 ± 1.2		−0.2 ± 1.3		31 (35.6%)		33 (37.9%)	
8. Nausea	OLL2716	1.0 ± 1.1	0.7 ± 1.0	0.7 ± 1.0	−0.3 ± 1.0	0.583	−0.3 ± 0.9	0.413	29 (33.3%)	0.873	30 (34.5%)	1.000
	Placebo	1.0 ± 1.0	0.8 ± 1.0	0.8 ± 0.9	−0.2 ± 1.0		−0.2 ± 1.2		31 (35.6%)		30 (34.5%)	
9. Belching	OLL2716	1.6 ± 1.2	1.4 ± 1.1	1.2 ± 1.1	−0.3 ± 1.1	0.653	−0.4 ± 1.0	0.818	33 (37.9%)	0.877	40 (46.0%)	1.000
	Placebo	1.7 ± 1.3	1.3 ± 1.1	1.3 ± 1.0	−0.3 ± 1.1		−0.4 ± 1.2		35 (40.2%)		40 (46.0%)	
10. Abdominal bloating	OLL2716	2.1 ± 1.2	1.7 ± 1.0	1.6 ± 1.2	−0.4 ± 1.2	0.108	−0.5 ± 1.3	0.638	43 (49.4%)	0.091	41 (47.1%)	0.649
	Placebo	2.0 ± 1.4	1.7 ± 1.3	1.4 ± 1.1	−0.2 ± 1.3		−0.6 ± 1.4		31 (35.6%)		45 (51.7%)	
PDS-like (1 and 2)	OLL2716	4.5 ± 1.3	3.8 ± 1.6	3.3 ± 1.8	−0.7 ± 1.5	0.455	−1.2 ± 1.7	0.689	43 (49.4%)	0.880	51 (58.6%)	0.757
	Placebo	4.6 ± 1.4	3.7 ± 1.8	3.2 ± 1.9	−0.9 ± 1.7		−1.4 ± 2.0		45 (51.7%)		54 (62.1%)	
EPS-like (4 and 5)	OLL2716	3.0 ± 2.1	2.0 ± 1.9	1.9 ± 1.9	−0.9 ± 2.1	0.028	−1.0 ± 1.9	0.073	52 (59.8%)	0.010	49 (56.3%)	0.068
	Placebo	2.4 ± 2.0	2.1 ± 1.8	1.9 ± 1.8	−0.3 ± 2.3		−0.5 ± 2.5		34 (39.1%)		36 (41.4%)	
FD-like (PDS-like and EPS-like)	OLL2716	7.5 ± 2.9	5.9 ± 3.0	5.2 ± 3.3	−1.6 ± 3.0	0.207	−2.3 ± 3.0	0.377	50 (57.5%)	0.129	52 (59.8%)	0.540
	Placebo	7.0 ± 2.8	5.7 ± 3.1	5.1 ± 3.4	−1.3 ± 3.3		−1.9 ± 4.0		39 (44.8%)		47 (54.0%)	

Changes in scores and number of participants with improved scores from baseline. Data are shown as mean ± standard deviation (n = 87 in the OLL2716 group and n = 87 in the placebo group). Intergroup comparison: [1] Wilcoxon rank sum test and [2] Fisher's exact test. Number of improved participants (%): (number of participants who improved/total number of participants evaluated) × 100. PDS, postprandial distress syndrome; EPS, epigastric pain syndrome; FD, functional dyspepsia. Sub-scales: PDS-like, "postprandial fullness" and "early satiety"; EPS-like, "epigastric pain" and "epigastric burning"; and FD-like, PDS- and EPS-like.

Table 3. Score change before ingestion and the number of participants with improved scores in the Short-form Nepean Dyspepsia Index (SF-NDI).

Items	Group	0 Week	6 Weeks	12 Weeks	Score Change before Ingestion (⊿)				Number of Participants with Improved Scores (%)			
					6 Weeks–0 Week	p Value [1]	12 Weeks–0 Week	p Value [1]	6 Weeks	p Value [2]	12 Weeks	p Value [2]
1. General emotional well-being	OLL2716	2.0 ± 0.6	1.7 ± 0.6	1.6 ± 0.6	−0.3 ± 0.7	0.084	−0.4 ± 0.7	0.822	28 (32.2%)	0.121	39 (44.8%)	0.760
	Placebo	2.0 ± 0.6	1.9 ± 0.6	1.6 ± 0.5	−0.1 ± 0.7		−0.4 ± 0.7		18 (20.7%)		36 (41.4%)	
2. Irritable, tense, or frustrated	OLL2716	2.2 ± 0.5	1.7 ± 0.6	1.6 ± 0.6	−0.5 ± 0.6	0.042	−0.6 ± 0.7	0.666	40 (46.0%)	0.061	47 (54.0%)	0.544
	Placebo	2.2 ± 0.5	1.9 ± 0.6	1.6 ± 0.6	−0.3 ± 0.7		−0.6 ± 0.7		27 (31.0%)		42 (48.3%)	
3. Fun (ability)	OLL2716	1.7 ± 0.7	1.3 ± 0.5	1.3 ± 0.5	−0.3 ± 0.8	0.353	−0.4 ± 0.8	0.457	34 (39.1%)	0.429	39 (44.8%)	0.279
	Placebo	1.6 ± 0.8	1.4 ± 0.6	1.3 ± 0.5	−0.2 ± 0.7		−0.3 ± 0.7		28 (32.2%)		31 (35.6%)	
4. Fun (enjoyment)	OLL2716	1.7 ± 0.7	1.3 ± 0.5	1.3 ± 0.5	−0.4 ± 0.8	0.284	−0.4 ± 0.8	0.909	38 (44.2%)	0.279	39 (45.3%)	0.760
	Placebo	1.7 ± 0.8	1.4 ± 0.7	1.3 ± 0.5	−0.3 ± 0.8		−0.4 ± 0.7		31 (35.6%)		37 (42.5%)	
5. Eat or drink (ability)	OLL2716	1.9 ± 0.7	1.6 ± 0.6	1.4 ± 0.6	−0.2 ± 0.7	0.215	−0.5 ± 0.7	0.226	29 (33.3%)	0.177	44 (50.6%)	0.287
	Placebo	1.8 ± 0.7	1.6 ± 0.6	1.4 ± 0.6	−0.1 ± 0.6		−0.4 ± 0.7		20 (23.0%)		36 (41.4%)	
6. Eating or drinking (enjoyment)	OLL2716	2.0 ± 0.7	1.5 ± 0.6	1.4 ± 0.6	−0.5 ± 0.9	0.031	−0.6 ± 0.8	0.150	42 (48.3%)	0.044	52 (59.8%)	0.171
	Placebo	1.9 ± 0.8	1.6 ± 0.6	1.4 ± 0.6	−0.2 ± 0.7		−0.5 ± 0.8		28 (32.2%)		42 (48.3%)	
7. Wondered (always)	OLL2716	1.7 ± 0.7	1.3 ± 0.5	1.2 ± 0.5	−0.4 ± 0.7	0.421	−0.4 ± 0.8	0.454	31 (35.6%)	0.63	38 (43.7%)	0.537
	Placebo	1.6 ± 0.8	1.4 ± 0.6	1.2 ± 0.5	−0.3 ± 0.8		−0.4 ± 0.9		27 (31.0%)		33 (37.9%)	
8. Thought (very serious illness)	OLL2716	1.3 ± 0.5	1.1 ± 0.4	1.1 ± 0.4	−0.1 ± 0.5	0.182	−0.2 ± 0.5	0.097	16 (18.4%)	0.403	19 (21.8%)	0.328
	Placebo	1.2 ± 0.4	1.2 ± 0.4	1.1 ± 0.3	0.0 ± 0.5		−0.1 ± 0.5		11 (12.6%)		13 (14.9%)	
9. Work or study (ability)	OLL2716	1.6 ± 0.6	1.5 ± 0.6	1.3 ± 0.5	−0.1 ± 0.7	0.620	−0.3 ± 0.7	0.677	24 (27.6%)	1.000	30 (34.5%)	1.000
	Placebo	1.6 ± 0.6	1.4 ± 0.7	1.3 ± 0.5	−0.1 ± 0.7		−0.3 ± 0.6		24 (27.6%)		29 (33.3%)	
10. Work or study (enjoyment)	OLL2716	1.6 ± 0.7	1.4 ± 0.6	1.3 ± 0.6	−0.2 ± 0.8	0.681	−0.3 ± 0.8	0.890	30 (34.5%)	0.747	31 (35.6%)	0.749
	Placebo	1.6 ± 0.7	1.4 ± 0.6	1.3 ± 0.5	−0.2 ± 0.8		−0.3 ± 0.7		27 (31.0%)		28 (32.2%)	
Tension (1 and 2)	OLL2716	4.1 ± 0.9	3.4 ± 1.1	3.1 ± 1.2	−0.7 ± 1.1	0.026	−1.0 ± 1.2	0.675	45 (51.7%)	0.032	54 (62.1%)	0.357
	Placebo	4.2 ± 1.0	3.8 ± 1.1	3.2 ± 1.1	−0.4 ± 1.3		−1.0 ± 1.2		30 (34.5%)		47 (54.0%)	
Interference with daily activities (3 and 4)	OLL2716	3.4 ± 1.2	2.7 ± 1.0	2.6 ± 1.0	−0.7 ± 1.4	0.262	−0.8 ± 1.4	0.575	42 (48.8%)	0.361	46 (53.5%)	0.288
	Placebo	3.3 ± 1.5	2.8 ± 1.3	2.5 ± 1.0	−0.5 ± 1.4		−0.8 ± 1.3		36 (41.4%)		39 (44.8%)	
Eating/drinking (5 and 6)	OLL2716	3.9 ± 1.3	3.2 ± 1.0	2.8 ± 1.1	−0.7 ± 1.4	0.057	−1.1 ± 1.5	0.095	45 (51.7%)	0.224	56 (64.4%)	0.217
	Placebo	3.6 ± 1.3	3.2 ± 1.1	2.8 ± 1.1	−0.4 ± 1.1		−0.9 ± 1.3		36 (41.4%)		47 (54.0%)	
Knowledge/control (7 and 8)	OLL2716	3.0 ± 1.0	2.5 ± 0.8	2.4 ± 0.8	−0.5 ± 1.0	0.240	−0.6 ± 1.1	0.223	37 (42.5%)	0.210	41 (47.1%)	0.358
	Placebo	2.9 ± 1.2	2.5 ± 0.9	2.4 ± 0.8	−0.3 ± 1.1		−0.5 ± 1.3		28 (32.2%)		34 (39.1%)	
Work/study (9 and 10)	OLL2716	3.2 ± 1.3	2.8 ± 1.1	2.6 ± 1.0	−0.3 ± 1.5	0.977	−0.6 ± 1.4	0.942	33 (37.9%)	0.875	36 (41.4%)	1.000
	Placebo	3.2 ± 1.3	2.9 ± 1.3	2.6 ± 0.9	−0.3 ± 1.4		−0.7 ± 1.2		31 (35.6%)		35 (40.2%)	

Changes in scores and number of participants with improved scores from baseline. Data are shown as mean ± standard deviation (n = 87 in the OLL2716 group and n = 87 in the placebo group). Intergroup comparison: [1] Wilcoxon rank sum test and [2] Fisher's exact test. Number of improved participants (%): (number of participants who improved/total number of participants evaluated) × 100. Sub-scales: tension, "general emotional well-being" and "irritable, tense, or frustrated"; interference with daily activities, "fun (ability)" and "fun (enjoyment)"; eating/drinking, "eat or drink (ability)" and "eating or drinking (enjoyment)"; knowledge/control, "wondered (always)" and "thought (very serious illness)"; and work/study, "work or study (ability)" and "work or study (enjoyment)".

Table 4. Score change before ingestion and the number of participants with improved scores in the Gastrointestinal Symptom Rating Scale (GSRS).

Items	Group	0 Week	6 Weeks	12 Weeks	Score Change before Ingestion (⊿)				Number of Participants with Improved Scores (%)			
					6 Weeks–0 Week	p Value [1]	12 Weeks–0 Week	p Value [1]	6 Weeks	p Value [2]	12 Weeks	p Value [2]
1. Abdominal pain	OLL2716	2.5 ± 1.1	2.0 ± 0.8	1.9 ± 0.8	−0.4 ± 1.2	0.369	−0.6 ± 1.1	0.143	37 (42.5%)	0.643	48 (55.2%)	0.288
	Placebo	2.2 ± 1.0	2.0 ± 1.0	1.8 ± 1.0	−0.2 ± 1.1		−0.4 ± 1.1		33 (37.9%)		40 (46.0%)	
2. Heartburn	OLL2716	2.1 ± 0.8	1.9 ± 0.8	1.7 ± 0.8	−0.2 ± 1.0	0.439	−0.4 ± 0.8	0.641	27 (31.0%)	0.522	38 (43.7%)	0.879
	Placebo	2.1 ± 1.0	1.8 ± 0.9	1.6 ± 0.8	−0.3 ± 1.0		−0.5 ± 0.9		32 (36.8%)		40 (46.0%)	
3. Acid regurgitation	OLL2716	2 ± 0.9	1.8 ± 0.8	1.6 ± 0.7	−0.2 ± 0.9	0.843	−0.4 ± 0.8	0.220	26 (29.9%)	0.736	42 (48.3%)	0.221
	Placebo	1.9 ± 0.9	1.7 ± 0.8	1.6 ± 0.8	−0.2 ± 0.9		−0.3 ± 1.0		23 (26.4%)		33 (37.9%)	
4. Sucking sensations in the epigastrium	OLL2716	2.1 ± 0.9	1.9 ± 0.9	1.7 ± 0.8	−0.2 ± 1.0	0.534	−0.4 ± 1.0	0.228	34 (39.1%)	0.340	38 (43.7%)	0.440
	Placebo	2.0 ± 0.8	1.8 ± 0.9	1.7 ± 0.9	−0.2 ± 0.9		−0.2 ± 0.9		27 (31.0%)		32 (36.8%)	
5. Nausea and vomiting	OLL2716	1.9 ± 1.0	1.6 ± 0.9	1.6 ± 0.7	−0.3 ± 1.2	0.498	−0.3 ± 0.8	0.701	33 (37.9%)	0.144	33 (37.9%)	0.635
	Placebo	1.9 ± 0.9	1.8 ± 0.8	1.7 ± 0.9	−0.1 ± 0.9		−0.3 ± 0.9		23 (26.4%)		29 (33.3%)	
6. Borborygmus	OLL2716	2.3 ± 1.1	2.4 ± 1.0	2.1 ± 1.0	0.0 ± 1.0	0.089	−0.3 ± 1.0	0.931	24 (27.6%)	0.147	34 (39.1%)	0.754
	Placebo	2.6 ± 1.2	2.3 ± 1.1	2.2 ± 1.1	−0.3 ± 1.1		−0.3 ± 1.2		34 (39.1%)		31 (35.6%)	
7. Abdominal distension	OLL2716	2.4 ± 1.0	2.1 ± 0.8	2.0 ± 0.8	−0.3 ± 1.1	0.507	−0.4 ± 1.0	0.654	35 (40.2%)	1.000	38 (43.7%)	0.761
	Placebo	2.4 ± 1.0	2.1 ± 0.9	1.9 ± 0.9	−0.4 ± 1.0		−0.5 ± 1.1		34 (39.1%)		41 (47.1%)	
8. Eructation	OLL2716	2.0 ± 0.9	1.8 ± 0.8	1.8 ± 0.9	−0.2 ± 0.9	0.671	−0.2 ± 0.9	0.830	27 (31.0%)	0.522	29 (33.3%)	0.528
	Placebo	2.1 ± 1.1	1.9 ± 0.8	1.8 ± 0.8	−0.3 ± 1.1		−0.3 ± 1.0		32 (36.8%)		34 (39.1%)	
9. Increased flatus	OLL2716	2.4 ± 1.0	2.2 ± 0.9	2.0 ± 0.8	−0.2 ± 1.0	0.821	−0.4 ± 1.0	0.605	32 (36.8%)	0.751	39 (44.8%)	0.646
	Placebo	2.5 ± 1.2	2.3 ± 1.1	2.1 ± 0.9	−0.2 ± 1.1		−0.4 ± 1.0		29 (33.3%)		35 (40.2%)	
10. Decreased passage of stools	OLL2716	2.4 ± 1.1	2.1 ± 1.0	1.8 ± 0.9	−0.3 ± 1.0	0.479	−0.6 ± 1.1	0.526	33 (37.9%)	0.877	42 (48.3%)	0.543
	Placebo	2.4 ± 1.4	2.0 ± 1.2	1.9 ± 1.0	−0.4 ± 1.1		−0.5 ± 1.2		35 (40.2%)		37 (42.5%)	
11. Increased passage of stools	OLL2716	2.1 ± 1.1	1.9 ± 1.0	1.6 ± 0.7	−0.1 ± 1.0	0.532	−0.4 ± 0.9	0.147	27 (31.0%)	0.747	34 (39.1%)	0.264
	Placebo	1.9 ± 1.1	1.7 ± 1.1	1.7 ± 1.0	−0.2 ± 1.0		−0.2 ± 1.1		30 (34.5%)		26 (29.9%)	
12. Loose stools	OLL2716	2.0 ± 1.0	1.9 ± 0.9	1.7 ± 0.7	−0.1 ± 0.9	0.813	−0.3 ± 0.8	0.617	26 (29.9%)	1.000	29 (33.3%)	1.000
	Placebo	1.9 ± 1.1	1.7 ± 0.8	1.6 ± 0.8	−0.2 ± 1.1		−0.3 ± 1.1		25 (28.7%)		28 (32.2%)	
13. Hard stools	OLL2716	2.2 ± 1.1	2.1 ± 1.0	1.8 ± 0.8	−0.1 ± 0.9	0.273	−0.4 ± 1.0	0.827	29 (33.3%)	0.347	33 (37.9%)	0.757
	Placebo	2.3 ± 1.2	1.9 ± 1.0	1.9 ± 1.0	−0.3 ± 1.1		−0.4 ± 1.2		36 (41.4%)		36 (41.4%)	
14. Urgent need for defecation	OLL2716	2.2 ± 1.0	2.2 ± 1.0	1.8 ± 0.8	0.0 ± 1.1	0.598	−0.4 ± 1.0	0.151	25 (28.7%)	1.000	34 (39.1%)	0.637
	Placebo	2.2 ± 1.2	2.2 ± 1.1	2.1 ± 1.1	0.0 ± 1.1		−0.1 ± 1.2		26 (29.9%)		30 (34.5%)	
15. Feeling of incomplete evacuation	OLL2716	2.4 ± 1.1	2.2 ± 0.9	1.9 ± 0.9	−0.2 ± 1.1	0.510	−0.5 ± 1.0	0.225	31 (35.6%)	0.630	41 (47.1%)	0.358
	Placebo	2.3 ± 1.0	2.2 ± 1.0	2.0 ± 0.9	−0.1 ± 0.9		−0.3 ± 1.0		27 (31.0%)		34 (39.1%)	
RS (2 and 3)	OLL2716	2.1 ± 0.8	1.9 ± 0.7	1.7 ± 0.7	−0.2 ± 0.8	0.780	−0.4 ± 0.7	0.762	31 (35.6%)	0.639	50 (57.5%)	0.648
	Placebo	2.0 ± 0.8	1.8 ± 0.7	1.6 ± 0.7	−0.3 ± 0.9		−0.4 ± 0.9		35 (40.2%)		46 (52.9%)	
AP (1, 4, and 5)	OLL2716	2.1 ± 0.8	1.8 ± 0.8	1.7 ± 0.7	−0.3 ± 0.9	0.251	−0.5 ± 0.8	0.126	44 (50.6%)	0.362	54 (62.1%)	0.169
	Placebo	2.0 ± 0.8	1.9 ± 0.8	1.7 ± 0.8	−0.2 ± 0.7		−0.3 ± 0.8		37 (42.5%)		44 (50.6%)	
IS (6, 7, 8, and 9)	OLL2716	2.3 ± 0.8	2.1 ± 0.7	2.0 ± 0.7	−0.2 ± 0.8	0.370	−0.3 ± 0.8	0.859	32 (36.8%)	0.219	48 (55.2%)	0.363
	Placebo	2.4 ± 0.9	2.1 ± 0.8	2.0 ± 0.8	−0.3 ± 0.8		−0.4 ± 0.9		41 (47.1%)		41 (47.1%)	

Table 4. *Cont.*

Items	Group	0 Week	6 Weeks	12 Weeks	Score Change before Ingestion (⊿)				Number of Participants with Improved Scores (%)			
					6 Weeks–0 Week	*p* Value [1]	12 Weeks–0 Week	*p* Value [1]	6 Weeks	*p* Value [2]	12 Weeks	*p* Value [2]
CS (10, 13, and 15)	OLL2716	2.4 ± 0.9	2.1 ± 0.8	1.8 ± 0.8	−0.2 ± 0.8	0.775	−0.5 ± 0.8	0.227	41 (47.1%)	0.879	54 (62.1%)	0.221
	Placebo	2.3 ± 1.0	2.1 ± 1.0	1.9 ± 0.9	−0.3 ± 0.8		−0.4 ± 0.9		43 (49.4%)		45 (51.7%)	
DS (11, 12, and 14)	OLL2716	2.1 ± 0.9	2.0 ± 0.8	1.7 ± 0.7	−0.1 ± 0.8	0.638	−0.4 ± 0.7	0.127	32 (36.8%)	1.000	45 (51.7%)	0.288
	Placebo	2.0 ± 1.0	1.9 ± 0.8	1.8 ± 0.8	−0.1 ± 0.7		−0.2 ± 0.9		31 (35.6%)		37 (42.5%)	
Upper GI (RS, AP, and IS)	OLL2716	2.2 ± 0.7	2.0 ± 0.6	1.8 ± 0.6	−0.2 ± 0.7	0.970	−0.4 ± 0.7	0.424	28 (32.2%)	0.271	47 (54.0%)	0.225
	Placebo	2.2 ± 0.8	2.0 ± 0.7	1.8 ± 0.7	−0.2 ± 0.7		−0.4 ± 0.7		36 (41.4%)		38 (43.7%)	
Lower GI (CS and DS)	OLL2716	2.2 ± 0.8	2.1 ± 0.7	1.8 ± 0.6	−0.1 ± 0.7	0.730	−0.4 ± 0.7	0.078	36 (41.4%)	0.533	52 (59.8%)	0.048
	Placebo	2.2 ± 0.9	2.0 ± 0.7	1.9 ± 0.7	−0.2 ± 0.6		−0.3 ± 0.8		31 (35.6%)		38 (43.7%)	
Over-all	OLL2716	2.2 ± 0.6	2.0 ± 0.6	1.8 ± 0.6	−0.2 ± 0.7	0.855	−0.4 ± 0.6	0.160	19 (21.8%)	0.229	39 (44.8%)	0.041
	Placebo	2.2 ± 0.7	2.0 ± 0.6	1.8 ± 0.7	−0.2 ± 0.6		−0.3 ± 0.7		27 (31.0%)		25 (28.7%)	

Changes in scores and number of participants with improved scores from baseline. Data are shown as mean ± standard deviation (n = 87 in the OLL2716 group and n = 87 in the placebo group). Intergroup comparison: [1] Wilcoxon rank sum test and [2] Fisher's exact test. Number of improved participants (%): (number of participants who improved/total number of participants evaluated) × 100. RS, reflux syndrome; AP, abdominal pain syndrome; IS, indigestion syndrome; CS, constipation syndrome; DS, diarrhea syndrome; GI, gastrointestinal. Sub-scales: RS, "heartburn" and "acid regurgitation"; AP, "abdominal pain", "sucking sensations in the epigastrium", and "nausea and vomiting"; IS, "borborygmus", "abdominal distension", "eructation", and "increased flatus"; CS, "decreased passage of stools", "hard stools", and "feeling of incomplete evacuation"; DS, "increased passage of stools", "loose stools", and "urgent need for defecation"; Upper GI, RS, AP, and IS; Lower GI, CS and DS.

Table 5. Score change before ingestion and the number of participants with improved scores in the Council on Nutrition Appetite Questionnaire-Japanese (CNAQ-J).

Items	Group	0 Week	6 Weeks	12 Weeks	Score Change before Ingestion (⊿)				Number of Participants with Improved Scores (%)			
					6 Weeks–0 Week	*p* Value [1]	12 Weeks–0 Week	*p* Value [1]	6 Weeks	*p* Value [2]	12 Weeks	*p* Value [2]
1. Appetite	OLL2716	3.3 ± 0.5	3.4 ± 0.5	3.4 ± 0.6	0.1 ± 0.5	0.393	0.1 ± 0.6	0.207	6 (6.9%)	0.307	7 (8.0%)	0.331
	Placebo	3.3 ± 0.6	3.4 ± 0.6	3.4 ± 0.6	0.0 ± 0.6		0.1 ± 0.6		11 (12.6%)		12 (13.8%)	
2. Feeling full	OLL2716	3.7 ± 0.5	3.7 ± 0.4	3.8 ± 0.4	0.1 ± 0.4	0.502	0.1 ± 0.5	0.781	5 (5.7%)	0.188	4 (4.6%)	0.535
	Placebo	3.7 ± 0.5	3.7 ± 0.4	3.8 ± 0.4	0.0 ± 0.5		0.1 ± 0.5		11 (12.6%)		7 (8.0%)	
3. Feeling hungry	OLL2716	3.0 ± 0.7	3.0 ± 0.8	3.1 ± 0.8	0.1 ± 0.7	0.570	0.1 ± 0.8	0.020	13 (14.9%)	0.685	13 (14.9%)	0.063
	Placebo	3.1 ± 0.8	3.2 ± 0.7	3.0 ± 0.7	0.0 ± 0.7		−0.1 ± 0.8		16 (18.4%)		24 (27.6%)	
4. Food tastes	OLL2716	3.6 ± 0.6	3.7 ± 0.6	3.7 ± 0.7	0.1 ± 0.6	0.271	0.1 ± 0.6	0.330	6 (6.9%)	0.432	9 (10.3%)	0.194
	Placebo	3.7 ± 0.6	3.7 ± 0.6	3.8 ± 0.7	0.1 ± 0.6		0.1 ± 0.7		10 (11.5%)		16 (18.4%)	
5. Food tastes compared to when younger	OLL2716	3.2 ± 0.6	3.3 ± 0.6	3.2 ± 0.6	0.0 ± 0.6	0.538	0.0 ± 0.7	0.873	9 (10.3%)	1.000	13 (14.9%)	1.000
	Placebo	3.3 ± 0.6	3.4 ± 0.7	3.3 ± 0.6	0.1 ± 0.6		0.0 ± 0.6		9 (10.3%)		12 (13.8%)	
6. Meal frequency a day	OLL2716	3.8 ± 0.4	3.8 ± 0.4	3.9 ± 0.4	0.0 ± 0.3	0.314	0.0 ± 0.4	0.162	3 (3.4%)	0.720	2 (2.3%)	1.000
	Placebo	3.9 ± 0.3	3.9 ± 0.4	3.9 ± 0.3	0.0 ± 0.4		0.0 ± 0.2		5 (5.7%)		2 (2.3%)	
7. Feeling sick or nauseated when eating	OLL2716	4.1 ± 0.7	4.3 ± 0.7	4.4 ± 0.7	0.2 ± 0.7	0.367	0.3 ± 0.7	0.210	11 (12.6%)	0.524	9 (10.3%)	0.194
	Placebo	4.2 ± 0.7	4.3 ± 0.7	4.3 ± 0.7	0.1 ± 0.9		0.1 ± 0.9		15 (17.2%)		16 (18,4%)	
8. Usual mood	OLL2716	3.4 ± 0.6	3.4 ± 0.6	3.4 ± 0.6	0.0 ± 0.5	0.087	0.0 ± 0.5	0.070	9 (10.3%)	0.794	7 (10.3%)	0.794
	Placebo	3.3 ± 0.5	3.4 ± 0.6	3.4 ± 0.5	0.1 ± 0.5		0.1 ± 0.5		7 (8.0%)		7 (8.0%)	
2 and 3	OLL2716	6.7 ± 0.9	6.8 ± 0.9	6.9 ± 1.0	0.1 ± 0.8	0.603	0.3 ± 1.0	0.045	–	–	–	–
	Placebo	6.9 ± 0.9	6.9 ± 0.8	6.9 ± 0.8	0.0 ± 0.9		0.0 ± 1.0		–	–	–	–
1, 2, and 3	OLL2716	9.9 ± 1.2	10.1 ± 1.2	10.3 ± 1.3	0.2 ± 1.0	0.531	0.4 ± 1.3	0.022	–	–	–	–
	Placebo	10.2 ± 1.3	10.3 ± 1.1	10.3 ± 1.1	0.1 ± 1.2		0.0 ± 1.3		–	–	–	–
Over-all	OLL2716	28.1 ± 2.5	28.7 ± 2.3	29.0 ± 2.5	0.6 ± 2.0	0.711	0.9 ± 2.4	0.144	–	–	–	–
	Placebo	28.6 ± 2.4	29.0 ± 2.5	29.0 ± 2.3	0.4 ± 2.2		0.4 ± 2.3		–	–	–	–

Changes in scores and number of participants with improved scores from baseline. Data are shown as mean ± standard deviation (n = 87 in the OLL2716 group and n = 87 in the placebo group). Intergroup comparison: [1] Wilcoxon rank sum test and [2] Fisher's exact test. Number of improved participants (%): (number of participants who improved/total number of participants evaluated) × 100. Sub-scales: pre- and post-meal satisfaction, "feeling full" and "feeling hungry"; eating satisfaction, "appetite", "feeling full", and "feeling hungry".

4. Discussion

In the present study, we investigated the effects of *L. paragasseri* OLL2716 intake on gastric discomfort and mental stress in healthy Japanese adults with gastric complaints, using four different questionnaires. An evaluation of the questionnaires showed that yogurt containing *L. paragasseri* OLL2716 helped relieve both gastric discomfort and mental stress. At the primary endpoint, epigastric pain and EPS-like symptoms in the OLL2716 group significantly improved compared with those in the placebo group (Table 2). The odds ratios for the improvement in gastric symptoms in the efficacy analysis population were investigated in an additional analysis. The relief odds ratios for the improvement in "epigastric pain" and "EPS-like symptoms" of the OLL2716 and placebo groups after 6 weeks of intake were 2.0 (95% CI, 1.1–3.7, $p = 0.022$) and 2.3 (95% CI, 1.3–4.3, $p = 0.007$), respectively. Similarly, a previous study [15] showed that yogurt containing *L. paragasseri* OLL2716 improved stomach pain caused by an empty stomach. These data strongly suggest that *L. paragasseri* OLL2716 intake reduces the risk of stomach upsets. Interestingly, *L. paragasseri* OLL2716 ameliorated FD in *H. pylori*-uninfected individuals [18]. Therefore, *L. paragasseri* OLL2716 may be a beneficial probiotic strain for the stomach in a wide range of healthy adults and patients. However, because a study on *H. pylori*-uninfected patients with FD showed an improvement in PDS symptoms [18], we hypothesized that PDS-like symptoms would also improve in the present study. The differences in the results could be due to the low baseline values for both PDS- and EPS-like symptoms because this study was conducted on healthy adults. Healthy adults were more responsive than unhealthy adults to changes in EPS-like symptoms because humans are generally more sensitive to pain.

At the secondary endpoint, *L. paragasseri* OLL2716 probably improved the stress associated with gastric disorders, based on the results of the SF-NDI. The SF-NDI is a common questionnaire used to assess the relationship between gastric complaints and the quality of life, including mental stress. In the present study, question items concerning stress, namely "tension" and sub-scales related to mental stress, at 6 weeks, were improved by the intake of the yogurt containing *L. paragasseri* OLL2716. An additional analysis was conducted because the intake of yogurt containing *L. paragasseri* OLL2716 may improve gastrointestinal symptoms and further enhance the quality of life, including mental stress. The relief odds ratios for improvement in "eating or drinking (enjoyment)" and "tension" of the OLL2716 and placebo groups after 6 weeks of intake were 2.0 (95% CI, 1.1–3.6, $p = 0.031$) and 2.0 (95% CI, 1.1–3.7, $p = 0.022$), respectively. The lack of statistical differences at 12 weeks could be due to the beneficial effects of placebo yogurt on the stomach in the long term; however, yogurt containing *L. paragasseri* OLL2716 was considered to be more effective earlier. To the best of our knowledge, these are the first subjective results indicating a relationship between the stomach and stress in *L. paragasseri* OLL2716-ingesting participants. Thus, *L. paragasseri* OLL2716 may ameliorate both gastric discomfort and mental stress in healthy adults with gastric complaints.

The GSRS is a popular questionnaire that is used to assess gastrointestinal symptoms. Whereas the change in GSRS scores was similar between the groups, the number of "Over-all" improved participants (total assessment) in the OLL2716 group at 12 weeks was improved compared with the placebo group. Compared with the placebo group, the odds ratios for improvement in "lower GI symptom" and "Over-all" in the OLL2716 group after 12 weeks of intake were 1.9 (95% CI, 1.0–3.5, $p = 0.035$) and 2.0 (95% CI, 1.1–3.8, $p = 0.029$), respectively. The results of the primary endpoint and GSRS evaluations suggest that *L. paragasseri* OLL2716 has a positive effect on the entire digestive tract. Otomi et al. reported that *L. paragasseri* OLL2716 improved GSRS scores [15] and suggested that *L. paragasseri* OLL2716 helped relieve upper gastrointestinal symptoms. However, the current study suggests that *L. paragasseri* OLL2716 alleviates upper and lower gastrointestinal symptoms. Further research is required to clarify whether *L. paragasseri* OLL2716 alleviates the lower gastrointestinal symptoms.

The CNAQ-J is commonly used to assess appetite. The changes at 12 weeks in the CNAQ-J score of feeling hungry, "pre- and post-meal satisfaction" and "eating satisfaction", were improved by the intake of the yogurt containing *L. paragasseri* OLL2716. Gastrointestinal signaling, such as neural pathways and released peptide hormones, has been reported to influence the feelings of satiety and hunger [29]. Additionally, acute mental stress suppresses appetite [30,31]. Therefore, in this study, it was considered possible that improvements in the gastrointestinal symptoms and stress led to the improvement in appetite.

The efficacy of *L. paragasseri* OLL2716 was not limited to gastrointestinal symptoms such as stomach pain and EPS-like symptoms but also affected stress and appetite. These improvements were considered to be due to the consumption of yogurt containing *L. paragasseri* OLL2716, which alleviated gastrointestinal symptoms such as stomach pain and subsequently led to reduced stress and improved appetite.

All previous clinical studies on *L. paragasseri* OLL2716 [15,18,32–34] were conducted with live bacteria in yogurt, leading us to consider that *L. paragasseri* OLL2716 is most effective when included in yogurt in the live bacterial state. Therefore, we chose yogurt as the formulation and set yogurt as the placebo. However, the control food was yogurt containing *Lactobacillus delbrueckii* subsp. *bulgaricus* and *Streptococcus thermophilus*, known as starter lactic acid bacteria, which are necessary for preparing yogurt. Therefore, considering that yogurt itself has a beneficial effect on the digestive tract, these effects may lead to an improvement in gastric symptoms. Additionally, yogurt is primarily composed of milk ingredients and contains α-lactalbumin, one of the whey proteins, and casein, both of which have been reported to have analgesic effects [35,36]. Given that the placebo yogurt contained these proteins in addition to the two types of lactic acid bacteria, it is possible that the placebo yogurt also demonstrated a certain degree of improvement. Nonetheless, it is considered to be highly significant that the continued intake of *L. paragasseri* OLL2716 resulted in even greater improvement.

We identified three limitations of this study: an unclear mechanism of action, the selection of participants primarily based on subjective measures, and the observation period. One limitation of this study is that although we found that the regular intake of yogurt containing *L. paragasseri* OLL2716 improved both gastric discomfort and mental stress in healthy adults with gastric complaints, the mechanism of action remains unclear. One of the presumed mechanisms for the beneficial effects of *L. paragasseri* OLL2716 is thought to be associated with the regulation of autonomic function because previous studies [3,15] have reported the improvement of stress markers, such as immunoglobulin A and amylase concentrations in saliva, also known as sympathetic markers. Moreover, in a previous study on *L. paragasseri* OLL2716, the main effects were observed in PDS symptoms [18]; however, in the present study, improvements were also observed in EPS-like symptoms such as stomach pain. Stomach pain is associated with peptic ulcers caused by stomach acid or *H. pylori* infection as well as gastritis caused due to medications such as NSAIDs, infections, alcohol, and gastroesophageal reflux disease [37–40]. The present study targeted healthy adults, excluding those with *H. pylori* infections, medication use, or excessive alcohol consumption; therefore, these factors were not thought to be the cause of stomach pain. Imbalances in autonomic nervous system function have also been reported as the causes of stomach pain [41]. Given that previous studies [3,15] have reported improvements in stress and sympathetic markers, the effects of *L. paragasseri* OLL2716 on the autonomic nervous system may also contribute to the improvement of stomach pain. However, as evaluations to support this finding were not conducted in the present study, further evidence of EPS-like symptoms is required.

Second, the selection of participants primarily based on subjective measures may have resulted in an inability to completely exclude participants with conditions such as gastroparesis, which could have prevented us from clearly demonstrating the differences between the OLL2716 and placebo groups. In this study, it was not feasible to conduct gastric emptying tests for all the participants because of scheduling constraints. However, in future studies, conducting such tests on selected participants may help clarify participant characteristics.

Regarding the third limitation, an increasing number of participants showed improvement after 6 and 12 weeks. Therefore, extending the intervention period might have led to more pronounced improvement effects. Future research should consider the duration of intake based on the evaluation criteria.

As suggested by this study, continuous intake of *L. paragasseri* OLL2716 may help improve gastric symptoms and alleviate stress, which in turn suggests that functional foods containing *L. paragasseri* OLL2716 could enhance the quality of life in healthy adults. Although the continuous intake of *L. paragasseri* OLL2716 has been reported to improve symptoms in patients with FD [18] and in those infected with *H. pylori* [32], its impact on the quality of life in these patients has not been investigated. Therefore, it is important to explore whether continuous intake of *L. paragasseri* OLL2716 could also contribute to improving the quality of life associated with symptoms in these patients, making this a topic for further research.

5. Conclusions

In conclusion, our findings suggest that regular intake of *L. paragasseri* OLL2716 may improve both gastric discomfort and mental stress in healthy adults with gastric complaints such as postprandial fullness or early satiety. Although one possible mechanism is the regulation of autonomic function, further research is required to clarify the mechanism of action of *L. paragasseri* OLL2716.

6. Patents

N.Y., K.K., A.N., and H.K. are the inventors of pending patents (Japanese Patent Application No. 2023–015847).

Supplementary Materials: The following supporting information can be downloaded at https://www.mdpi.com/article/10.3390/nu16183188/s1: Table S1. CONSORT 2010 checklist of information included for randomized trial.

Author Contributions: Conceptualization and methodology, N.Y., K.K. and H.K.; investigation, N.Y. and K.K.; additional analysis, A.S.; writing—original draft preparation, N.Y., A.N. and K.H.; writing—review and editing, N.Y., A.N., T.T., A.S. and K.H.; project administration, A.N. and H.K. All authors have read and agreed to the published version of the manuscript.

Funding: This research and the APC was funded by Meiji Co., Ltd.

Institutional Review Board Statement: The study was conducted according to the guidelines of the Declaration of Helsinki and approved by the Ethics Committee of Meiji Co., Ltd. (Tokyo, Japan) (Review Number: 193; date of approval: 8 October 2020) and Ageo Central Second Hospital (Saitama, Japan) (Review Number: GI1–2001; date of approval: 9 November 2020).

Informed Consent Statement: Informed consent was obtained from all the participants involved in this study.

Data Availability Statement: Data supporting the results of this study are available from the corresponding author upon reasonable request. However, data for the test food analysis result are not publicly available due to include confidential product specifications.

Acknowledgments: We thank the staff of APO PLUS STATION Co., Ltd. for supporting this study as a contract research organization. We would like to thank Takaya Kitamura at Meiji Co., Ltd. for monitoring this study. We would like to thank Hisanobu Noji, Takuya Takabayashi, and Rina Inomata at Meiji Co., Ltd. for supplying the test food products.

Conflicts of Interest: N.Y., A.N., A.S., T.T. and K.H. are employees of Meiji Co., Ltd. K.K. and H.K. are employees of Meiji Holdings Co., Ltd.

References

1. Parkman, H.P.; Hasler, W.L.; Fisher, R.S. American Gastroenterological Association technical review on the diagnosis and treatment of gastroparesis. *Gastroenterology* **2004**, *127*, 1592–1622. [CrossRef] [PubMed]
2. Browning, K.N. Stress-induced modulation of vagal afferents. *Neurogastroenterol. Motil.* **2019**, *31*, e13758. [CrossRef] [PubMed]
3. Ohtsu, T.; Haruma, K.; Ide, Y.; Takagi, A. The effect of continuous intake of *Lactobacillus gasseri* OLL2716 on mild to moderate delayed gastric emptying: A randomized controlled study. *Nutrients* **2021**, *13*, 1852. [CrossRef] [PubMed]
4. Vijayvargiya, P.; Jameie-Oskooei, S.; Camilleri, M.; Chedid, V.; Erwin, P.J.; Murad, M.H. Association between delayed gastric emptying and upper gastrointestinal symptoms: A systematic review and meta-analysis. *Gut* **2019**, *68*, 804–813. [CrossRef]
5. Travagli, R.A.; Anselmi, L. Vagal neurocircuitry and its influence on gastric motility. *Nat. Rev. Gastroenterol. Hepatol.* **2016**, *13*, 389–401. [CrossRef]
6. Holtmann, G.; Talley, N.J. The stomach–brain axis. *Best Pract. Res. Clin. Gastroenterol.* **2014**, *28*, 967–979. [CrossRef]
7. Rebollo, I.; Devauchelle, A.D.; Béranger, B.; Tallon-Baudry, C. Stomach-brain synchrony reveals a novel, delayed-connectivity resting-state network in humans. *eLife* **2018**, *7*, e33321. [CrossRef]
8. Konturek, P.C.; Brzozowski, T.; Konturek, S.J. Stress and the gut: Pathophysiology, clinical consequences, diagnostic approach, and treatment options. *J. Physiol. Pharmacol.* **2011**, *62*, 591–599.
9. Aro, P.; Talley, N.J.; Agréus, L.; Johansson, S.E.; Bolling-Sternevald, E.; Storskrubb, T.; Ronkainen, J. Functional dyspepsia impairs quality of life in the adult population. *Aliment. Pharmacol. Ther.* **2011**, *33*, 1215–1224. [CrossRef]
10. Talley, N.J.; Locke, G.R., 3rd; Lahr, B.D.; Zinsmeister, A.R.; Tougas, G.; Ligozio, G.; Rojavin, M.A.; Tack, J. Functional dyspepsia, delayed gastric emptying, and impaired quality of life. *Gut* **2006**, *55*, 933–939. [CrossRef]
11. Jones, M.P.; Tack, J.; Van Oudenhove, L.; Walker, M.M.; Holtmann, G.; Koloski, N.A.; Talley, N.J. Mood and anxiety disorders precede development of functional gastrointestinal disorders in patients but not in the population. *Clin. Gastroenterol. Hepatol.* **2017**, *15*, 1014–1020. [CrossRef] [PubMed]
12. Koloski, N.A.; Jones, M.; Kalantar, J.; Weltman, M.; Zaguirre, J.; Talley, N.J. The brain–gut pathway in functional gastrointestinal disorders is bidirectional: A 12-year prospective population-based study. *Gut* **2012**, *61*, 1284–1290. [CrossRef] [PubMed]
13. Koloski, N.A.; Jones, M.; Talley, N.J. Evidence that independent gut-to-brain and brain-to-gut pathways operate in the irritable bowel syndrome and functional dyspepsia: A 1-year, population-based prospective study. *Aliment. Pharmacol. Ther.* **2016**, *44*, 592–600. [CrossRef] [PubMed]
14. Black, C.J.; Drossman, D.A.; Talley, N.J.; Ruddy, J.; Ford, A.C. Functional gastrointestinal disorders: Advances in understanding and management. *Lancet* **2020**, *396*, 1664–1674. [CrossRef]
15. Otomi, K.; Ymaguchi, T.; Watanabe, S.; Kobayashi, A.; Kobayashi, H.; Hashiguchi, N. Effects of yogurt containing *Lactobacillus gasseri* OLL2716 on autonomic nerve activities and physiological functions. *Health* **2015**, *7*, 397–405. [CrossRef]
16. Gomi, A.; Yamaji, K.; Watanabe, O.; Yoshioka, M.; Miyazaki, K.; Iwama, Y.; Urita, Y. *Bifidobacterium bifidum* YIT 10347 fermented milk exerts beneficial effects on gastrointestinal discomfort and symptoms in healthy adults: A double-blind, randomized, placebo-controlled study. *J. Dairy Sci.* **2018**, *101*, 4830–4841. [CrossRef]
17. Sakamoto, I.; Igarashi, M.; Kimura, K.; Takagi, A.; Miwa, T.; Koga, Y. Suppressive effect of *Lactobacillus gasseri* OLL 2716 (LG21) on *Helicobacter pylori* infection in humans. *J. Antimicrob. Chemother.* **2001**, *47*, 709–710. [CrossRef]
18. Ohtsu, T.; Takagi, A.; Uemura, N.; Inoue, K.; Sekino, H.; Kawashima, A.; Uchida, M.; Koga, Y. The ameliorating effect of *Lactobacillus gasseri* OLL2716 on functional dyspepsia in *Helicobacter pylori*-uninfected individuals: A randomized, controlled study. *Digestion* **2017**, *96*, 92–102. [CrossRef]
19. Stanghellini, V.; Chan, F.K.; Hasler, W.L.; Malagelada, J.R.; Suzuki, H.; Tack, J.; Talley, N.J. Gastroduodenal disorders. *Gastroenterology* **2016**, *150*, 1380–1392. [CrossRef]
20. Tack, J.; Talley, N.J.; Camilleri, M.; Holtmann, G.; Hu, P.; Malagelada, J.R.; Stanghellini, V. Functional gastroduodenal disorders. *Gastroenterology* **2006**, *130*, 1466–1479. [CrossRef]
21. Veldhuyzen van Zanten, S.J.; Chiba, N.; Armstrong, D.; Barkun, A.N.; Thomson, A.B.; Mann, V.; Escobedo, S.; Chakraborty, B.; Nevin, K. Validation of a 7-point Global Overall Symptom scale to measure the severity of dyspepsia symptoms in clinical trials. *Aliment. Pharmacol. Ther.* **2006**, *23*, 521–529. [CrossRef]
22. Iwakiri, R.; Tominaga, K.; Furuta, K.; Inamori, M.; Furuta, T.; Masuyama, H.; Kanke, K.; Nagahara, A.; Haruma, K.; Kinoshita, Y.; et al. Randomised clinical trial: Rabeprazole improves symptoms in patients with functional dyspepsia in Japan. *Aliment. Pharmacol. Ther.* **2013**, *38*, 729–740. [CrossRef] [PubMed]
23. Talley, N.J.; Verlinden, M.; Jones, M. Quality of life in functional dyspepsia: Responsiveness of the Nepean Dyspepsia Index and development of a new 10-item short form. *Aliment. Pharmacol. Ther.* **2001**, *15*, 207–216. [CrossRef] [PubMed]
24. Arslan, G.; Lind, R.; Olafsson, S.; Florvaag, E.; Berstad, A. Quality of life in patients with subjective food hypersensitivity: Applicability of the 10-item short form of the Nepean Dyspepsia Index. *Dig. Dis. Sci.* **2004**, *49*, 680–687. [CrossRef] [PubMed]
25. Kulich, K.R.; Madisch, A.; Pacini, F.; Piqué, J.M.; Regula, J.; Van Rensburg, C.J.; Ujszászy, L.; Carlsson, J.; Halling, K.; Wiklund, I.K. Reliability and validity of the Gastrointestinal Symptom Rating Scale (GSRS) and Quality of Life in Reflux and Dyspepsia (QOLRAD) questionnaire in dyspepsia: A six-country study. *Health Qual. Life Outcomes* **2008**, *6*, 12. [CrossRef] [PubMed]
26. Revicki, D.A.; Wood, M.; Wiklund, I.; Crawley, J. Reliability and validity of the Gastrointestinal Symptom Rating Scale in patients with gastroesophageal reflux disease. *Qual. Life Res.* **1998**, *7*, 75–83. [CrossRef]

27. Tokudome, Y.; Okumura, K.; Kumagai, Y.; Hirano, H.; Kim, H.; Morishita, S.; Watanabe, Y. Development of the Japanese version of the Council on Nutrition Appetite Questionnaire and its simplified versions, and evaluation of their reliability, validity, and reproducibility. *J. Epidemiol.* **2017**, *27*, 524–530. [CrossRef]
28. Wilson, M.M.G.; Thomas, D.R.; Rubenstein, L.Z.; Chibnall, J.T.; Anderson, S.; Baxi, A.; Diebold, M.R.; Morley, J.E. Appetite assessment: Simple appetite questionnaire predicts weight loss in community-dwelling adults and nursing home residents. *Am. J. Clin. Nutr.* **2005**, *82*, 1074–1081. [CrossRef]
29. Tack, J.; Verbeure, W.; Mori, H.; Schol, J.; van den Houte, K.; Huang, I.H.; Balsiger, L.; Broeders, B.; Colomier, E.; Scarpellini, E.; et al. The gastrointestinal tract in hunger and satiety signalling. *United Eur. Gastroenterol. J.* **2021**, *9*, 727–734. [CrossRef]
30. Nakamura, C.; Ishii, A.; Matsuo, T.; Ishida, R.; Yamaguchi, T.; Takada, K.; Uji, M.; Yoshikawa, T. Neural effects of acute stress on appetite: A magnetoencephalography study. *PLoS ONE* **2020**, *15*, e0228039. [CrossRef]
31. Petrowski, K.; Wintermann, G.B.; Joraschky, P.; Päßler, S. Chewing after stress: Psychosocial stress influences chewing frequency, chewing efficacy, and appetite. *Psychoneuroendocrinology* **2014**, *48*, 64–76. [CrossRef] [PubMed]
32. Kimura, K. Health benefits of probiotics: Probiotics for *Helicobacter pylori* infection. *Food Sci. Technol. Res.* **2004**, *10*, 1–5. [CrossRef]
33. Takagi, A.; Yanagi, H.; Ozawa, H.; Uemura, N.; Nakajima, S.; Inoue, K.; Kawai, T.; Ohtsu, T.; Koga, Y. Effects of *Lactobacillus gasseri* OLL2716 on *Helicobacter pylori*-associated dyspepsia: A multicenter, randomized, double-blind, controlled trial. *Gastroenterol. Res. Pract.* **2016**, *2016*, 7490452. [CrossRef] [PubMed]
34. Koga, Y.; Ohtsu, T.; Kimura, K.; Asami, Y. Probiotic *L. gasseri* strain (LG21) for the upper gastrointestinal tract acting through improvement of indigenous microbiota. *BMJ Open Gastroenterol.* **2019**, *6*, e000314. [CrossRef] [PubMed]
35. Yamaguchi, M.; Yoshida, K.; Uchida, M. Novel functions of bovine milk-derived α-lactalbumin: Anti-nociceptive and anti-inflammatory activity caused by inhibiting cyclooxygenase-2 and phospholipase A_2. *Biol. Pharm. Bull.* **2009**, *32*, 366–371. [CrossRef]
36. Carrillo, W.; Monteiro, K.M.; Spindola, H.; Ramos, M.; de Carvalho, J.E. Antiulcerative and antinociceptive activities of casein and whey proteins. *J. Med. Food* **2018**, *21*, 605–611. [CrossRef]
37. Lukic, S.; Mijac, D.; Filipovic, B.; Sokic-Milutinovic, A.; Tomasevic, R.; Krstic, M.; Milosavljevic, T. Chronic abdominal pain: Gastroenterologist approach. *Dig. Dis.* **2022**, *40*, 181–186. [CrossRef]
38. Bailey, J.M. Gastrointestinal conditions: Peptic ulcer disease. *FP Essent.* **2024**, *540*, 16–23.
39. Scida, S.; Russo, M.; Miraglia, C.; Leandro, G.; Franzoni, L.; Meschi, T.; De'Angelis, G.L.; Di Mario, F. Relationship between *Helicobacter pylori* infection and GERD. *Acta Biomed.* **2018**, *89*, 40–43. [CrossRef]
40. Leung, W.K.; Graham, D.Y. Ulcer and gastritis. *Endoscopy* **2001**, *33*, 8–15. [CrossRef]
41. Mayer, E.A. The neurobiology of stress and gastrointestinal disease. *Gut* **2000**, *47*, 861–869. [CrossRef]

nutrients

MDPI

Article

Persimmon Fiber-Rich Ingredients Promote Anti-Inflammatory Responses and the Growth of Beneficial Anti-Inflammatory Firmicutes Species from the Human Colon

Lucía López-Bermudo [1,2,†], Bryan Moreno-Chamba [3,4,†], Julio Salazar-Bermeo [3,4], Nicholas J. Hayward [5], Amanda Morris [5], Gary J. Duncan [5], Wendy R. Russell [5], Antonio Cárdenas [1,2], Ángeles Ortega [1,2], Blanca Escudero-López [1,2], Genoveva Berná [1,2], Nuria Martí Bruña [3], Sylvia H. Duncan [5], Madalina Neacsu [5,*] and Franz Martin [1,2,*]

1 Andalusian Center of Molecular Biology and Regenerative Medicine (CABIMER), Pablo de Olavide University, University of Seville, CSIC, 41092 Seville, Spain; lucia.lopez@cabimer.es (L.L.-B.); maortega@upo.es (Á.O.); besclop@upo.es (B.E.-L.); gberamo@upo.es (G.B.)
2 Biomedical Research Network on Diabetes and Related Metabolic Diseases (CIBERDEM), Instituto de Salud Carlos III, 28029 Madrid, Spain
3 Institute of Research, Development and Innovation in Sanitary Biotechnology of Elche, Miguel Hernández University, 03202 Elche, Spain; bryan.morenoc@umh.es (B.M.-C.)
4 Institute of Food Engineering for Development, Universitat Politècnica de València, 46022 Valencia, Spain
5 Rowett Institute, University of Aberdeen, Foresterhill, Aberdeen AB25 2ZD, UK
* Correspondence: m.neacsu@abdn.ac.uk (M.N.); fmarber@upo.es (F.M.)
† These authors contributed equally to this work.

Abstract: Persimmon fruit processing-derived waste and by-products, such as peels and pomace, are important sources of dietary fiber and phytochemicals. Revalorizing these by-products could help promote circular nutrition and agricultural sustainability while tackling dietary deficiencies and chronic diseases. In this study, fiber-rich fractions were prepared from the by-products of Sharoni and Brilliant Red persimmon varieties. These fractions were quantified for their phenolic composition and assessed for their ability to promote the growth of beneficial human colonic Firmicutes species and for their in vitro anti-inflammatory potential. Gallic and protocatechuic acids, delphinidin, and cyanidin were the main phenolics identified. *Faecalibacterium prausnitzii* strains showed significantly higher growth rates in the presence of the Brilliant Red fraction, generating more than double butyrate as a proportion of the total short-chain fatty acids (39.5% vs. 17.8%) when compared to glucose. The fiber-rich fractions significantly decreased the inflammatory effect of interleukin-1β in Caco-2 cells, and the fermented fractions (both from Sharoni and Brilliant Red) significantly decreased the inflammatory effect of interleukin-6 and tumor necrosis factor-α in the RAW 264.7 cells. Therefore, fiber-rich fractions from persimmon by-products could be part of nutritional therapies as they reduce systemic inflammation, promote the growth of beneficial human gut bacteria, and increase the production of beneficial microbial metabolites such as butyrate.

Keywords: dietary fiber; *Faecalibacterium prausntizii*; *Eubacterium eligens*; pectins; anti-inflammatory activity; butyrate

1. Introduction

Persimmon (*Diospyros kaki* Thunb.), belonging to the Ebanaceae family, is a fruit that originated in China [1]. Due to its taste, nutrient content, bioactive compounds, and health benefits [2], its consumption has increased in recent years, with more than 4 million tons/year produced [3] in countries like China, South Korea, Japan, and Spain. The astringent varieties like 'Red Brilliant' and 'Triumph or Sharoni' are the most commercially cultivated [4], in Spain and Israel [5]. The persimmon processing leads to the generation of several by-products, such as peels, pomace, hulls, and leaves [6], which could deliver zero-waste persimmon production, promoting circular nutrition [7]. Furthermore, the development of nutraceuticals and functional food ingredients could tackle global dietary deficiencies/malnutrition and diseases by delivering important nutrients like dietary fiber and bioactive phytochemicals [8,9].

The main waste produced by persimmon industrialization is the peel and pulp, which are separated during juice production [10], which serves as a potential valuable source for phytochemicals and polysaccharides. Persimmon by-products are rich in carotenoids such as lutein, zeaxanthin, β-cryptoxanthin, β-carotene, α-carotene, and lycopene, as well as polyphenols like gallic acid, fumaric acid, epigallocatechin, and catechin [11,12]. Although some research has been carried out studying the bioactive composition of persimmon by-products [1], bonded phenolics and carotenoids to dietary fiber and indigestible compounds hold promise to be fully explored and utilized.

Persimmon by-product processing involves different methods, such as the use of solvents or the application of technologies such as ultrasound, to obtain bioactive compounds. Fermentation, an ancient technology, has been applied for shelf-life enhancement, organoleptic improvement, and the production of certain metabolites [13,14]. It also shows potential for bioactive extraction from food materials, influencing the discovery of new bioactive compounds, the improvement of health benefits, and even an increase in the bioaccessibility and bioavailability of some compounds [13,14], which is a key aspect during the gastrointestinal digestion of complex macromolecules like dietary fibers.

The dietary fibers consumed in our diets usually escape digestion by bacteria in the upper intestinal tract and are then colonized by bacteria in the colon [15]. The human colon hosts a dense and varied collection of microorganisms. Molecular profiling of the gut microbiota shows that the healthy human colon hosts two bacterial phyla, the Firmicutes and Bacteroidetes, usually making up the majority of the total microbiota [16] and the composition and metabolic output of the human gut microbiota have a major impact on health and disease [17]. Both the micronutrient and macronutrient content of diets, including fiber content and composition, are important drivers in modulating the composition of the gut microbiota and activities [18], some of which can have an impact on wider health, including the brain [19].

Certain colonic anaerobes possess key fiber degrading enzymes [15] and cross feed to other members of the microbiota in the large intestine [20]. Moreover, primary and secondary metabolites produced by the fermentation of selective compounds in dietary fiber may lead to the stimulation of anti-inflammatory response pathways, playing a role in the modulation of severe gastrointestinal (GI) disorders like ulcerative colitis [21]. Furthermore, the interaction of dietary polysaccharides and gut microbiota supports the modulation of the gut barrier, which is associated with a stimulation of the immune system and a reduction in pro-inflammatory cytokines, especially in the exposure to external molecules like pathogenic bacteria or their endotoxins [22]. In this sense, dietary fiber, throughout its interaction with gut microbiota, can be implied in the modulation of the cellular adaptative, increasing its importance to be consumed. Through their receptors in the immune cells, metabolites originating from the gut microbiota influence the metabolism of the immune cells. These compounds from various microbes have an anti-inflammatory activity. Gut microbiota-derived metabolites can inhibit inflammatory cells and improve the differentiation and activity of immunosuppressive cells [23]. Persimmon tannins [24] and polysaccharides [25] modify the microbiota balance in rats and mice and act as anti-inflammatory and antioxidant agents [26].

The average intake of fiber in the UK is 10–15 g despite the recommendation to maintain health being 30 g fiber per day [27]. Plant cell walls are a valuable source of carbon for gut bacteria and are primarily composed of cellulose, hemicellulose, and pectin. Pectins are structurally highly complex, and degradation of pectin-rich fibers requires that pectin degraders possess an array of glycosyl hydrolase and polysaccharide lyases [28]. One Firmicutes species, *Eubacterium* (now *Lachnospira*) *eligens*, has been reported to have a wide-ranging collection of specialist pectin-degrading enzymes and has been reported to be efficient at degrading pectins whilst other species, such as *Faecalibacterium prausnitzii* which is one of the most abundant bacteria in a healthy human colon, is likely to be more specialized and possess the ability to utilize more specific components of pectin [28,29].

The main products formed by fermentation by the human gut microbiota are the short-chain fatty acids, including the three main acids acetate, propionate, and butyrate. In particular, the latter product, butyrate, is the main end product of *Faecalibacterium prausnitzii* and is generated by these bacteria utilizing acetate, generated by certain bacteria, such as *Eubacterium eligens*, to form butyrate via the butyryl CoA:acetate CoA transferase route. Butyrate is the major energy source for colonocytes, possesses anti-inflammatory activity, and regulates apoptosis [30]. Metabolism of dietary fibers also results in the release of plant polyphenols, which are largely responsible for giving fruits and vegetables their color and can also have potent anti-inflammatory activities [31].

This study aims to revalorize persimmon by-products by producing fiber-rich ingredients, characterize their phenolic composition, and assess their ability to promote beneficial, understudied bacteria, in particular representative Firmicutes species, from the human colon. Furthermore, the study assessed the in vitro anti-inflammatory potential of the persimmon fiber-rich ingredients.

2. Materials and Methods

2.1. Preparations of Soluble Fiber-Rich Fractions from the Persimmon Fruits

Two varieties of fresh persimmon fruits (Red Brilliant and Sharoni varieties) were purchased from a local market in Elche, Spain. The fruits were washed and cut into 2–3 cm wedges to be further processed using an industrial fruit juice machine to separate juice and solid by-products. The by-products, made up of pulp and peels, were vacuum dried at 60 °C and stored at 20 °C for further treatment (Figure 1). Batches of the by-product of each persimmon variety were hydrated with water (1:10 *w*/*v*) at room temperature prior to fermentation or hydrolysis treatments.

Fermentation treatment: A total of 1 L of each suspension containing by-products from each variety was inoculated with 1 mL of fresh suspensions (10^7 CFU/mL) of each *Streptococcus salivarius* subsp. *thermophilus* CECT 7207 and *Lactobacillus casei* CECT 475. The suspensions were incubated at 37 °C for 24 h at 150 rpm in a benchtop incubator shaker. After incubation, the suspensions were filtered, discarding the liquid. The solid fractions were vacuum dried at 60 °C overnight. The powder obtained from each sample was further subjected to hydrolysis.

Hydrolysis treatment: Each suspension containing by-products from each variety, fermented and non-fermented, was exposed to alkaline hydrolysis by adjusting the pH of each suspension to 12 with 5 M NaOH (color change of suspension from clear to deep dark) using a pH meter (Hanna Instruments edge®, Daselab S.L., Valencia, Spain). The mixtures were heated at 40 °C for 24 h under constant shaking (150 rpm) in a benchtop incubator shaker; then, the pH was lowered to 2.5 with 5 M HCl (acidic hydrolysis, color change from deep dark to red). The suspensions were filtered, and the liquid fractions were freeze dried (LyoQuest, Telstar, Barcelona, Spain) to obtain persimmon water-soluble fractions of Red Brilliant (RB) and Sharoni (SH) without fermentation, and water-soluble fermented persimmon fractions of Red Brilliant (RBF) and Sharoni (SHF).

2.2. Chemicals and Reagents

Standards and reagents: Standards for the free sugar analysis, including glucose, fructose, sucrose, rhamnose, fucose, arabinose, xylose, mannose, galactose, galacturonic acid, glucuronic acid, and maltose, were purchased from Sigma-Aldrich (Dorset, UK) and Thermo-Fisher Scientific (UK). Standards for the anthocyanin aglycones (anthocyanins) analysis, including delphinidin (>95%), cyanidin (>95%), pelargonidin (undeclared purity), and peonidin (>96.5%), were all purchased from Sigma-Aldrich (Dorset, UK) and malvidin (>95%) from Phytolab, Germany. The aglycone standard, petunidin, was purchased from ChemFaces (Wuhan, China) at a purity of >95%. All the phenolic standards were purchased from Sigma-Aldrich (Gillingham, UK), Phytolab (Vestenbergsgreuth, Germany) or synthesized, as described previously [32]. General reagents were purchased from Sigma-Aldrich (Dorset, UK) and Fischer Scientific (Loughborough, UK). Reagents used for hydrolysis to obtain persimmon fiber-rich fractions were purchased from Sigma-Aldrich (Madrid, Spain), while strains used to ferment persimmon by-products to obtain fermented persimmon fractions were purchased from Spanish Type Culture Collection (CECT, Valencia, Spain).

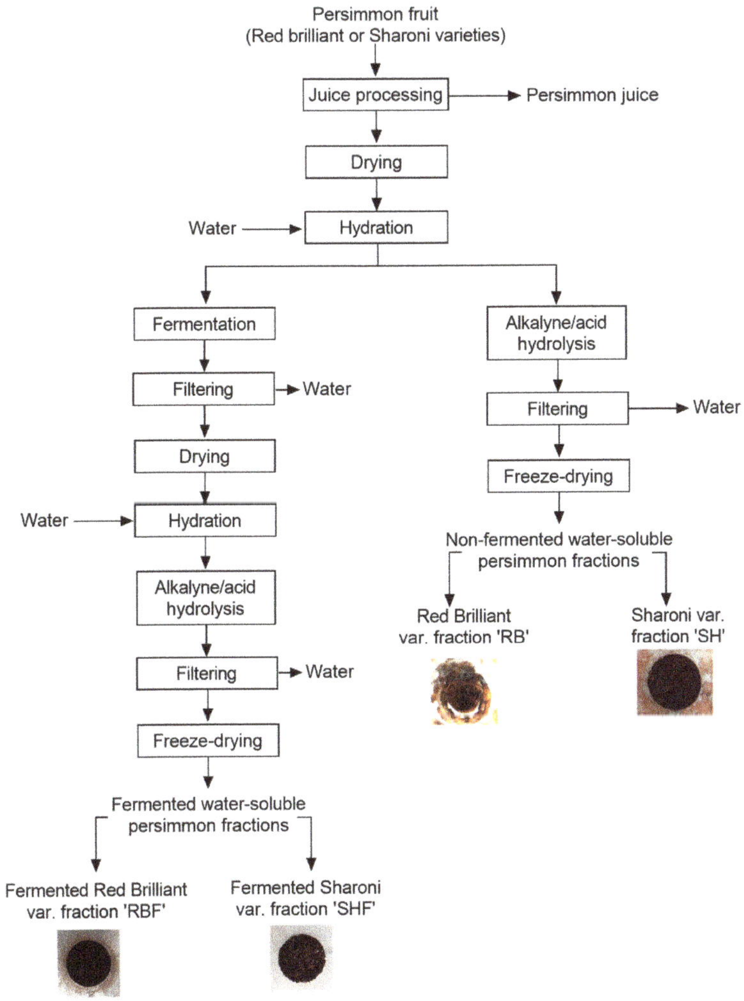

Figure 1. Diagram for processing of the persimmon fruit by-products to obtain soluble fiber-rich fractions through alkaline/acid hydrolysis or fermentation.

2.3. *Analysis of Free Sugars Composition of Persimmon Fiber-Rich Fractions*

The quantification of mono- and disaccharides from the persimmon fraction was performed using a 1260 Infinity HPLC from Agilent (Wokingham, UK) equipped with a RI detection and an Asahipak NH2P-50 4E (5 µm; 25 cm × 0.46 cm), (Shodex, Japan) column connected to an Asahipak NH2P-50G 4A pre-column (4.6 mm × 10 mm, Shodex, Japan). Persimmon fractions (approx. 0.05 g, n = 3) were dissolved in 250 µL of water and filtered using 0.2 µm filters prior to the HPLC analysis using an isocratic solvent program consisting of 70% acetonitrile at a constant flow of 1 mL/min. Quantification of the free sugars was performed using external calibration curves using validated standards for each sugar analyzed.

2.4. Analysis of Anthocyanin Composition of Persimmon Fiber-Rich Fractions

Quantification of the anthocyanidin (anthocyanin aglycones) content from persimmon fiber-rich fractions was performed by extraction and hydrolysis methods adapted from Zhang et al. [33]. Briefly, samples of persimmon (0.05 g, n = 3) were extracted with methanol–water–hydrochloric acid (ratio of 50:33:17; *v*/*v*/*v*; 3 mL) three consecutive times, and the supernatants and the pellet were combined and hydrolyzed at 100 °C for 60 min. Hydrolyzed samples were then immediately cooled to room temperature, filtered using 0.2 μm filters, and analyzed using a 1260 Infinity HPLC from Agilent (Wokingham, UK) with a Synergi 4 μm Polar-RP 80A (250 × 4.6 mm) column with a Polar-RP 4 × 3 mm pre-column from Phenomenex (Macclesfield, UK). The HPLC system was equipped with a DAD detector, and spectra were recorded between 200 and 700 nm. The chromatograms were monitored at 530 nm. For the HPLC separations, the following solvents were used: A: formic acid (2.125%) and B: acetonitrile/methanol (85:15, *v*/*v*) in an isocratic program using 18% B for 40 min at 1 mL/min, as described by Zhang et al. [33]. The column temperature was held at 35 °C. The separation and quantification of anthocyanins were performed using external standardization, as described previously [34].

2.5. Analysis of Other Phenolic Molecules Composition of Persimmon Fiber-Rich Fractions

Phenolic molecule extraction used an existing method [35], which has been further amended. Briefly, the persimmon samples (approx. 0.1 g dry weight; n = 3) were suspended in HCl (0.2 M; 3 mL) and then extracted three consecutive times into ethyl acetate (5 mL). The three ethyl acetate extracts were combined and dried before being dissolved with methanol (0.5 mL), which represented the "free fraction" and was stored at −70 °C prior to LC-MS analysis.

The remaining aqueous fraction's pH was adjusted for the alkaline hydrolysis using NaOH (4 M) and stirred at room temperature for 4 h under nitrogen. The pH was reduced afterwards to 2 with HCl (6 M), and samples were extracted again into ethyl acetate (3 × 5 mL). The three ethyl acetate extracts were combined and dried before being dissolved with methanol (0.5 mL), which represented the 'alkaline-bound fraction'.

The pH of the aqueous fraction was then adjusted again using HCl (6M) for the acid hydrolysis and incubated at 95 °C for 30 min. The samples were cooled to room temperature and extracted three times with ethyl acetate (5 mL each). The three ethyl acetate extracts were combined and dried being dissolved with methanol (0.5 mL), which represented the 'acid-bound fraction'. The alkaline-bound and acid-bound fractions were combined and represented as bound fractions analyzed using LC-MS/MS analysis.

To prepare samples for LC-MS analysis, an aliquot of free and respectively bound fractions dissolved in methanol, as described above, was mixed with Internal Standard 1 (IS1) for negative-mode mass spectrometry (13C benzoic acid) and Internal Standard 2 (IS2) for positive-mode mass spectrometry (2-amino-3,4,7,8-tetramethylimidazo [4,5-*f*]quinoxaline).

The liquid chromatography separation of phenolic metabolites was performed on an Agilent 1100 LC-MS system using a Zorbax Eclipse 5 μm, 150 mm × 4 mm column from Agilent Technologies (Wokingham, UK) as described elsewhere [36]. All persimmon sample extracts prepared as mentioned above were screened for phenolic acids and derivatives, flavonoids, and lignan metabolites. Three gradient solvents were used to separate the different categories of metabolites, with the mobile-phase solvents being water containing 0.1% acetic acid (A) and acetonitrile containing 0.1% acetic acid (B). In all cases, the flow rate was 300 μL/min with an injection volume of 5 μL. The LC eluent was directed, without splitting, into an ABI 3200 triple-quadrupole mass spectrometer (Applied Biosystems, Warrington, UK) fitted with a turbo-ion spray source. All the metabolites were quantified using multiple reaction monitoring.

2.6. Preparation of Bacterial Cultures and Assessment of Growth on Persimmon Fiber-Rich Fractions Using a Microtitre Plate Assay

Six bacterial strains were tested for their ability to utilize soluble persimmon fiber-rich fractions as growth substrates using the microtiter plate method in triplicate (technical replicates) as described previously [28].

Bacterial strains included two strains that represented the two phylogroups of *Faecalibacterium prausnitzii* (A2–165, M21/2), *Coprococcus species* L2-50 held by the Rowett Institute, Aberdeen; *Eubacterium eligens* (DSM3376 = ATCC27750); and *Bifidobacterium bifidum* (DSM20456) were purchased from Deutsche Sammlung von Mikroorganismen und Zellkulturen (DSMZ) (Braunschweig, Germany); and *Bacteroides thetaiotaomicron* (ATCC 5482) was from the American type culture collection (ATCC) (Manassas, VA, USA). The strains for growth studies were pre-prepared by culturing on M2 medium containing 0.2% (*w*/*v*) of each glucose, cellobiose, and soluble potato starch [37] for 20–24 h at 37 °C under a stream of CO_2 gas. Sterile microtiter plates were maintained in an anaerobic cabinet (Don Whitley, Shipley, UK) for 24 h prior to adding 200 μL of anaerobic pre-reduced basal M2 medium containing either 0.2% (*w*/*v*) of glucose or persimmon fractions as the single growth substrate. The wells were then inoculated with overnight cultures grown anaerobically in M2GSC medium in triplicate for each substrate or basal medium as a control. The plates were covered with Q-optical seals (Bio-Rad, UK), and then the plates were incubated for 24 h at 37 °C in a Tecan Safire 2 microplate reader (Tecan Group Ltd., Männerdorf, Switzerland), with optical readings at 650 nm taken every hour following low-speed shaking for 5 s, as described previously [28]. For the mixed bacterial culture studies and assessment of growth on persimmon fibers using a microtiter plate assay, the strains were prepared and mixed in equal amounts and processed, as described.

2.7. Growth Rate Determinations

Growth rate calculations were performed using at least three values during the mid-exponential growth phase, as described previously [38]. Briefly, semi-log values of the growth from the microtiter plates were used to determine the growth rates per hour.

2.8. Quantification of Short-Chain Fatty Acid Analysis

Shor- chain fatty acid (SCFA) formation was determined in culture supernatants by gas chromatography, as described previously [39]. Briefly, following derivatization of the samples using N- tert-butyldimethylsilyl-N-methyltrifluoroacetamide, the samples were analyzed using a Hewlett–Packard gas chromatograph fitted with a fused silica capillary column with helium as the carrier gas. The lower detection limit for each of the acids was 0.2 mM.

2.9. Mammalian Cell Culture

The cell lines human colorectal adenocarcinoma (Caco-2), human hepatoma (HepG2), murine macrophage (RAW 264.7), and 3T3-L1 mouse embryo fibroblast were obtained from ATCC (American Type Culture Collection, Gaithersburg, MD, USA).

All cell lines were usually grown in culture T-flasks in a CO_2-incubator at 37 °C with 5% CO_2. The culture medium was Dulbecco's Modified Eagle Medium (DMEM) (Sigma Aldrich, Madrid, Spain) supplemented with 0.1 mM of non-essential amino acids (Gibco, Thermo Fisher Scientific Inc., Waltham, MA, USA), 100 U/mL of penicillin/0.1 g/mL of streptomycin (Gibco), 10% fetal bovine serum (FBS) (Hyclone Laboratories Inc., Logan, UT, USA), and 2 mM L-glutamine (Gibco). The culture medium was refreshed every 24–48 h. At a confluency of 70–80% (every 2–3 days), cells were split at a ratio of 1:3–1:5 using trypsin-EDTA (Gibco). For all the experiments, cell passages 10–30 were used.

In the case of 3T3-L1, they were maintained in a pre-adipocyte state using the above-mentioned splitting and feeding protocol. To obtain the adipocyte phenotype, cells were cultured to 85–90% confluency and maintained for 3 days in the previously indicated feeding media. Then, the medium was completely changed to the differentiation media.

This medium included DMEM supplemented with 10% FBS, 1 μg/mL of insulin (Sigma Aldrich, Madrid, Spain), 0.25 μM dexamethasone (DEX) (Sigma-Aldrich, Madrid, Spain), and 0.5 mM 3-Isobutyl-1-methylxanthine (IBMX) (Sigma-Aldrich). This medium was left on the cells for 48 h. Afterwards, the cells were cultured for another 5 days in post-differentiation media containing insulin without DEX and IBMX. Fresh insulin was added to media each day of feeding. This post-differentiation medium was added to the cells until the end of the experiment. Adipocytes were used 7–10 days after differentiation.

2.10. Persimmon Fiber-Rich Fractions Preparation for Cell Culture Studies

Individual stock solutions of the four fractions (RB, RBF, SH, and SHF) at 250 mg/mL were prepared in dimethyl sulfoxide (DMSO) used as a vehicle (Thermo Fisher Scientific Inc.). Different concentrations of the stock solutions were prepared in the corresponding media (100, 250, and 500 μg/mL), with the final DMSO concentrations being 0.04%, 0.1%, and 0.2%, respectively. A concentration of 10% (*v*/*v*) DMSO was used as a control for the maximal cytotoxicity.

2.11. Viability and Cytotoxicity Assays

Cells were seeded on a 96-well plate (Thermo Fisher Scientific Inc.) at 5×10^4 (HepG2), 2×10^4 (Caco-2), 10×10^3 cells/well (adipocytes derived from 3T3-L1), and 10×10^3 cells/well (RAW 264.7) and incubated for 24 h at 37 °C and 5% CO_2. The cell culture medium was then removed, and the cells were exposed, for 24 h and 48 h, to 100 μL of cell culture medium (control), 10% DMSO in medium, the three different concentrations of the four fractions (100, 250, and 500 μg/mL) in the complete medium, or different DMSO concentrations equivalent to the amount of vehicle in the corresponding assayed fractions (0.04%, 0.1%, and 0.2%). Empty wells and sample controls, containing the medium with no cells to determine the baseline signal, were included in all assays.

For the human cell lines HepG2 and Caco-2, the Alamar Blue assay was used to determine the effect of the extracts on cell viability. This assay incorporates an oxidation–reduction (REDOX) indicator, assessing the mitochondrial ability to reduce resazurin into the fluorescent product resorufin [40]. Briefly, 100 μL of Alamar Blue working solution was prepared by mixing cell culture medium with a stock solution of resazurin sodium salt (5 mg/mL) (Panreac AppliChem, Barcelona, Spain) in phosphate-buffered saline (PBS) in a 10:1 ratio and then added to each well. Cells were then incubated for 2 h at 37 °C before measuring fluorescence.

For adipocytes and RAW 264.7 cells, the Live/Dead Viability/Cytotoxicity Kit for mammalian cells (Thermo Fisher Scientific Inc.) was used. This 2-color fluorescence cell viability assay is based on 2 probes: calcein AM, which identifies live cells, and ethidium homodimer-1 (EthD-1), which detects dead cells. Consequently, live cells are stained green and dead cells are stained red. The cell viability and cytotoxicity were evaluated in line with the manufacturer's instruction. Briefly, after incubations with the different concentrations of the four fractions, the supernatant was removed, and the cells were washed with PBS twice. Moreover, samples of live and dead cells were prepared for controls. To prepare dead cell controls, live cells were treated with 0.1% saponin for 10 min. The following controls were prepared: (i) all live cells labeled only with EthD-1; (ii) all live cells labeled only with calcein AM; (iii) all dead cells labeled only with EthD-1; (iv) all dead cells labeled only with calcein AM; (v) one cell-free control labeled with EthD-1; and another cell-free control labeled with calcein AM. After determining the optimal dye concentrations, cell culture medium with calcein AM (2 uM) and EthD-1 (4 uM) was added to each well and incubated for another 30 min. The supernatant was then removed, and cells were washed with PBS twice. Live and dead cell numbers were calculated from standard curves with known numbers of live and dead cells.

The results were averaged over 4 different independent experiments with 5 replicates per experiment. Fluorescence was measured with a multimodal Varioskan Lux spectrophotometer (Thermo Fisher Scientific Inc.). For the alamar blue assay, the excitation and emission wavelengths were 530 and 590 nm, respectively. For the Live/Dead kit, the excitation and emission filters were 485/530 nm (live cells) and 530/645 nm (dead cells). The results indicated the percentage of viable cells in relation to the control of each compound after background fluorescence was subtracted. This was calculated as the number of live cells divided by the number of total cells at each time point.

2.12. *Inflammatory Cytokine Tests*

Caco-2 and RAW 264.7 cells were seeded in 12- and 24-well plates at a density of 2×10^5 and 10×10^3 cells/well, respectively, and allowed to adhere for 22 h before initiating different treatments. To analyze the anti-inflammatory effect of the fractions, inflammation was induced by treating Caco-2 cells with 25 ng/mL interleukin 1β (IL-1β) (R&D Systems, Inc., Minneapolis, MN, USA) and RAW 264.7 cells with 1 μg/mL lipopolysaccharide (LPS) (Sigma-Aldrich). IL-1β and LPS were diluted in complete media from a 10 μg/mL stock and 1 mg/mL stock solution in PBS, respectively.

The protocol was to culture the cells with media containing 100 μg/mL of the different four fractions (100 μg/mL) for 2 h at 37 °C. Next, cells were cultured with the inflammatory stimulus for 24 h in the presence or absence of the same concentrations of the different four fractions in the completed media. Non-stimulated Caco-2 and RAW 264.7 cells were cultivated in parallel with the four different fractions (100 μg/mL) in complete media for 24 h and used as controls. Additional controls were cells cultivated for 24 h with complete media, media with vehicle (DMSO 0.04%), or media with IL-1β (25 ng/mL) or LPS (1 μg/mL). After treatments, the extracellular media were collected, centrifuged at 1500 rpm to remove debris, and kept at −20 °C until analysis.

Interleukin 6 (IL-6) and tumor necrosis factor α (TNF-α) were quantified using a sandwich ELISA method (R&D Systems, Inc.) according to the manufacturers' instructions. Fluorescence values were recorded with a microplate reading Varioskan Lux spectrophotometer (Thermo Fisher Scientific Inc.) at 450 nm for IL-6 and 520 nm for TNF-α.

2.13. *Cellular Antioxidant Enzyme Activity Assays*

HepG2 cells were seeded in 12-well plates at a density of 1×10^6 cells/well in complete medium and allowed to adhere for 24 h before initiating different treatments. After incubating cells for 24 h at 37 °C with each of the four different fractions (100 μg/mL) in complete media, oxidative stress was induced by cultivating the cells with 500 μM H_2O_2 for another 2 h at 37 °C, either in the presence or absence of the same concentrations of the different four fractions in complete media. Cells cultivated in parallel with the four different fractions (100 μg/mL) in complete media were used as controls. Cells incubated in complete media and then added 500 μM H_2O_2 for 2 h at 37 °C were treated as the model group. Additional controls were cells cultivated in complete media or in media with vehicle (DMSO 0.04%).

After treatment, cells were then lysed in an ice-cold solubilization buffer (10 mM Tris ClH pH 7.5, 150 mM ClNa, 0.1 mM EDTA) and a supernatant was collected by centrifuging at 10,000× *g* for 15 min at 4 °C. Samples were frozen at −20 °C until the enzyme activity assay.

Superoxide dismutase (SOD), catalase (CAT), and glutathione reductase (GR) activities were determined using colorimetric and fluorescent activity kits (Thermo Fisher Scientific Inc.) according to the manufacturers' instructions. Lectures were recorded with a microplate reading Varioskan Lux spectrophotometer (Thermo Fisher Scientific Inc.) at 450 nm for SOD, 560 nm for CAT, and 390 nm excitation and 510 nm emission for GR. Results were expressed as U/mL for SOD and CAT and mU/mL for GR.

2.14. *Statistical Analysis*

All the analyses were performed in triplicate, and the results are presented as the mean ± standard deviation (SD). The significance differences between persimmon sample composition (in terms of varieties and preparation treatments) were assessed using a paired

t test (significance level $p < 0.05$) with a significance threshold of * $p < 0.05$, ** $p < 0.01$, and *** $p < 0.001$. The phytochemicals profiles measured by LC-MS/MS were determined by principal component analysis (PCA), unit variance (UV)-scaled using SIMCA 14.1 (Umetrics, Cambridge, UK). For the rest of the results presented, statistical comparisons were performed using the two-tailed unpaired t test, U-Mann–Whitney test, and the Kruskal–Wallis test. p-values less than 0.05 were considered statistically significant.

The SPSS Version 29.0 software package (IBM, Armonk, NY, USA) was used for statistical analysis. GraphPad Prism 10.0 (GraphPad Software Inc, San Diego, CA, USA) was used to graph the data.

3. Results

The persimmon fruit by-products from varieties 'Red Brilliant' and 'Sharoni' were further processed to obtain soluble fiber-rich fractions (Figure 1) to determine their potential functionalities.

3.1. Composition of the Soluble Persimmon Fiber-Rich Fractions

The free monosaccharides compositions of Brilliant Red (RB), Sharoni (SH), RBF, and SHF fractions are shown in Table 1. RB fraction had significantly ($p < 0.05$) higher levels of fructose and glucose ($p < 0.001$) than SH fraction. Moreover, as might be expected, fermented fractions had significantly ($p < 0.001$) lower levels of fructose and glucose than non-fermented fractions.

Table 1. Free monosaccharides content in persimmon RB, SH, RBF, and SHF fractions.

Compound (g/100 g)	Fiber-Rich Persimmon Fraction			
	RB	**SH**	**RBF**	**SHF**
Fructose	18.47 ± 0.10 *	8.62 ± 0.92	0 ***	0 ***
Glucose	20.31 ± 1.23 ***	4.83 ± 0.48	0 ***	0 ***

RB: Red Brilliant fraction; RBF: Red Brilliant fermented fraction; SH: Sharoni fraction; SHF: Sharoni fermented fraction. Values are expressed as g/100 g dry weight. Mean ± SD (n = 3). * $p < 0.05$; *** $p < 0.001$.

Principal component analysis (PCA) analysis of the persimmon metabolites measured by LCMS showed that the phytochemicals profile of the molecules extracted in the free fractions of the RB and RBF was different than the one from SH and SHF. Furthermore, this analysis also indicates that the profile of the bound phytochemicals, those extracted following the hydrolysis (alkaline and acid), is also different that the one extracted in the free fraction, but the profile of the bound phytochemicals for RB, RBF, SH, and SHF is similar (Figure 2A).

Gallic acid and protocatechuic acid were the most abundant molecules measured by LC-MS/MS in the persimmon samples, they are mostly found in free forms in the RB and RBF and more equally distributed between free and bound forms in the SH and SHF samples (Figure 2C–F). The fermentation had a significant effect on the extractability of the phytochemicals, especially for the Brilliant Red variety of persimmon, incrementing the number of molecules extractable (Figure 2B–D); the sum of free molecules measured by LCMS in RB before fermentation rises from 63 mg to 133 mg after the fermentation, and the bound molecules from 6 to 47 mg, respectively. The Sharoni variety seems to be richer in the phytochemicals measured by LCMS than the Brilliant Red variety; however, as the fermentation increases the extractability of the molecules, the RBF fraction is richer in extractable phytochemicals (Figure 2G,H).

The HPLC analysis demonstrated that the persimmon fiber-rich products contain anthocyanins in abundance, also that the fermentation significantly increased ($p < 0.01$) the amount of extractable delphinidin and cyanidin in both varieties of persimmon (Figure 3), and these anthocyanins are more abundant in the samples from the Sharoni variety.

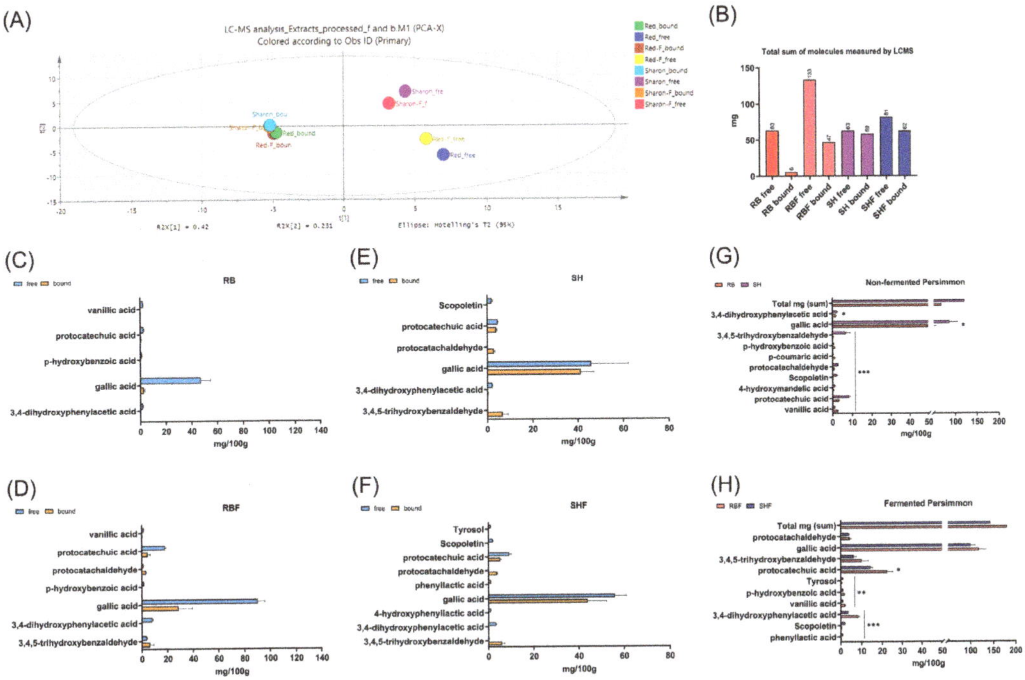

Figure 2. Composition of the soluble persimmon fiber-rich fractions. (**A**) The principal component analysis (PCA) of all plant metabolites measured by LCMS in free and bound fractions from persimmon samples. (**B**) The total phytochemicals (free and bound) content (in mg) obtained by summing the individual plant metabolites measured by LCMS. (**C–F**) The most abundant molecules (above 1 mg/100 g), measured by LCMS in free and bound fractions (as mean ± SD, n = 3) in the persimmon fiber-rich fractions. (**G,H**) Ten most abundant plant metabolites (as mean ± SD, n = 3 summing the bound and free extractable molecules) measured in the persimmon samples, where RB and RBF are fiber-rich persimmon fractions from Brilliant Red variety, fermented (F) or not; respectively, SH and SHF are fiber-rich persimmon fractions from Sharoni variety, fermented (F) or not; * $p < 0.05$, ** $p < 0.01$, *** $p < 0.001$. RB: Red Brilliant; RBF: Red Brilliant fermented; SH: Sharoni; SHF: Sharoni fermented.

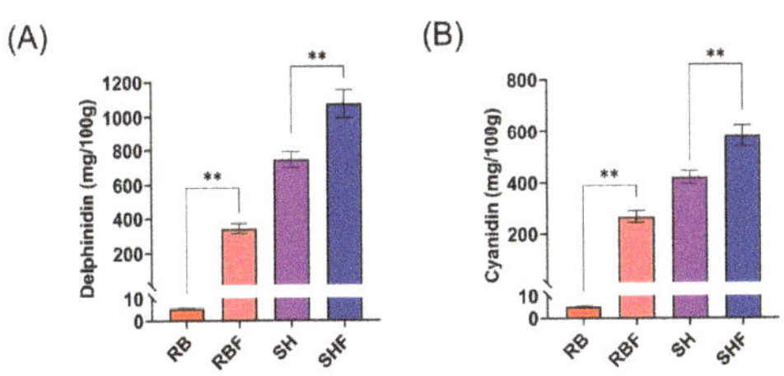

Figure 3. Values of anthocyanins in soluble persimmon fiber-rich fractions. The anthocyanins (**A**) delphinidin and (**B**) cyanidin in mg/100 g persimmon sample (as mean ± SD, n = 3) measured by HPLC, where RB, RBF, are fiber-rich persimmon fractions from Brilliant Red variety, fermented (F) or not; respectively, SH, SHF are fiber-rich persimmon fractions from Sharoni variety, fermented (F) or not; ** $p < 0.01$. RB: Red Brilliant fraction; RBF: Red Brilliant fermented fraction; SH: Sharoni fraction; SHF: Sharoni fermented fraction.

3.2. Effects of Persimmon Fiber-Rich Fractions on Cell Viability

The effects of the four fractions on cell viability were studied in four cell lines (two of human and two of murine origin) after 24 and 48 h of culture with the fractions. Three concentrations of the fractions were tested (100, 250 and 500 μg/mL). In human cell lines Alamar Blue assay was used. For murine cell lines, the Live/Dead Viability/Cytotoxicity Kit was employed.

Firstly, none of the different DMSO concentrations equivalent to the amount of vehicle in the corresponding assayed fractions (0.04%, 0.1% and 0.2%) exerted any effect on cell viability.

In all cell lines (Figure 4A–D), the culture with the fractions at 100 μg/mL for 24 h did not affect cell viability. In the 48-hour culture, there was only a significant decrease ($p < 0.05$) in viability with RBF and SHF fractions and only for the Caco-2 cell line (Figure 4D).

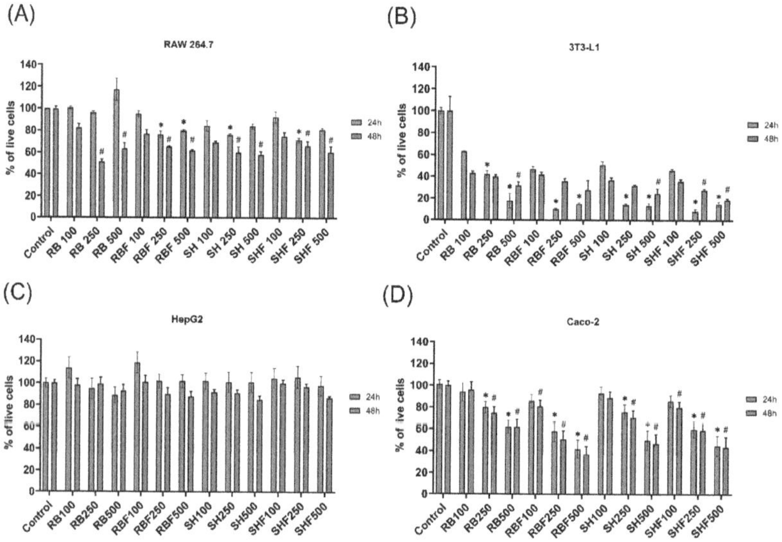

Figure 4. Effect of the different persimmon fiber-rich fractions on the viability on murine and human culture cell lines over 24 and 48 h culture incubations. (**A**) Effect of the fractions on the RAW 264.7 cell line. (**B**) Effect of the fractions on the 3T3-L1 cell line differentiated to adipocytes. (**C**) Effect of the fractions on the HepG2 cell line. (**D**) Effect of the fractions on the Caco-2 cell line. Viability was expressed as a percentage relative to the control (cells cultured without fraction). Blue bars indicate a 24 h culture with the fractions and red bars indicate 48 h culture. The concentrations of the fractions were 100, 250, and 500 μg/mL. Values are expressed as the mean ± SD (n = 4). * $p < 0.05$ fractions versus control at 24 h. # $p < 0.05$ fractions versus control at 48 h. RB: Red Brilliant fraction; RBF: Red Brilliant fermented fraction; SH: Sharoni fraction; SHF: Sharoni fermented fraction.

In the RAW 264.7 cell line (Figure 4A), only the RB fraction, at any concentration and during 24 h of culture, did not affect cell viability. In the rest of the culture conditions (type of fraction and duration of culture), there was a significant ($p < 0.05$) decrease in cell viability.

For the HepG2 cell line (Figure 4C), none of the concentrations of the four fractions affected cell viability at 24 h or 48 h of culture with the fractions. At the other extreme was the adipocyte cell line (Figure 4B), where any concentration above 100 μg/mL of all fractions significantly ($p < 0.05$) decreased viability.

In the Caco-2 cell line (Figure 4D), when cultured at concentrations above 100 μg/mL, the four fractions significantly ($p < 0.05$) decreased cell viability at 24 and 48 h of culture.

3.3. Anti-Inflammatory Effects of Persimmon Fiber-Rich Fractions

The anti-inflammatory effects were evaluated for the four fractions, but only at concentrations with no cytotoxic effects observed (100 µg/mL). One murine (RAW 264.7) and one human (Caco-2) cell line were studied. Two conditions were tested: (i) previous culture with the fractions for 2 h and then addition of the proinflammatory stimulus for 24 h but without the fractions, and (ii) culture of the fractions and the proinflammatory stimulus together during 26 h. None of the four fractions had proinflammatory effects in the two cell lines, nor did 0.04% DMSO.

In the Caco-2 cell line, all the fractions and in both conditions tested significantly ($p < 0.05$) decreased the inflammatory effect of IL-1β (Figure 5C).

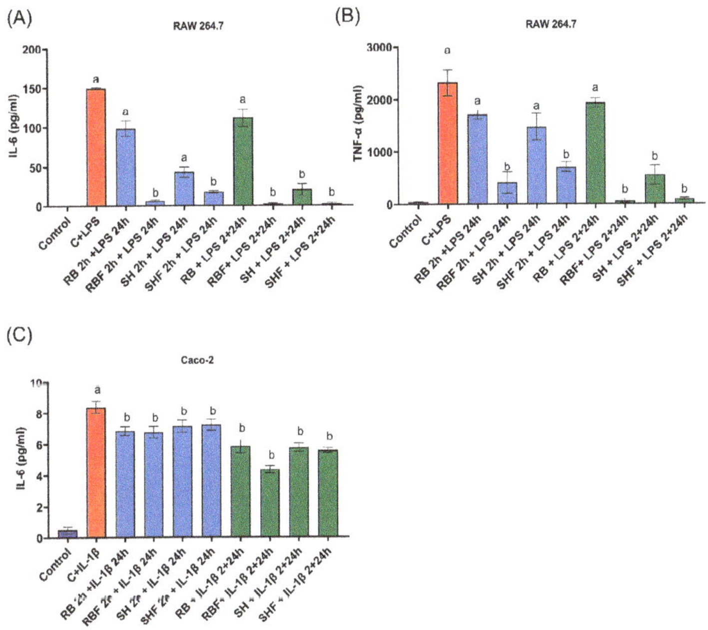

Figure 5. Effects of the different persimmon fiber-rich fractions on inflammation, in murine and human cell lines. (**A**) Effect of the fractions on IL-6 release, in RAW 264.7 cell line. (**B**) Effect of the fractions on TNF-α release, in RAW 264.7 cell line. (**C**) Effect of the fractions on IL-6 release, in Caco-2 cell line. The red bar represents the inflammatory stimulation condition with LPS or IL-1β. The blue bars are the culture conditions with the addition of the fractions before the induction of inflammation and then the addition of LPS (1 µg/mL) or IL-1β (25 ng/mL) with the absence of the fractions. The green bars indicate the simultaneous culture condition of the fractions and LPS (1 µg/mL) or IL-1β (25 ng/mL). The concentration of fraction used was 100 µg/mL. All cultures lasted 24 h. Values are expressed as the mean ± SD (n = 7). Different letters indicate a significant difference ($p < 0.05$) versus induction of inflammation with LPS or IL-1β. RB: Red Brilliant fraction; RBF: Red Brilliant fermented fraction; SH: Sharoni fraction; SHF: Sharoni fermented fraction. LPS: lipopolysaccharide.

In the RAW 264.7 cell line, RBF and SHF fractions significantly decreased ($p < 0.05$) IL 6 (Figure 5A) and TNF-α (Figure 5B) LPS-induced production in the two conditions tested. The RB fraction had no anti-inflammatory effects. In the case of SH, there was a significant ($p < 0.05$) reduction of IL-6 and TNF-α production in the assay condition of simultaneously culture of fractions and proinflammatory stimulus.

3.4. *Effects of Persimmon Fiber-Rich Fractions on the Activity of Antioxidant Enzymes*

The antioxidant activity of the four fractions was analyzed in the human line HepG2 at the concentration of the fractions at which viability was not affected (100 µg/mL) and in the same two culture conditions already mentioned in the previous section.

None of the four fractions had any effect on antioxidant enzyme activity in the HepG2 cell lines, nor did 0.04% DMSO, in the absence of any oxidative stress stimulus.

The antioxidant activity of SOD, CAT, and GR enzymes was studied. SOD (Figure 6A) and GR (Figure 6C) enzymes significantly ($p < 0.05$) increased their activity after induction of oxidative damage with H_2O_2 when HepG2 cells were cultured with the different fractions and for the two conditions tested. There was only one exception, and that was the cultur.e of SH fractions together with H_2O_2 (Figure 6C). In the case of CAT enzyme activity (Figure 6B), there was a significant ($p < 0.05$) increase in its activity for all fractions in the condition of previous culture with the fractions. When RBF and SH fractions and H_2O_2 were cultured simultaneously, there was no increase in the activity of the antioxidant enzyme.

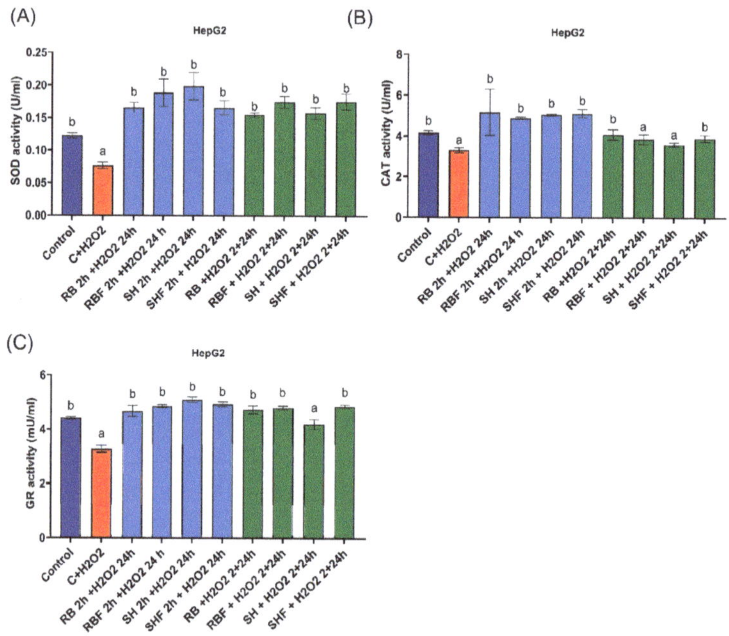

Figure 6. Effects of different persimmon fiber-rich fractions on antioxidant enzyme activity. (**A**) Effect of fractions on SOD activity, in HepG2 line. (**B**) Effect of fractions on CAT activity, in HepG2 line. (**C**) Effect of fractions on GR activity, in HepG2 cell line. The red bar represents the pro-oxidant stimulation condition with H_2O_2 (500 µM). The blue bars are the culture conditions with the addition of the fractions before the induction of oxidative stress and then the addition of H_2O_2 with the absence of the fractions. The green bars indicate the simultaneous culture condition of the fractions and H_2O_2. The concentration of fraction used was 100 µg/mL. All cultures lasted 24 h. Values are expressed as the mean ± SD (n = 7). Different letters indicate a significant difference ($p < 0.05$) versus induction of oxidation with H_2O_2. RB: Red Brilliant fraction; RBF: Red Brilliant fermented fraction; SH: Sharoni fraction; SHF: Sharoni fermented fraction. SOD: superoxide dismutase; CAT: catalase; GR: glutathione reductase.

3.5. *Effects of Persimmon Fiber-Rich Fractions on Bacterial Growth*

All strains showed variable, although significant, growth ($p < 0.01$) with the RB fraction (Figure 7), while no significant growth was detected with RBF, SH, and SHF fractions. *Faecalibacterium prausnitzii* strains (A2-165 and M21/2) showed a significantly ($p < 0.05$

and $p < 0.01$, respectively) higher growth rate with RB fraction as a carbohydrate source than with glucose (Figure 7B,C). For the rest of the species, except for *Eubacterium eligens* (Figure 7D), the growth rate with RB fraction as a carbohydrate source was significantly ($p < 0.01$, $p < 0.001$ and $p < 0.01$, respectively) lower than glucose (Figure 7A,E,F).

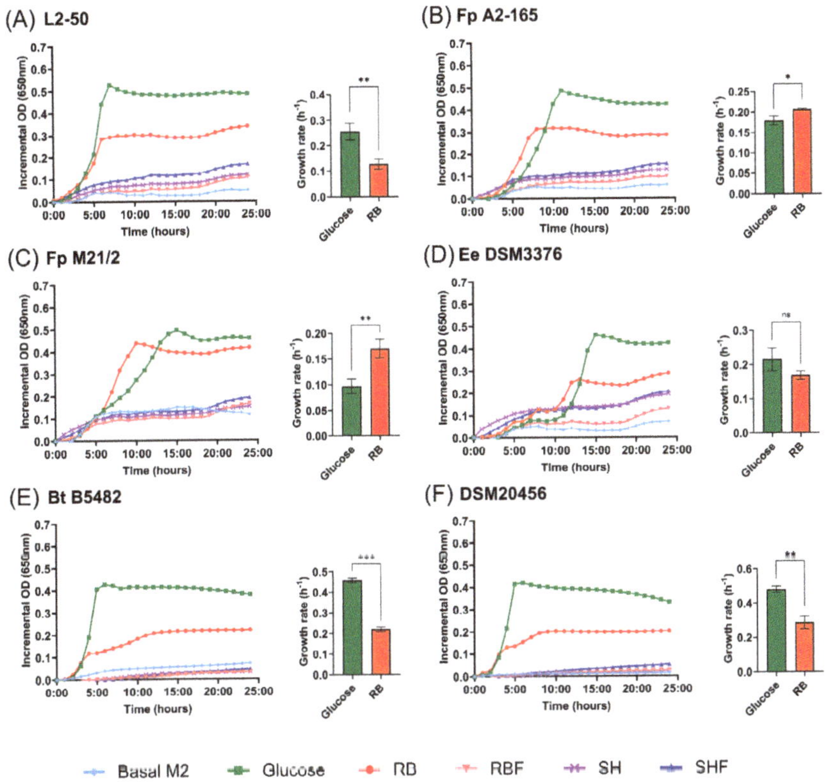

Figure 7. Growth stimulation by RB, RBF, SH, and SHF fiber-rich fractions and comparison of growth rates (h^{-1}) between glucose and RB fraction in different bacterial strains. (**A**) *Coprococcus* species L2-50; (**B**) *Faecalibacterium prausnitzii* A2–165; (**C**) *Faecalibacterium prausnitzii* M21/2; (**D**) *Eubacterium eligens* DSM3376; (**E**) *Bacteroides thetaiotaomicron* B5482; (**F**) *Bifidobacterium bifidum* DSM20456. The final fraction concentration was 0.2%. Growth stimulation in microtiter plates values were expressed as mean OD650 values. Values in growth rate were expressed as mean ± SD (n = 3). Basal M2 medium contains no added carbohydrate source. * $p < 0.05$; ** $p < 0.01$; *** $p < 0.001$. ns: not significant. RB: Red Brilliant fraction; RBF: Red Brilliant fermented fraction; SH: Sharoni fraction; SHF: Sharoni fermented fraction.

The mix of three Firmicutes strains (*Eubacterium eligens* DSM3376, *Faecalibacterium prausnitzii* A2-165, and *Faecalibacterium prausnitzii* M21/2) and one Bacteroidetes strain (*Bacteroides thetaiotaomicron* B5482) showed equal growth rate with RB fraction and glucose (Figure 8A). In the mix of bacteria, fermentation of the RB fraction resulted in the production of formate, acetate, propionate, butyrate, and succinate at the end of the incubation (Figure 8B), while glucose fermentation resulted in the production of the same SCFA products as the RB fraction plus lactate (Figure 8B). The butyrate level was higher (no significant difference) in RB compared to glucose, while lactate (approx. 5 mM) was only detected when the bacteria were grown on glucose, which is likely due to more rapid growth. The ratio of butyrate as a proportion of the total SCFA was higher ($p < 0.01$) on RB compared to glucose (Figure 8C).

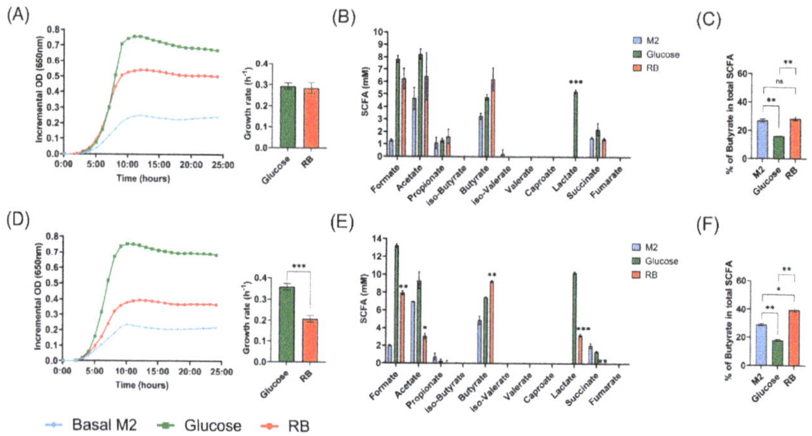

Figure 8. Growth stimulation and short-chain fatty acids production of a mix of three Firmicutes and one Bacteroidetes strain (**A–C**) and a mix of two Firmicutes, one Bacteroidetes and one Bifidobacterium strain (**D–F**) by glucose and RB fraction. (**A**) Growth rate of three Firmicutes (*Eubacterium eligens* DSM3376; *Faecalibacterium prausnitzii* A2–165; *Faecalibacterium prausnitzii* M21/2) and one Bacteroidetes (*B. thetaiotaomicron* B5482) strains; and (**B**) short-chain fatty acids (SCFA) profile in the medium after a 24 h culture by glucose and RB fraction (n = 2); and (**C**) percentage of butyrate in total SCFA (n = 2). (**D**) Growth rate of two Firmicutes (*Eubacterium eligens* DSM3376, *Faecalibacterium prausnitzii* M21/2), one Bacteroidetes (*B. thetaiotaomicron* B5482) and one Bifidobacteria (*Bifidobacterium bifidum* DSM20456) strains, (**E**) SCFA profile in the medium after 24 h culture by glucose and RB fraction (n = 2) and (**F**) percentage of butyrate in total SCFA (n = 2). Growth stimulation in microtiter plates values were expressed as mean OD650 values. Values in growth rate were expressed as mean ± SD (n = 10). Basal M2 medium contains no added carbohydrate source. mM= millimolar. * $p < 0.05$; ** $p < 0.01$; *** $p < 0.001$. SCFA: short-chain fatty acids; RB: Red Brilliant fraction.

In a separate experiment with a more complex mix of bacteria, a mix of two Firmicutes strains (*Eubacterium eligens* DSM3376 and *Faecalibacterium prausnitzii* M21/2), one *Bifidobacterium* strain (*Bifidobacterium bifidum* DSM20456), and one Bacteroidetes strain (*Bacteroides thetaiotaomicron* B5482) showed growth with RB fraction, although the growth rate was significantly ($p < 0.001$) lower than glucose (Figure 8D). In the mix of bacteria, fermentation of the RB fraction resulted in the production of formate, acetate, butyrate, and lactate (Figure 8E). Glucose fermentation produced the same SCFA as the RB fraction plus propionate and succinate (Figure 8E). The latter two products are likely to only come from the fermentation activity of the *Bacteroides thetaiotaomicron* strain B5482. In Figure 8E, the butyrate level was significantly higher ($p < 0.01$) in RB compared to glucose. In addition, the percentage of butyrate as a proportion of the total SCFA in RB was twice as much ($p < 0.01$) as in glucose (Figure 8F).

Overall, the RB fermentation results in a more beneficial fermentation profile compared to that of glucose.

4. Discussion

The concentrations of bioactive phytochemicals, such as delphinidin and cyanidin, found in the persimmon fiber-rich fractions, specifically the fermented fractions from the Sharoni variety, are similar to those found in berries [41]. Delphinidin and cyanidin health-promoting benefits such as antioxidant, anti-cancer, anti-inflammatory, cardioprotective and anti-hypertensive are well documented by several reviews [42–44]. Berries such as bilberries are particularly rich in anthocyanins and are consumed because of their high content of these bioactives. We also observed that the fermentation used to prepare the persimmon fractions significantly increased the extractability of several bioactive phytochemicals, such as anthocyanins and phenolic acids, suggesting that these molecules could therefore be more bioavailable in vivo and therefore exert health benefits. In fact, the anti-inflammatory effects exerted by the persimmon fiber-rich fractions in murine and human cell lines were higher in both fermented fractions and in the Sharoni variety, which had significantly higher levels of anthocyanins. Regarding the antioxidant activity, we found that all persimmon fiber-rich fractions had antioxidant activity in a human liver cell line. Inflammation and oxidative stress are the most common features of chronic diseases. Functional food ingredients, such as fermented fiber-rich persimmon fractions rich in anthocyanins, could therefore be used in the design of functional foods as part of nutritional therapies to prevent the development of chronic diseases.

The extracts used in the study were rich in pectin [45]. We found it remarkable that the persimmon pectin fiber-rich fractions in cultured cell lines did not show cytotoxic effects unless they were used at high, non-physiological, and not nutritionally relevant concentrations higher than 100 ug/mL. This agrees with studies showing that pectin can reduce in vitro tumoral cell proliferation and viability in different cell lines from human and murine origins at the same concentrations [46,47]. Several signaling pathways have been proposed for pectin in vitro, with anti-cancer activity being the most relevant to the DNA and mitochondrial damage, the increase in the production of reactive oxygen species, higher levels of apoptosis, cell cycle arrest, and the interference with extracellular matrix proteins [46,47]. Our data show that the most significant effects of the extracts occur in a murine fibroblast cell line and a human colon cancer line. The studies indicate that pectin can reduce in vitro cell viability in some tumor lines more than in others. This occurs regardless of whether the lines are of murine or human origin. No effects of pectin have been described in healthy cells. These results are encouraging as they guarantee their use for the development of food ingredients without having detrimental adverse events in healthy cells and with anti-cancer properties that decrease tumor cell proliferation. Although our results and those of other studies are based on in vitro studies, they open the possibility to further analyze the use of extracts as possible adjuvant therapies against some tumors.

It is, however, important to determine whether industrial by-products from the fruit industry can be used to promote the growth and metabolic output of bacteria in the human colon to promote health as this will help to drive a greener economy and reduce waste by-products from the food industry. The selectivity of pectin-rich food by products, such as those derived from persimmon fruits, for gut bacteria suggest that these fruit by-products could be used as prebiotics. The selectivity of pectin and pectic-oligosaccharides as a growth substrate was demonstrated by Chung et al. [28]. They showed that *Eubacterium eligens* and *Faecalibacterium prausnitzii* utilized pectin and pectic-oligosaccharides as growing substrate. These bacteria showed also to grow very well on model short oligomers (with degree of polymerization of 4 and 5) substrates. Others also showed that a range of whole fruits and vegetables also promoted these important bacterial species [48]. The current study revealed that two *F. prausnitzii* strains (A2-165 and M21/2) grew more rapidly on the none pre-fermented BR substrate than on glucose. One of the other Firmicutes strains of *E. eligens* (DSM3766) showed no significant difference in growth rates between the two substrates, while all the other strains tested achieved lower growth rates on the BR substrate when compared to glucose. When several of the strains belonging to the Bacteroidetes and Firmicutes phyla were mixed together and compared from the growth and fermentation of the none pre-fermented BR compared to glucose, butyrate was one of the major fermentation products formed on the BR substrate, suggesting that this fruit by-product may be a valuable substrate to promote the growth of health-promoting bacteria from the human colon. Moreover, in additional incubations, which included a bacterial strain from a third phylum, *B. bifidum*, it was still evident that butyrate was one of the main fermentation products. These findings are extremely important because of the strains in the mix that only *F. prausnitzii* can form butyrate, and this therefore demonstrates that this bacterium can compete very well for the BR substrate within the strain mixes. *F. prausnitzii* is recognized as one of the most abundant bacterial species in the healthy human colon, in part due to its ability to produce butyrate and also because it has been shown to attenuate the development of colitis in mouse models [49,50]. These bacteria help maintain the integrity of the intestinal mucosa, release molecules with anti-inflammatory properties, and contribute to the maintenance of a healthy microbiota [51]. *F. prausnitzii* also possesses the capacity to induce the anti-inflammatory cytokine interleukin 10 (IL-10) [52], as does the important key pectin degrader *E. eligens* [28]. Butyrate is the major energy source for colonocytes and may modulate carcinogenesis through its effects on proliferation, differentiation, and apoptosis in the gut, as well as stimulation of the immune system [53]. Butyrate also promotes blood flow and gut motility in the colon, which are important for digestion [54]. Furthermore, butyrate supports wound healing processes that are needed to repair epithelial injury [55]. Moreover, butyrate can prevent and treat diet-induced insulin resistance [56]. The crosstalk between the microbiome and host immunity regulates inflammatory status and has an important role in the prevention of chronic non-communicable diseases and health maintenance [57].

The persimmon fruit by-products tested in this study contain a heterogeneous family of phytochemicals with anti-inflammatory, antioxidant, and anti-cancer effects. These effects are most probably due to their richness in anthocyanins. Moreover, the pectin fiber-rich fraction inhibited "in vitro" tumoral cell line proliferation. In addition, the unfermented fractions clearly supported growth of important human colonic bacterial species with respect to promoting the growth of anti-inflammatory and butyrate-producing species that are likely to promote colonic and general health. Thus, fiber-rich fractions, obtained from persimmon fruit by-products can be used to generate foods with healthier properties to be considered as part of new therapeutic resources to promote health. The "in vivo" functionality of new therapeutic foods containing persimmon fiber-rich fractions will be the scope of further investigations.

Author Contributions: Conceptualization, N.M.B., S.H.D., M.N. and F.M.; methodology, L.L.-B., B.M.-C., J.S.-B., N.J.H., A.M., G.J.D., A.C., Á.O. and B.E.-L.; validation, W.R.R., N.M.B., S.H.D., M.N. and F.M.; formal analysis, L.L.-B., B.M.-C., Á.O., B.E.-L., G.B., N.M.B., S.H.D., M.N. and F.M.; investigation, L.L.-B., B.M.-C., J.S.-B., N.J.H., A.M., G.J.D., A.C., Á.O. and B.E.-L.; resources, M.N. and F.M.; data curation, L.L.-B., G.B., S.H.D., M.N. and F.M.; writing—original draft preparation, L.L.-B., N.M.B., S.H.D., M.N. and F.M.; visualization, N.M.B., S.H.D., M.N. and F.M.; supervision, M.N. and F.M.; project administration, F.M.; funding acquisition, F.M. All authors have read and agreed to the published version of the manuscript.

Funding: This research was funded by the Scottish Government's Rural and Environment Science and Analytical Services Division (RESAS) Strategic Research Programme (grant number RI-B6-1) and the Spanish Ministerio de Ciencia e Innovación (grant number PID2020-116731RB-C21 to F.M.). JS-B was a recipient of a predoctoral fellowship from Miguel Hernández University (Ayudas a la contratación de personal investigador en formación 2022), Spain. BM-C was a recipient of the Mobility Scholarship from Santander Open Access (Movilidad Internacional UMH 2023—2da convocatoria; 02706/2023), Spain. LL-B was a recipient of an FPU contract (FPU19/04497) from the Ministerio de Educación y Formación Profesional (Spain). LL-B's stay at The Rowett Institute was supported by a mobility fellowship from the Training Program of the Centro de Investigación Biomédica en Red de Diabetes y Enfermedades Metabólicas Asociadas (CIBERDEM), which is funded by the Instituto de Salud Carlos III (Spain).

Institutional Review Board Statement: Not applicable.

Informed Consent Statement: Not applicable.

Data Availability Statement: The original contributions presented in the study are included in the article, further inquiries can be directed to the corresponding authors.

Acknowledgments: We thank Jose Moral Rodríguez Plan de Garantía Juvenil (Ministerio de Trabajo y Economía Social, Spain) for his help and technical support with the cell cultures.

Conflicts of Interest: The authors declare no conflicts of interest.

References

1. Matheus, J.R.V.; de Andrade, C.J.; Miyahira, R.F.; Fai, A.E.C. Persimmon (*Diospyros Kaki* L.): Chemical Properties, Bioactive Compounds and Potential Use in the Development of New Products—A Review. *Food Rev. Int.* **2022**, *38*, 384–401. [CrossRef]
2. Pérez-Burillo, S.; Oliveras, M.J.; Quesada, J.; Rufián-Henares, J.A.; Pastoriza, S. Relationship between Composition and Bioactivity of Persimmon and Kiwifruit. *Food Res. Int.* **2018**, *105*, 461–472. [CrossRef] [PubMed]
3. FAO World Food and Agriculture-Statistical Yearbook. Available online: http://www.fao.org/faostat/es/#data/QC (accessed on 1 July 2024).
4. Conesa, C.; Laguarda-Miró, N.; Fito, P.; Seguí, L. Evaluation of Persimmon (*Diospyros Kaki* Thunb. Cv. Rojo Brillante) Industrial Residue as a Source for Value Added Products. *Waste Biomass Valorization* **2020**, *11*, 3749–3760. [CrossRef]
5. MAPA (Ministerio de Agricultura, P. y A. MAPA (Ministerio de Agricultura, Pesca y Alimentación). Superficies y Producciones Anuales de Cultivos-Anuario de Estadística. Available online: https://www.mapa.gob.es/es/estadistica/temas/estadisticas-agrarias/agricultura/superficies-producciones-anuales-cultivos/ (accessed on 1 July 2024).
6. Sharma, S.K.; Bansal, S.; Mangal, M.; Dixit, A.K.; Gupta, R.K.; Mangal, A.K. Utilization of Food Processing By-Products as Dietary, Functional, and Novel Fiber: A Review. *Crit. Rev. Food Sci. Nutr.* **2016**, *56*, 1647–1661. [CrossRef]
7. Lee, J.; Kim, M.; Jung, J.; Heo, J.W.; Lee, K.H.; Kim, S.; Son, H.; Chun, Y.; Yoo, H.Y. Valorization of Persimmon Calyx, an Industrial Biowaste, as a Potential Resource for Antioxidant Production. *Environ. Technol. Innov.* **2023**, *30*, 103038. [CrossRef]
8. Baiano, A. Recovery of Biomolecules from Food Wastes—A Review. *Molecules* **2014**, *19*, 14821–14842. [CrossRef] [PubMed]
9. Torres-León, C.; Ramírez-Guzman, N.; Londoño-Hernandez, L.; Martinez-Medina, G.A.; Díaz-Herrera, R.; Navarro-Macias, V.; Alvarez-Pérez, O.B.; Picazo, B.; Villarreal-Vázquez, M.; Ascacio-Valdes, J.; et al. Food Waste and Byproducts: An Opportunity to Minimize Malnutrition and Hunger in Developing Countries. *Front. Sustain. Food Syst.* **2018**, *2*, 52. [CrossRef]
10. Gea-Botella, S.; Agulló, L.; Martí, N.; Martínez-Madrid, M.C.; Lizama, V.; Martín-Bermudo, F.; Berná, G.; Saura, D.; Valero, M. Carotenoids from Persimmon Juice Processing. *Food Res. Int.* **2021**, *141*, 109882. [CrossRef] [PubMed]
11. González, C.; Hernando, I.; Moraga, G. In vitro and in vivo digestion of persimmon and derived products: A Review. *Foods* **2021**, *10*, 3083. [CrossRef]
12. Direito, R.; Rocha, J.; Sepodes, B.; Eduardo-Figueira, M. From *Diospyros kaki* L. (Persimmon) phytochemical profile and health impact to new product perspectives and waste valorization. *Nutrients* **2021**, *13*, 3283. [CrossRef]
13. Hur, S.J.; Lee, S.Y.; Kim, Y.-C.; Choi, I.; Kim, G.-B. Effect of Fermentation on the Antioxidant Activity in Plant-Based Foods. *Food Chem.* **2014**, *160*, 346–356. [CrossRef] [PubMed]
14. Sadh, P.K.; Kumar, S.; Chawla, P.; Duhan, J.S. Fermentation: A Boon for Production of Bioactive Compounds by Processing of Food Industries Wastes (By-Products). *Molecules* **2018**, *23*, 2560. [CrossRef] [PubMed]
15. Flint, H.J.; Scott, K.P.; Louis, P.; Duncan, S.H. The Role of the Gut Microbiota in Nutrition and Health. *Nat. Rev. Gastroenterol. Hepatol.* **2012**, *9*, 577–589. [CrossRef] [PubMed]
16. Hou, K.; Wu, Z.-X.; Chen, X.-Y.; Wang, J.-Q.; Zhang, D.; Xiao, C.; Zhu, D.; Koya, J.B.; Wei, L.; Li, J.; et al. Microbiota in Health and Diseases. *Signal Transduct. Target. Ther.* **2022**, *7*, 135. [CrossRef] [PubMed]
17. Cho, I.; Blaser, M.J. The Human Microbiome: At the Interface of Health and Disease. *Nat. Rev. Genet.* **2012**, *13*, 260–270. [CrossRef] [PubMed]
18. Johnson, A.J.; Vangay, P.; Al-Ghalith, G.A.; Hillmann, B.M.; Ward, T.L.; Shields-Cutler, R.R.; Kim, A.D.; Shmagel, A.K.; Syed, A.N.; Walter, J.; et al. Daily Sampling Reveals Personalized Diet-Microbiome Associations in Humans. *Cell Host Microbe* **2019**, *25*, 789–802.e5. [CrossRef]
19. Cryan, J.F.; Mazmanian, S.K. Microbiota–Brain Axis: Context and Causality. *Science (1979)* **2022**, *376*, 938–939. [CrossRef] [PubMed]
20. David, L.A.; Maurice, C.F.; Carmody, R.N.; Gootenberg, D.B.; Button, J.E.; Wolfe, B.E.; Ling, A.V.; Devlin, A.S.; Varma, Y.; Fischbach, M.A.; et al. Diet Rapidly and Reproducibly Alters the Human Gut Microbiome. *Nature* **2014**, *505*, 559–563. [CrossRef]
21. Carlson, J.L.; Erickson, J.M.; Lloyd, B.B.; Slavin, J.L. Health Effects and Sources of Prebiotic Dietary Fiber. *Curr. Dev. Nutr.* **2018**, *2*, nzy005. [CrossRef]
22. Cani, P.D.; Bibiloni, R.; Knauf, C.; Waget, A.; Neyrinck, A.M.; Delzenne, N.M.; Burcelin, R. Changes in Gut Microbiota Control Metabolic Endotoxemia-Induced Inflammation in High-Fat Diet–Induced Obesity and Diabetes in Mice. *Diabetes* **2008**, *57*, 1470–1481. [CrossRef]
23. Wang, J.; Zhu, N.; Su, X.; Gao, Y.; Yang, R. Gut-Microbiota-Derived Metabolites Maintain Gut and Systemic Immune Homeostasis. *Cells* **2023**, *12*, 793. [CrossRef] [PubMed]
24. Zhu, W.; Lin, K.; Li, K.; Deng, X.; Li, C. Reshaped Fecal Gut Microbiota Composition by the Intake of High Molecular Weight Persimmon Tannin in Normal and High-Cholesterol Diet-Fed Rats. *Food Funct.* **2018**, *9*, 541–551. [CrossRef] [PubMed]

25. Yang, Z.; Xu, M.; Li, Q.; Wang, T.; Zhang, B.; Zhao, H.; Fu, J. The Beneficial Effects of Polysaccharide Obtained from Persimmon (*Diospyros Kaki* L.) on the Proliferation of Lactobacillus and Gut Microbiota. *Int. J. Biol. Macromol.* **2021**, *182*, 1874–1882. [CrossRef] [PubMed]
26. Hwang, K.C.; Shin, H.Y.; Kim, W.J.; Seo, M.S.; Kim, H. Effects of a High-Molecular-Weight Polysaccharides Isolated from Korean Persimmon on the Antioxidant, Anti-Inflammatory, and Antiwrinkle Activity. *Molecules* **2021**, *26*, 1600. [CrossRef]
27. Dong, W.; Yang, Z. Association of Dietary Fiber Intake with Myocardial Infarction and Stroke Events in US Adults: A Cross-Sectional Study of NHANES 2011–2018. *Front. Nutr.* **2022**, *9*, 936926. [CrossRef] [PubMed]
28. Chung, W.S.F.; Walker, A.W.; Louis, P.; Parkhill, J.; Vermeiren, J.; Bosscher, D.; Duncan, S.H.; Flint, H.J. Modulation of the Human Gut Microbiota by Dietary Fibres Occurs at the Species Level. *BMC Biol.* **2016**, *14*, 3. [CrossRef]
29. Lindstad, L.J.; Lo, G.; Leivers, S.; Lu, Z.; Michalak, L.; Pereira, G.V.; Røhr, Å.K.; Martens, E.C.; McKee, L.S.; Louis, P.; et al. Human Gut *Faecalibacterium prausnitzii* Deploys a Highly Efficient Conserved System to Cross-Feed on β-Mannan-Derived Oligosaccharides. *mBio* **2021**, *12*, e0362820. [CrossRef] [PubMed]
30. Louis, P.; Hold, G.L.; Flint, H.J. The Gut Microbiota, Bacterial Metabolites and Colorectal Cancer. *Nat. Rev. Microbiol.* **2014**, *12*, 661–672. [CrossRef]
31. Human Microbiome Project Consortium. Structure, Function and Diversity of the Healthy Human Microbiome. *Nature* **2012**, *486*, 207–214. [CrossRef]
32. Russell, W.R.; Burkitt, M.J.; Scobbie, L.; Chesson, A. Radical Formation and Coupling of Hydroxycinnamic Acids Containing 1,2-Dihydroxy Substituents. *Bioorganic Chem.* **2003**, *31*, 206–215. [CrossRef]
33. Zhang, Z.; Kou, X.; Fugal, K.; McLaughlin, J. Comparison of HPLC Methods for Determination of Anthocyanins and Anthocyanidins in Bilberry Extracts. *J. Agric. Food Chem.* **2004**, *52*, 688–691. [CrossRef]
34. Neacsu, M.; Christie, J.S.; Duncan, G.J.; Vaughan, N.J.; Russell, W.R. Buckwheat, Fava Bean and Hemp Flours Fortified with Anthocyanins and Other Bioactive Phytochemicals as Sustainable Ingredients for Functional Food Development. *Nutraceuticals* **2022**, *2*, 150–161. [CrossRef]
35. Russell, W.R.; Scobbie, L.; Labat, A.; Duthie, G.G. Selective Bio-availability of Phenolic Acids from Scottish Strawberries. *Mol. Nutr. Food Res.* **2009**, 53. [CrossRef]
36. Neacsu, M.; Vaughan, N.J.; Perri, V.; Duncan, G.J.; Walker, R.; Coleman, M.; Russell, W.R. Nutritional and Chemical Profiling of UK-Grown Potato Bean (*Apios americana* Medik) Reveal Its Potential for Diet Biodiversification and Revalorisation. *J. Food Compos. Anal.* **2021**, *98*, 103821. [CrossRef]
37. Miyazaki, K.; Martin, J.C.; Marinsek-Logar, R.; Flint, H.J. Degradation and Utilization of Xylans by the Rumen Anaerobe *Prevotella bryantii* (Formerly *P. ruminicola* subsp. *Brevis) B14*. *Anaerobe* **1997**, *3*, 373–381. [CrossRef]
38. Pirt, S. *Principles of Microbe and Cell Cultivation*; Blackwell Scientific: Oxford, UK, 1975.
39. Richardson, A.J.; Calder, A.G.; Stewart, C.S.; Smith, A. Simultaneous Determination of Volatile and Non-Volatile Acidic Fermentation Products of Anaerobes by Capillary Gas Chromatography. *Lett. Appl. Microbiol.* **1989**, *9*, 5–8. [CrossRef]
40. Rampersad, S.N. Multiple applications of Alamar Blue as an indicator of metabolic function and cellular health in cell viability bioassays. *Sensors* **2012**, *12*, 12347–12360. [CrossRef]
41. Witol, A.; Lemire, M.; Dudonné, S.; Walshe-Roussel, B.; Desjardins, Y.; Cuerrier, A.; Harris, C.S. Profiling the Phenolic Acids, Flavonoids and Tannins in Skunk Currants (*Ribes glandulosum*) of Northern Québec, Canada. *J. Berry Res.* **2018**, *8*, 119–127. [CrossRef]
42. da Silva, A.P.G.; Zia, S.; John, O.D.; de Souza, M.C.; da Silva, L.C.; Sganzela, W.G. Delphinidin: Sources, Biosynthesis, Bioavailability, Bioactivity, and Pharmacology. In *Handbook of Dietary Flavonoids*; Springer International Publishing: Cham, Switzerland, 2023; pp. 1–31.
43. Husain, A.; Chanana, H.; Khan, S.A.; Dhanalekshmi, U.M.; Ali, M.; Alghamdi, A.A.; Ahmad, A. Chemistry and Pharmacological Actions of Delphinidin, a Dietary Purple Pigment in Anthocyanidin and Anthocyanin Forms. *Front. Nutr.* **2022**, *9*, 746881. [CrossRef]
44. Mattioli, R.; Francioso, A.; Mosca, L.; Silva, P. Anthocyanins: A Comprehensive Review of Their Chemical Properties and Health Effects on Cardiovascular and Neurodegenerative Diseases. *Molecules* **2020**, *25*, 3809. [CrossRef]
45. Salazar-Bermeo, J.; Moreno-Chamba, B.; Martínez-Madrid, M.C.; Saura, D.; Valero, M.; Martí, N. Potential of Persimmon Dietary Fiber Obtained from Byproducts as Antioxidant, Prebiotic and Modulating Agent of the Intestinal Epithelial Barrier Function. *Antioxidants* **2021**, *10*, 1668. [CrossRef]
46. Emran, T.B.; Islam, F.; Mitra, S.; Paul, S.; Nath, N.; Khan, Z.; Das, R.; Chandran, D.; Sharma, R.; Lima, C.M.G.; et al. Pectin: A Bioactive Food Polysaccharide with Cancer Preventive Potential. *Molecules* **2022**, *27*, 7405. [CrossRef]
47. Palko-Łabuz, A.; Maksymowicz, J.; Sobieszczańska, B.; Wikiera, A.; Skonieczna, M.; Wesołowska, O.; Środa-Pomianek, K. Newly Obtained Apple Pectin as an Adjunct to Irinotecan Therapy of Colorectal Cancer Reducing *E. coli* Adherence and β-Glucuronidase Activity. *Cancers* **2021**, *13*, 2952. [CrossRef]
48. Lordan, C.; Thapa, D.; Ross, R.P.; Cotter, P.D. Potential for Enriching Next-Generation Health-Promoting Gut Bacteria through Prebiotics and Other Dietary Components. *Gut Microbes* **2020**, *11*, 1–20. [CrossRef]
49. Sokol, H.; Pigneur, B.; Watterlot, L.; Lakhdari, O.; Bermúdez-Humarán, L.G.; Gratadoux, J.-J.; Blugeon, S.; Bridonneau, C.; Furet, J.-P.; Corthier, G.; et al. *Faecalibacterium prausnitzii* Is an Anti-Inflammatory Commensal Bacterium Identified by Gut Microbiota Analysis of Crohn Disease Patients. *Proc. Natl. Acad. Sci. USA* **2008**, *105*, 16731–16736. [CrossRef]

50. Leylabadlo, H.E.; Ghotaslou, R.; Feizabadi, M.M.; Farajnia, S.; Moaddab, S.Y.; Ganbarov, K.; Khodadadi, E.; Tanomand, A.; Sheykhsaran, E.; Yousefi, B.; et al. The Critical Role of *Faecalibacterium prausnitzii* in Human Health: An Overview. *Microb. Pathog.* **2020**, *149*, 104344. [CrossRef]
51. Alameddine, J.; Godefroy, E.; Papargyris, L.; Sarrabayrouse, G.; Tabiasco, J.; Bridonneau, C.; Yazdanbakhsh, K.; Sokol, H.; Altare, F.; Jotereau, F. *Faecalibacterium prausnitzii* Skews Human DC to Prime IL10-Producing T Cells Through TLR2/6/JNK Signaling and IL-10, IL-27, CD39, and IDO-1 Induction. *Front. Immunol.* **2019**, *10*, 143. [CrossRef]
52. Singh, V.; Lee, G.; Son, H.; Koh, H.; Kim, E.S.; Unno, T.; Shin, J.-H. Butyrate Producers, "The Sentinel of Gut": Their Intestinal Significance with and beyond Butyrate, and Prospective Use as Microbial Therapeutics. *Front. Microbiol.* **2023**, *13*, 1103836. [CrossRef]
53. Klepinina, L.; Klepinin, A.; Truu, L.; Chekulayev, V.; Vija, H.; Kuus, K.; Teino, I.; Pook, M.; Maimets, T.; Kaambre, T. Colon cancer cell differentiation by sodium butyrate modulates metabolic plasticity of Caco-2 cells via alteration of phosphotransfer network. *PLoS ONE* **2021**, *16*, e0245348. [CrossRef]
54. Mitsui, R.; Ono, S.; Karaki, S.; Kuwahara, A. Neural and non-neural mediation of propionate-induced contractile responses in the rat distal colon. *Neurogastroenterol. Motil.* **2005**, *17*, 585–594. [CrossRef]
55. Hou, Y.; Wang, L.; Yi, D.; Ding, B.; Chen, X.; Wang, Q.; Zhu, H.; Liu, Y.; Yin, Y.; Gong, J.; et al. Dietary supplementation with tributyrin alleviates intestinal injury in piglets challenged with intrarectal administration of acetic acid. *Br. J. Nutr.* **2014**, *111*, 1748–1758. [CrossRef]
56. Gao, Z.; Yin, J.; Zhang, J.; Ward, R.E.; Martin, R.J.; Lefevre, M.; Cefalu, W.T.; Ye, J. Butyrate improves insulin sensitivity and increases energy expenditure in mice. *Diabetes* **2009**, *58*, 1509–1517. [CrossRef]
57. Zheng, D.; Liwinski, T.; Elinav, E. Interaction between microbiota and immunity in health and disease. *Cell Res.* **2020**, *30*, 492–506. [CrossRef]

nutrients

MDPI

Article

Effect of Intake of Bifidobacteria and Dietary Fiber on Resting Energy Expenditure: A Randomized, Placebo-Controlled, Double-Blind, Parallel-Group Comparison Study

Yuhei Baba [1,*], Naoki Azuma [2], Yasuo Saito [2], Kazuma Takahashi [2], Risa Matsui [2] and Tsuyoshi Takara [3]

1 Dairy Business Division, Ezaki Glico Co., Ltd., 4-6-5 Utajima, Nishiyodogawa-ku, Osaka 555-8502, Japan
2 R&D Laboratory, Ezaki Glico Co., Ltd., 4-6-5 Utajima, Nishiyodogawa-ku, Osaka 555-8502, Japan; naoki.azuma@glico.com (N.A.); yasuo.saito@glico.com (Y.S.); kazuma.takahashi@glico.com (K.T.); risa.matsui@glico.com (R.M.)
3 Medical Corporation Seishinkai Takara Clinic, 9F Taisei Bldg., 2-3-2 Higashi-gotanda, Shinagawa-ku, Tokyo 142-0022, Japan; t-takara@takara-clinic.com
* Correspondence: yuhei.baba@glico.com; Tel.: +81-80-9596-0792

Abstract: *Bifidobacterium animalis* subsp. *lactis* GCL2505 in combination with inulin has been shown to have several health benefits, including an improvement in the intestinal microbiota and a reduction in human visceral fat. Previous studies have suggested that the visceral fat reduction of GCL2505 and inulin may be achieved by improving daily energy expenditure. This parallel, placebo-controlled, randomized, double-blind study was conducted to evaluate the effects of GCL2505 and inulin on resting energy expenditure (REE) in overweight or mildly obese Japanese adults (n = 44). Participants ingested 1×10^{10} colony forming units of GCL2505 and 5.0 g of inulin daily for 4 weeks. REE score at week 4 was set as the primary endpoint. At week 4, the REE score of the GCL2505 and inulin group was significantly higher than that of the placebo group, with a difference of 84.4 kcal/day. In addition, fecal bifidobacteria counts were significantly increased in the GCL2505 and inulin group. Our results indicated that the intake of GCL2505 and inulin improves energy balance, which is known to be a major factor of obesity, by modulating the microbiota in the gut. This is the first report to demonstrate the effects of probiotics and dietary fiber on REE in humans.

Keywords: *Bifidobacterium animalis* subsp. *lactis*; probiotics; inulin; visceral fat; metabolic syndrome; gut microbiota; resting energy expenditure; obesity

1. Introduction

Obesity is a state resulting from the excessive accumulation of fat and can have harmful effects on health. The World Health Organization diagnostic criteria defined a body mass index (BMI) of $\geq$25 kg/m^2 as overweight and $\geq$30 kg/m^2 as obese; in 2022, 43% of adults aged $\geq$ 18 years worldwide were overweight and 16% were obese [1]. Being overweight or obese has caused an estimated 5 million deaths in 2019 through non-communicable diseases such as cardiovascular disease, diabetes, cancer, neurological diseases, chronic respiratory diseases, and gastrointestinal diseases [2]. Furthermore, if no action is taken, economic losses and healthcare costs due to being overweight or obese are projected to reach $3 trillion annually by 2030 [3], making being overweight or obese one of the greatest public health crises of the 21st century.

Being overweight or obese is primarily caused by an imbalance between energy intake and energy expenditure. Obesity is commonly treated by caloric restriction to lower energy intake as well as exercise to increase energy expenditure. However, with industrialization and urbanization, the widespread use of trains and automobiles has led to reduced amounts of physical activity. Furthermore, it is known that 6 out of 10 obese people live in low- and middle-income countries [4]. Poverty often causes people to select cheap, high-calorie foods [1]. These social factors are some of the reasons for the increase in overweight and obese individuals.

Resting energy expenditure (REE) accounts for the largest portion of the amount of energy a person expends each day (approximately 60%), followed by physical activity (approximately 30%) and diet-induced thermogenesis (DIT; approximately 10%) [5]. It has been reported that the main determinants of DIT heat production are the energy content of the diet and the proportion of protein and alcohol, and it is difficult to control energy expenditure by regulating DIT because energy intake rises as DIT increases [6]. In addition, obese individuals are reported to be less physically active than non-obese individuals [7]. Therefore, REE has attracted attention because efforts can be made to reasonably improve energy consumption. REE consists mainly of respiration, visceral activity, and temperature maintenance. It is known that body temperature is maintained primarily through non-shivering thermogenesis specific to brown adipose tissue (BAT) [8]. A previous study reported that cold stimulation and intake of some capsinoids increased energy expenditure and decreased body fat in participants with low BAT activity [9]. Furthermore, it has been demonstrated not only in experimental animals such as mice but also in humans that BAT dysfunction contributes to obesity [10,11]. These previous studies have suggested that BAT contributes to the elimination of overweight and obesity by increasing energy expenditure through non-shivering thermogenesis.

In addition to cold stimulation and capsinoids, the effects of short-chain fatty acids (SCFAs) on BAT activity and energy expenditure have attracted attention. SCFAs are saturated aliphatic organic acids consisting of one to six carbons, of which acetate (C2), propionate (C3), and butyrate (C4) are the most abundant ($\geq$95%). The relationship between SCFAs and BAT activity has been studied, and animal studies have reported that the administration of butyric acid activates BAT and improves energy expenditure [12], while administration of acetic acid enhances the expression of genes related to BAT function and increases oxygen consumption, an indicator of energy expenditure [13]. Furthermore, human studies have shown that the infusion of a mixture of SCFAs into the intestine improves lipid oxidation and REE compared with the placebo [14]. These previous studies have suggested a possible effect of SCFAs on energy expenditure via the activation of BAT; the most reasonable way to synthesize SCFAs is fermentation in the body by intestinal bacteria [15]. Non-digestible carbohydrates (dietary fiber) are fermented to produce energy for bacterial growth, and SCFAs are produced as the main end product [16]. Therefore, probiotics, which produce SCFAs in the gut, prebiotics, which are capitalized by intestinal bacteria, and synbiotics, which are a complex of probiotics and prebiotics, are suitable materials for increasing the level of SCFAs in the gut. Previous animal studies have shown that probiotics and prebiotics support the BAT-mediated enhancement of REE. The combined administration of the probiotic strain *Bifidobacterium adolescentis* 2016_7_2 and a high-fat diet leads to a decrease in the respiratory quotient (RQ) and an increase in the expression of Ucp-1 in BAT [17], and the intake of the prebiotic caffeoylquinic acid improves energy expenditure with the help of the microbiota [18]. In humans, however, there have been no reports of improved energy metabolism by probiotics, prebiotics, or synbiotics, although there are reports of enhanced fat oxidation with the administration of oligopeptides [19] and 24 g of inulin [20].

Bifidobacterium animalis subsp. *lactis* GCL2505, commercially named "BifiX" in Japan, is a probiotic strain that grows in the gut that passes through and was originally isolated from the feces of healthy adults [21,22]. Previous animal experiments have revealed that SCFAs produced in the intestine by GCL2505 have anti-metabolic syndrome effects such as improving glucose tolerance and suppressing visceral fat accumulation [23] and affect host metabolic homeostasis, including the enhancement of glucose tolerance and suppression of body fat accumulation, via G protein-coupled receptor 43, which is also known as the SCFA receptor [24]. Clinical studies have reported that the intake of GCL2505 was associated with improved cognitive function [25] and vascular endothelial function [26] when administered in combination with inulin [27], a typical prebiotic material. Furthermore, we recently conducted a clinical trial to evaluate the efficacy of GCL2505 and inulin on obesity [28]. In healthy adult men and women with a BMI of $\geq$23 kg/m^2 and <30, 12 weeks of GCL2505 and inulin intake significantly decreased visceral and total fat area, increased the total number of bifidobacteria, and decreased the levels of several lipid markers. Therefore, it was suggested that the combined intake of GCL2505 and inulin improves the intestinal environment and reduces abdominal fat related to the SCFA-mediated pathway. Because GCL2505 can proliferate in the gut, it may contribute to the increase in SCFA levels in the gut, thereby exerting anti-obesity effects.

In other words, the effect on visceral fat caused by the combined intake of GCL2505 (a high producer of SCFAs) and inulin is thought to be due to the increased presence of SCFAs in the gut, but the mechanism of action has not been clarified. From the previous studies, it was considered that the visceral-fat-reduction effect of GCL2505 and inulin may be achieved by improving daily energy expenditure. Therefore, the present study conducted a parallel, placebo-controlled, randomized, double-blind study to evaluate the effects of intake of the synbiotics GCL2505 and inulin on REE in overweight or mildly obese Japanese adults.

2. Materials and Methods

2.1. Participants

This study was approved by the Ethics Committee of Medical Corporation Seishinkai Takara Clinic on 20 September 2023 (approval number: 2309-00178-0078-3BTC). Written informed consent was obtained from all participants in accordance with the Declaration of Helsinki. Participants were Japanese men and women between the ages of 25 and 61 years at the time of consent who satisfied the inclusion criteria, did not satisfy any of the exclusion criteria, and were deemed eligible by the study investigator. The inclusion criteria were as follows: (1) in good health, (2) BMI between 25 kg/m^2 and 30 kg/m^2, (3) body fat percentage of at least 15% in men and 25% in women, and (4) the top 40 participants with the lowest resting energy metabolism among those satisfying selection criteria (1) through (3) and not satisfying the exclusion criteria. The exclusion criteria were as follows: (1) undergoing treatment for or a history of malignant tumor, heart failure, or myocardial infarction, (2) having a pacemaker or implantable cardioverter-defibrillator, (3) undergoing treatment for arrhythmia, liver damage, kidney damage, cerebrovascular disease, rheumatism, diabetes, dyslipidemia, hypertension, or other chronic diseases, (4) consuming foods for specified health uses or foods with functional claims, (5) taking pharmaceuticals (including herbal medicines) or supplements, (6) having allergies (to pharmaceuticals or food related to the tested food), (7) pregnant, lactating, or intending to become pregnant during the study period, (8) participation in other clinical trials during the 28 days prior to the date of consent, (9) took antibiotics during the 28 days prior to the date of consent; and (10) deemed ineligible by the principal investigator.

2.2. Management of Participants

Participants were managed as follows: (1) During the study period, the ingestion or non-ingestion of test drinks and the occurrence of menstruation (for women only) were recorded daily in a logbook provided by the contract research institute. (2) The physical condition of the study participants was ascertained by interview at the time of their visit to the hospital. (3) Dietary intake was examined 3 and 2 days before each test day. A Calorie and Nutrition Diary (CAND) was used for the dietary survey [29]. The participants were asked to submit their CAND at the time of each examination visit. (4) For breakfast and lunch on the day before each test, participants consumed the specified prescribed diet. Participants spent the evening at their designated accommodation facility and consumed the prescribed dinner no later than 12 h before the start of the test the next day. Participants were asked to avoid eating and drinking anything other than the prescribed diet and were only allowed to drink water. (5) Participants were asked to ensure compliance with the following points during study participation: (a) from the date of obtaining consent to participate in the study until the final test (4 weeks post-test), avoid binge eating and drinking and maintain their previous lifestyle; (b) if any change in physical condition occurs during the study period, immediately contact the sponsoring clinical research organization and ask for instructions on what to do next; (c) consume the test drinks according to the prescribed dosage and administration, at an intake rate of at least 80%; (d) during the test period, avoid, to the extent possible, consuming foods for specified health use, foods with functional claims, fermented foods such as yogurt, kimchi, and natto, and other foods/beverages with possible functional properties; (e) avoid alcohol consumption and excessive exercise from 3 days before each examination until the end of the examination on the same day; (f) refrain from consuming caffeine-rich beverages (energy drinks, coffee, etc.) for 3 days before the test, and avoid consuming caffeine-containing beverages on the day before the test.

2.3. Test Foods

The test foods were a dairy drink containing inulin (Orafti GR; Beneo GmbH, Mannheim, Germany) and GCL2505 (active drink) or a placebo. The active drink was made by diluting the fermented dairy drink in which the bifidobacteria count was measured with a non-fermented dairy drink containing the same ingredients to stabilize the bifidobacteria count. The active drink contained 5.0 g of inulin and 1×10^{10} colony-forming units of GCL2505 per 100 g. The placebo was prepared with the same ingredients as the active drink, with the addition of food-grade acetic acid and lactic acid to adjust flavor and pH; the basic ingredients were skim milk powder, fructose dextrose, sucrose, yeast extract, acidifier, stabilizer, and flavoring. The nutritional details of the test foods are shown in Table 1.

Table 1. Nutritional details of the test drinks.

	Active	Placebo
Energy, kcal/100 g	60.2	47.7
Moisture, g/100 g	82.3	87.0
Protein, g/100 g	2.8	2.8
Fat, g/100 g	0.1	0.1
Carbohydrate, g/100 g	14.9	9.1
Ash, g/100 g	1.1	1.1

The active drink contained 5.0 g of inulin and 1.0×10^{10} colony-forming units of GCL2505.

2.4. Experimental Design

The study was a randomized, placebo-controlled, double-blind, parallel-group study. Participants who satisfied the eligibility criteria at the time of the screening test were assigned by the allocation manager to either the active or placebo group at a 1:1 ratio, using an allocation table generated by the open-source software program R (ver. 4.2.1); the algorithm used block random allocation with a random block size of seven. For the sample size, the final target number of participants was set at 20, based on our previous study of resting energy expenditure with GCL2505 and inulin (UMIN000050836, unpublished). The participant selection process is shown in Figure 1. In this study, 81 participants were screened. After screening, 44 participants were eligible; 22 were assigned to the active group and 22 were assigned to the placebo group. The doses of GCL2505 and inulin were determined based on previous studies, respectively [28,30]. Participants in the active and placebo groups consumed 100 g of test foods once daily for 4 weeks. Both the participants and observers were blinded to group allocation for the duration of the study. Double blinding was accomplished by labeling the test foods with an identification number only. The identification numbers of the active and placebo drinks were kept strictly confidential and were not disclosed until the allocation manager sent out the allocation list after the study was completed. The allocation manager generated the allocation order based on the identification numbers of the test foods provided and created an allocation list and an emergency key. The emergency key was sealed in an envelope for each study participant, and the envelope was stamped with an allotment seal and sealed. After the study was completed and the data were fixed, the allocation manager confirmed that the allocation list and emergency key had not been opened, and the identification numbers of the test foods were revealed. The primary outcome was REE at week 4. Secondary outcomes included REE at week 2, RQ, carbohydrate oxidation, fat oxidation, body weight, BMI, body fat percentage, and muscle mass at weeks 2 and 4, as well as fecal SCFAs (formic acid, acetic acid, lactic acid, propionic acid, n-butyric acid, iso-butyric acid, succinic acid, n-valeric acid, and iso-valeric acid) and the number of bifidobacteria in feces at week 4. The study was conducted at the Medical Corporation Seishinkai Takara Clinic (Tokyo, Japan) from October to December 2023 by Orthomedico Inc., a contract research organization (Tokyo, Japan), and was registered with the University Hospital Medical Information Network Clinical Trials Registry as UMIN000052435. This article conforms to the Consolidated Standards of Reporting Trials (CONSORT) 2010 guidelines (Supplementary Materials, Table S1).

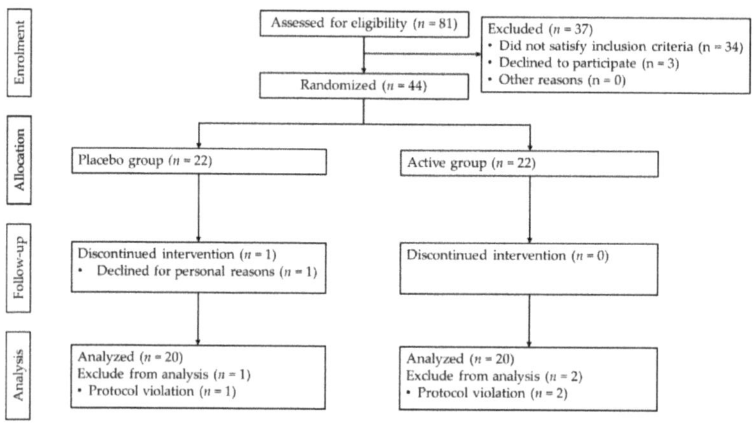

Figure 1. Flowchart of participant selection.

2.5. Indirect Calorimetry

Participants' oxygen uptake (VO_2) and carbon dioxide production (VCO_2) were measured using a respiratory gas analyzer (AE310S; Minato Medical Science, Osaka, Japan) in the morning of the test day. Participants were instructed to rest in their assigned accommodations from the evening before the test and to consume their assigned dinner in the accommodations at least 12 h before the start of the test the next day. Measurements were taken in a resting sitting position at a comfortable room temperature; VO_2 and VCO_2 were recorded continuously for 15 min. REE and RQ were calculated by the following equations:

$$\text{REE (kcal/day)} = [3.9 \times VO_2\ (\text{mL/min}) + 1.1 \times VCO_2\ (\text{mL/min})] \times 1.44 \quad (1)$$

$$RQ = VCO_2/VO_2 \quad (2)$$

2.6. Anthropometric Measurements and Body Composition

Body weight and height were measured in units of 0.1 kg and 0.1 cm, respectively, with the participant standing. BMI was calculated by dividing weight (kg) by the square of height (m).

2.7. Fecal Samples

Fecal samples were submitted on weeks 0 and 4. Fecal samples were handled according to previously described procedures [28] and promptly transported to the Kyoto Institute of Nutrition and Pathology (Kyoto, Japan) by refrigerated transport at temperatures below −15 °C.

2.8. Fecal Short-Chain Fatty Acids

Concentrations of SCFAs in feces were determined using ion-exclusion high-performance liquid chromatography (HPLC) according to the procedure of Morishima et al. [31]. Specifically, 0.3 g of feces was placed into a 1.5 mL microtube. To a suspension consisting of feces and 0.6 mL distilled water, 0.09 mL of 12% perchloric acid was added and allowed to stand on ice for 3 min after suspension. Samples were then centrifuged (15,000× *g*, 10 min, 4 °C), and the collected supernatant was filtered through a 0.45 µm COSMONICE® Filter W (water-based: Nacalai Tesque, Inc., Kyoto, Japan) before being subjected to analysis. An LC-10ADVP pump, CDD-10A VP conductometer, Shim-Pack SCR-102(H) Column (8.0 mm × 30 cm × 2 columns), and CTO-20AC column heater module (all manufactured by Shimadzu Corporation, Kyoto, Japan) were used to measure the analytical concentrations of SCFAs. Distilled water for HPLC (Fujifilm Wako Pure Chemical Corporation, Osaka, Japan) containing 5 mM *p*-toluenesulfonic acid was prepared as the mobile phase and filtered through a 0.45 µm cellulose acetate membrane filter (Toyo Roshi Kaisha, Ltd., Tokyo, Japan) for use in the measurements. The post-column pH buffering solution was distilled water with 5 mM *p*-toluenesulfonic acid, 20 mM Bis-Tris, and 100 µM EDTA (free acid) added. Mobile phase and pH buffering solution were supplied at a flow rate of 0.8 mL/min each. The column temperature was set at 45 °C. Components were identified by the Kyoto Institute of Nutrition and Pathology (Kyoto, Japan), using a CBM-20A data module (Shimadzu Corporation, Kyoto, Japan).

2.9. Fecal DNA Extraction

Bacterial DNA was extracted from fecal samples, according to the procedure of Tourlousse et al. [32]. Specifically, 0.2 g of fecal sample, 700 µL of FE1 buffer, and 10 µL of RNase were added to a tube containing beads, and a bead-beating homogenizer (FastPrep-24; MP Biomedicals, Irvine, CA, USA) was run at 6 m/s for 1 min to destroy the cells. This process was repeated three times, during which the samples were held at room temperature for 5 min. The samples were then centrifuged at 12,000× *g* for 15 min with 90 µL of FE2 buffer added. The supernatant (up to 500 µL) was collected and mixed with FB buffer and isopropanol, each at 0.4× the volume of the supernatant obtained. Finally, the sample was loaded onto a spin column and washed according to the manufacturer's instructions. The purified DNA was eluted from the column by 50 µL of Tris-EDTA buffer (pH 8.0).

2.10. Fecal Bifidobacteria

Real-time polymerase chain reaction (PCR) was performed with reference to Tanaka et al., using genus-specific primers capable of detecting *Bifidobacterium* spp., including GCL2505 [33]. The primer sequences were as follows: *Bifidobacterium* spp. sense primer, 5′-GATTCTGGCTCAGGATGAACGC-3′; *Bifidobacterium* spp. antisense primer, 5′-CTGATAGGACGCGACCCCAT-3′. Each PCR reaction mixture contained 20 pmol of each primer, 5 µL of SYBR® premix Ex taq (Takara Bio, Kusatsu, Japan), and 1 µL of DNA solution. This procedure was performed by the Kyoto Institute of Nutrition and Pathology (Kyoto, Japan).

2.11. Statistical Analysis

All measurement data are presented as means and standard deviations (SD) or 95% confidence intervals (CIs). All statistical analyses were performed using IBM® SPSS® Statistics 23 (IBM Corp., Armonk, NY, USA). p-value < 0.05 was used as the threshold for determining significant differences. Missing data were treated as missing values and no proxy values were used. Unpaired t-tests were used to assess baseline at study entry between participants in both groups and dietary bias during the study period. The participants' indirect calorimetry and body parameters during the study period were compared between groups, using a mixed-effects model for repeated measures, based on the restricted maximum likelihood method. The mean structure of models other than BMI assumed a fully unstructured variance–covariance matrix for error terms, including time, group (active or placebo), sex, baseline values, baseline BMI values, and interactions between time and group, and between baseline values and time. The mean structure of models for BMI included baseline value, time point, group (active or placebo), sex, interaction between time point and group, and interaction between baseline value and group, and assumed a fully unstructured variance–covariance matrix in the error term. Satterthwaite's method was used to estimate the degrees of freedom. The difference between groups at each time point was calculated as the difference in the marginal estimated means. Fecal SCFAs concentrations were statistically analyzed by covariate analysis adjusted for baseline (week 0). Fecal bifidobacteria were compared within each group by paired *t*-test and between the two groups by a covariate analysis adjusted for the baseline (week 0).

3. Results

3.1. Subjects (Analysis Target Population)

At the start of the study, a significant difference in plasma glucose values between the active and placebo groups was observed but was deemed acceptable because it was within the reference range. There were no differences in the baseline characteristics of other participant data between the two groups (Table 2). By the end of the study, one participant from the placebo group withdrew for personal reasons. After the completion of the entire study, three participants whose consumption rate of the test food was less than 80% were excluded according to study participant management criteria (5)-(c) (n = 1 from the placebo group and n = 2 from the active). Finally, 40 participants were analyzed, with 20 in the placebo group and 20 in the active group. There were no reported harms or unintended effects in either group.

Table 2. Participant characteristics at baseline.

	Active Group	Placebo Group	*p*-Value
Age, years	47.75 (11.07)	47.75 (9.68)	1.00
Female, *n* (%)	16 (80.00)	16 (80.00)	1.00
Height, cm	160.18 (6.44)	157.50 (6.45)	0.20
Body weight, kg	68.61 (8.37)	66.86 (6.21)	0.46
Body mass index, kg/m^2	26.65 (1.47)	26.90 (1.02)	0.53
Systolic blood pressure, mmHg	120.40 (18.19)	123.95 (15.50)	0.51
Diastolic blood pressure, mmHg	79.95 (13.22)	80.90 (10.32)	0.80
White blood cell count, /μL	6130.00 (1475.80)	6000.00 (1069.68)	0.75
Red blood cell count, $\times 10^4$/μL	470.00 (31.98)	463.90 (38.73)	0.59
Hemoglobin, g/dL	13.77 (1.31)	13.64 (1.26)	0.75
Hematocrit, %	44.18 (3.45)	43.65 (3.29)	0.62
Platelet count, $\times 10^4$/μL	26.20 (5.01)	29.27 (5.84)	0.08
Total serum protein, g/dL	7.06 (0.37)	7.09 (0.33)	0.79
Aspartate aminotransferase, U/L	21.20 (11.73)	22.35 (12.84)	0.77
Alanine aminotransferase, U/L	22.50 (19.03)	22.95 (17.02)	0.94
Total bilirubin, mg/dL	0.68 (0.33)	0.69 (0.28)	0.96
γ-Glutamyl transpeptidase, U/L	35.05 (50.11)	26.80 (15.07)	0.49
Blood urea nitrogen, mg/dL	12.93 (3.66)	11.29 (1.71)	0.08
Creatinine, mg/dL	0.70 (0.13)	0.68 (0.07)	0.55
Uric acid, mg/dL	5.29 (1.10)	5.64 (1.25)	0.35
Sodium (Na), mEq/L	142.25 (2.12)	141.60 (2.01)	0.33
Chlorine (Cl), mEq/L	102.60 (1.27)	102.25 (2.02)	0.52
Potassium (K), mEq/L	4.15 (0.23)	4.07 (0.28)	0.33
Serum amylase, U/L	80.40 (52.48)	75.50 (26.09)	0.71
Total cholesterol, mg/dL	215.80 (45.89)	225.75 (34.27)	0.44
HDL cholesterol, mg/dL	57.65 (12.52)	55.70 (13.67)	0.64
LDL cholesterol, mg/dL	131.20 (40.90)	139.20 (34.01)	0.51
Triglycerides, mg/dL	118.30 (46.71)	134.20 (59.54)	0.35
Glucose, mg/dL	88.40 (5.78)	93.05 (6.21)	0.02
HbA1c (NGSP), %	5.41 (0.21)	5.47 (0.31)	0.48
Compliance rate of the test sample, % *	99.1 (3.5)	102.7 (7.1)	-

All data are presented as mean (SD). Differences between placebo and active groups were assessed by unpaired *t*-test. * Indicates compliance rate of the test sample, excluding participants who dropped out.

3.2. Dietary Composition

Nutrients ingested by the participants were calculated from the food records for 2 days before REE measurement (Table 3). No statistically significant differences were observed between the two groups in energy, protein, fat, carbohydrate, and dietary fiber. Participants were required to eat the designated dinner at the provided accommodations no later than 12 h before the start of the next day's test, thereby ensuring that the content of the meal the day before the REE measurement did not affect the test results. Accordingly, it was concluded that dietary content did not affect the results of this study.

Table 3. Changes in dietary composition in the active (n = 20) and placebo (n = 20) groups during the intervention period.

		0 Week	2 Weeks	4 Weeks
Energy, kcal	Active	2430.02 (1999.86, 2860.17)	2201.98 (1821.45, 2582.50)	2270.44 (1886.14, 2654.74)
	Placebo	2234.22 (1971.28, 2497.17)	2236.50 (1957.50, 2515.50)	1943.19 (1624.80, 2261.58)
	p-value	0.45	0.89	0.21
Protein, g	Active	102.14 (82.54, 121.73)	87.80 (70.33, 105.28)	93.64 (75.36, 111.93)
	Placebo	90.81 (78.16, 103.46)	91.12 (77.40, 104.83)	78.38 (64.20, 92.57)
	p-value	0.35	0.77	0.20
Fat, g	Active	93.48 (73.39, 113.56)	79.80 (60.79, 98.81)	85.86 (66.31, 105.40)
	Placebo	80.75 (68.49, 93.01)	79.75 (67.23, 92.27)	64.54 (52.08, 77.00)
	p-value	0.30	1.00	0.08
Carbohydrate, g	Active	280.39 (234.10, 326.68)	270.57 (233.45, 307.69)	265.11 (230.95, 299.27)
	Placebo	272.41 (242.12, 302.70)	275.72 (238.19, 313.25)	250.93 (211.18, 290.67)
	p-value	0.78	0.85	0.60
Dietary fiber, g	Active	15.42 (11.42, 19.42)	14.06 (10.65, 17.48)	14.73 (11.74, 17.71)
	Placebo	15.09 (12.56, 17.62)	14.82 (12.15, 17.48)	14.69 (11.24, 18.13)
	p-value	0.89	0.74	0.99

All data are presented as mean (95% CIs). Differences between the placebo and active groups were assessed by unpaired *t*-test.

3.3. Indirect Calorimetry

The REE score at week 4 (the primary endpoint) of the active group (1376.5 ± 272.8 kcal/day) was greater than that of the placebo group (1303.2 ± 188.1 kcal/day), and a significant difference was confirmed (p = 0.042 by repeated measurements analysis using a linear mixed model). In addition, the REE score in the active group at week 2 (1435.9 ± 195.2 kcal/day) was also statistically higher than in the placebo group (1345.5 ± 231.6 kcal/day) (p = 0.002 by repeated measurements analysis using a linear mixed model). In contrast, no significant differences were observed between the two groups in RQ, carbohydrate oxidation, or lipid oxidation (Table 4).

Table 4. Changes in indirect calorimetry in the active (n = 20) and placebo (n = 20) groups during the intervention period.

		0 Week	2 Weeks	4 Weeks
Resting energy expenditure, kcal/day	Active	1326.8 (1262.9, 1390.7)	1435.9 (1464.2, 1565.5) *	1376.5 (1393.0, 1517.6) *
	Placebo	1325.1 (1257.3, 1392.8)	1345.5 (1362.5, 1463.5)	1303.2 (1308.7, 1433.0)
	Difference between groups	1.8 (−94.4, 97.9)	101.8 (39.2, 164.4)	84.4 (3.2, 165.7)
	p-value	0.971	0.002	0.042

Table 4. *Cont.*

		0 Week	2 Weeks	4 Weeks
Respiratory quotient	Active	0.83 (0.81, 0.85)	0.83 (0.81, 0.85)	0.85 (0.82, 0.87)
	Placebo	0.83 (0.82, 0.84)	0.83 (0.81, 0.84)	0.84 (0.80, 0.86)
	Difference between groups	−0.01 (−0.03, 0.02)	0.00 (−0.02, 0.03)	0.01 (−0.02, 0.05)
	p-value	0.649	0.695	0.429
Carbohydrate oxidation amount, mg/min	Active	133.2 (111.57, 154.77)	153.1 (138.6, 178.4)	163.1 (138.7, 198.5)
	Placebo	139.5 (123.95, 155.00)	142.0 (121.9, 162.6)	144.9 (115.0, 175.3)
	Difference between groups	−6.3 (−33.9, 21.3)	16.2 (−9.1, 41.6)	23.4 (−17.2, 64.0)
	p-value	0.645	0.202	0.250
Fat oxidation amount, mg/min	Active	82.4 (72.74, 92.02)	85.6 (83.6, 100.9)	76.3 (73.8, 92.4)
	Placebo	79.0 (71.01, 87.08)	80.2 (78.8, 96.0)	74.6 (72.5, 91.1)
	Difference between groups	3.3 (−9.6, 16.3)	4.8 (−6.3, 15.9)	1.3 (−10.8, 13.5)
	p-value	0.606	0.382	0.826

All data are presented as mean (95% CIs). Differences between the placebo and active groups were assessed by the mixed-effects model for repeated measures. * $p < 0.05$.

3.4. Fecal Bifidobacteria

The quantified number of bifidobacteria in feces was converted into logarithmic values and compared (Figure 2). Inter-group comparison at week 4 revealed that the total number of bifidobacteria was significantly increased in the active group (11.5 ± 0.9 log cells/g feces) compared with the placebo (11.3 ± 1.2 log cells/g feces) ($p = 0.037$ by analysis of covariance with baseline values as covariates). Intra-group comparison revealed a statistically significant increase in the total number of bifidobacteria in the active group at week 4 compared with that at week 0 (10.5 ± 2.0 log cells/g feces) ($p = 0.013$ by paired *t*-test). In contrast, the number of fecal bifidobacteria in the placebo group did not change during the study period (week 0: 11.3 ± 1.0 log cells/g feces).

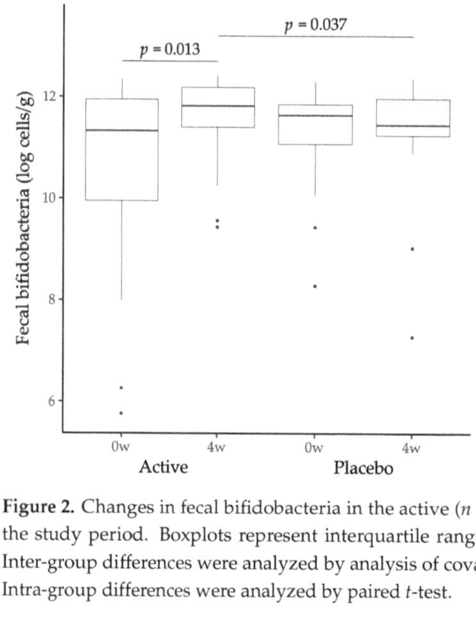

Figure 2. Changes in fecal bifidobacteria in the active (n = 20) and placebo (n = 20) groups during the study period. Boxplots represent interquartile range (25–75%) with median shown in black. Inter-group differences were analyzed by analysis of covariance with baseline values as covariates. Intra-group differences were analyzed by paired *t*-test.

3.5. Fecal Short-Chain Fatty Acids

The fecal concentration of propionic acid in the active group (15.4 ± 6.0 mmol/kg wet feces) at week 4 was statistically lower than in the placebo group (20.9 ± 9.2 mmol/kg wet feces) ($p = 0.015$ by analysis of covariance with baseline values as covariates). In all the items except propionic acid, there were no statistically significant differences between the active group and the placebo during the study period (Table 5).

Table 5. Changes in fecal SCFA concentrations in the active ($n = 20$) and placebo ($n = 20$) groups during the intervention period.

		0 Week	4 Weeks
Formic acid, mmol/kg wet feces	Active	0.0 (0.0, 0.0)	0.0 (−0.2, 0.2)
	Placebo	0.0 (0.0, 0.0)	0.2 (0.0, 0.5)
	p-value	-	0.156
Acetic acid, mmol/kg wet feces	Active	55.6 (43.7, 67.5)	42.4 (34.7, 50.2)
	Placebo	49.6 (40.3, 58.9)	52.9 (45.1, 60.7)
	p-value	0.444	0.063
Lactic acid, mmol/kg wet feces	Active	0.4 (−0.4, 1.2)	0.0 (−0.1, 0.1)
	Placebo	0.3 (0.0, 0.6)	0.0 (0.0, 0.1)
	p-value	0.759	0.335
Propionic acid, mmol/kg wet feces	Active	19.0 (14.4, 23.6)	15.4 (12.1, 18.5) *
	Placebo	18.6 (13.8, 23.4)	20.9 (17.8, 24.2)
	p-value	0.918	0.015
n-Butyric acid, mmol/kg wet feces	Active	9.5 (7.3, 11.7)	7.4 (5.6, 9.2)
	Placebo	7.9 (5.3, 10.6)	8.6 (6.8, 10.4)
	p-value	0.377	0.378
iso-Butyric acid, mmol/kg wet feces	Active	0.9 (0.1, 1.7)	0.3 (−0.2, 1.0)
	Placebo	0.0 (0.0, 0.0)	0.6 (0.0, 1.2)
	p-value	0.051	0.612
Succinic acid, mmol/kg wet feces	Active	1.0 (0.0, 2.1)	1.4 (−0.1, 2.9)
	Placebo	0.7 (0.2, 1.1)	0.4 (−1.0, 1.9)
	p-value	0.547	0.356
n-Valeric acid, mmol/kg wet feces	Active	0.8 (−0.3, 1.9)	0.4 (−0.1, 1.0)
	Placebo	0.3 (−0.3, 0.9)	0.3 (−0.2, 0.9)
	p-value	0.449	0.876
iso-Valeric acid, mmol/kg wet feces	Active	1.0 (0.1, 2.0)	0.4 (−0.2, 1.0)
	Placebo	0.0 (0.0, 0.0)	0.6 (−0.1, 1.1)
	p-value	0.051	0.808

All data are presented as mean (95% CIs). Differences between the placebo and active groups were assessed by analysis of covariance with baseline values as covariates. * $p < 0.05$.

3.6. Anthropometric Parameters

Body weight, BMI, body fat percentage, and muscle mass were measured during the study period (Table 6). There were no statistically significant differences between the active and placebo groups.

Table 6. Changes in anthropometric parameters in the active ($n = 20$) and placebo ($n = 20$) groups during the intervention period.

		0 Week	2 Weeks	4 Weeks
Body weight, kg	Active	68.6 (64.9, 72.3)	69.0 (67.8, 69.3)	68.9 (67.6, 69.2)
	Placebo	66.9 (64.1, 69.6)	67.1 (67.6, 68.9)	67.1 (67.5, 69.0)
	p-value	0.459	0.526	0.726

Table 6. *Cont.*

		0 Week	2 Weeks	4 Weeks
Body mass index, kg/m^2	Active	26.6 (26.0, 27.3)	26.8 (26.8, 27.3)	26.7 (26.7, 27.2)
	Placebo	26.9 (26.5, 27.3)	27.0 (26.7, 27.2)	27.0 (26.7, 27.2)
	p-value	0.528	0.753	0.965
Body fat rate, %	Active	37.6 (35.1, 40.2)	37.9 (37.0, 38.9)	37.7 (35.5, 40.0)
	Placebo	37.4 (34.6, 40.2)	37.3 (36.5, 38.5)	35.6 (33.6, 38.1)
	p-value	0.903	0.291	0.213
Muscle mass, kg	Active	40.5 (37.3, 43.7)	40.5 (40.1, 41.1)	40.6 (39.1, 42.2)
	Placebo	39.7 (36.7, 42.6)	39.9 (40.3, 41.2)	41.1 (40.4, 43.5)
	p-value	0.714	0.540	0.239

All data are presented as mean (95% CIs). Differences between the placebo and active group were assessed by the mixed-effects model for repeated measures.

4. Discussion

We investigated the effect of consuming a dairy drink containing *Bifidobacterium animalis* subsp. *lactis* GCL2505 and inulin on REE in healthy adults. The results showed that the active group had a statistically significantly higher REE score for the primary outcome compared with the placebo group and an increased number of fecal bifidobacteria. Because we have previously shown that GCL2505 and inulin reduce visceral and body fat area, the present results were considered to support earlier findings.

GCL2505 and inulin, which have been shown to inhibit fat accumulation [28], are expected to contribute to the prevention of weight gain and obesity. Obesity is a state of excessive fat accumulation resulting from an imbalance between energy intake and energy expenditure [34]. In other words, continuous intake of GCL2505 and inulin, which has the effect of reducing fat in humans, may cause a rebalancing of energy levels. Thus, the effect of GCL2505 and inulin on REE, which significantly affects the amount of energy a person consumes per day, was verified in this study. The mechanism by which probiotics suppress obesity has been investigated in several studies, and it was reported that the suppressive effect of *Lactobacillus gasseri* SBT2055 on visceral fat accumulation was due to inhibitory effects on the absorption and promotion of lipid excretion in the intestinal tract [35]. Inhibitory effects of *Bifidobacterium breve* B-3 [36] and *Lactobaclillus paracasei* subsp. *paracasei* F199 [37] on fat accumulation in adipose tissue have also been reported. *Lactobacillus gasseri* BNR17 is one of the few bacteria shown to improve energy metabolism. Animal studies have also suggested that the effects of BNR17 on visceral fat accumulation and abdominal circumference reduction depend on increased expression of genes related to fatty acid metabolism [38]. Meanwhile, in the present study, the direct effect of intake of bifidobacteria and dietary fiber on REE was confirmed in clinical trials. REE is strictly determined by summing the metabolic rates of body tissues [39]; however, indirect calorimetry, which is simple, noninvasive, and highly accurate for measurement [40,41], was used in the present study. To date, no studies have demonstrated an ameliorative effect of probiotics on REE, and the reasons for this are not clear. However, for REE to be measured in this study, participants were provided accommodations the day before at the testing facility and were kept in a strictly controlled environment. Given the study design, along with the high ability of GCL2505 and inulin to improve SCFAs levels in the gut, the present study may be the first to demonstrate the effect of probiotics and soluble dietary fiber on REE in a clinical trial. REE (the primary endpoint of this study) was correlated with basal metabolic rate [42], suggesting that the increase in REE may also indicate an increase in the participants' basal metabolic rate.

It was hypothesized that the increase in REE with the intake of GCL2505 and inulin was achieved by a mechanism of action comprising the following two steps. In Step 1, the intake of GCL2505 and inulin increases bifidobacteria in the gut and increases the levels of SCFAs. In a previous animal study, ingestion of GCL2505 alone contributed to an increase in the number of bifidobacteria in the feces and a concomitant increase in acetic acid levels in the feces and blood [24]. In addition, previous clinical studies reported that GCL2505 and inulin increased the total number of bifidobacteria in feces [43] and that intake of inulin increased SCFAs such as acetic acid via an increase in bifidobacteria [44]. These results support our hypothesis. However, in the present study, we were unable to detect an increase in the levels of SCFAs in the gut as a result of the intervention, and there was no significant difference between the two groups in terms of acetic and butyric acid concentrations in the feces at week 4. Propionate levels in the active group were significantly lower than in the placebo. This discrepancy can be explained by the detectable stability of SCFAs in the gut. The amount of SCFAs present in the feces was known to be highly influenced by stool retention time and other factors [45], with large variations between individuals and between days. Another clinical trial with increased stool collection points and a crossover study design was conducted to confirm the effect of GCL2505 and inulin on the levels of SCFAs in the gut, and the positive impact of the intervention was confirmed (UMIN000050924). In Step 2, the increased levels of SCFAs may activate BAT via the activation of its receptor, G protein-coupled receptor 41 (GPR41), as well as sympathetic nerve stimulation, which, in turn, promotes REE. As mentioned above, previous studies in animals have reported that the administration of acetic acid and butyric acid increases energy expenditure via the activation of brown adipocytes [12,13] and that SCFAs contribute to energy regulation via GPR41 [46]. In addition, SCFAs have been shown to promote the expression of GPR41 and 43 in adipose tissue as well as the differentiation of adipocytes into beige adipocytes, which have the same function as brown adipocytes [47]. In clinical trials, colonic infusion of SCFAs promoted increased lipid oxidation and REE [14]. An observational study in humans revealed a positive correlation between the relative abundance of the genus Bifidobacterium and BAT activity [48], together with the possibility that this effect occurs in an SCFA-dependent manner. These previous studies suggested that SCFAs are likely to have an effect on BAT and REE. Thus, based on the hypotheses thus far, it is speculated that intake of GCL2505 and inulin may increase the concentration of SCFAs in the gut by increasing the total number of bifidobacteria in the gut, thereby stimulating the sympathetic nervous system via GPR41 and activating BAT to improve energy expenditure.

RQ, which is the ratio of carbohydrate to fat oxidation [49], did not change in either the active or placebo groups during the study period. Brooks et al. showed that the energy supply from carbohydrates (glycogen and glucose) increased with increasing exercise intensity, thereby causing the RQ values to change [50]. These previous studies showed that RQ was influenced by environmental factors. Therefore, the stability of the RQ observed in the present study can be attributed to a highly accurate test design that was set up to suppress the influence of environmental factors. The group difference between the active and placebo groups based on the marginal estimated means of REE at week 4 was 84.4 kcal/day (Table 4). It has been reported that the amount of energy deficit required to reduce body weight by 1 kg is around 7400–7700 kcal [51,52]. Based on this concept, the theoretical value of the weight difference between the groups during the current intervention period of 28 days was calculated to be about 0.3 kg. It was considered that the change in values was too small to confirm intervention-induced changes in the body weight and BMI of participants in the active group during this study. However, because the effect on REEs has been confirmed, it is expected that continued intake of GCL2505 and inulin would have a long-term optimizing effect on body weight and BMI. Furthermore, in this study, no aspects of BAT activity were measured, such as body temperature, density of BAT, or cold-induced thermogenesis. In addition, although total bifidobacteria counts were quantified by qPCR, the change in intestinal microbiota was not comprehensively analyzed by 16S amplicon sequence analysis or shotgun metagenomic sequencing. Therefore, further studies are needed to determine the impact of bacteria other than bifidobacteria and whether they influenced the changes in BAT activity or the improvement in REE. Moreover, hypotheses regarding the mechanism may be fully discussed by recruiting a sufficient number of subjects, measuring SCFA accurately while minimizing individual differences and daily variability, and combining this with BAT activity measurements.

Metabolic syndrome, which is the simultaneous development of insulin resistance, obesity, atherosclerosis, and several metabolic diseases represented by dyslipidemia and hypertension, is one of the major public health threats of our time [53]. It is known that the diseases contributing to metabolic syndrome start from obesity, and the overview is sometimes described by the concept of the "metabolic domino effect" [54]. Meanwhile, it has become clear that intestinal bacteria play a pivotal role in maintaining host energy metabolism homeostasis [55–57], and probiotics, one of the most promising materials for improving the intestinal environment, are expected to contribute to solving this problem. The ability of probiotics to reach the intestine alive is considered very important for their health benefits to the host. We previously reported that GCL2505 not only reaches the intestinal tract alive after ingestion but also has the characteristic of proliferating in the intestinal tract [21] and has an excellent ability to increase the number of bifidobacteria in the gut [22]. Clinical trials have demonstrated that GCL2505 is effective in reducing visceral fat [58], and animal studies have shown that this effect is due to the high production of SCFAs via the intestinal viability and intestinal proliferation of GCL2505 [59]. In addition, GCL2505 in combination with inulin has been shown in clinical trials to reduce body fat [28], improve vascular endothelial function [26], and improve cognitive function [25]. The effects on cognitive function have been suggested to be due to the anti-inflammatory effects of SCFAs, and further studies are expected. Although a series of research studies have proven the effectiveness of GCL2505 and inulin, the REE improvement effect has been newly revealed in this study. Thus, we can expect that the effects of GCL2505 and inulin on the diseases that constitute metabolic syndrome will contribute to preventing the onset and progression of these diseases in a domino-like fashion, thereby reducing or eliminating the risk of metabolic syndrome.

5. Conclusions

The administration of GCL2505 and inulin improved REE by increasing the levels of bifidobacteria in the gut. These results suggest that the intake of GCL2505 and inulin improves the energy balance, which is the underlying cause of obesity. These results might also suggest the underlying mechanism of the anti-obesity effects of GCL2505 and inulin, and this is the first report of the effects of probiotics and dietary fiber on REE.

Supplementary Materials: The following supporting information can be downloaded at: https://www.mdpi.com/article/10.3390/nu16142345/s1, Table S1: CONSORT 2010 checklist.

Author Contributions: Conceptualization, Y.B.; methodology, Y.B.; validation, Y.B., N.A., Y.S. and K.T.; formal analysis, Y.B.; investigation, Y.B., N.A., Y.S., K.T. and R.M.; writing—original draft preparation, Y.B., N.A., Y.S., K.T. and R.M.; writing—review and editing, Y.B., N.A., Y.S., K.T. and R.M.; visualization, Y.B.; supervision, T.T. All authors have read and agreed to the published version of the manuscript.

Funding: This study was funded by Ezaki Glico Co., Ltd., Japan, and there was no external funding.

Institutional Review Board Statement: The study was conducted in accordance with the Declaration of Helsinki and approved by the Ethics Committee of Medical Corporation Seishinkai Takara Clinic (approval date: 20 September 2023; approval number: 2309-00178-0078-3BTC).

Informed Consent Statement: Informed consent was obtained from all participants involved in the study.

Data Availability Statement: The datasets generated during the present study and/or analyzed during the current study are available from the responsible author upon reasonable request.

Acknowledgments: We would like to express our sincere appreciation to the team members involved in the production of the test foods for this research.

Conflicts of Interest: The authors declare that this study received funding from Ezaki Glico Co., Ltd. Y.B., N.A., Y.S., K.T. and R.M. are employees of Ezaki Glico Co., Ltd. The other author, T.T., reports no conflicts of interest in this work.

References

1. World Health Organization. Obesity and Overweight. Available online: https://www.who.int/news-room/fact-sheets/detail/obesity-and-overweight (accessed on 25 April 2024).
2. GBD 2019 Risk Factors Collaborators. Global burden of 87 risk factors in 204 countries and territories, 1990–2019: A systematic analysis for the Global Burden of Disease Study 2019. *Lancet* **2020**, *396*, 1223–1249. [CrossRef] [PubMed]
3. Okunogbe, A.; Nugent, R.; Spencer, G.; Powis, J.; Ralston, J.; Wilding, J. Economic impacts of overweight and obesity: Current and future estimates for 161 countries. *BMJ Glob. Health* **2022**, *7*, e009773. [CrossRef] [PubMed]
4. Rocha, T.X.; Melson, E.; Zamora, J.; Fernandez-Felix, B.M.; Arlt, W.; Thangaratinam, S. Sex-specific obesity and cardiometabolic disease risks in low- and middle-income countries: A meta-analysis involving 3,916,276 individuals. *J. Clin. Endocrinol. Metab.* **2024**, *109*, 1145–1153. [CrossRef] [PubMed]
5. Müller, M.J.X.; Geisler, C. From the past to future: From energy expenditure to energy intake to energy expenditure. *Eur. J. Clin. Nutr.* **2017**, *71*, 358–364. [CrossRef] [PubMed]
6. Westerterp, K.R. Diet induced thermogenesis. *Nutr. Metab.* **2004**, *1*, 5. [CrossRef] [PubMed]
7. Ravussin, E. Physiology. A NEAT way to control weight? *Science* **2005**, *307*, 530–531. [CrossRef] [PubMed]
8. Lee, P.; Smith, S.; Linderman, J.; Courville, A.B.; Brychta, R.J.; Dieckmann, W.; Werner, C.D.; Chen, K.Y.; Celi, F.S. Temperature-acclimated brown adipose tissue modulates insulin sensitivity in humans. *Diabetes* **2014**, *63*, 3686–3698. [CrossRef] [PubMed]
9. Yoneshiro, T.; Aita, S.; Matsushita, M.; Kayahara, T.; Kameya, T.; Kawai, Y.; Iwanaga, T.; Saito, M. Recruited brown adipose tissue as an antiobesity agent in humans. *J. Clin. Investig.* **2013**, *123*, 3404–3408. [CrossRef] [PubMed]
10. Kajimura, S.; Saito, M. A new era in brown adipose tissue biology: Molecular control of brown fat development and energy homeostasis. *Annu. Rev. Physiol.* **2014**, *76*, 225–249. [CrossRef]
11. Saito, M.; Yoneshiro, T.; Matsushita, M. Activation and recruitment of brown adipose tissue by cold exposure and food ingredients in humans. *Best Pract. Res. Clin. Endocrinol. Metab.* **2016**, *30*, 537–547. [CrossRef]
12. Gao, Z.; Yin, J.; Zhang, J.; Ward, R.E.; Martin, R.J.; Lefevre, M.; Cefalu, W.T.; Ye, J. Butyrate improves insulin sensitivity and increases energy expenditure in mice. *Diabetes* **2009**, *58*, 1509–1517. [CrossRef] [PubMed]
13. Hanatani, S.; Motoshima, H.; Takaki, Y.; Kawasaki, S.; Igata, M.; Matsumura, T.; Kondo, T.; Senokuchi, T.; Ishii, N.; Kawashima, J.; et al. Acetate alters expression of genes involved in beige adipogenesis in 3T3-L1 cells and obese KK-Ay mice. *J. Clin. Biochem. Nutr.* **2016**, *59*, 207–214. [CrossRef] [PubMed]
14. Canfora, E.E.; van der Beek, C.M.; Jocken, J.W.E.; Goossens, G.H.; Holst, J.J.; Olde Damink, S.W.M.; Lenaerts, K.; Dejong, C.H.C.; Blaak, E.E. Colonic infusions of short-chain fatty acid mixtures promote energy metabolism in overweight/obese men: A randomized crossover trial. *Sci. Rep.* **2017**, *7*, 2360. [CrossRef] [PubMed]
15. Tremaroli, V.; Bäckhed, F. Functional interactions between the gut microbiota and host metabolism. *Nature* **2012**, *489*, 242–249. [CrossRef] [PubMed]
16. Flint, H.J.; Duncan, S.H.; Scott, K.P.; Louis, P. Links between diet, gut microbiota composition and gut metabolism. *Proc. Nutr. Soc.* **2015**, *74*, 13–22. [CrossRef] [PubMed]
17. Wang, B.; Kong, Q.; Cui, S.; Li, X.; Gu, Z.; Zhao, J.; Zhang, H.; Chen, W.; Wang, G. *Bifidobacterium adolescentis* isolated from different hosts modifies the intestinal microbiota and displays differential metabolic and immunomodulatory properties in mice fed a high-fat diet. *Nutrients* **2021**, *13*, 1017. [CrossRef] [PubMed]
18. Liu, Y.; Zhong, X.; Lin, S.; Xu, H.; Liang, X.; Wang, Y.; Xu, J.; Wang, K.; Guo, X.; Wang, J.; et al. *Limosilactobacillus reuteri* and caffeoylquinic acid synergistically promote adipose browning and ameliorate obesity-associated disorders. *Microbiome* **2022**, *10*, 226. [CrossRef] [PubMed]
19. Jin, A.; Kan, Z.; Tan, Q.; Shao, J.; Han, Q.; Chang, Y.; An, N.; Yi, M. Supplementation with food-derived oligopeptides promotes lipid metabolism in young male cyclists: A randomized controlled crossover trial. *J. Int. Soc. Sports Nutr.* **2023**, *20*, 2254741. [CrossRef] [PubMed]
20. van der Beek, C.M.; Canfora, E.E.; Kip, A.M.; Gorissen, S.H.M.; Olde Damink, S.W.M.; van Eijk, H.M.; Holst, J.J.; Blaak, E.E.; Dejong, C.H.C.; Lenaerts, K. The prebiotic inulin improves substrate metabolism and promotes short-chain fatty acid production in overweight to obese men. *Metabolism* **2018**, *87*, 25–35. [CrossRef] [PubMed]
21. Ishizuka, A.; Tomizuka, K.; Aoki, R.; Nishijima, T.; Saito, Y.; Inoue, R.; Ushida, K.; Mawatari, T.; Ikeda, T. Effects of administration of *Bifidobacterium animalis* subsp. *lactis* GCL2505 on defecation frequency and bifidobacterial microbiota composition in humans. *J. Biosci. Bioeng.* **2012**, *113*, 587–591. [CrossRef]
22. Tanaka, Y.; Takami, K.; Nishijima, T.; Aoki, R.; Mawatari, T.; Ikeda, T. Short- and long-term dynamics in the intestinal microbiota following ingestion of *Bifidobacterium animalis* subsp. *lactis* GCL2505. *Biosci. Microbiota Food Health* **2015**, *34*, 77–85. [CrossRef] [PubMed]
23. Aoki, R.; Kamikado, K.; Suda, W.; Takii, H.; Mikami, Y.; Suganuma, N.; Hattori, M.; Koga, Y. A proliferative probiotic *Bifidobacterium* strain in the gut ameliorates progression of metabolic disorders via microbiota modulation and acetate elevation. *Sci. Rep.* **2017**, *7*, 43522. [CrossRef] [PubMed]

24. Horiuchi, H.; Kamikado, K.; Aoki, R.; Suganuma, N.; Nishijima, T.; Nakatani, A.; Kimura, I. *Bifidobacterium animalis* subsp. *lactis* GCL2505 modulates host energy metabolism via the short-chain fatty acid receptor GPR43. *Sci. Rep.* **2020**, *10*, 4158. [CrossRef] [PubMed]
25. Azuma, N.; Mawatari, T.; Saito, Y.; Tsukamoto, M.; Sampei, M.; Iwama, Y. Effect of continuous ingestion of bifidobacteria and dietary fiber on improvement in cognitive function: A randomized, double-blind, placebo-controlled trial. *Nutrients* **2023**, *15*, 4175. [CrossRef] [PubMed]
26. Azuma, N.; Saito, Y.; Nishijima, T.; Aoki, R.; Nishihira, J. Effect of daily ingestion of *Bifidobacterium* and dietary fiber on vascular endothelial function: A randomized, double-blind, placebo-controlled, parallel-group comparison study. *Biosci. Biotechnol. Biochem.* **2023**, *88*, 86–96. [CrossRef] [PubMed]
27. Mensink, M.A.; Frijlink, H.W.; van der Voort Maarschalk, K.; Hinrichs, W.L.J. I Inulin, a flexible oligosaccharide I: Review of its physicochemical characteristics. *Carbohydr. Polym.* **2015**, *130*, 405–419. [CrossRef] [PubMed]
28. Baba, Y.; Saito, Y.; Kadowaki, M.; Azuma, N.; Tsuge, D. Effect of continuous ingestion of bifidobacteria and inulin on reducing body fat: A randomized, double-blind, placebo-controlled, parallel-group comparison study. *Nutrients* **2023**, *15*, 5025. [CrossRef]
29. Suzuki, N.; Baba, A.; Kakinuma, T.; Sano, Y.; Tanaka, M.; Ouchi, S.; Watanabe, T.; Yamamoto, K. A novel dietary questionnaire: The Calorie and Nutrition Diary (CAND). *New Food Indust.* **2019**, *61*, 721–732.
30. Kolida, S.; Meyer, D.; Gibson, G.R. A Double-Blind Placebo-Controlled Study to Establish the Bifidogenic Dose of Inulin in Healthy Humans. *Eur. J. Clin. Nutr.* **2007**, *61*, 1189–1195. [CrossRef] [PubMed]
31. Morishima, S.; Aoi, W.; Kawamura, A.; Kawase, T.; Takagi, T.; Naito, Y.; Tsukahara, T.; Inoue, R. Intensive, prolonged exercise seemingly causes gut dysbiosis in female endurance runners. *J. Clin. Biochem. Nutr.* **2021**, *68*, 253–258. [CrossRef] [PubMed]
32. Tourlousse, D.M.; Narita, K.; Miura, T.; Sakamoto, M.; Ohashi, A.; Shiina, K.; Matsuda, M.; Miura, D.; Shimamura, M.; Ohyama, Y.; et al. Validation and standardization of DNA extraction and library construction methods for metagenomics-based human fecal microbiome measurements. *Microbiome* **2021**, *9*, 95. [CrossRef] [PubMed]
33. Tanaka, K.; Nakamura, Y.; Terahara, M.; Yanagi, T.; Nakahara, S.; Furukawa, O.; Tsutsui, H.; Inoue, R.; Tsukahara, T.; Koshida, S. Poor bifidobacterial colonization is associated with late provision of colostrum and improved with probiotic supplementa-tion in low birth weight infants. *Nutrients* **2019**, *11*, 839. [CrossRef] [PubMed]
34. Carneiro, I.P.; Elliott, S.A.; Siervo, M.; Padwal, R.; Bertoli, S.; Battezzati, A.; Prado, C.M. Is obesity associated with altered energy expenditure? *Adv. Nutr.* **2016**, *7*, 476–487. [CrossRef] [PubMed]
35. Ogawa, A.; Kobayashi, T.; Sakai, F.; Kadooka, Y.; Kawasaki, Y. Lactobacillus gasseri SBT2055 suppresses fatty acid release through enlargement of fat emulsion size in vitro and promotes fecal fat excretion in healthy Japanese subjects. *Lipids Health Dis.* **2015**, *14*, 20. [CrossRef] [PubMed]
36. Kondo, S.; Xiao, J.-Z.; Satoh, T.; Odamaki, T.; Takahashi, S.; Sugahara, H.; Yaeshima, T.; Iwatsuki, K.; Kamei, A.; Abe, K. Antiobesity effects of *Bifidobacterium breve* strain B-3 supplementation in a mouse model with high-fat diet-induced obesity. *Biosci. Biotechnol. Biochem.* **2010**, *74*, 1656–1661. [CrossRef] [PubMed]
37. Aronsson, L.; Huang, Y.; Parini, P.; Korach-André, M.; Håkansson, J.; Gustafsson, J.-Å.; Pettersson, S.; Arulampalam, V.; Rafter, J. Decreased fat storage by *Lactobacillus paracasei* is associated with increased levels of angiopoietin-like 4 protein (ANGPTL4). *PLoS ONE* **2010**, *5*, e13087. [CrossRef] [PubMed]
38. Kim, J.; Yun, J.M.; Kim, M.K.; Kwon, O.; Cho, B. *Lactobacillus gasseri* BNR17 supplementation reduces the visceral fat accumulation and waist circumference in obese adults: A randomized, double-blind, placebo-controlled trial. *J. Med. Food* **2018**, *21*, 454–461. [CrossRef] [PubMed]
39. Nelson, K.M.; Weinsier, R.L.; Long, C.L.; Schutz, Y. Prediction of resting energy expenditure from fat-free mass and fat mass. *Am. J. Clin. Nutr.* **1992**, *56*, 848–856. [CrossRef] [PubMed]
40. Frings-Meuthen, P.; Henkel, S.; Boschmann, M.; Chilibeck, P.D.; Alvero Cruz, J.R.; Hoffmann, F.; Möstl, S.; Mittag, U.; Mulder, E.; Rittweger, N.; et al. Resting energy expenditure of master athletes: Accuracy of predictive equations and primary determinants. *Front. Physiol.* **2021**, *12*, 641455. [CrossRef] [PubMed]
41. Frączek, B.; Grzelak, A.; Klimek, A.T. Analysis of daily energy expenditure of elite athletes in relation to their sport, the measurement method and energy requirement norms. *J. Hum. Kinet.* **2019**, *70*, 81–92. [CrossRef]
42. Compher, C.; Frankenfield, D.; Keim, N.; Roth-Yousey, L.; Evidence Analysis Working Group. Best practice methods to apply to measurement of resting metabolic rate in adults: A systematic review. *J. Am. Diet. Assoc.* **2006**, *106*, 881–903. [CrossRef] [PubMed]
43. Anzawa, D.; Mawatari, T.; Tanaka, Y.; Yamamoto, M.; Genda, T.; Takahashi, S.; Nishijima, T.; Kamasaka, H.; Suzuki, S.; Kuriki, T. Effects of synbiotics containing *Bifidobacterium animalis* subsp. *lactis* GCL2505 and inulin on intestinal bifidobacteria: A randomized, placebo-controlled, crossover study. *Food Sci. Nutr.* **2019**, *7*, 1828–1837. [CrossRef] [PubMed]
44. Birkeland, E.; Gharagozlian, S.; Birkeland, K.I.; Valeur, J.; Måge, I.; Rud, I.; Aas, A.-M. Prebiotic effect of inulin-type fructans on faecal microbiota and short-chain fatty acids in type 2 diabetes: A randomised controlled trial. *Eur. J. Nutr.* **2020**, *59*, 3325–3338. [CrossRef] [PubMed]
45. Müller, M.; Hermes, G.D.A.; Canfora, E.E.; Smidt, H.; Masclee, A.A.M.; Zoetendal, E.G.; Blaak, E.E. Distal colonic transit is linked to gut microbiota diversity and microbial fermentation in humans with slow colonic transit. *Am. J. Physiol. Gastrointest. Liver Physiol.* **2020**, *318*, G361–G369. [CrossRef] [PubMed]

46. Kimura, I.; Inoue, D.; Maeda, T.; Hara, T.; Ichimura, A.; Miyauchi, S.; Kobayashi, M.; Hirasawa, A.; Tsujimoto, G. Short-chain fatty acids and ketones directly regulate sympathetic nervous system via G protein-coupled receptor 41 (GPR41). *Proc. Natl. Acad. Sci. USA* **2011**, *108*, 8030–8035. [CrossRef]
47. Lu, Y.; Fan, C.; Li, P.; Lu, Y.; Chang, X.; Qi, K. Short chain fatty acids prevent high-fat-diet-induced obesity in mice by regulating G protein-coupled receptors and gut microbiota. *Sci. Rep.* **2016**, *6*, 37589. [CrossRef]
48. Ortiz-Alvarez, L.; Acosta, F.M.; Xu, H.; Sanchez-Delgado, G.; Vilchez-Vargas, R.; Link, A.; Plaza-Díaz, J.; Llamas, J.M.; Gil, A.; Labayen, I.; et al. Fecal microbiota composition is related to brown adipose tissue 18F-fluorodeoxyglucose uptake in young adults. *J. Endocrinol. Investig.* **2023**, *46*, 567–576. [CrossRef] [PubMed]
49. Frayn, K.N. Calculation of substrate oxidation rates in vivo from gaseous exchange. *J. Appl. Physiol.* **1983**, *55*, 628–634. [CrossRef] [PubMed]
50. Brooks, G.A. Importance of the "crossover" concept in exercise metabolism. *Clin. Exp. Pharmacol. Physiol.* **1997**, *24*, 889–895. [CrossRef]
51. Wishnofsky, M. Caloric equivalents of gained or lost weight. *Am. J. Clin. Nutr.* **1958**, *6*, 542–546. [CrossRef] [PubMed]
52. Webster, J.D.; Hesp, R.; Garrow, J.S. The composition of excess weight in obese women estimated by body density, total body water and total body potassium. *Hum. Nutr. Clin. Nutr.* **1984**, *38*, 299–306. [PubMed]
53. Huang, P.L. A comprehensive definition for metabolic syndrome. *Dis. Model. Mech.* **2009**, *2*, 231–237. [CrossRef] [PubMed]
54. Itoh, H. What is 'metabolic domino effect'?—New concept in lifestyle-related diseases. *Nihon Rinsho. Jpn. J. Clin. Med.* **2003**, *61*, 1837–1843.
55. Turnbaugh, P.J.; Ley, R.E.; Mahowald, M.A.; Magrini, V.; Mardis, E.R.; Gordon, J.I. An obesity-associated gut microbiome with increased capacity for energy harvest. *Nature* **2006**, *444*, 1027–1031. [CrossRef] [PubMed]
56. Ley, R.E.; Turnbaugh, P.J.; Klein, S.; Gordon, J.I. Human gut microbes associated with obesity. *Nature* **2006**, *444*, 1022–1023. [CrossRef] [PubMed]
57. Ridaura, V.K.; Faith, J.J.; Rey, F.E.; Cheng, J.; Duncan, A.E.; Kau, A.L.; Griffin, N.W.; Lombard, V.; Henrissat, B.; Bain, J.R.; et al. Gut microbiota from twins discordant for obesity modulate metabolism in mice. *Science* **2013**, *341*, 1241214. [CrossRef] [PubMed]
58. Takahashi, S.; Anzawa, D.; Takami, K.; Ishizuka, A.; Mawatari, T.; Kamikado, K.; Sugimura, H.; Nishijima, T. Effect of *Bifidobacterium animalis* ssp. *lactis* GCL2505 on visceral fat accumulation in healthy Japanese adults: A randomized controlled trial. *Biosci. Microbiota Food Health* **2016**, *35*, 163–171. [CrossRef]
59. Aoki, R.; Tsuchida, S.; Arai, Y.; Ohno, K.; Nishijima, T.; Mawatari, T.; Mikami, Y.; Ushida, K. Effect of *Bifidobacterium animalis* subsp. *lactis* GCL2505 on the physiological function of intestine in a rat model. *Food Sci. Nutr.* **2016**, *4*, 782–790. [CrossRef] [PubMed]

nutrients

MDPI

Article

Faecalibacterium prausnitzii Supplementation Prevents Intestinal Barrier Injury and Gut Microflora Dysbiosis Induced by Sleep Deprivation

Xintong Wang [1], Yixuan Li [1], Xifan Wang [2], Ran Wang [1], Yanling Hao [1], Fazheng Ren [1,3], Pengjie Wang [1,3,*] and Bing Fang [1,*]

1 Key Laboratory of Precision Nutrition and Food Quality, Department of Nutrition and Health, China Agricultural University, Beijing 100083, China; xtwang@cau.edu.cn (X.W.); liyixuan@cau.edu.cn (Y.L.); wangran@cau.edu.cn (R.W.); haoyl@cau.edu.cn (Y.H.); renfazheng@cau.edu.cn (F.R.)
2 Department of Obstetrics and Gynecology, Columbia University, New York, NY 10032, USA; wangxfan@126.com
3 Food Laboratory of Zhongyuan, Luohe 462000, China
* Correspondence: wpj1019@cau.edu.cn (P.W.); bingfang@cau.edu.cn (B.F.)

Abstract: Sleep deprivation (SD) leads to impaired intestinal barrier function and intestinal flora disorder, especially a reduction in the abundance of the next generation of probiotic *Faecalibacterium prausnitzii* (*F. prausnitzii*). However, it remains largely unclear whether *F. prausnitzii* can ameliorate SD-induced intestinal barrier damage. A 72 h SD mouse model was used in this research, with or without the addition of *F. prausnitzii*. The findings indicated that pre-colonization with *F. prausnitzii* could protect against tissue damage from SD, enhance goblet cell count and MUC2 levels in the colon, boost tight-junction protein expression, decrease macrophage infiltration, suppress pro-inflammatory cytokine expression, and reduce apoptosis. We found that the presence of *F. prausnitzii* helped to balance the gut microbiota in SD mice by reducing harmful bacteria like *Klebsiella* and *Staphylococcus*, while increasing beneficial bacteria such as *Akkermansia*. Ion chromatography analysis revealed that *F. prausnitzii* pretreatment increased the fecal butyrate level in SD mice. Overall, these results suggested that incorporating *F. prausnitzii* could help reduce gut damage caused by SD, potentially by enhancing the intestinal barrier and balancing gut microflora. This provides a foundation for utilizing probiotics to protect against intestinal illnesses.

Keywords: sleep deprivation; *Faecalibacterium prausnitzii*; intestinal barrier function; intestinal microflora; short chain fatty acids

1. Introduction

Sleep is an essential physiological process of the body, but with the acceleration of social rhythm, there is still a 1/3 of the world's population that suffer from the insufficient sleep phenomenon [1]. Sleep loss can damage the body's cognitive function, gastrointestinal digestion, absorption function and endocrine homeostasis [2–4]. Recently, research has found that sleep deprivation (SD) induces intestinal barrier dysfunction [5]; the prevention of disease-causing substances and bacteria from entering the bloodstream is the primary function of the intestinal mucosal barrier [6]. MUC2, the main mucin that forms gels in the intestines, plays a key role in creating the mucus barrier [7]. Tight-junction proteins like claudins, zona occludens, and occludin are responsible for preserving the integrity of the intestinal epithelial barrier [8]. The weakening of the intestinal epithelial barrier is marked by a reduction in mucous layer thickness, disturbance in the distribution of tight-junction proteins, and compromised intestinal integrity resulting in heightened intestinal permeability. This allows toxins to move into the lamina propria, triggering an exaggerated immune response from the host's immune cells [9]. Since SD can seriously affect the intestinal health of the body, it is of great significance and great demand to find a healthy and green substance or method to alleviate the adverse effects of insomnia.

Recently, more and more studies have shown that severe intestinal flora imbalances are related to SD, so the research and application of microecological agents are receiving more attention [10]. Previous research has shown that SD can disrupt the balance of intestinal microorganisms in mice, resulting in a reduction in beneficial bacteria like *Faecalibacterium* [11]. When the intestinal barrier becomes more permeable, it is unable to effectively block out the microorganisms in the surroundings [12]. This leads to a significant amount of detrimental bacteria and their byproducts passing through the intestinal barrier during times of stress, subsequently impacting intestinal epithelial cells and immune cells in a chain reaction [13]. Nevertheless, certain research has indicated that adjusting and replenishing the equilibrium of gut bacteria could improve gut health and decrease long-lasting inflammation [14,15]. Therefore, preserving the balance of gut bacteria and improving the strength of the intestinal barrier could offer a novel approach to preventing and treating intestinal damage caused by SD.

Faecalibacterium is recognized as a key bacterium in maintaining a healthy gut, and fluctuations in its levels may indicate an imbalance in the human gut microbiota, making it a potential contender for future probiotics [16]. In patients with inflammatory bowel disease (IBD), there is a reduction in the richness of two mucosa-associated *Faecalibacterium* phylotypes [17]. In a recent study using multicenter amplicon sequencing data, Chen et al. found that, within a cohort of 708 individuals (354 irritable bowel syndrome (IBS) patients and 354 healthy controls), the genus *Faecalibacterium* was one of the depleted taxa in IBS patients [18]. Furthermore, *Faecalibacterium* abundance was found to be higher in healthy individuals than in individuals with mild Alzheimer's disease patients, in whom it was positively correlated with cognitive performance [19]. It can be seen that *Faecalibacterium* is associated with a variety of human diseases. SCFAs are among the most thoroughly investigated bacterial metabolites; *Faecalibacterium* is considered the main producer of butyrate [20]. Butyrate serves as a crucial energy source for cells in the intestines and is essential for regulating intestinal peristalsis, maintaining epithelial barrier function, and supporting the immune system's health [21,22]. In addition, butyrate is involved in the enhancement of the colon barrier by increasing the synthesis of tight-junction proteins and the production of antimicrobial peptides [23]. In chronic kidney disease, *Faecalibacterium* could restore the disturbed intestinal microflora and drive its metabolite butyrate to play renal protection [24]. The potential of *Faecalibacterium* to enhance intestinal health in sleep deprived mice is uncertain, and further research is needed to understand the mechanism behind its effects.

In this study, we constructed an acute SD model for 72 consecutive hours and administered *Faecalibacterium prausnitzii* (*F. prausnitzii*) with gavage before SD. We analyzed the intestinal flora and intestinal barrier in mice through the identification of colon inflammation markers, LPS, proteins related to tight junctions, and 16s rDNA sequencing. Our results showed that *F. prausnitzii* could improve intestinal barrier damage and intestinal microflora disturbance induced by SD. Its metabolite butyrate has participated in the improvement of *F. prausnitzii* on intestinal damage.

2. Materials and Methods

2.1. Animal and Experimental Design

Forty-eight male ICR mice, 8 weeks old, from Vital River Laboratory Animal Technology Co., Ltd. in Beijing, China, were housed in cages under standard environmental conditions with a temperature of 21 ± 1 °C and relative humidity of 50 ± 10%. All animal experiments were performed at the SPF animal room, China Agricultural University. They were kept on a 14 h light/10 h dark cycle, with lights on at 7:00 a.m. and off at 9:00 p.m. The mice were provided with ad libitum access to food and water. All mice used in the experiments were housed in groups of four per cage under the same environmental and husbandry conditions. The mice were randomly divided into four groups: (1) control group (CON, $n = 12$); (2) sleep deprivation group (SD, $n = 12$); (3) *F. prausnitzii* colonization group (FP group, $n = 12$); and (4) sleep deprivation with *F. prausnitzii* colonization group (SD + FP group, $n = 12$). The number of mice for each group was determined based on similar experiments from the literature [25].

To construct a pseudo-sterile mouse model, all mice were given water with 1 g/kg ampicillin (A102048, Aladdin, Shanghai, China), 100 mg/kg gentamicin (G100391, Aladdin, Shanghai, China), 0.5 g/kg neomycin (N412785, Aladdin, Shanghai, China), 0.5 g/kg vancomycin (V105495, Aladdin, Shanghai, China), and 10 mg/kg erythromycin (E105345, Aladdin, Shanghai, China) for 10 days. A new solution of antibiotics was made daily to ensure its effectiveness. Mice (FP and SD + FP) received an oral gavage of 10^8 CFU of *F. prausnitzii* (in 0.2 mL PBS) at 8 a.m. The CON and SD groups received a 0.2 mL PBS vehicle via oral gavage. After 14 days of inoculation, mice in the SD group and SD + FP group were subjected to SD for 72 h. A modified multi-platform water environment method was utilized to establish the SD mouse model [11]. Four mice were housed in one cage, and their body weight was recorded daily.

At 8 a.m. on the day the experiment concluded, every mouse was euthanized through anesthesia with chloral hydrate. The mouse colon contents were gathered and frozen at −80 °C for microbial analysis. One part of the colon was preserved in 4% paraformaldehyde for morphological examination, while another section was stored at −80 °C for RT-PCR and ELISA testing.

2.2. Probiotic Culture

The probiotic *F. prausnitzii* (DSM 17677) was obtained from DSMZ. *F. prausnitzii* probiotics were introduced into the Modified Reinforced Clostridial Broth Medium (MZMD039B, Ningbo Mingzhou Biotechnology Co., Ltd., Ningbo, China) and incubated at 37 °C for 48 h in an anaerobic setting. After centrifuging the probiotic culture, the supernatant was removed and the bacterial precipitates were then mixed with 20% glycerol. Plate counting was used to determine the bacterial content, and the bacterial solution was diluted to a final concentration of 10^8 CFU (0.2 mL). Mice were treated via oral gavage with 200 μL of either *F. prausnitzii* suspension or anaerobic PBS.

2.3. Hematoxylin and Eosin (H&E) Staining

Freshly isolated colon samples were fixed with 4% paraformaldehyde overnight. After dehydration, the tissues were encased in paraffin and cut into slices that were 5 μm thick. Colon tissue sections were stained with H&E (G1120, Solarbio, Beijing, China) and evaluated histologically based on established criteria from a previous publication [26]. Each tissue section was microscopically assessed for the loss of epithelial surface, destruction of crypt and infiltration of immunocytes (each category scored from 0 to 4). The total scores indicate the cumulative pathological score recorded for every individual sample.

2.4. AB-PAS Staining

Colon tissue sections were treated with alcian blue and periodic acid-Schiff (AB-PAS) stains following the instructions provided by the manufacturer (G1285, Solarbio, Beijing, China). The average number of goblet cells in 30 randomly selected intact crypts per group was determined by calculating goblet cell counts.

2.5. Immunohistochemical Staining

Paraffin sections were treated with rabbit anti-ZO-1 (1:200, 21773-1-AP, Proteintech, Chicago, IL, USA), rabbit anti-mucin-2 (MUC2) (1:2000, 27675-1-AP, Proteintech, Chicago, IL, USA), rabbit anti-occludin (1:700, GB111401-100, Servicebio, Wuhan, China), rabbit anti-cleaved caspase-3 (1:200, 9661, CST, Boston, MA, USA), and mouse anti-F4/80 (1:1000, GB11027, Servicebio, Wuhan, China) primary antibodies overnight at 4 °C. Afterward, the portions were washed with 0.01M PBS (pH 7.4) and then exposed to biotinylated goat anti-rabbit IgG (1:300, GB23303, Servicebio, Wuhan, China) or rabbit anti-mouse IgG

(1:300, GB23301, Servicebio, Wuhan, China) for a duration of 2 h at ambient temperature. Afterward, the portions were dyed with hematoxylin and placed on a mount. In every instance, slides lacking the main antibody were analyzed for comparison. Cells showing immunoreactivity exhibited a yellow-brown stain in their cytoplasm. Positive cells in five cross-sections were randomly selected for each sample, and at least 30 fields were counted for each group. The integrated optical density (IOD) was measured by using ImageJ software (version 4.0.2; Scion Corp., Frederick, MD, USA)

2.6. Real-Time Reverse Transcription-Polymerase Chain Reaction (RT-PCR)

Six colon samples were used to extract total RNA with the FastPure® Cell/Tissue Total RNA Isolation Kit V2 (RC112; Vazyme, Nanjing, China). The HiScript III All-in-one RT SuperMix Perfect for qPCR (R333; Vazyme, Nanjing, China) was utilized to synthesize the initial cDNA strand. The ChamQ Universal SYBR qPCR Master Mix (Q711; Vazyme, Nanjing, China) was utilized for RT-PCR amplification. Every sample underwent two rounds of testing. Supplementary Table S1 contains the list of RT-PCR primers.

2.7. Enzyme-Linked Immunosorbent Assay (ELISA)

The colon tissues were collected for the detection of inflammatory factors, such as tumor necrosis factor alpha (TNF-α) (MK2868A; Meike, Yancheng, China) and interleukin-6 (IL-6) (MK5737A; Meike, Yancheng, China), and lipopolysaccharide (LPS) (MK3418A; Meike, Yancheng, China) concentrations using a competitive ELISA. The tests were conducted in compliance with the guidelines provided by the manufacturer. Every sample underwent two rounds of testing. Both the intra-assay and inter-assay coefficients of variation (CV) were below 15%.

2.8. Gut Microbiota Analysis

Samples of feces were gathered and frozen at −80 degrees Celsius for analysis using high-throughput DNA sequencing. Mouse feces were used to extract total DNA with the FastDNATM SPIN Kit from MP Bio in the Irvine, CA, USA. The Thermo Scientific Nanodrop 2000 (Waltham, MA, USA) was utilized to assess the strength and cleanliness of the samples. PCR was used to amplify DNA templates with specific primers 338F (ACTCC-TACGGGAGGCAGCAG) and 806R (GGACTACHVGGGTWTCTAAT). PCR products were identified using 2% agarose gel electrophoresis, and were then purified and quantified using the qubit 4.0 system (Thermo Fisher Scientific, Waltham, MA, USA). Per the standard protocol of Majorbio Bio-Pharm Technology Co., Ltd. in Shanghai, China, the amplicons were combined in equal amounts for sequencing on the Illumina MiSeq PE300 platform. The analysis of data was conducted on the Majorbio cloud platform (cloud.majorbio.com, accessed on 6 November 2023).

2.9. Detection of SCFAs Using Ion Chromatography

Ion chromatography was used to determine the levels of short-chain fatty acids in feces as per the previously studied technique [27]. Briefly, around 25 mg of feces was mixed on a spin with 4 mL of sterile deionized water. Next, the specimen was spun at 8000 times the force of gravity at a temperature of 4 degrees Celsius for a duration of 10 min. After collection, the supernatant underwent filtration using a 0.22 mm mesh and was then analyzed for SCFAs content using an ion chromatography system (DIONEX ICS-3000, Thermo Fisher Scientific, Waltham, MA, USA).

2.10. Statistical Analysis

The information was presented as the average plus standard error and assessed with GraphPad Prism version 9 from GraphPad Software located in La Jolla, CA, USA. The data were analyzed with a normal distribution test and homogeneity test of variance. Group variances were compared using a one-way ANOVA, with Tukey's post hoc test conducted afterwards. The Kruskal–Wallis test was used to examine variations in microbial composition among various groups at the genus level. Any p-values less than 0.05 were deemed to be statistically significant.

3. Results

3.1. *F. prausnitzii* Colonization Alleviated Intestinal Mucosal Barrier Disruption Induced by SD

A mouse model was created to simulate intestinal barrier damage induced by SD, with or without colonization by *F. prausnitzii*, to investigate the potential beneficial effects of *F. prausnitzii* on the condition (Figure 1A). As illustrated in Figure 1B, we noticed a difference in the physical appearance of the mice, specifically in terms of their weight, across the CON, FP, SD, and SD + FP groups. Following a 72 h SD period, the weight of sleep-deprived mice was notably less than that of the control group. HE staining revealed that the control group exhibited normal tissues, while the SD group showed a breakdown of the epithelial barrier and invasion of inflammatory cells. However, pre-treatment with *F. prausnitzii* attenuated the above intestinal damage induced by SD (Figure 1C,F). Histological scoring revealed a significant increase in the SD group (66.1%, $p < 0.001$) compared to the CON group. However, colonization of *F. prausnitzii* alleviated the damage in the SD + FP group compared with the SD group (35.6%, $p = 0.04$).

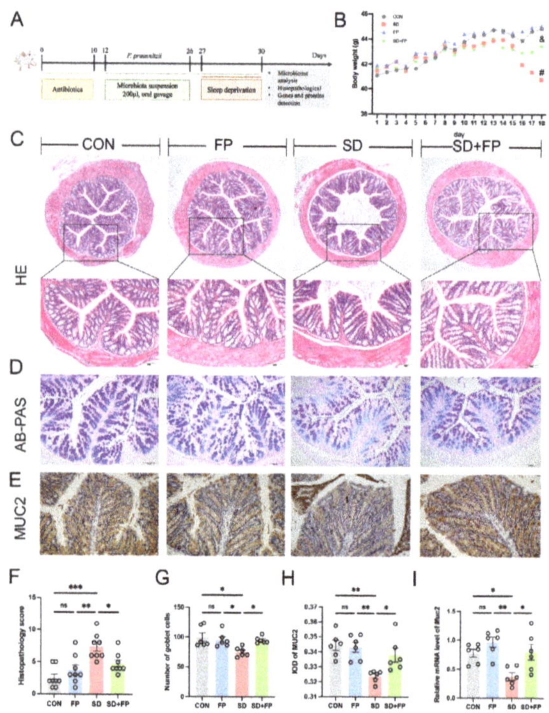

Figure 1. *Faecalibacterium prausnitzii* colonization alleviated intestinal mucosal barrier disruption induced by SD. (**A**) Diagram showing the layout of the experiment. (**B**) Body weight. (**C**) Hematoxylin and eosin (H&E) staining, Bar = 100 μm. (**D**) Alcian blue, and periodic acid-Schiff (AB-PAS) staining, Bar = 100 μm. (**E**) Representative captures of immunohistochemical of MUC2 in colon, Bar = 100 μm. (**F**) Histologic scores ($n = 8$). (**G**) The number of goblet cells per crypt ($n = 6$). (**H**) IOD of MUC2 in the intestinal tissue from each treatment group ($n = 6$). (**I**) The mRNA levels of Muc2 in the colon ($n = 6$). One-way ANOVA was utilized to evaluate variances. The study included a control group (CON), a group colonized with *Faecalibacterium prausnitzii* (FP), a group subjected to sleep deprivation (SD), and a group experiencing sleep deprivation with *Faecalibacterium prausnitzii* colonization (SD + FP).

The outcome indicates the average value plus or minus the standard error. *, $p < 0.05$; **, $p < 0.01$; ***, $p < 0.001$ compared with the control group. ns, non-significance. # $p < 0.05$ vs. the CON group; & $p < 0.05$ vs. the SD group.

AB-PAS staining was utilized to further identify the quantity of goblet cells. In the SD group, there was a notable decrease in the number of goblet cells in the colon (24.4%, $p = 0.01$) when compared to the control group (Figure 1D,G). Following the colonization of *F. prausnitzii*, there was a notable rise in the number of goblet cells in the colon within the SD + FP group (26.2%, $p = 0.04$) in contrast to the SD group (Figure 1D,G). MUC2, which is released by goblet cells, plays a significant role in safeguarding the intestinal epithelium. Immunohistochemistry results showed that the expression of MUC2 was significantly decreased in the SD group, compared with the CON group (5.9%, $p = 0.002$). In contrast, after *F. prausnitzii* colonization, the expression of MUC2 was significantly increased (4.4%, $p = 0.04$) in the SD + FP group compared with the SD group (Figure 1E,H). Additionally, it was discovered that the mRNA levels of Muc2 in the colon of sleep-deprived mice were decreased compared to the control mice ($p = 0.03$), but the colonization of *F. prausnitzii* significantly increased the expression of this gene ($p = 0.02$) (Figure 1I).

Subsequently, the presence of tight-junction proteins was examined through immunohistochemistry and RT-PCR analyses. In the SD group, there was a significant decrease in the levels of ZO-1 ($p = 0.008$) and occludin ($p = 0.05$) tight-junction proteins compared to the CON group, as shown in Figure 2A–D. Conversely, following the colonization of *F. prausnitzii*, there was a notable increase in the expression of ZO-1 ($p = 0.007$) and occludin ($p = 0.03$) in the SD + FP group when compared to the SD group (Figure 2A–D). Furthermore, alterations in colonic tight-junction proteins were noted at the mRNA level in sleep-deprived mice. Specifically, the mRNA levels of colonic ZO-1, Cloudin-1, and occludin decreased after SD, but the presence of *F. prausnitzii* improved these markers (Figure 2E–G). Together, these data suggested that SD leads to the disruption of intestinal barrier integrity and decreased numbers of goblet cells but could be improved by *F. prausnitzii* colonization.

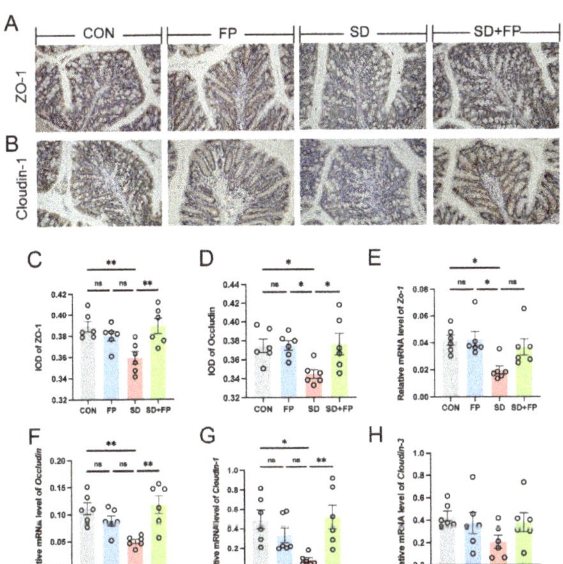

Figure 2. *Faecalibacterium prausnitzii* colonization increased the expression of tight-junction proteins in sleep-deprived mice. (**A**) Representative captures of immunohistochemical of ZO-1 in colon,

Bar = 100 μm. (**B**) Representative captures of immunohistochemical of occludin in colon, Bar = 100 μm. (**C**) IOD of ZO-1 in the intestinal tissue from each treatment group (n = 6). (**D**) IOD of occludin in the intestinal tissue from each treatment group (n = 6). (**E–H**) The mRNA levels of *ZO-1*, *occludin*, *Cloudin-1*, and *Cloudin-3* in the colon (n = 6). One-way ANOVA was utilized to evaluate variances. The study included a control group (CON), a group colonized with *Faecalibacterium prausnitzii* (FP), a group subjected to sleep deprivation (SD), and a group experiencing sleep deprivation with *Faecalibacterium prausnitzii* colonization (SD + FP). The outcome indicates the average value plus or minus the standard error. *, $p < 0.05$; **, $p < 0.01$ compared with the control group. ns, non-significance.

3.2. The Colonization of F. prausnitzii Suppressed the Production of Inflammatory Cytokines in SD Mice

We assessed alterations in inflammatory markers and macrophages in the colon to examine the inflammation response triggered by SD. By utilizing F4/80 as a marker for macrophages, we confirmed that there was a 32.6% increase in the amount of F4/80-positive cells in the colon of the SD group compared to the CON group (Figure 3A,B). We also observed a significant increase in IL-6 (41.3%, p = 0.04), and TNF-α (50.6%, p = 0.009) levels in the SD group compared with the CON group (Figure 3C,D). Nevertheless, the establishment of *F. prausnitzii* notably inhibited the stimulation of macrophages and reduced the levels of inflammatory molecules. Specifically, there was a decrease in the quantity of F4/80-positive cells in the colon of the SD + FP group compared to the SD group, with a significance level of p = 0.02. We also observed a significant decrease in IL-6 (38.3%, p = 0.005) and TNF-α (33.7%, p = 0.008) levels in sleep-deprived mice with *F. prausnitzii* colonization compared with SD mice. Additionally, the expression of pro-inflammatory cytokines (Tnf-α, Il-1β, Il-6) and F4/80 mRNA decreased in sleep-deprived mice following the introduction of *F. prausnitzii* (Figure 3F–I). Interestingly, we also observed a significant increase in LPS in the colon in the sleep-deprived mice versus the control mice, while the colonization of *F. prausnitzii* reversed this change (Figure 3E). These results suggested that the colonization of *F. prausnitzii* could improve the intestinal inflammation caused by SD.

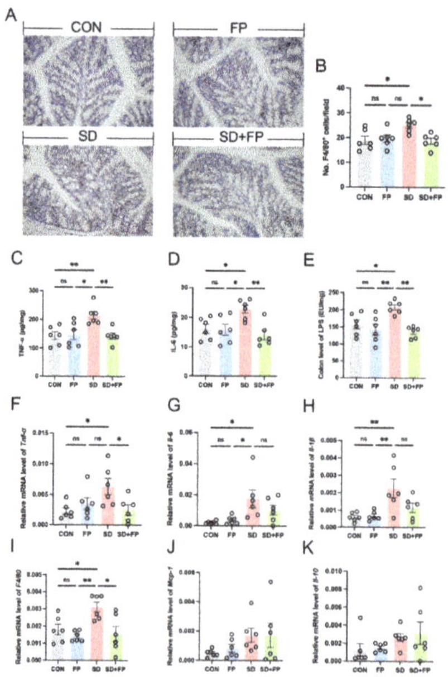

Figure 3. *Faecalibacterium prausnitzii* colonization inhibited the expression of inflammatory cytokines in sleep-deprived mice. (**A**) Representative captures of immunohistochemical of F4/80 in colon,

Bar = 100 μm. (**B**) The quantification of F4/80+ cells ($n = 6$). (**C**,**D**) The levels of cytokines (TNF-α and IL-6) in the colon ($n = 6$). (**E**) The levels of LPS in the colon ($n = 6$). (**F**–**K**) The mRNA levels of Tnf-α, Il-6, Il-1β, F4/80, Mcp-1 and Il-10 in the colon ($n = 6$). One-way ANOVA was utilized to evaluate variances. The study included a control group (CON), a group colonized with *Faecalibacterium prausnitzii* (FP), a group subjected to sleep deprivation (SD), and a group experiencing sleep deprivation with *Faecalibacterium prausnitzii* colonization (SD + FP). The outcome indicates the average value plus or minus the standard error. *, $p < 0.05$; **, $p < 0.01$ compared with the control group. ns, non-significance.

3.3. Colonization of *F. prausnitzii* Mitigated Gut Microbiota Dysbiosis Induced by SD in Mice

To examine how the presence of *F. prausnitzii* affects the composition of the gut microbiota in mice subjected to SD, 16s rDNA gene sequencing was performed. In total, 24 samples were collected from four groups of mice ($n = 6$) and subsequently sequenced to generate V3–V4 16s rDNA gene profiles. The Venn diagram indicated that SD treatment or *F. prausnitzii* colonization resulted in different microbial changes in mice (Figure 4B). β-diversity analysis measures the level of dissimilarity among various microbial communities (Figure 4D–F). The Bray–Curtis PCoA analysis indicated distinct group separation, with PC1, PC2, and PC3 explaining 25.65%, 20.37%, and 13.60% of the variation, respectively (Adonis, $p = 0.001$, R2 = 0.6698, Figure 4E). UPGMA analysis indicated that the SD + FP group exhibited a closer relationship with the CON group compared to the SD group, further supporting the findings of PCoA (Figure 4C). Verrucomicrobiota, Firmicutes, and Proteobacteria were the most common microorganisms at the phylum level (Figure 4G). At the genus level, *Akkermansia, Blautia, Escherichia-Shigella, Parasutterella, Lachnoclostridium, Coprobacillus*, and *Klebsiella* were the predominant floras (Figure 4H). LDA and LEfSe were used to determine the particular bacterial phyla linked to CON, FP, SD, and SD + FP groups, to pinpoint the key taxa that could account for the variations among the groups. Additionally, the LEfSe analysis revealed 52 taxa biomarkers in three groups, each identified with an LDA score greater than 3 and a *p*-value less than 0.05 (Figure 5A,B). *Akkermansia* was the most abundant bacterium at the genus level, and its content in the SD + FP group was the highest among all groups. Specifically, the relative abundance of *Akkermansia* was higher in SD + FP group than in SD group ($p = 0.07$, Figure 5C). Furthermore, the comparative prevalence of *Klebsiella* ($p < 0.001$, LDA score = 4.49, Figure 5D), *Enterobacter* ($p = 0.002$, LDA score = 3.78, Figure 5E), *Empedobacter* ($p < 0.001$, LDA score = 3.29, Figure 5F), *Proteus* ($p < 0.001$, LDA score = 3.86, Figure 5G), *Staphylococcus* ($p = 0.01$, LDA score = 3.35, Figure 5H), *Comamonas* ($p < 0.001$, LDA score = 3.29, Figure 5I), and *Acinetobacter* ($p < 0.001$, LDA score = 3.53, Figure 5J) was notably elevated in the SD group compared to the CON, FP, and SD + FP group ($p < 0.05$).

3.4. *F. prausnitzii* Colonization Alleviated Colon SCFAs Reduction in SD Mice

We assessed the concentrations of short-chain fatty acids in the feces of various treated mice using ion chromatography, as these are crucial metabolites produced by Faecalibacterium. The results showed that the level of fecal butyrate was decreased in the SD group compared with the CON group. In addition, the level of fecal butyrate was higher in the FP group than in the SD group. However, after colonization with *F. prausnitzii* in sleep-deprived mice, the level of fecal butyrate was significantly increased ($p = 0.009$) compared to the sleep-deprived mice (Figure 6C). No notable variances were observed in the levels of acetate in the feces (Figure 6A), propionate (Figure 6B), and valerate (Figure 6D) among the four groups ($p > 0.05$).

3.5. *F. prausnitzii* Colonization Reduced Intestinal Apoptosis Level Induced by SD

The inflammatory response is accompanied by the excessive apoptosis of cells engaged in inflammation. The SD group showed a notable increase in cleaved caspase-3 expression compared to the CON group ($p = 0.04$). However, the colonization with *F. prausnitzii* reversed these changes. The protein levels of cleaved caspase-3 were lower in the SD + FP group compared to the SD group, with a significance level of $p = 0.03$ (Figure 6E,F). In addition, we quantified the mRNA expression of pro-apoptotic Bax and anti-apoptotic Bcl-2. The findings indicated that SD decreased the expression of the anti-apoptotic protein Bcl-2, but colonization with *F. prausnitzii* reversed these effects (Figure 6G). No significant changes were detected for Bax levels in different treated groups (Figure 6H).

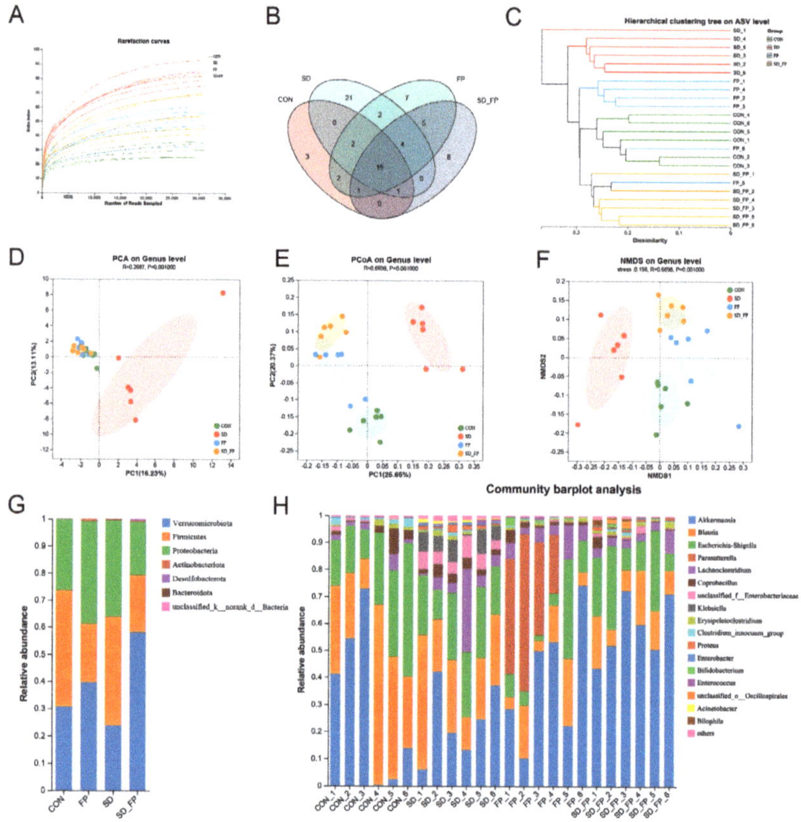

Figure 4. Composition of the colonic microbiota in mice. (**A**) Rarefaction curves. (**B**) Venn diagram. (**C**) Unweighted pair-group method with arithmetic mean (UPGMA) analysis (at the genus level). (**D**) Principal component analysis (PCA). (**E**) PCoA score plot. (**F**) Nonmetric multidimensional scaling (NMDS) score plot based on the binary_jaccard distance plot based on the ASV of the gut microbe. (**G**) Relative abundances of gut microbiota at the phylum level. (**H**) Relative abundances of gut microbiota at the genus level. The study included a control group (CON), a group colonized with *Faecalibacterium prausnitzii* (FP), a group subjected to sleep deprivation (SD), and a group experiencing sleep deprivation with *Faecalibacterium prausnitzii* colonization (SD + FP).

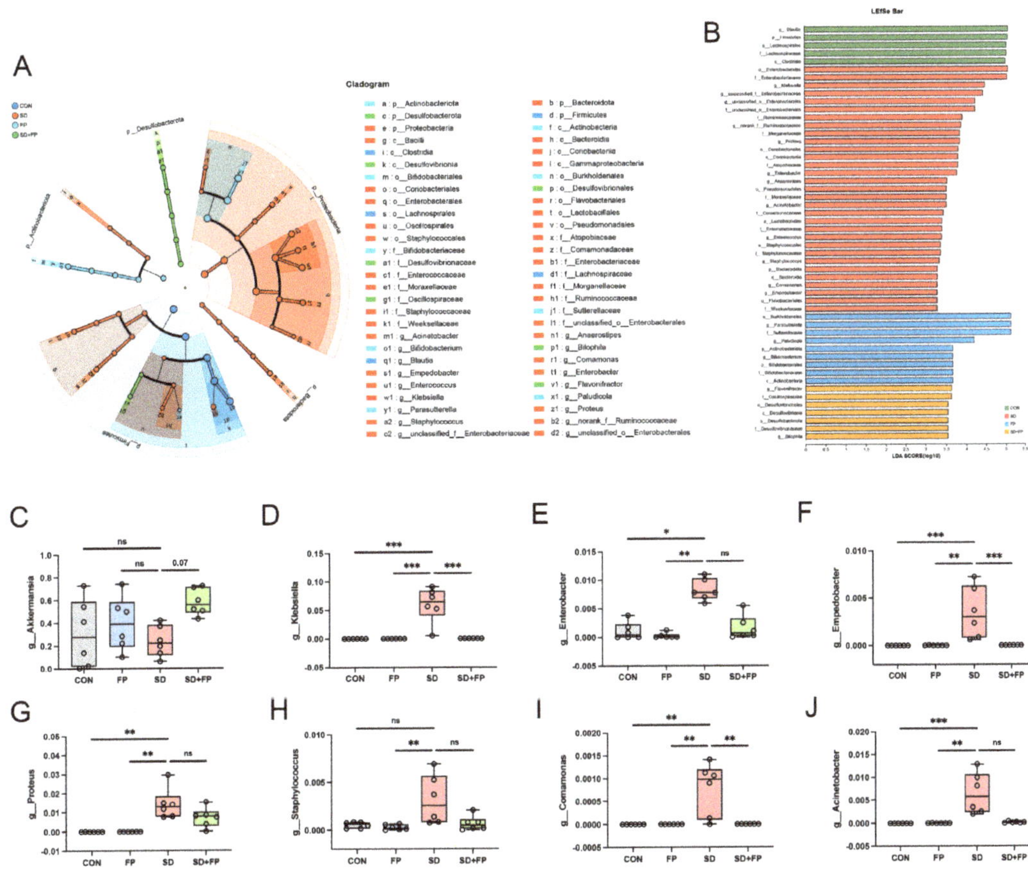

Figure 5. *Faecalibacterium prausnitzii* colonization inhibited the colonic microbial dysbiosis induced by sleep deprivation. (**A**) Taxonomic cladogram obtained from LEfSe sequence analysis in the colon. Biomarker taxa are highlighted by colored circles and shaded areas. (**B**) The diameter of each circle reflects the abundance of that taxon in the community. A cutoff value of 3 was used for LDA. (**C–J**) Relative abundance of *g__Akkermansia*, *g__Klebsiella*, *g__Enterobacter*, *g__Enterobacter*, *g__Proteus*, *g__Staphylococcus*, *g__Comamonas*, *g__Acinetobacter* in the colon microbiota based on the LefSe results. The study included a control group (CON), a group colonized with *Faecalibacterium prausnitzii* (FP), a group subjected to sleep deprivation (SD), and a group experiencing sleep deprivation with *Faecalibacterium prausnitzii* colonization (SD + FP). The outcome indicates the average value plus or minus the standard error. *, $p < 0.05$; **, $p < 0.01$; ***, $p < 0.001$ compared with the control group. ns, non-significance.

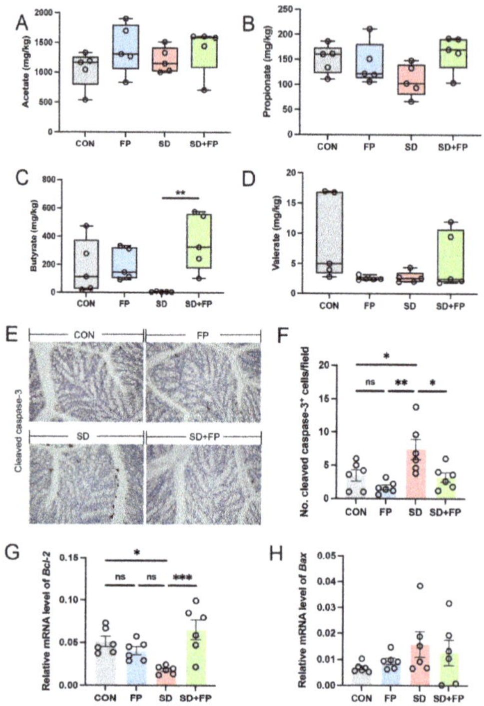

Figure 6. *Faecalibacterium prausnitzii* colonization reversed decrease in SCFAs and increase in intestinal apoptosis induced by sleep deprivation. (**A**) The colon acetate concentration ($n = 5$). (**B**) The colon propionate concentration ($n = 5$). (**C**) The colon butyrate concentration ($n = 5$). (**D**) The colon valerate concentration ($n = 5$). (**E**) Representative captures of immunohistochemical of cleaved caspase-3 in colon, Bar = 100 µm. (**F**) The quantification of cleaved caspase-3+ cells ($n = 6$). (**G**,**H**) The mRNA levels of Bax and Bcl-2 in the colon ($n = 6$). One-way ANOVA was utilized to evaluate variances. The study included a control group (CON), a group colonized with *Faecalibacterium prausnitzii* (FP), a group subjected to sleep deprivation (SD), and a group experiencing sleep deprivation with *Faecalibacterium prausnitzii* colonization (SD + FP). The outcome indicates the average value plus or minus the standard error. *, $p < 0.05$; **, $p < 0.01$; ***, $p < 0.001$ compared with the control group. ns, non-significance.

4. Discussion

Sleep is the most essential physiological function of the human body, which is both important and complex. Lack of sleep can cause problems with the digestive system, immune system, metabolism, and circulation [28–30]. Our previous research has shown that SD could affect the intestinal barrier in mice, leading to changes in gut bacteria, including a decrease in beneficial *Faecalibacterium* in the colon [11]. *Faecalibacterium*, as the next generation of probiotics, has diverse functional effects such as anti-inflammatory effects, antioxidant and antimicrobial [31]. Furthermore, it was recently found that *Faecalibacterium* could ameliorate renal dysfunction in patients with chronic kidney disease partly through the butyrate-mediated GPR43 signaling in the kidney [24]. Uncertainty remains regarding the potential of *F. prausnitzii* colonization to repair intestinal barrier damage and restore intestinal microflora balance in sleep-deprived mice. In this study, by constructing an SD mouse model with *F. prausnitzii* intervention, it was found that SD caused intestinal microflora disturbance and intestinal permeability increase in mice, which then led to the introduction of toxic substances into the intestinal cavity to induce inflammation and the programmed death of colon epithelial cells. The preimplantation of *F. prausnitzii* remodeled the disturbed intestinal microflora in sleep-deprived mice, improved overtransition inflammation and apoptosis, enhanced intestinal integrity, and ultimately improved intestinal function (Figure 7).

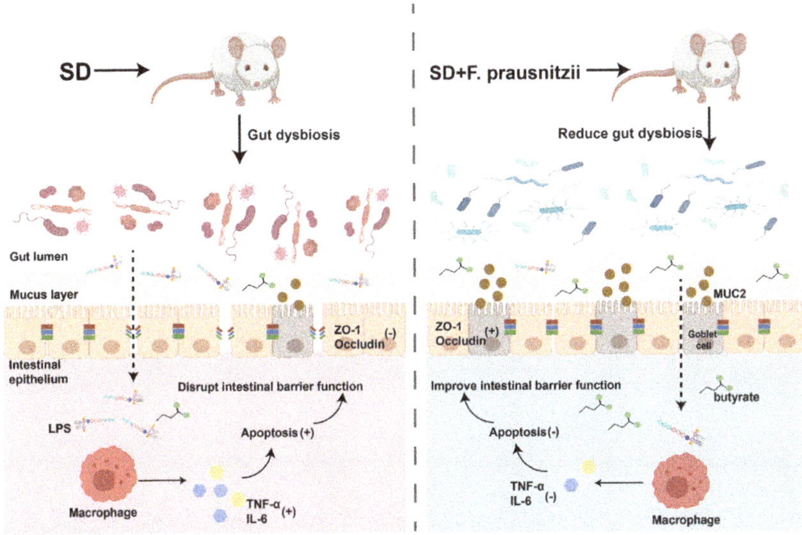

Figure 7. Schematic diagram of the potential mechanism by which *Faecalibacterium prausnitzii* ameliorates intestinal injury induced by sleep deprivation. A feasible mechanism is that the colonization of *Faecalibacterium prausnitzii* ameliorates impaired intestinal barrier function in SD mice through regulating intestinal inflammation, programmed death, and the gut microbiota. SD: sleep deprivation; LPS: lipopolysaccharide; ZO-1: tight-junction protein zonula occluden-1; TNF-α: tumor necrosis factor-alpha; IL-6; interleukin-6. (+); promote; (-); inhibit.

Lack of sleep may impact the gut tissue in both structure and performance [32]. In this study, we found that acute SD caused the disruption of colonic epithelial cell morphology and function in mice. Through HE staining, it could be observed that there is inflammatory cell infiltration and intestinal epithelial cell damage in the colon of SD mice, and the colonization of *F. prausnitzii* could improve it. Goblet cells produce mucin glycoproteins that create a protective mucus barrier to prevent bacterial infiltration [33,34]. The main mucin is found in the intestines. The deletion of the MUC2 gene destroyed the intestinal mucus barrier and disturbed intestinal symbiotic flora, which made mice sensitive to the colonization of Citrobacter muris [35]. Goblet cells and MUC2 protein expression were significantly decreased in the SD group compared to the control group but were restored to normal levels with *F. prausnitzii* colonization. Maintaining the integrity of the intestinal epithelial cell barrier is dependent on the consistent presence of tight-junction protein. Exposure to stress or infection can disrupt the regulation of tight-junction protein expression in the body, leading to an increase in the size of the intestinal epithelial paracellular space. This can expose lamina propria immune cells to harmful bacteria, ultimately causing inflammation in the intestines [36,37]. A previous study found that the supernatant of *F. prausnitzii* enhances the intestinal barrier function by affecting paracellular permeability and may thereby attenuate the severity of DSS-induced colitis in mice [38]. In sleep-deprived mice, the levels of ZO-1, occludin, and Claudin-1, which are colonic proteins related to tight junctions, were found to be notably decreased. However, the presence of *F. prausnitzii* was able to bring back the expression of these tight-junction proteins.

Additionally, we assessed alterations in the intestinal microflora composition in the colon contents of mice following various treatments. Overall, the alpha-diversity and beta-diversity results showed that the microbiomes of the SD + FP group were similar to the CON group, while the SD group was less similar to the other groups, suggesting that the colonization of *F. prausnitzii* improved the disturbed intestinal microflora of SD mice. During the research, we noticed a rise in the proportion of *Klebsiella*, *Proteus*, *Staphylococcus*, and *Enterococcus* in sleep-deprived mice. *Klebsiella*, a prevalent pathogen, can be found in the respiratory system and gut of humans and can produce different harmful elements like adhesins, capsular polysaccharides, siderophores, and lipopolysaccharides to induce the body's immune response [39]. A prior investigation discovered that *Klebsiella pneumoniae* triggered IBD by activating caspase-11-mediated IL18 in the cells lining the intestines [40]. *Proteus* is a common intestinal symbiotic bacterium, which is considered a potential pathogenic bacterium [41]. The relative abundance of *Proteus* was found to be significantly increased in MPTP-induced Parkinson's disease [42]. *Staphylococcus* is a Gram-positive bacterium that could cause pneumonia, pseudomembranous colitis, pericarditis, sepsis, and other acute and chronic infections [43]. In general, we have observed the increase in various conditional bacteria in sleep-deprived mice. Conversely, following colonization by *F. prausnitzii*, there has been a notable reduction in the proportion of detrimental bacteria in the intestines of sleep-deprived mice. It shows that *F. prausnitzii* has corrected the disordered intestinal microflora caused by SD. Additionally, there was a notable rise in the proportion of *Akkermansia* in sleep-deprived mice following the colonization of *F. prausnitzii*. *Akkermansia* is a mucin-degrading bacteria in the gut. The addition of *Akkermansia* led to an increase in the quantity of goblet cells that produce mucin in the mice [44]. In addition, in the mucous layer, there are still some bacteria that cannot degrade mucus, but use carbon and nitrogen generated by mucin degradation by *Akkermansia*; these microorganisms include *F. prausnitzii*, *Rothella*, etc. Studies have shown that when Akkermansia and Faecalibacterium are co-cultured, they exhibit synergistic growth and produce butyrate [45]. Overall, our findings indicated that the presence of *F. prausnitzii* can restore the balance of gut bacteria in sleep-deprived mice by increasing the levels of beneficial bacteria and reducing the levels of harmful bacteria.

Additionally, there was a notable rise in the proportion of various detrimental microbes in the intestines of sleep-deprived mice, particularly *Klebsiella*. The research discovered that *klebsiella* pneumoniae can stimulate the generation of fully developed IL18 in colon epithelial cells and gut organoids, leading to colitis and enhancing DSS-induced colitis [40]. Following SD, our findings indicated a notable rise in the presence of LPS in the colon, along with an observed increase in gut permeability. Therefore, we speculated that the excess LPS produced by the increased harmful bacteria may cross the intestinal barrier and enter the lamina propria of the colon. TLR4 is a receptor for bacterial LPS in response to LPS-induced inflammation [46]. The activation of NF-κB leads to the production of different inflammatory mediators and cytokines [47,48]. Likewise, we noted that the presence of *F. prausnitzii* was able to decrease macrophage infiltration and lower the amount of pro-inflammatory cytokines in the colon. Similar to our results, *F. prausnitzii* generates butyrate to support Th17/Treg equilibrium and improve colorectal colitis by blocking histone deacetylase 1 [49]. Hence, restoring gut immune balance with Faecalibacterium could improve intestinal damage caused by SD.

Faecalibacterium is a major butyrate-producing bacteria and most of its functions are based on its metabolites. As a result, we also assessed the concentration of short-chain fatty acids present in the colon contents. Similar to our previous findings, the content of butyrate in the colon contents of sleep-deprived mice has decreased significantly [50]. Interestingly, the content of butyrate in sleep-deprived mice has increased significantly after pre-colonization with *Faecalibacterium*. It could be seen that the colonization of *Faecalibacterium* has significantly increased the content of butyrate, its metabolite. However, alterations in various other short-chain fatty acids (acetate, propionate, and valerate) did not show statistical significance among the different treatment groups. The research has shown that butyrate enhances the production of mucus in the goblet cells of the intestines, reinforces the chemical defense, and boosts the synthesis of tight-junction protein, which helps restore the compromised physical barrier [51]. Moreover, butyrate exhibits a beneficial anti-inflammatory impact on both intestinal epithelial cells and immune cells [52,53]. In prior research, it was noted that *Prevotellaceae* generates butyrate to reduce cardiotoxicity associated with PD-1/PD-L1 inhibitors through the PPARα-CYP4X1 pathway in macrophages located in the colon [54]. Hence, we hypothesized that the positive impact of *F. prausnitzii* colonization on intestinal barrier impairment in sleep-deprived mice is due to the significant generation of its metabolite butyrate. According to the previous results, we have observed the reduction in goblet cells, and the colonization of *F. prausnitzii* could alleviate them to a certain extent. Studies have shown that butyrate can regulate inflammatory response by inhibiting the NF-κB pathway, effectively inhibiting the apoptosis pathway, and ultimately improving intestinal barrier function [55]. Hence, we speculated that the presence of *F. prausnitzii* has a beneficial effect in decreasing colon cell apoptosis. In sleep-deprived mice, the colon exhibited elevated levels of cleaved caspase-3 protein and reduced levels of Bcl-2 anti-apoptotic factor mRNA, as indicated by our findings. Unlike sleep-deprived mice, no significant apoptosis has been observed in sleep-deprived mice colonized by *F. prausnitzii*.

Based on the worldwide human health problem of sleep loss, this study first explored the relationship between *F. prauznitzi* and sleep deficiency from the perspective of probiotics. In addition, this study evaluated the beneficial effects of colonization of *F. prauznitzi* on intestinal damage in sleep-deprived mice from multiple aspects of the intestinal barrier, including the mechanical barrier, immune barrier and microbial barrier. More importantly, through the detection of intestinal short-chain fatty acids, the relationship between *F. prauznitzii* and butyrate was found, which provided the direction for further exploration of the mechanism of *F. prauznitzii*. There are some limitations to our study. The animal model we constructed in this study was used to simulate acute SD, but a large number of people also face chronic sleep restriction in their daily lives. Therefore, in the follow-up research, it is necessary to expand the breadth of sleep research and more comprehensive research on the beneficial effects of *F. prauznitzii* on organismal health. In addition, the results of this study are based on animal models, and while our results provide valuable insights into the role of *F. prauznitzii* in alleviating intestinal damage in sleep-deprived mice, further research is needed to apply these results to human health.

F. prauznitzii, a health-related human intestinal bacterium with reduced levels in patients with a variety of metabolic diseases and IBD, is considered a next-generation probiotic with therapeutic potential. However, *F. prausnitzii* is strictly anaerobic, and two challenges need to be addressed before it can be used in human subjects. The first is that bacteria need to be produced on a large scale under strict anaerobic conditions, and the second is that adequate hypoxia conditions must be maintained throughout the culture process (such as centrifugation, filtration or lyophilization) [56]. At present, some studies have carried out preliminary exploration, using the symbiotic relationship between *F. prausnitzii* and *Desulfovibrio piger*, a sulfate-reducing bacterium, which can be used for the large-scale production of *F. prausnitzii* and improve storage stability [20]. Therefore, exploring the health effect of *F. prausnitzii* in the human body and transforming it for industry is the focus of future research. In addition, as scientists become more and more clear about the classification of *F. prausnitzii*, exploring and verifying the mechanism of action of *F. prausnitzii* in humans or animals is also the focus of future research.

5. Conclusions

The findings of our study indicated that *F. prausnitzii*, when used as probiotics, had a positive impact on alleviating the intestinal damage induced by SD. The positive impact of *F. prausnitzii* could be due to its ability to prevent SD-induced inflammation in the intestines, reduce dysbiosis in the gut microbiome, and promote the function of the intestinal barrier by potentially involving its metabolite butyrate in the process. These results provided new evidence for the protective mechanisms of *F. prausnitzii* against intestinal injury, thereby revealing the potential values of *F. prausnitzii* in SD-related body damage and having significant implications for *F. prausnitzii* as probiotics.

Supplementary Materials: The following supporting information can be downloaded at: https://www.mdpi.com/article/10.3390/nu16081100/s1, Table S1: Primers for real-time PCR.

Author Contributions: X.W. (Xintong Wang): Conceptualization, methodology, formal analysis, writing—original draft, visualization; Y.L.: conceptualization, formal analysis, validation, supervision; X.W. (Xifan Wang): methodology; R.W.: methodology; Y.H.: methodology; F.R.: resources, supervision, project administration, funding acquisition; P.W. and B.F.: conceptualization, funding acquisition, project administration, resources, writing—review and editing. All authors have read and agreed to the published version of the manuscript.

Funding: This research was funded by the National Key R&D Program of China (No. 2022YFF1100100) and the 111 project of the Education Ministry of China (No. B18053).

Institutional Review Board Statement: The study was conducted in accordance with the Declaration of Helsinki and approved by the Animal Welfare and Ethics Committee of China Agricultural University (Ethical No. CAU201709112, approval date: 11 September 2017).

Informed Consent Statement: Not applicable.

Data Availability Statement: Data are contained within the article.

Conflicts of Interest: The authors declare no conflicts of interest.

References

1. Kroese, F.M.; Evers, C.; Adriaanse, M.A.; de Ridder, D.T.D. Bedtime procrastination: A self-regulation perspective on sleep insufficiency in the general population. *J. Health Psychol.* **2016**, *21*, 853–862. [CrossRef] [PubMed]
2. Zamore, Z.; Veasey, S.C. Neural consequences of chronic sleep disruption. *Trends Neurosci.* **2022**, *45*, 678–691. [CrossRef]
3. Neroni, B.; Evangelisti, M.; Radocchia, G.; Di Nardo, G.; Pantanella, F.; Villa, M.P.; Schippa, S. Relationship between sleep disorders and gut dysbiosis: What affects what? *Sleep. Med.* **2021**, *87*, 1–7. [CrossRef] [PubMed]
4. Zeng, Y.; Zhang, Z.; Liang, S.; Chang, X.; Qin, R.; Chen, H.; Guo, L. Paternal sleep deprivation induces metabolic perturbations in male offspring via altered LRP5 DNA methylation of pancreatic islets. *J. Pineal Res.* **2023**, *74*, e12863. [CrossRef]
5. Fang, D.; Xu, T.; Sun, J.; Shi, J.; Li, F.; Yin, Y.; Wang, Z.; Liu, Y. Nicotinamide Mononucleotide Ameliorates Sleep Deprivation-Induced Gut Microbiota Dysbiosis and Restores Colonization Resistance against Intestinal Infections. *Adv. Sci.* **2023**, *10*, e2207170. [CrossRef]
6. Martel, J.; Chang, S.H.; Ko, Y.F.; Hwang, T.L.; Young, J.D.; Ojcius, D.M. Gut barrier disruption and chronic disease. *Trends Endocrinol. Metab.* **2022**, *33*, 247–265. [CrossRef]
7. Cani, P.D.; Jordan, B.F. Gut microbiota-mediated inflammation in obesity: A link with gastrointestinal cancer. *Nat. Rev. Gastroenterol. Hepatol.* **2018**, *15*, 671–682. [CrossRef] [PubMed]
8. Gong, Y.; Xia, W.; Wen, X.; Lyu, W.; Xiao, Y.; Yang, H.; Zou, X. Early inoculation with caecal fermentation broth alters small intestine morphology, gene expression of tight junction proteins in the ileum, and the caecal metabolomic profiling of broilers. *J. Anim. Sci. Biotechnol.* **2020**, *11*, 8. [CrossRef]
9. Carloni, S.; Bertocchi, A.; Mancinelli, S.; Bellini, M.; Erreni, M.; Borreca, A.; Braga, D.; Giugliano, S.; Mozzarelli, A.M.; Manganaro, D.; et al. Identification of a choroid plexus vascular barrier closing during intestinal inflammation. *Science* **2021**, *374*, 439–448. [CrossRef]
10. Chen, H.; Wang, C.; Bai, J.; Song, J.; Bu, L.; Liang, M.; Suo, H. Targeting microbiota to alleviate the harm caused by sleep deprivation. *Microbiol. Res.* **2023**, *275*, 127467. [CrossRef]
11. Gao, T.; Wang, Z.; Dong, Y.; Cao, J.; Lin, R.; Wang, X.; Yu, Z.; Chen, Y. Role of melatonin in sleep deprivation-induced intestinal barrier dysfunction in mice. *J. Pineal Res.* **2019**, *67*, e12574. [CrossRef] [PubMed]
12. Horowitz, A.; Chanez-Paredes, S.D.; Haest, X.; Turner, J.R. Paracellular permeability and tight junction regulation in gut health and disease. *Nat. Rev. Gastroenterol. Hepatol.* **2023**, *20*, 417–432. [CrossRef] [PubMed]
13. Duan, J.; Matute, J.D.; Unger, L.W.; Hanley, T.; Schnell, A.; Lin, X.; Krupka, N.; Griebel, P.; Lambden, C.; Sit, B.; et al. Endoplasmic reticulum stress in the intestinal epithelium initiates purine metabolite synthesis and promotes Th17 cell differentiation in the gut. *Immunity* **2023**, *56*, 1115–1131.e9. [CrossRef] [PubMed]
14. Pontarollo, G.; Kollar, B.; Mann, A.; Khuu, M.P.; Kiouptsi, K.; Bayer, F.; Brandão, I.; Zinina, V.V.; Hahlbrock, J.; Malinarich, F.; et al. Commensal bacteria weaken the intestinal barrier by suppressing epithelial neuropilin-1 and Hedgehog signaling. *Nat. Metab.* **2023**, *5*, 1174–1187. [CrossRef] [PubMed]
15. Doré, E.; Joly-Beauparlant, C.; Morozumi, S.; Mathieu, A.; Lévesque, T.; Allaeys, I.; Duchez, A.C.; Cloutier, N.; Leclercq, M.; Bodein, A.; et al. The interaction of secreted phospholipase A2-IIA with the microbiota alters its lipidome and promotes inflammation. *JCI Insight* **2022**, *7*, e152638. [CrossRef]
16. Lopez-Siles, M.; Duncan, S.H.; Garcia-Gil, L.J.; Martinez-Medina, M. Faecalibacterium prausnitzii: From microbiology to diagnostics and prognostics. *ISME J.* **2017**, *11*, 841–852. [CrossRef] [PubMed]

17. Lopez-Siles, M.; Martinez-Medina, M.; Abellà, C.; Busquets, D.; Sabat-Mir, M.; Duncan, S.H.; Aldeguer, X.; Flint, H.J.; Garcia-Gil, L.J. Mucosa-associated Faecalibacterium prausnitzii phylotype richness is reduced in patients with inflammatory bowel disease. *Appl. Environ. Microbiol.* **2015**, *81*, 7582–7592. [CrossRef] [PubMed]
18. Chen, H.; Ou, R.; Tang, N.; Su, W.; Yang, R.; Yu, X.; Zhang, G.; Jiao, J.; Zhou, X. Alternation of the gut microbiota in irritable bowel syndrome: An integrated analysis based on multicenter amplicon sequencing data. *J. Transl. Med.* **2023**, *21*, 117. [CrossRef]
19. Ueda, A.; Shinkai, S.; Shiroma, H.; Taniguchi, Y.; Tsuchida, S.; Kariya, T.; Kawahara, T.; Kobayashi, Y.; Kohda, N.; Ushida, K.; et al. Identification of Faecalibacterium prausnitzii strains for gut microbiome-based intervention in Alzheimer's-type dementia. *Cell Rep. Med.* **2021**, *2*, 100398. [CrossRef]
20. Khan, M.T.; Dwibedi, C.; Sundh, D.; Pradhan, M.; Kraft, J.D.; Caesar, R.; Tremaroli, V.; Lorentzon, M.; Bäckhed, F. Synergy and oxygen adaptation for development of next-generation probiotics. *Nature* **2023**, *620*, 381–385. [CrossRef]
21. Zhang, L.; Liu, C.; Jiang, Q.; Yin, Y. Butyrate in Energy Metabolism: There Is Still More to Learn. *Trends Endocrinol. Metab.* **2021**, *32*, 159–169. [CrossRef] [PubMed]
22. Liang, L.; Liu, L.; Zhou, W.; Yang, C.; Mai, G.; Li, H.; Chen, Y. Gut microbiota-derived butyrate regulates gut mucus barrier repair by activating the macrophage/WNT/ERK signaling pathway. *Clin. Sci.* **2022**, *136*, 291–307. [CrossRef] [PubMed]
23. Song, Q.; Cheng, S.W.; Zou, J.; Li, K.S.L.; Cheng, H.; Wai Lau, D.T.; Han, Q.; Yang, X.; Shaw, P.C.; Zuo, Z. Role of gut microbiota on regulation potential of Dendrobium officinale Kimura & Migo in metabolic syndrome: In-vitro fermentation screening and in-vivo verification in db/db mice. *J. Ethnopharmacol.* **2024**, *321*, 117437. [PubMed]
24. Li, H.B.; Xu, M.L.; Xu, X.D.; Tang, Y.Y.; Jiang, H.L.; Li, L.; Xia, W.J.; Cui, N.; Bai, J.; Dai, Z.M.; et al. Faecalibacterium prausnitzii Attenuates CKD via Butyrate-Renal GPR43 Axis. *Circ. Res.* **2022**, *131*, e120–e134. [CrossRef] [PubMed]
25. Pan, W.; Zhao, J.; Wu, J.; Xu, D.; Meng, X.; Jiang, P.; Shi, H.; Ge, X.; Yang, X.; Hu, M.; et al. Dimethyl itaconate ameliorates cognitive impairment induced by a high-fat diet via the gut-brain axis in mice. *Microbiome* **2023**, *11*, 30.
26. Ma, L.; Shen, Q.; Lyu, W.; Lv, L.; Wang, W.; Yu, M.; Yang, H.; Tao, S.; Xiao, Y. Clostridium butyricum and Its Derived Extracellular Vesicles Modulate Gut Homeostasis and Ameliorate Acute Experimental Colitis. *Microbiol. Spectr.* **2022**, *10*, e0136822. [CrossRef]
27. Wu, J.; Lin, Z.; Wang, X.; Zhao, Y.; Zhao, J.; Liu, H.; Johnston, L.J.; Lu, L.; Ma, X. Limosilactobacillus reuteri SLZX19-12 Protects the Colon from Infection by Enhancing Stability of the Gut Microbiota and Barrier Integrity and Reducing Inflammation. *Microbiol. Spectr.* **2022**, *10*, e0212421. [CrossRef]
28. Palmer, C.A.; Bower, J.L.; Cho, K.W.; Clementi, M.A.; Lau, S.; Oosterhoff, B.; Alfano, C.A. Sleep loss and emotion: A systematic review and meta-analysis of over 50 years of experimental research. *Psychol. Bull.* **2023**. [CrossRef]
29. Hyun, M.K.; Baek, Y.; Lee, S. Association between digestive symptoms and sleep disturbance: A cross-sectional community-based study. *BMC Gastroenterol.* **2019**, *19*, 34. [CrossRef] [PubMed]
30. Sang, D.; Lin, K.; Yang, Y.; Ran, G.; Li, B.; Chen, C.; Li, Q.; Ma, Y.; Lu, L.; Cui, X.Y.; et al. Prolonged sleep deprivation induces a cytokine-storm-like syndrome in mammals. *Cell* **2023**, *186*, 5500–5516.e21. [CrossRef]
31. Touch, S.; Godefroy, E.; Rolhion, N.; Danne, C.; Oeuvray, C.; Straube, M.; Galbert, C.; Brot, L.; Alonso Salgueiro, I.; Chadi, S.; et al. Human CD4+CD8α+ Tregs induced by Faecalibacterium prausnitzii protect against intestinal inflammation. *JCI Insight* **2022**, *7*, e154722. [CrossRef] [PubMed]
32. Li, N.; Tan, S.; Wang, Y.; Deng, J.; Wang, N.; Zhu, S.; Tian, W.; Xu, J.; Wang, Q. Akkermansia muciniphila supplementation prevents cognitive impairment in sleep-deprived mice by modulating microglial engulfment of synapses. *Gut Microbes.* **2023**, *15*, 2252764. [CrossRef] [PubMed]
33. Furter, M.; Sellin, M.E.; Hansson, G.C.; Hardt, W.D. Mucus Architecture and Near-Surface Swimming Affect Distinct Salmonella Typhimurium Infection Patterns along the Murine Intestinal Tract. *Cell Rep.* **2019**, *27*, 2665–2678.e2663. [CrossRef] [PubMed]
34. McLoughlin, K.; Schluter, J.; Rakoff-Nahoum, S.; Smith, A.L.; Foster, K.R. Host Selection of Microbiota via Differential Adhesion. *Cell Host Microbe.* **2016**, *19*, 550–559. [CrossRef] [PubMed]
35. Birchenough, G.M.H.; Schroeder, B.O.; Sharba, S.; Arike, L.; Recktenwald, C.V.; Puértolas-Balint, F.; Subramani, M.V.; Hansson, K.T.; Yilmaz, B.; Lindén, S.K.; et al. Muc2-dependent microbial colonization of the jejunal mucus layer is diet sensitive and confers local resistance to enteric pathogen infection. *Cell Rep.* **2023**, *42*, 112084. [CrossRef] [PubMed]
36. Turpin, W.; Lee, S.H.; Raygoza Garay, J.A.; Madsen, K.L.; Meddings, J.B.; Bedrani, L.; Power, N.; Espin-Garcia, O.; Xu, W.; Smith, M.I.; et al. Increased Intestinal Permeability Is Associated with Later Development of Crohn's Disease. *Gastroenterology* **2020**, *159*, 2092–2100.e2095. [CrossRef] [PubMed]
37. Pedersen, T.K.; Brown, E.M.; Plichta, D.R.; Johansen, J.; Twardus, S.W.; Delorey, T.M.; Lau, H.; Vlamakis, H.; Moon, J.J.; Xavier, R.J.; et al. The CD4(+) T cell response to a commensal-derived epitope transitions from a tolerant to an inflammatory state in Crohn's disease. *Immunity* **2022**, *55*, 1909–1923.e6. [CrossRef] [PubMed]
38. Carlsson, A.H.; Yakymenko, O.; Olivier, I.; Håkansson, F.; Postma, E.; Keita, A.V.; Söderholm, J.D. Faecalibacterium prausnitzii supernatant improves intestinal barrier function in mice DSS colitis. *Scand. J. Gastroenterol.* **2013**, *48*, 1136–1144. [CrossRef] [PubMed]
39. Martin, R.M.; Bachman, M.A. Colonization, Infection, and the Accessory Genome of Klebsiella pneumoniae. *Front. Cell Infect. Microbiol.* **2018**, *8*, 4. [CrossRef]
40. Zhang, Q.; Su, X.; Zhang, C.; Chen, W.; Wang, Y.; Yang, X.; Liu, D.; Zhang, Y.; Yang, R. Klebsiella pneumoniae Induces Inflammatory Bowel Disease Through Caspase-11-Mediated IL18 in the Gut Epithelial Cells. *Cell Mol. Gastroenterol. Hepatol.* **2023**, *15*, 613–632. [CrossRef]

41. Wasfi, R.; Hamed, S.M.; Amer, M.A.; Fahmy, L.I. Proteus mirabilis Biofilm: Development and Therapeutic Strategies. *Front. Cell Infect. Microbiol.* **2020**, *10*, 414. [CrossRef]
42. Zhou, X.; Lu, J.; Wei, K.; Wei, J.; Tian, P.; Yue, M.; Wang, Y.; Hong, D.; Li, F.; Wang, B.; et al. Neuroprotective Effect of Ceftriaxone on MPTP-Induced Parkinson's Disease Mouse Model by Regulating Inflammation and Intestinal Microbiota. *Oxid. Med. Cell Longev.* **2021**, *2021*, 9424582. [CrossRef] [PubMed]
43. Howden, B.P.; Giulieri, S.G.; Wong Fok Lung, T.; Baines, S.L.; Sharkey, L.K.; Lee, J.Y.H.; Hachani, A.; Monk, I.R.; Stinear, T.P. Staphylococcus aureus host interactions and adaptation. *Nat. Rev. Microbiol.* **2023**, *21*, 380–395. [CrossRef] [PubMed]
44. Cani, P.D.; Depommier, C.; Derrien, M.; Everard, A.; de Vos, W.M. Akkermansia muciniphila: Paradigm for next-generation beneficial microorganisms. *Nat. Rev. Gastroenterol. Hepatol.* **2022**, *19*, 625–637. [CrossRef] [PubMed]
45. Lukovac, S.; Belzer, C.; Pellis, L.; Keijser, B.J.; de Vos, W.M.; Montijn, R.C.; Roeselers, G. Differential modulation by Akkermansia muciniphila and Faecalibacterium prausnitzii of host peripheral lipid metabolism and histone acetylation in mouse gut organoids. *mBio* **2014**, *5*, e01438-14. [CrossRef] [PubMed]
46. Liu, J.; Kang, R.; Tang, D. Lipopolysaccharide delivery systems in innate immunity. *Trends Immunol.* **2024**. [CrossRef] [PubMed]
47. Lai, J.L.; Liu, Y.H.; Liu, C.; Qi, M.P.; Liu, R.N.; Zhu, X.F.; Zhou, Q.G.; Chen, Y.Y.; Guo, A.Z.; Hu, C.M. Indirubin Inhibits LPS-Induced Inflammation via TLR4 Abrogation Mediated by the NF-kB and MAPK Signaling Pathways. *Inflammation* **2017**, *40*, 1–12. [CrossRef]
48. Ju, M.; Liu, B.; He, H.; Gu, Z.; Liu, Y.; Su, Y.; Zhu, D.; Cang, J.; Luo, Z. MicroRNA-27a alleviates LPS-induced acute lung injury in mice via inhibiting inflammation and apoptosis through modulating TLR4/MyD88/NF-κB pathway. *Cell Cycle* **2018**, *17*, 2001–2018. [CrossRef]
49. Zhou, L.; Zhang, M.; Wang, Y.; Dorfman, R.G.; Liu, H.; Yu, T.; Chen, X.; Tang, D.; Xu, L.; Yin, Y.; et al. Faecalibacterium prausnitzii Produces Butyrate to Maintain Th17/Treg Balance and to Ameliorate Colorectal Colitis by Inhibiting Histone Deacetylase 1. *Inflamm Bowel Dis.* **2018**, *24*, 1926–1940. [CrossRef]
50. Gao, T.; Wang, Z.; Dong, Y.; Cao, J.; Chen, Y. Melatonin-Mediated Colonic Microbiota Metabolite Butyrate Prevents Acute Sleep Deprivation-Induced Colitis in Mice. *Int. J. Mol. Sci.* **2021**, *22*, 11894. [CrossRef]
51. Singh, V.; Lee, G.; Son, H.; Koh, H.; Kim, E.S.; Unno, T.; Shin, J.H. Butyrate producers, "The Sentinel of Gut": Their intestinal significance with and beyond butyrate, and prospective use as microbial therapeutics. *Front. Microbiol.* **2022**, *13*, 1103836. [CrossRef]
52. Fu, X.; Liu, Z.; Zhu, C.; Mou, H.; Kong, Q. Nondigestible carbohydrates, butyrate, and butyrate-producing bacteria. *Crit. Rev. Food Sci. Nutr.* **2019**, *59*, S130–S152. [CrossRef]
53. Recharla, N.; Geesala, R.; Shi, X.Z. Gut Microbial Metabolite Butyrate and Its Therapeutic Role in Inflammatory Bowel Disease: A Literature Review. *Nutrients* **2023**, *15*, 2275. [CrossRef]
54. Chen, Y.; Liu, Y.; Wang, Y.; Chen, X.; Wang, C.; Chen, X.; Yuan, X.; Liu, L.; Yang, J.; Zhou, X. Prevotellaceae produces butyrate to alleviate PD-1/PD-L1 inhibitor-related cardiotoxicity via PPARα-CYP4X1 axis in colonic macrophages. *J. Exp. Clin. Cancer Res.* **2022**, *41*, 1. [CrossRef]
55. Couto, M.R.; Gonçalves, P.; Magro, F.; Martel, F. Microbiota-derived butyrate regulates intestinal inflammation: Focus on inflammatory bowel disease. *Pharmacol. Res.* **2020**, *159*, 104947. [CrossRef]
56. Martín, R.; Rios-Covian, D.; Huillet, E.; Auger, S.; Khazaal, S.; Bermúdez-Humarán, L.G.; Sokol, H.; Chatel, J.M.; Langella, P. Faecalibacterium: A bacterial genus with promising human health applications. *FEMS Microbiol. Rev.* **2023**, *47*, fuad039. [CrossRef]

nutrients

MDPI

Article

Effects of *Lacticaseibacillus paracasei* Strain Shirota on Daytime Performance in Healthy Office Workers: A Double-Blind, Randomized, Crossover, Placebo-Controlled Trial

Hiroko Kikuchi-Hayakawa [1,*], Hiroshi Ishikawa [1], Kazunori Suda [1,2], Yusuke Gondo [1], Genki Hirasawa [1], Hayato Nakamura [1], Mai Takada [1], Mitsuhisa Kawai [1] and Kazunori Matsuda [1,*]

[1] Yakult Central Institute, 5-11 Izumi, Kunitachi-shi, Tokyo 186-8650, Japan

[2] Yakult Honsha European Research Center for Microbiology VOF, Technologiepark 94 bus 3, 9052 Ghent, Belgium

* Correspondence: hiroko-hayakawa@yakult.co.jp (H.K.-H.); kazunori-matsuda@yakult.co.jp (K.M.)

Abstract: *Lacticaseibacillus paracasei* strain Shirota (LcS) modulates psychological homeostasis via the gut–brain axis. To explore the possible efficacy of LcS for improving daytime performance, we conducted a double-blind, randomized, crossover, placebo-controlled study of 12 healthy office workers with sleep complaints. The participants received fermented milk containing viable LcS (daily intake of 1×10^{11} colony-forming units) and non-fermented placebo milk, each for a 4-week period. In the last week of each period, the participants underwent assessments of their subjective mood and measurements of physiological state indicators via an electroencephalogram (EEG) and heart rate variability in the morning and afternoon. The attention score in the afternoon as assessed by the visual analog scale was higher in the LcS intake period than in the placebo intake period ($p = 0.041$). Theta power on EEG measured at rest or during an auditory oddball task in the afternoon was significantly lower in the LcS period than in the placebo period ($p = 0.025$ and 0.009, respectively). The change rate of theta power was associated with the change in attention score. Treatment-associated changes were also observed in heart rate and the sympathetic nerve activity index. These results indicate that LcS has possible efficacy for improving daytime performance, supported by observations of the related physiological state indicators.

Keywords: probiotics; daytime performance; electroencephalograms; heart rate

1. Introduction

Exposure to psychological stress causes various physical and mental disorders via the nervous and endocrine systems, such as depression, stomach pain, headaches, loss of appetite, diarrhea, fatigue, and insomnia [1–3]. Occupational stress and sleep disorders are associated with presenteeism, which refers to productivity losses from employees working with poor health conditions [4,5]. In Japan, costs corresponding to lost productivity due to presenteeism account for 80% of companies' total health-related costs and are higher than the costs of absenteeism [6].

One of the symptoms of mental stress is disordered sleep, which has been found to increase cognitive task errors [7] and the risk of accidents [8]. This is because sleep disturbances can cause fatigue, drowsiness, decreased alertness, and lack of motivation, affecting daytime performance [7]. These mood indicators are commonly assessed by self-reported questionnaires. However, physiological state indicators such as electroencephalography (EEG) and heart rate variability (HRV) have been used as surrogate measures of daytime performance. EEG is a non-invasive technique for recording the brain's electrical activity and reflects changes in mental status and cognitive function [9,10]. The EEG frequency bands beta (13–30 Hz), alpha (8–13 Hz), theta (4–8 Hz), and delta (0–4 Hz) are commonly used for studying brain activities. The P300 wave is an event-related potential (ERP) component elicited by the target stimulus in an oddball task occurring at around the 300 ms latency region and is commonly used to estimate cognitive performance. HRV analysis provides information about the status of the autonomic nervous system [11]. High-frequency (HF) components at 0.15–0.4 Hz reflect parasympathetic nerve activity, whereas the ratio of HF to low-frequency (LF) (0.04–0.15 Hz) components reflects sympathetic nerve activity [12].

The digestive system is highly regulated by the brain and state of mind, such as anxiety, depression, and fear, which greatly affect its functions. The brain and gut exchange information in both directions through the nervous, endocrine, and immune systems and are closely influenced by each other, forming a relationship called the gut–brain axis [13]. In recent years, the concept of a microbiota–gut–brain axis has been gaining attention as studies of stress [14–16] and mental illness [17–19] have produced a growing body of evidence that the gut microbiota is involved in the gut–brain axis. In addition, probiotics that contribute to the health of the host, either directly or through the improvement of the intestinal environment, are expected to exert positive effects via the gut–brain axis [20–23].

Lacticaseibacillus paracasei strain Shirota (LcS; formerly named the *Lactobacillus casei* strain Shirota [24]) is a widely used probiotic strain, which was selected from the microbial collection of Dr. Minoru Shirota, the founder of Yakult Honsha Co., Ltd. (Tokyo, Japan) LcS has a long history of safe use as a food material for over 80 years, and the United States Food and Drug Administration has accredited it as Generally Recognized As Safe (GRAS) [25]. LcS has been revealed in numerous clinical studies to have functions such as bowel movement normalization [26,27], infection prevention [28], and immunomodulation [29]. It has also been found to act on the nervous and endocrine systems. We previously reported that an LcS intervention in medical students preparing for an academic exam prevented the onset of physical symptoms and attenuated a stress-induced rise in salivary cortisol [30–32]. In addition, it was effective in maintaining the quality of sleep under academic stress conditions by suppressing the decline in deep sleep and shortening sleep latency, as measured by both subjective questionnaires and sleep EEG measures [33].

To explore the possible efficacy of LcS for improving daytime performance, we conducted a double-blind, randomized, crossover, placebo-controlled study in healthy participants with sleep complaints. We assessed their subjective mood using perceived mood questionnaires and measured physiological indicators via EEG and HRV.

2. Materials and Methods

2.1. Test Beverages

Fermented milk containing at least 1.0×10^{11} colony-forming units of LcS (strain no.: YIT 9029) per bottle (100 mL) was used as the test beverage. The nutritional content per bottle was as follows: 1.5 g of protein, 0.1 g of fat, 14.1 g of carbohydrates, and 63.0 kcal calories. The probiotic strain was obtained from the Culture Collection Research Laboratory of Yakult Central Institute, Tokyo, Japan. Since there is insufficient evidence for the efficacy of any probiotic strain on daytime performance, we decided to use an inert placebo, non-fermented milk, to evaluate the contribution of both LcS and its metabolites to any health benefits for the endpoint. The nutritional content, color, flavor, taste, and pH of the non-fermented placebo milk were adjusted to match the test beverage. The placebo was made from the same ingredients as the test beverage, except that lactic acid was added to match the acidity [31], and the identical plastic bottle was used for storage and the provision of the test beverage and the placebo. The beverages were hand-delivered to each participant weekly and stored at 0–10 °C.

2.2. Study Design and Participants

This study was conducted in Kunitachi-shi, Tokyo, from August to December 2021 as a double-blind, randomized, crossover, placebo-controlled study. The study protocol was registered at the UMIN (University Hospital Medical Information Network) Clinical

Trials Registry (ID: UMIN000044852). The study comprised a 3-week pre-intervention period, a first 4-week intervention period, a 4-week washout period, and a second 4-week intervention period (Figure 1). Healthy male and female office workers were recruited for this study. Because little information was available on the impact of probiotics on daytime performance, the present study is seen as an exploratory study in this field, and no formal sample size calculation was performed. We set the sample size to 12 in total by considering the handling capability of the EEG system, which was to be used only on weekdays during a specified period. There are several reports that cognitive abilities decline with age [34–36] and lack of sleep [37,38]. It has also been reported that the reaction times of the go/no-go task decline with age older compared to 40 years [35]. Based on these observations, office workers between 40 and 59 years old who were aware of impaired sleep quality were chosen as the participants. Those meeting any of the following criteria were excluded from the study: a history of sleep-related illnesses, use of medicines or health foods that could influence sleep, smoking habits, allergies to milk or soy, placement of an artificial cardiac pacemaker, skin disorders on the face or head, severe dry eye symptoms, and participation in other clinical trials during the study period. Finally, 5 males and 7 females were enrolled and were divided into the following two groups: one group was treated with the placebo and LcS in that order, and the other group was treated in the reverse order. Simple randomization was performed by assigning random numbers from a random number table to the groups. The codes of the treatments were assigned by personnel not involved in the study, and participants and those giving treatment or assessing outcomes were blinded to the treatment allocation.

	Pre-intervention period (3 weeks)	Intervention period I	Washout period (4 weeks)	Intervention period II
Placebo first		Placebo (4 weeks)		LcS (4 weeks)
LcS first		LcS (4 weeks)		Placebo (4 weeks)

Figure 1. Study design. LcS, *Lacticaseibacillus paracasei* strain Shirota.

2.3. *Questionnaires*

The daytime mood, including perceived sleepiness, physical fatigue, motivation, attention, and optimism, was surveyed using a visual analog scale (VAS) ranging from 0 mm for "severely unfavorable" daytime performance to 100 mm for "supremely favorable". Sleepiness was also evaluated on a 9 point Likert scale using the Karolinska Sleepiness Scale (KSS) [39]. Participants completed the questionnaires at an arbitrary time in the morning (8:00–11:00) and in the afternoon (15:00–18:00) every day for 5 weekdays during the last week of each period. Information on sleep habits and stress levels was also collected via retrospective self-assessments for the previous 1-month period before intervention using the Pittsburgh Sleep Quality Index (PSQI) [40] and the Japanese version of the Perceived Stress Scale (JPSS) [41], respectively.

2.4. *Daytime EEG and HRV Measurement*

EEG and HRV measurements were conducted in a shielded room during the last week of each period: in the pre-intervention period and in intervention periods I and II. Participants underwent measurements on any day from Tuesday to Thursday in the morning (8:00–11:00) and in the afternoon (15:00–18:00). Each measurement required about 1 h, including preparation and cleanup, and the day and time of each individual's measurements were matched between the experimental periods. We instructed the participants not to consume alcohol for at least 24 h before the measurement and caffeinated beverages on the day of measurement and confirmed the participants' adherence to the instructions before the measurement.

EEG signals were recorded with Ag/AgCl electrodes placed on the scalp at the Fz, Cz, and Pz positions according to the International 10–20 System. Two electrodes placed on both ears were regarded as a reference for the scalp electrodes, and two forehead electrodes served as the ground and system reference. To aid in the elimination of data obtained during eye movements or blinks, an electrooculogram (EOG) was recorded with an electrode placed 2 cm lateral to the lateral canthus of the left eye and 2 cm above the upper edge of the left orbit. Electrodes were also attached to the left arm to monitor the electrocardiogram (ECG) for HRV analysis. EEG and ECG were recorded using a Polymate® biological signal recording device (Miyuki Giken Co., Ltd., Tokyo, Japan) with a sampling rate of 1000 Hz; the impedances of all electrodes were kept below 10 kΩ.

The procedure for each measurement is shown in Figure 2. The participants performed the whole task seated on an armchair, fixating their eyes on a point directly in front of them at a 0.5 m distance. During a 5 min task period, participants performed an auditory oddball task with a 2-tone paradigm. The procedures were compliant with the guidelines of the Japanese Society of Clinical Neurophysiology [42], and a target tone (2000 Hz) and a standard tone (1000 Hz) were randomly emitted every 2.0 s with a 100 ms duration for each, 150 times in total, with appearance rates of 20% and 80%, respectively, using the Stimuli Output Sequencer Program v. 2009-Jan. (NoruPro Light Systems, Inc., Tokyo, Japan). The participants responded by pressing a button only for the target stimulus, and their reaction time was calculated. Finally, EEG was recorded with the eyes closed for 2 min to estimate the basal resting state.

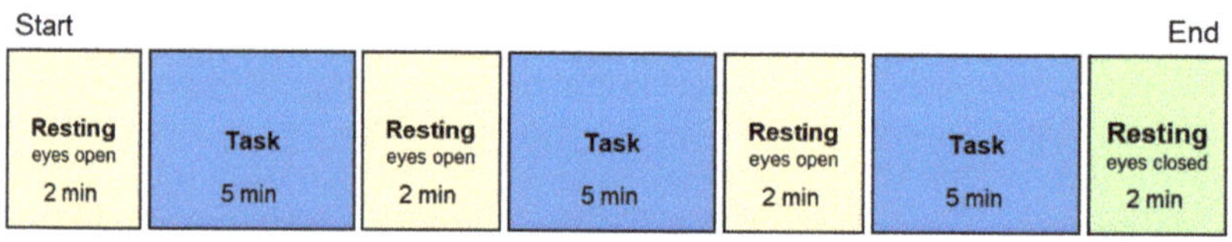

Figure 2. Experimental procedure for electroencephalogram and heart rate variability measurement.

2.5. Analysis of EEG and HRV

The MATLAB R2018b mathematical analysis software (MathWorks, Inc., Natick, MA, USA) was used to analyze spontaneous brain waves [43]. Fourier transformation was performed for data removed using a 0.5–30-Hz band-pass filter, and power spectrum analysis was performed in each frequency band (theta wave, 4–8 Hz; alpha wave, 8–13 Hz; beta wave, 13–30 Hz). The event-related potential (ERP) P300 was obtained and analyzed via the arithmetic mean method using the EP Travel Light software v. 2009-Feb. (NoruPro Light Systems, Inc.). When P300a and P300b were observed, P300b was selected, and when three or more peaks were observed, it was considered inconclusive. An analysis program based on the MemCal method using the RR Interval Analysis software v. 2009-Feb. (NoruPro Light Systems, Inc.) was used to analyze heart rate and HRV. LF components at 0.04–0.15 Hz, reflecting sympathetic and parasympathetic nerve activity, and HF components at 0.15–0.4 Hz, reflecting parasympathetic nerve activity, were analyzed, and the ratio of the LF/HF parameters was calculated as an index of sympathetic nerve activity [12].

2.6. Statistical Analysis

The order effect and time effect of the crossover test were calculated from repeated-measures analysis of variance in the pre-intervention period, intervention period I, and intervention period II. Variables were subjected to the assessment of treatment efficacy if they did not have any significant order and time effects ($p \geq 0.05$). The pairwise comparison between the placebo and LcS intake periods was performed using the Wilcoxon rank sum test. Fisher's exact test was used to compare the proportions in participant attributes. To assess statistical associations between variables, Spearman's rank correlation coefficient (ρ) was calculated for differences in the data between the placebo and LcS periods. Spearman's rank correlation coefficient (ρ) was also calculated for the data at the pre-intervention period. Due to the exploratory character of the study and the clear definition of a single primary outcome parameter, no correction for multiplicity testing was applied. The significance level was set at 0.05, with less than 0.05 considered significant. The EZR and R v. 4.2.0

statistical analysis software (R Foundation for Statistical Computing, Vienna, Austria) were used for the analysis.

3. Results

3.1. Characteristics of Participants

The participants were 12 healthy male and female office workers who consented to participate and were divided into the following two groups: one group received the placebo treatment first, while the other group received the LcS treatment first. There were no significant differences between the two groups in terms of their age, height, weight, JPSS, PSQI, and number of complaints about sleep (Table 1). The rate of compliance for test beverage consumption was calculated as the percentage of the actual intake over a defined period, and the results were 99.7% in the placebo treatment and 100% in the LcS one. No adverse events related to the test beverage intake were identified. Blinding was confirmed at the end of the study based on responses when the participants were asked to guess which beverage they received first.

Table 1. Characteristics of the participants.

	Placebo First	LcS First	Total
Number	2 males, 4 females	3 males, 3 females	5 males, 7 females
Age (years)	47.2 ± 7.8	45.5 ± 5.9	46.3 ± 6.6
Height (cm)	161.75 ± 7.04	166.50 ± 9.89	164.13 ± 8.55
Body weight (kg)	54.66 ± 12.08	58.16 ± 13.22	56.42 ± 12.21
Body mass index	20.71 ± 3.16	20.80 ± 2.89	20.76 ± 2.88
JPSS	18.2 ± 7.5	25.5 ± 11.2	21.8 ± 9.8
PSQI	6.7 ± 1.6	7.3 ± 1.8	7.0 ± 1.7
Complaints about sleep			
Sleep latency	1/6	0/6	1/12
Wake after sleep onset	5/6	4/6	9/12
Waking up too early	3/6	3/6	6/12
Sleep quality	1/6	2/6	3/12
Sleepiness on rising	2/6	2/6	4/12
Daytime sleepiness	3/6	4/6	7/12
Sleep duration	2/6	2/6	4/12

Values are expressed as the means ± standard deviation (quantitative variables) or proportions (complaints about sleep). *Lacticaseibacillus paracasei* strain Shirota (LCS); JPSS, Japanese Perceived Stress Score; PSQI, Pittsburgh Sleep Quality Index.

3.2. Perceived Mood

Daytime mood was assessed using the KSS and VAS items asking about perceived sleepiness, fatigue, motivation, attention, and optimism. The scores were generally better during the LcS intake period than during the placebo intake period in both the morning and afternoon (Figure 3). The VAS score of attention in the afternoon was significantly higher during the LcS intake period than during the placebo intake period (Figure 3e, $p = 0.041$).

3.3. EEG

There was no significant difference in the alpha, beta, or beta/alpha ratio between the treatments, both during the resting state and during the tasks (Figure 4). Theta power (μV^2/min) observed during the resting state with the eyes open and during the oddball task in the afternoon was significantly lower during the LcS intake period than during the placebo intake period ($p = 0.025$ and 0.009, respectively).

Table S1 shows the button-press reaction time and the ERP P300 latency to the target stimulus in auditory oddball tasks. The reaction time during the LcS intake period tended to be shorter when compared to the placebo intake period in both the morning and the afternoon, although this difference was not statistically significant. The efficacy assessment of the ERP P300 latency was available only in the morning measurements because, for the afternoon measurements, we were able to determine the P300 in only 8 participants for both intervention periods (10 in the LcS intake period and 8 in the placebo). The morning P300 latency was shorter during the LcS intake period than during the placebo intake period.

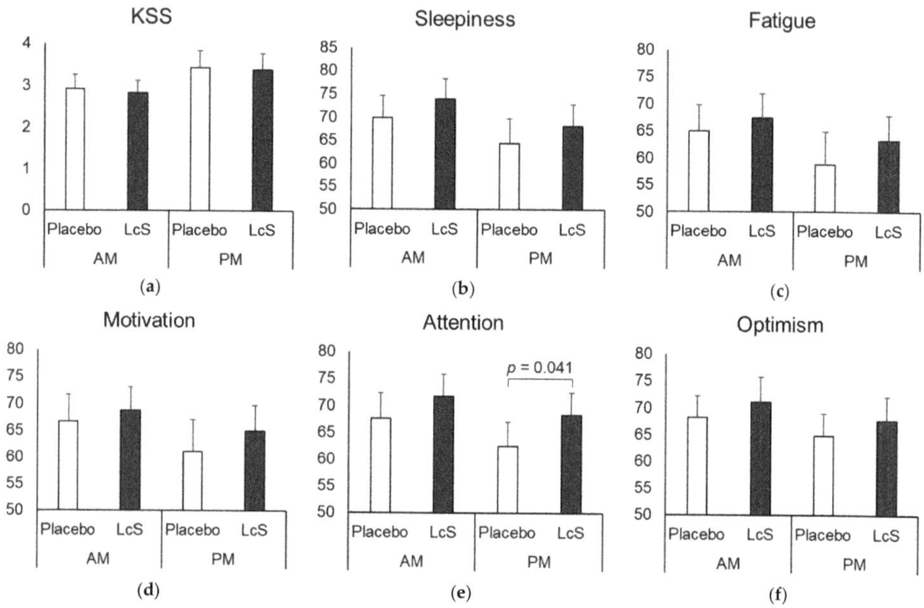

Figure 3. Impact of *Lacticaseibacillus paracasei* strain Shirota (LcS) intake on daytime mood. Participants completed the questionnaires in the morning (AM) and in the afternoon (PM) every day for 5 weekdays during the last week of each 4-week placebo and LcS intervention period. Sleepiness was evaluated with the Karolinska Sleepiness Scale (KSS) (**a**). Perceived sleepiness (**b**), physical fatigue (**c**), motivation (**d**), attention (**e**), and optimism (**f**) were also surveyed using a visual analog scale (VAS) from 0 mm for "severely unfavorable" daytime performance to 100 mm for "supremely favorable". Values are expressed as the means + standard error (n = 12). The Wilcoxon rank sum test was used for the comparison between treatments.

3.4. HRV

Table 2 shows the heart rate and autonomic nerve indices calculated from HRV. Heart rate was lower during the LcS intake period than during the placebo intake period, regardless of the time (AM/PM) and activity (resting/task). No significant difference was observed in either the LF/HF ratio (an index of sympathetic nerve activity) or the HF component (parasympathetic nerve activity), except for the LF/HF ratio during the afternoon oddball task.

3.5. Relationship between the Perceived Mood and Physiological Parameters

We analyzed the relationship between the perceived mood and physiological indices using the differences between the placebo and LcS interventions (LcS-placebo, Figure 5a). The LcS/placebo ratio was used in EEG data. In the morning, heart rate during the resting state with the eyes closed was associated with the KSS (ρ = 0.636; 95% CI 0.193 to 0.880), sleepiness (−0.545; −0.905 to 0.139), motivation (−0.699; −0.914 to −0.156), fatigue (−0.441; −0.871 to 0.169), attention (−0.474; −0.862 to 0.319), and optimism (−0.730; −0.939 to −0.291). In the afternoon, LF/HF during the resting state with eyes closed was associated with the KSS (0.687; 0.256 to 0.889), motivation (−0.531; −0.978 to 0.162), fatigue (−0.552; −0.927 to 0.028), attention (−0.538; −0.865 to 0.036), and optimism (−0.552; −0.900 to 0.068). As for EEG, the theta wave during the resting state with the eyes closed was associated with the KSS

(0.493; −0.228 to 0.919) and attention (−0.497; −0.913 to 0.235). The relationships between the theta wave, heart rate, LF/HF, and attention are also shown graphically (Figure 5b–d). These relationships indicate that physiological indices reflect subjective indices.

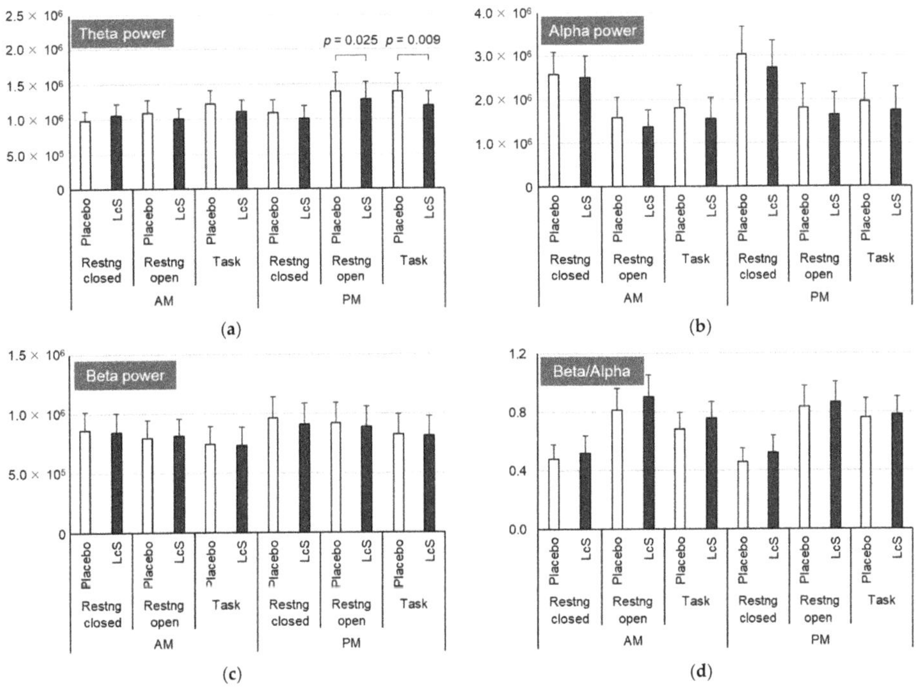

Figure 4. Impact of *Lacticaseibacillus paracasei* strain Shirota (LcS) intake on electroencephalography (EEG). EEG was measured during the last week of each 4-week placebo and LcS intervention period, and the power spectrum (μV^2/min) of theta waves (4–8 Hz) (**a**), alpha waves (8–13 Hz) (**b**), beta waves (13–30 Hz) (**c**), and the beta/alpha ratio (**d**) were calculated during the resting state (eyes open and closed) and during the task both in the morning (AM) and the afternoon (PM). Values are expressed as the means + standard error (n = 12). The Wilcoxon rank sum test was used for the comparison between treatments.

Table 2. Impact of LcS intake on the following HRV metrics: heart rate and autonomic nerve indices.

	Time	Treatment	HR (bpm)	LF/HF	HF (ms^2)
Resting (Eyes closed)	AM	Placebo	67.8 ± 3.1	4.13 ± 1.20	321.1 ± 105.9
		LcS	64.7 ± 3.0 *	3.60 ± 1.12	200.4 ± 37.1
	PM	Placebo	68.0 ± 3.0	3.52 ± 1.45	390.2 ± 215.1
		LcS	66.9 ± 4.0	5.28 ± 3.16	279.3 ± 103.7
Resting (Eyes open)	AM	Placebo	67.5 ± 2.5	6.69 ± 1.86	307.9 ± 129.4
		LcS	66.5 ± 2.8	4.73 ± 1.07	295.5 ± 90.8
	PM	Placebo	68.5 ± 2.7	5.67 ± 1.50	348.0 ± 142.2
		LcS	67.0 ± 3.4	6.59 ± 2.50	352.5 ± 96.4
Task	AM	Placebo	67.4 ± 2.6	4.55 ± 1.33	296.5 ± 63.7
		LcS	65.9 ± 2.9	3.32 ± 1.00	430.2 ± 198.4
	PM	Placebo	68.0 ± 3.1	4.64 ± 1.32	420.9 ± 203.2
		LcS	66.4 ± 3.4	3.24 ± 0.87 *	289.8 ± 70.8

Heart rate variability (HRV) was measured during the last week of each 4-week placebo and *Lacticaseibacillus paracasei* strain Shirota (LcS) intervention period, and heart rate (HR), the ratio of low-frequency (LF) components to high-frequency (HF) components, and HF were calculated in the resting state (with the eyes closed and open) and during the auditory oddball task. Values are expressed as the means ± standard error (n = 12). The Wilcoxon rank sum test was used for the comparison between treatments (*, $p < 0.050$). AM, morning; PM, afternoon.

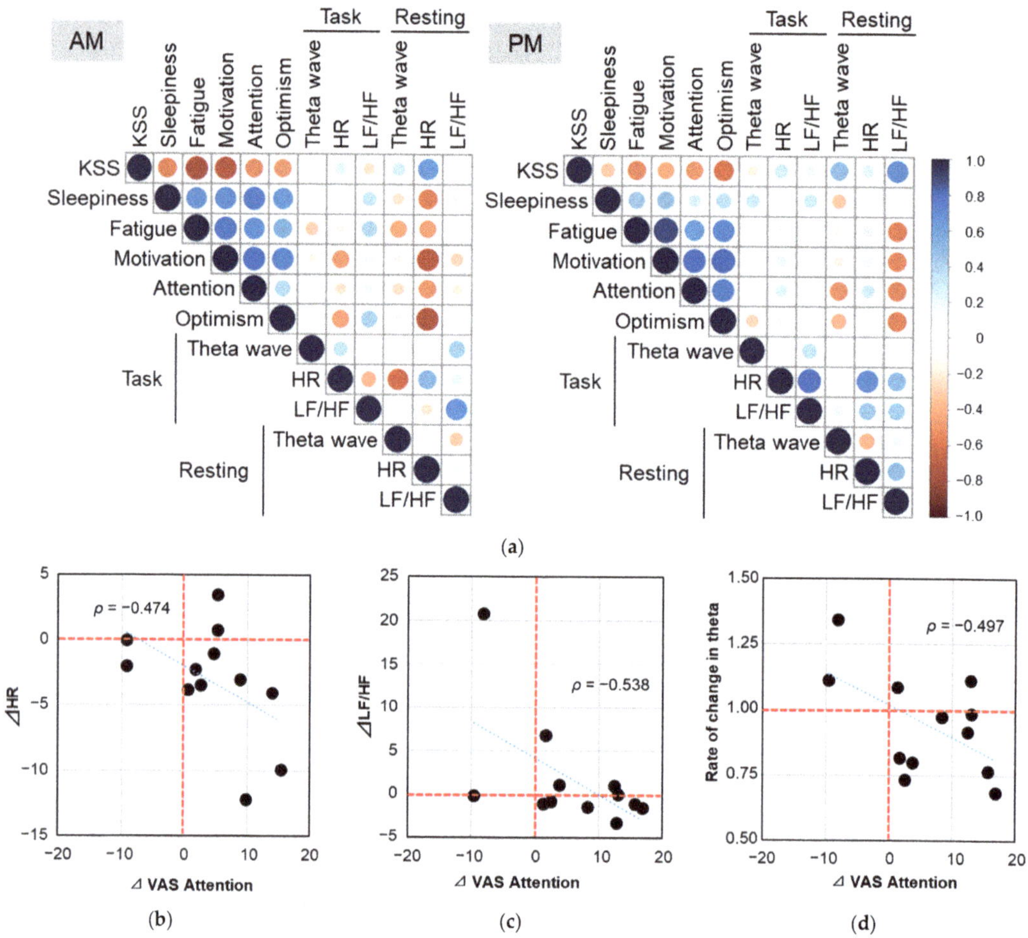

Figure 5. Correlation analysis between perceived daytime mood and physiological indices. The correlation matrix was based on Spearman's rank correlation coefficient (ρ) between the perceived mood and physiological indices using the difference in values of 12 participants between the placebo and *Lacticaseibacillus paracasei* strain Shirota (LcS) intervention periods (**a**). Color intensity and the size of circles are proportional to correlation coefficients. Relationships between the visual analog scale (VAS) attention score and morning heart rate (HR), (**b**) afternoon ratio of low-frequency (LF) components to high-frequency (HF) components (LF/HF), (**c**) and afternoon electroencephalogram theta power (**d**). Plots from 12 participants and the fitted linear trendline (dotted blue line) are shown, and data lying on dashed red lines indicate no difference between the periods. AM, morning; KSS, Karolinska Sleepiness Scale; PM, afternoon.

4. Discussion

We investigated whether LcS affects the daytime performance of healthy office workers with sleep complaints and observed that LcS intake suppresses the decline in perceived attention in the afternoon. This result is considered to be supported by the fact that other VAS items (sleepiness, fatigue, motivation, and optimism) tended to be better during the LcS intake period. All items of VAS were lower in the afternoon than in the morning, suggesting an overall decline in performance in the afternoon. The VAS score of attention in the afternoon was significantly higher during the LcS intake period than during the placebo intake period. The VAS attention score in the afternoon during the LcS intake period was similar to that of the placebo intake period in the morning, indicating that the effect size of LcS was comparable to the range of diurnal variation.

During the LcS intake period, the EEG theta power that appeared during the resting state (eyes open) and when performing tasks in the afternoon was significantly lower than that during the placebo intake period. Correlation analysis showed that the changes in daytime attention with LcS intervention were associated with the changes in theta power in the resting state (eyes closed) in the afternoon ($\rho = -0.497$). These results indicate that LcS increases the arousal level both subjectively and objectively. Spontaneous brain waves are constantly observed, and the degree of arousal is estimated by the intensity of beta, alpha, theta, and delta waves (classified by frequency). Alpha and beta waves are observed in the wakeful state [9], and theta waves increase when the degree of wakefulness is reduced. The beta/alpha ratio during wakefulness has also been widely used as an index of attention and stress in previous reports [44]. The beta/alpha value tended to be higher during LcS intake than with the placebo, supporting an increased arousal level, although this difference was not statistically significant.

The shortened latency of the ERP P300 also supports the results, indicating improved perceived attention. ERP P300 latency is believed to indicate the attentional state of the brain. Prolonged P300 latency is related to aging [34], depression [45], and dementia [46]. The ingestion of polyunsaturated fatty acids [47] or caffeine [48] can shorten P300 latency. However, the ERP P300 peak measured in the afternoon was not detectable in this study, suggesting that P300 characterizes brain traits under good conditions but may not be suitable for assessing brain status under poor conditions. For example, a decrease in P300's amplitude has been reported during sleep deprivation [49]. However, the fact that the LcS intervention showed a tendency for a shorter button-press reaction time in our task provides data supporting increased attention.

Heart rate during the LcS intake period was lower than that during the placebo intake period, regardless of the time of day or if the individual was resting/performing a task, with a significant difference in the resting heart rate (with eyes closed) in the morning. Moreover, changes in heart rate with closed eyes in the morning were correlated with improved mood (positively for the KSS, negatively for motivation and optimism). A reduction in heart rate induced by LcS intervention has also been observed after the intake of a hot water extract of LcS [50], which is consistent with our results. The LF/HF values during the afternoon oddball task were significantly lower during the LcS intake period. The reduction in heart rate and LF/HF with LcS in the present study supports the finding that LcS affects the autonomic nervous system because HRV has been considered a product of emotional response or stress exerted via the autonomic nervous system. These neurological responses are consistent with the ability of LcS to attenuate a stress-induced rise in cortisol in the academic test stress model [30–32]. In an animal study, the intragastric administration of LcS stimulated gastric vagal afferent activity in a dose-dependent manner [32], and it has been shown to suppress stress-induced sympathetic nerve activity via signal transmission through the gastric vagal afferent [51,52]. The results of these animal and human experiments suggest that LcS acts via the autonomic nervous system to improve performance. LcS may also affect mood because several probiotics have been reported to normalize anxiety-like behavior via the vagus nerve [53,54].

Only individuals with sleep complaints were included in this study, and stress was not taken into consideration. The stress levels of the participants enrolled in this study were not very high in terms of JPSS values. LcS has been shown to prevent the worsening of stress and sleep quality under academic stress [30–33], but the effect on stress and sleep may be mild in people with low-stress levels, as in this study. When we analyzed the relationship between perceived mood and other parameters in the pre-intervention period, a high negative correlation was confirmed between the stress state and perceived mood in the morning and afternoon, and the degree of correlation was stronger than that with sleep (Table S2). Our results are consistent with reports that mental and physical stress and sleep disturbances directly and indirectly affect job performance [4,55]. The Japanese Ministry of Health, Labor, and Welfare has reported that half of the Japanese population lives with worry or stress and that one-fourth of working-age people have sleep complaints [56]. The effects of LcS on work performance may be captured more clearly by targeting participants with high-stress levels.

The present study must be seen as an exploratory study in this field, and thus, no formal sample size calculation was performed. We set the sample size to 12, considering the handling capability of the EEG system, but this may be insufficient for making strong conclusions. To eliminate this limitation, we adopted a crossover study design. The crossover design has high power and statistical efficiency, and it is possible to obtain an estimate with the same level of accuracy as a parallel design, even with a smaller number of participants [57]. We confirmed that the crossover trial was conducted properly without any time or order effect. Another concern was the selection of a placebo and its impact when used in the crossover study. We selected non-fermented milk as a placebo, the nutritional content, color, flavor, taste, and pH of which were adjusted to match the test beverage, but it was practically impossible to make the characteristics of the non-fermented milk have complete correspondence with the fermented one, considering the wide variety of metabolites produced by microbial fermentation. Blinding was assured by confirming that the participants could not guess at the end of the study which treatment they received first. However, it cannot be concluded that subtle differences in flavor and texture between the test beverage and the placebo did not have any impact on the study outcomes. We believe that the current study basically provides reliable results on the effect of LcS on daytime performance, and further studies using a randomized, placebo-controlled, double-blind, parallel-group study design with a larger sample size are needed to address such limitations.

5. Conclusions

Our results suggest that LcS intervention improves mood when measured as daytime performance indices compared with the placebo. Furthermore, the daytime perceived mood indices were associated with physiological parameters such as EEG and autonomic nerve activity data.

Supplementary Materials: The following supporting information can be downloaded at: https://www.mdpi.com/article/10.3390/nu15245119/s1, Table S1: Effect of LcS on reaction time and ERP P300 latency in the auditory oddball task. Table S2: Correlation of the daytime mood with other indicators during the pre-intervention period.

Author Contributions: Conceptualization, H.K.-H. and K.M.; Data curation, H.K.-H.; Formal analysis, H.K.-H. and K.M.; Investigation, H.K.-H., H.I., H.N., K.S., Y.G., G.H. and M.T.; Project administration, M.K. and K.M.; Resources, H.K.-H. and H.I.; Supervision, M.K. and K.M.; Validation, H.I.; Visualization, H.K.-H. and K.M.; Writing—original draft, H.K.-H. and K.M.; Writing—review and editing, H.K.-H. and K.M. All authors have read and agreed to the published version of the manuscript.

Funding: This research received no external funding.

Institutional Review Board Statement: The present study was conducted according to the guidelines of the Declaration of Helsinki in its most recent version and approved by the Institutional Review Board of Chiyoda Paramedical Care Clinic, Tokyo, Japan (protocol code: YKH21C4, date of approval: 9 July 2021).

Informed Consent Statement: Written informed consent was obtained from all participants involved in this study.

Data Availability Statement: The data are not available due to the nature of this research. Participants of this study did not agree for their data to be shared publicly.

Acknowledgments: We appreciate Satoshi Matsubara and Kan Shida for their helpful advice. We also thank Osamu Watanabe, Daisuke Nozaki and the members of their teams for the provision of the test beverages.

Conflicts of Interest: H.K.-H., H.I., Y.G., G.H., M.T., M.K. and K.M. are employees of the Yakult Central Institute, Yakult Honsha Co., Ltd. K.S. and H.N. were the employees at the time of this work. H.K.-H., H.I. and K.M. are inventors of a pending patent related to this work (patent applicant: Yakult Honsha Co., Ltd., Tokyo, Japan; application number: Japanese Patent Application No. 2023-152247; status of application: application).

References

1. Selye, H. The general adaptation syndrome and the diseases of adaptation. *J. Clin. Endocrinol. Metab.* **1946**, *6*, 117–230. [CrossRef]
2. Charmandari, E.; Tsigos, C.; Chrousos, G. Endocrinology of the stress response. *Annu. Rev. Physiol.* **2005**, *67*, 259–284. [CrossRef] [PubMed]
3. Chu, B.; Marwaha, K.; Sanvictores, T.; Ayers, D. Physiology, Stress Reaction. In *StatPearls*; StatPearls Publishing: Treasure Island, FL, USA, 2023.
4. Furuichi, W.; Shimura, A.; Miyama, H.; Seki, T.; Ono, K.; Masuya, J.; Inoue, T. Effects of job stressors, stress response, and sleep disturbance on presenteeism in office workers. *Neuropsychiatr. Dis. Treat.* **2020**, *16*, 1827–1833. [CrossRef] [PubMed]
5. Rosekind, M.R.; Gregory, K.B.; Mallis, M.M.; Brandt, S.L.; Seal, B.; Lerner, D. The cost of poor sleep: Workplace productivity loss and associated costs. *J. Occup. Environ. Med.* **2010**, *52*, 91–98. [CrossRef] [PubMed]
6. Ministry of Economy. Trade and Industry of Japan Corporate Health Management Guidebook: Recommendation for Health Promotion through Cooperation and Cooperation. Revised 1st Edition. Available online: https://www.meti.go.jp/policy/mono_info_service/healthcare/kenkokeiei-guidebook2804.pdf (accessed on 24 February 2023).
7. Lo, J.C.; Groeger, J.A.; Santhi, N.; Arbon, E.L.; Lazar, A.S.; Hasan, S.; von Schantz, M.; Archer, S.N.; Dijk, D.-J. Effects of partial and acute total sleep deprivation on performance across cognitive domains, individuals and circadian phase. *PLoS ONE* **2012**, *7*, e45987. [CrossRef]
8. Abe, T.; Komada, Y.; Nishida, Y.; Hayashida, K.; Inoue, Y. Short sleep duration and long spells of driving are associated with the occurrence of Japanese drivers' rear-end collisions and single-car accidents. *J. Sleep Res.* **2010**, *19*, 310–316. [CrossRef]
9. Kaida, K.; Takahashi, M.; Akerstedt, T.; Nakata, A.; Otsuka, Y.; Haratani, T.; Fukasawa, K. Validation of the Karolinska sleepiness scale against performance and EEG variables. *Clin. Neurophysiol.* **2006**, *117*, 1574–1581. [CrossRef] [PubMed]
10. Wutzl, B.; Golaszewski, S.M.; Leibnitz, K.; Langthaler, P.B.; Kunz, A.B.; Leis, S.; Schwenker, K.; Thomschewski, A.; Bergmann, J.; Trinka, E. Narrative Review: Quantitative EEG in disorders of consciousness. *Brain Sci.* **2021**, *11*, 697. [CrossRef]
11. Von Rosenberg, W.; Chanwimalueang, T.; Adjei, T.; Jaffer, U.; Goverdovsky, V.; Mandic, D.P. Resolving ambiguities in the LF/HF ratio: LF-HF scatter plots for the categorization of mental and physical stress from HRV. *Front. Physiol.* **2017**, *8*, 360. [CrossRef]
12. Sawada, Y.; Ohtomo, N.; Tanaka, Y.; Tanaka, G.; Yamakoshi, K.; Terachi, S.; Shimamoto, K.; Nakagawa, M.; Satoh, S.; Kuroda, S.; et al. New technique for time series analysis combining the maximum entropy method and non-linear least squares method: Its value in heart rate variability analysis. *Med. Biol. Eng. Comput.* **1997**, *35*, 318–322. [CrossRef]
13. Al Omran, Y.; Aziz, Q. The brain-gut axis in health and disease. *Adv. Exp. Med. Biol.* **2014**, *817*, 135–153. [CrossRef]
14. Holdeman, L.V.; Good, I.J.; Moore, W.E. Human fecal flora: Variation in bacterial composition within individuals and a possible effect of emotional stress. *Appl. Environ. Microbiol.* **1976**, *31*, 359–375. [CrossRef] [PubMed]
15. Bailey, M.T.; Coe, C.L. Maternal separation disrupts the integrity of the intestinal microflora in infant rhesus monkeys. *Dev. Psychobiol.* **1999**, *35*, 146–155. [CrossRef]
16. Sudo, N.; Chida, Y.; Aiba, Y.; Sonoda, J.; Oyama, N.; Yu, X.-N.; Kubo, C.; Koga, Y. Postnatal microbial colonization programs the hypothalamic-pituitary-adrenal system for stress response in mice. *J. Physiol.* **2004**, *558*, 263–275. [CrossRef] [PubMed]
17. Son, J.S.; Zheng, L.J.; Rowehl, L.M.; Tian, X.; Zhang, Y.; Zhu, W.; Litcher-Kelly, L.; Gadow, K.D.; Gathungu, G.; Robertson, C.E.; et al. Comparison of fecal microbiota in children with autism spectrum disorders and neurotypical siblings in the simons simplex collection. *PLoS ONE* **2015**, *10*, e0137725. [CrossRef]
18. Aizawa, E.; Tsuji, H.; Asahara, T.; Takahashi, T.; Teraishi, T.; Yoshida, S.; Ota, M.; Koga, N.; Hattori, K.; Kunugi, H. Possible association of *Bifidobacterium* and *Lactobacillus* in the gut microbiota of patients with major depressive disorder. *J. Affect. Disord.* **2016**, *202*, 254–257. [CrossRef]
19. Zheng, P.; Zeng, B.; Zhou, C.; Liu, M.; Fang, Z.; Xu, X.; Zeng, L.; Chen, J.; Fan, S.; Du, X.; et al. Gut microbiome remodeling induces depressive-like behaviors through a pathway mediated by the host's metabolism. *Mol. Psychiatry* **2016**, *21*, 786–796. [CrossRef] [PubMed]
20. Huang, R.; Wang, K.; Hu, J. Effect of probiotics on depression: A systematic review and meta-analysis of randomized controlled trials. *Nutrients* **2016**, *8*, 483. [CrossRef]
21. Martínez-González, A.E.; Andreo-Martínez, P. Prebiotics, probiotics and fecal microbiota transplantation in autism: A systematic review. *Rev. Psiquiatr. Salud Ment.* **2020**, *13*, 150–164. [CrossRef]
22. Li, W.; Guo, J.; Shen, Y.; Huang, L.; Leng, B.; Fan, D.; Shui, L.; Chen, C. Probiotics, prebiotics, and synbiotics for the treatment of dementia: Protocol for a systematic review. *Medicine* **2020**, *99*, e18608. [CrossRef]

23. Otaka, M.; Kikuchi-Hayakawa, H.; Ogura, J.; Ishikawa, H.; Yomogida, Y.; Ota, M.; Hidese, S.; Ishida, I.; Aida, M.; Matsuda, K.; et al. Effect of *Lacticaseibacillus paracasei* Strain Shirota on Improvement in Depressive Symptoms, and Its Association with Abundance of Actinobacteria in Gut Microbiota. *Microorganisms* **2021**, *9*, 1026. [CrossRef] [PubMed]
24. Zheng, J.; Wittouck, S.; Salvetti, E.; Franz, C.M.A.P.; Harris, H.M.B.; Mattarelli, P.; O'Toole, P.W.; Pot, B.; Vandamme, P.; Walter, J.; et al. A taxonomic note on the genus *Lactobacillus*: Description of 23 novel genera, emended description of the genus *Lactobacillus* Beijerinck 1901, and union of *Lactobacillaceae* and *Leuconostocaceae*. *Int. J. Syst. Evol. Microbiol.* **2020**, *70*, 2782–2858. [CrossRef] [PubMed]
25. U.S. Food and Drug Administration. Agency Response Letter GRAS Notice No. GRN 000429. Available online: https://www.fda.gov/downloads/Food/IngredientsPackagingLabeling/GRAS/NoticeInventory/UCM309143.pdf (accessed on 30 October 2023).
26. Matsumoto, K.; Takada, T.; Shimizu, K.; Moriyama, K.; Kawakami, K.; Hirano, K.; Kajimoto, O.; Nomoto, K. Effects of a probiotic fermented milk beverage containing *Lactobacillus casei* strain Shirota on defecation frequency, intestinal microbiota, and the intestinal environment of healthy individuals with soft stools. *J. Biosci. Bioeng.* **2010**, *110*, 547–552. [CrossRef] [PubMed]
27. Nagata, S.; Asahara, T.; Wang, C.; Suyama, Y.; Chonan, O.; Takano, K.; Daibou, M.; Takahashi, T.; Nomoto, K.; Yamashiro, Y. The effectiveness of *Lactobacillus* beverages in controlling infections among the residents of an aged care facility: A randomized placebo-controlled double-blind trial. *Ann. Nutr. Metab.* **2016**, *68*, 51–59. [CrossRef] [PubMed]
28. Shida, K.; Sato, T.; Iizuka, R.; Hoshi, R.; Watanabe, O.; Igarashi, T.; Miyazaki, K.; Nanno, M.; Ishikawa, F. Daily intake of fermented milk with *Lactobacillus casei* strain Shirota reduces the incidence and duration of upper respiratory tract infections in healthy middle-aged office workers. *Eur. J. Nutr.* **2017**, *56*, 45–53. [CrossRef]
29. Nagao, F.; Nakayama, M.; Muto, T.; Okumura, K. Effects of a fermented milk drink containing *Lactobacillus casei* strain Shirota on the immune system in healthy human subjects. *Biosci. Biotechnol. Biochem.* **2000**, *64*, 2706–2708. [CrossRef]
30. Kato-Kataoka, A.; Nishida, K.; Takada, M.; Suda, K.; Kawai, M.; Shimizu, K.; Kushiro, A.; Hoshi, R.; Watanabe, O.; Igarashi, T.; et al. Fermented milk containing *Lactobacillus casei* strain Shirota prevents the onset of physical symptoms in medical students under academic examination stress. *Benef. Microbes* **2016**, *7*, 153–156. [CrossRef]
31. Kato-Kataoka, A.; Nishida, K.; Takada, M.; Kawai, M.; Kikuchi-Hayakawa, H.; Suda, K.; Ishikawa, H.; Gondo, Y.; Shimizu, K.; Matsuki, T.; et al. Fermented milk containing *Lactobacillus casei* strain Shirota preserves the diversity of the gut microbiota and relieves abdominal dysfunction in healthy medical students exposed to academic stress. *Appl. Environ. Microbiol.* **2016**, *82*, 3649–3658. [CrossRef]
32. Takada, M.; Nishida, K.; Kataoka-Kato, A.; Gondo, Y.; Ishikawa, H.; Suda, K.; Kawai, M.; Hoshi, R.; Watanabe, O.; Igarashi, T.; et al. Probiotic *Lactobacillus casei* strain Shirota relieves stress-associated symptoms by modulating the gut–brain interaction in human and animal models. *Neurogastroenterol. Motil.* **2016**, *28*, 1027–1036. [CrossRef]
33. Takada, M.; Nishida, K.; Gondo, Y.; Kikuchi-Hayakawa, H.; Ishikawa, H.; Suda, K.; Kawai, M.; Hoshi, R.; Kuwano, Y.; Miyazaki, K.; et al. Beneficial effects of *Lactobacillus casei* strain Shirota on academic stress-induced sleep disturbance in healthy adults: A double-blind, randomised, placebo-controlled trial. *Benef. Microbes* **2017**, *8*, 153–162. [CrossRef]
34. Van Dinteren, R.; Arns, M.; Jongsma, M.L.A.; Kessels, R.P.C. P300 development across the lifespan: A systematic review and meta-analysis. *PLoS ONE* **2014**, *9*, e87347. [CrossRef] [PubMed]
35. Kropotov, J.; Ponomarev, V.; Tereshchenko, E.P.; Müller, A.; Jäncke, L. Effect of aging on ERP components of cognitive control. *Front. Aging Neurosci.* **2016**, *8*, 69. [CrossRef]
36. Hellerstedt, R.; Moccia, A.; Brunskill, C.M.; Bowman, H.; Bergström, Z.M. Aging reduces EEG markers of recognition despite intact performance: Implications for forensic memory detection. *Cortex* **2021**, *140*, 80–97. [CrossRef] [PubMed]
37. Ganesan, S.; Magee, M.; Stone, J.E.; Mulhall, M.D.; Collins, A.; Howard, M.E.; Lockley, S.W.; Rajaratnam, S.M.W.; Sletten, T.L. The impact of shift work on sleep, alertness and performance in healthcare workers. *Sci. Rep.* **2019**, *9*, 4635. [CrossRef] [PubMed]
38. Lima, N.C.; Kirov, R.; de Almondes, K.M. Impairment of executive functions due to sleep alterations: An integrative review on the use of P300. *Front. Neurosci.* **2022**, *16*, 906492. [CrossRef] [PubMed]
39. Akerstedt, T.; Gillberg, M. Subjective and objective sleepiness in the active individual. *Int. J. Neurosci.* **1990**, *52*, 29–37. [CrossRef] [PubMed]
40. Buysse, D.J.; Reynolds, C.F.; Monk, T.H.; Berman, S.R.; Kupfer, D.J. The Pittsburgh Sleep Quality Index: A new instrument for psychiatric practice and research. *Psychiatry Res.* **1989**, *28*, 193–213. [CrossRef]
41. Iwahashi, S.; Tanaka, Y.; Fukudo, S.; Hongo, M. The development of the Japanese version of the perceived stress scale. *Jpn. Soc. Psycosom. Med.* **2002**, *42*, 459–466. [CrossRef]
42. Shimokouch, M.; Kuroiw, Y.; Tobimatsu, S.; Ichikawa, G.; Yoshikawa, H.; Yamakawa, T.; Kakimoto, R.; Shibazaki, H.; Ozaki, I.; Sonoo, M.; et al. Guidelines for evoked potentials. *Jpn. J. Electroencephalogr. Electromyogr.* **1998**, *26*, 185–200.
43. Ono, Y. Frequency analysis of spontaneous brain waves. In *Introduction to Biological Processing with MATLAB*; Corona Publishing Co., Ltd.: Tokyo, Japan, 2018; pp. 24–43. ISBN 978-4-339-07245-7.
44. Takahashi, T.; Murata, T.; Hamada, T.; Omori, M.; Kosaka, H.; Kikuchi, M.; Yoshida, H.; Wada, Y. Changes in EEG and autonomic nervous activity during meditation and their association with personality traits. *Int. J. Psychophysiol.* **2005**, *55*, 199–207. [CrossRef]
45. Vandoolaeghe, E.; van Hunsel, F.; Nuyten, D.; Maes, M. Auditory event related potentials in major depression: Prolonged P300 latency and increased P200 amplitude. *J. Affect. Disord.* **1998**, *48*, 105–113. [CrossRef]

46. Yamaguchi, S.; Tsuchiya, H.; Yamagata, S.; Toyoda, G.; Kobayashi, S. Event-related brain potentials in response to novel sounds in dementia. *Clin. Neurophysiol.* **2000**, *111*, 195–203. [CrossRef]
47. Tokuda, H.; Sueyasu, T.; Kontani, M.; Kawashima, H.; Shibata, H.; Koga, Y. Low doses of long-chain polyunsaturated fatty acids affect cognitive function in elderly Japanese men: A randomized controlled trial. *J. Oleo Sci.* **2015**, *64*, 633–644. [CrossRef]
48. Pan, J.; Takeshita, T.; Morimoto, K. Acute caffeine effect on repeatedly measured P300. *Environ. Health Prev. Med.* **2000**, *5*, 13–17. [CrossRef] [PubMed]
49. Morris, A.M.; So, Y.; Lee, K.A.; Lash, A.A.; Becker, C.E. The P300 event-related potential. The effects of sleep deprivation. *J. Occup. Med.* **1992**, *34*, 1143–1152. [PubMed]
50. Nakajima, K.; Hata, Y.; Osono, Y.; Hamura, M.; Kobayashi, S.; Watanuki, M. Antihypertensive Effect of Extracts of *Lactobacillus casei* in Patients with Hypertension. *J. Clin. Biochem. Nutr.* **1995**, *18*, 181–187. [CrossRef]
51. Tanida, M.; Imanishi, K.; Akashi, H.; Kurata, Y.; Chonan, O.; Naito, E.; Kunihiro, S.; Kawai, M.; Kato-Kataoka, A.; Shibamoto, T. Injection of *Lactobacillus casei* strain Shirota affects autonomic nerve activities in a tissue-specific manner, and regulates glucose and lipid metabolism in rats. *J. Diabetes Investig.* **2014**, *5*, 153–161. [CrossRef] [PubMed]
52. Tanida, M.; Takada, M.; Kato-Kataoka, A.; Kawai, M.; Miyazaki, K.; Shibamoto, T. Intragastric injection of *Lactobacillus casei* strain Shirota suppressed spleen sympathetic activation by central corticotrophin-releasing factor or peripheral 2-deoxy-d-glucose in anesthetized rats. *Neurosci. Lett.* **2016**, *619*, 114–120. [CrossRef] [PubMed]
53. Bravo, J.A.; Forsythe, P.; Chew, M.V.; Escaravage, E.; Savignac, H.M.; Dinan, T.G.; Bienenstock, J.; Cryan, J.F. Ingestion of *Lactobacillus* strain regulates emotional behavior and central GABA receptor expression in a mouse via the vagus nerve. *Proc. Natl. Acad. Sci. USA* **2011**, *108*, 16050–16055. [CrossRef] [PubMed]
54. Bercik, P.; Park, A.J.; Sinclair, D.; Khoshdel, A.; Lu, J.; Huang, X.; Deng, Y.; Blennerhassett, P.A.; Fahnestock, M.; Moine, D.; et al. The anxiolytic effect of *Bifidobacterium longum* NCC3001 involves vagal pathways for gut-brain communication. *Neurogastroenterol. Motil.* **2011**, *23*, 1132–1139. [CrossRef]
55. Ishibashi, Y.; Shimura, A. Association between work productivity and sleep health: A cross-sectional study in Japan. *Sleep Health* **2020**, *6*, 270–276. [CrossRef] [PubMed]
56. Ministry of Health, Labour and Welfare of Japan Comprehensive Survey of Living Conditions. 2019. Available online: https://www.mhlw.go.jp/toukei/list/20-21.html (accessed on 24 February 2023).
57. Dwan, K.; Li, T.; Altman, D.G.; Elbourne, D. CONSORT 2010 statement: Extension to randomised crossover trials. *BMJ* **2019**, *366*, l4378. [CrossRef] [PubMed]

nutrients

MDPI

Article

Effect of Continuous Ingestion of Bifidobacteria and Inulin on Reducing Body Fat: A Randomized, Double-Blind, Placebo-Controlled, Parallel-Group Comparison Study

Yuhei Baba [1,*], Yasuo Saito [2], Mei Kadowaki [2], Naoki Azuma [2] and Daisuke Tsuge [3]

[1] Dairy Business Division, Ezaki Glico Co., Ltd., 4-6-5 Utajima, Nishiyodogawa-Ku, Osaka 555-8502, Japan
[2] R&D Laboratory, Ezaki Glico Co., Ltd., 4-6-5 Utajima, Nishiyodogawa-Ku, Osaka 555-8502, Japan; yasuo.saito@glico.com (Y.S.); mei.kadowaki@glico.com (M.K.); naoki.azuma@glico.com (N.A.)
[3] Shinagawa Season Terrace Health Care Clinic, Shinagawa Season Terrace (5F), 1-2-70 Konan, Minato-Ku, Tokyo 108-0075, Japan; shibaura@sempos.or.jp
* Correspondence: yuhei.baba@glico.com; Tel.: +81-80-9596-0792

Abstract: *Bifidobacterium animalis* subsp. *lactis* GCL2505 has been shown to have several positive health effects, including improved defecation frequency and reduced visceral fat. It is known that combined intake of GCL2505 and inulin increases the total number of bifidobacteria compared with ingestion of GCL2505 alone. This randomized, double-blind, placebo-controlled, parallel-group study was conducted to confirm that consumption of GCL2505 and inulin reduces abdominal fat (n = 120). Participants consumed a test beverage containing 1×10^{10} colony-forming units of GCL2505 per 100 g and 2.0 g of inulin per 100 g for 12 weeks. A change in the visceral fat area (VFA) was set as the primary endpoint. There were significant reductions in VFA and total fat area. The intervention significantly increased the total number of bifidobacteria and affected the levels of several lipid markers. Regression analysis of bifidobacteria and measured parameters showed that total bifidobacteria correlated with VFA and body mass index (BMI), while endogenous bifidobacteria and *Bifidobacterium animalis* subsp. *lactis* correlated only with BMI, suggesting that increases in both contributed to the decrease in VFA. These results suggest that combined intake of GCL2505 and inulin improves the intestinal environment and reduces abdominal fat in association with the SCFA-mediated pathway.

Keywords: *Bifidobacterium animalis* subsp. *lactis*; inulin; synbiotics; abdominal fat; overweight; gut microbacteria

1. Introduction

The World Health Organization defines overweight and obesity as abnormal or excessive fat accumulation that has the potential to negatively impact health. Obesity is a risk factor for lifestyle diseases such as hypertension, dyslipidemia and diabetes, as well as non-communicable diseases such as cardiovascular diseases (mainly heart disease and stroke), musculoskeletal diseases (especially osteoarthritis) and some cancers. The worldwide prevalence of obesity nearly tripled between 1975 and 2016, with more than 650 million adults becoming obese. Overweight and obesity are linked to more deaths worldwide compared with underweight [1]. Thus, the obesity epidemic is one of the greatest public health challenges of the twenty-first century. Abdominal visceral fat accumulation (abdominal obesity) is a form of obesity related to environmental factors such as diet and physical inactivity and is also an underlying component of metabolic syndrome, which is a risk factor for coronary heart disease, hypertension, type-2 diabetes and impaired glucose tolerance. Furthermore, accumulation of abdominal visceral fat is considered to have greater negative health implications compared with obesity in general [2]. Caloric restriction and exercise are commonly used to treat obesity. For severe obesity, bariatric surgery and pharmacotherapy may also be used, but these have issues such as invasiveness and continuity. Therefore, there is a need for treatments such as functional foods, which do not require major lifestyle changes and are easier to apply to daily life [3].

The fundamental cause of overweight and obesity is an energy imbalance between calories ingested and calories expended [1]. In addition, it has been reported that the gut microbiota and obesity are closely related, with the composition and diversity of gut microbiota altered in overweight and diabetes [4]. Accordingly, probiotics and prebiotics may offer one approach to treating overweight and obesity by regulating the gut microbiota. Probiotics are defined as "live microorganisms that, when administered in adequate amounts, confer a health benefit on the host" [5]. Lactic acid bacteria (especially *Lactobacillus* spp.) and bifidobacteria have been reported as probiotics in the treatment of obesity. Consumption of *Lactobacillus gasseri* SBT2055 has been shown to reduce body weight, body fat percentage, body fat mass, visceral fat mass, subcutaneous fat mass, waist circumference, hip circumference, waist-to-hip ratio (WHR) and triglycerides (TG) [6,7], while consumption of strain BNR17 has been shown to reduce visceral fat mass [8]. Consumption of *Lactiplantibacillus plantarum* DAD-13 has been shown to reduce body weight and body mass index (BMI) [9], while strain K50 has been shown to reduce total cholesterol (TC) and TG [10], and strain LMT-1-40 has been shown to reduce body fat mass and abdominal VFA and to change insulin-related parameters [11]. In addition, it has been reported that consumption of *Bifidobacterium breve* B-3 decreases body fat mass, body fat percentage and abdominal VFA [12]. Consumption of *Bifidobacterium animalis* subsp. *lactis* CECT8145 decreases BMI, waist circumference and waist circumference/height [13], while consumption of strain GCL 2505 decreases visceral fat [14]. Furthermore, several meta-analyses have suggested an association of probiotics with glucose metabolism, lipid metabolism, body fat mass, body weight, visceral adipose tissue and hepatic adiposity [15–17]. Meanwhile, prebiotics are nonviable food components that confer a health benefit on the host and are associated with modulation of the gut microbiota [18]. Multiple meta-analyses [19–21] have shown that prebiotics reduce body weight, BMI, body fat, fasting glucose, insulin and fasting TG. However, the efficacy of a single probiotic depends on various factors, including diet and indigenous bacteria [22–24]. In addition, the effect of prebiotics is influenced by the gut microbiota of the individual and its sugar capitalization [24,25]. For these reasons, there may be limitations to using single probiotics or prebiotics to treat obesity in diverse populations. The combination of probiotics and prebiotics is called synbiotics [26], and it has been reported that intake of synbiotics increases gut bifidobacteria [27,28]; even in subjects with low probiotic strain growth, total bifidobacteria counts increased due to an increase in endogenous bifidobacteria [29]. Furthermore, several animal studies have reported that intake of synbiotics acts synergistically against obesity [30,31]. Clinical trials investigating the health benefits of synbiotics in a variety of areas found the following benefits. For intestinal health, there was a reduction in abdominal pain frequency [32] as well as an overall improvement in symptoms of irritable bowel syndrome [33]. For obesity and metabolic diseases, reductions in body weight, BMI [34] and body fat percentage [35] as well as improvements in inflammatory markers [36] have been reported. Improvements in the stress response [37] as well as reductions in tension and drowsiness [38] have also been reported. Therefore, it may be beneficial to consider a synbiotic or multi-strain probiotic approach with a view toward providing benefits to a broader population.

Bifidobacterium animalis subsp. *lactis*, a probiotic strain, is commonly used in fermented dairy products and has shown numerous health benefits related to gastrointestinal and immune health [39,40]. *Bifidobacterium animalis* subsp. *lactis* GCL2505 is a probiotic strain isolated from the feces of healthy adults that can grow in the gut [41,42] and it is used in fermented milk products in Japan. In clinical trials, daily consumption of fermented milk containing 8×10^{10} colony-forming units (CFU) of GCL2505, which contains higher numbers of bifidobacteria compared with regular fermented milk, reduced abdominal VFA [14]. Also, combined intake of GCL2505 and inulin was shown to be more effective in increasing the total number of bifidobacteria compared with ingestion of GCL2505 alone [29]. Based on these findings, this study evaluated the effects of the synbiotic intake of GCL2505 and inulin on abdominal fat accumulation in overweight Japanese adults in a placebo-controlled, randomized, double-blind, parallel-group study.

2. Materials and Methods

2.1. Participants

Participants were Japanese men and women between the ages of 20 and 65 years at the time of consent, who satisfied the inclusion criteria, did not satisfy any of the exclusion criteria and were deemed eligible to participate by the principal investigator. The inclusion criteria were as follows: (1) BMI of 23 or higher and less than 30 at the screening test; (2) able to abstain from alcohol for 2 days before each measurement; and (3) fully informed of the purpose and content of the study, deemed to have the capacity to consent, volunteered of their own accord to participate in the study based on a thorough understanding of the purpose and content of the study and provided written informed consent to participate in the study. Exclusion criteria were as follows: (1) regularly taking medications that affect obesity, hyperlipidemia or lipid metabolism; (2) treatment for severe kidney disease, heart disease, respiratory disease, endocrine disease, diabetes or other illness (excluding transient illnesses such as colds); (3) unable to stop intake of health foods or supplements that affect obesity, hyperlipidemia or lipid metabolism; (4) unable to restrict the intake of foods that might affect the intestinal microbiota; (5) use of antibiotics within 1 month prior to the start of the study; (6) regularly use intestinal drugs and laxatives (including strong laxatives); (7) history of digestive surgery (excluding appendectomy); (8) history of allergy to any of the study food ingredients; (9) current or former drug or alcohol dependence; (10) presence of metal (e.g., surgical implants) that precludes computed tomography (CT) scans of the measurement site; (11) implanted medical devices such as cardiac pacemakers and cardioverter-defibrillators; (12) claustrophobia; (13) pregnant or lactating, or expecting to become pregnant during the study; (14) participation in research involving the ingestion of other foods or the use of pharmaceuticals, the application of cosmetics or pharmaceuticals or participation in other research while participating in this study; and (15) deemed ineligible by the principal investigator.

2.2. Test Foods

The test products were a dairy drink (active drink) containing inulin (Orafti GR; BENEO GmbH, Mannheim, Germany) and GCL2505 or placebo. The active drink contained 1×10^{10} CFU of GCL2505 and 2.0 g of inulin per 100 g. The placebo was prepared using the same ingredients as the active drink, with the addition of food-grade acetic acid and lactic acid to adjust the flavor and pH; the basic ingredients were skim milk powder, fructose, dextrose, sucrose, yeast extract, acidifier, stabilizer and flavoring. The nutritional details of the test products are shown in Table 1.

Table 1. Nutritional details of the test drinks.

Parameter	Placebo Group	Active Group
Energy, kcal/100 g	47.0	52.0
Moisture, g/100 g	87.0	84.8
Protein, g/100 g	2.8	2.8
Fat, g/100 g	0.1	0.1
Carbohydrate, g/100 g	9.0	11.2
Ash, g/100 g	1.1	1.1

The active drink contained 2.0 g inulin and 1.0×10^{10} colony-forming units GCL2505.

2.3. Experimental Design

This was a randomized, placebo-controlled, double-blind, parallel-group study. Participants were randomized by computer-generated randomization into two groups (1:1), with age at screening, sex, body weight, body fat rate and VFA serving as stratification factors for randomization in block sizes of four. The controller (allocation manager) assigned the two groups to the test drink intake group and the control food intake group. For the sample size, the final target number of subjects was set at 60, referring to previous reports on visceral fat reduction with probiotics [14,43,44]. Participants in the active and placebo groups consumed 100 g of dairy beverage once daily for 12 weeks. Both the participants and observers were blinded to the group allocation for the duration of the study. Double blinding was accomplished by labeling the test drink with only an identification number. The change in VFA between weeks 0 and 12 was set as the primary endpoint. The secondary endpoints were VFA between weeks 0 and 8, subcutaneous fat area (SFA) between weeks 0, 8 and 12, total fat area (TFA) between weeks 0, 8 and 12, body weight, BMI, body fat rate, waist circumference, hip circumference, WHR, TC, low-density lipoprotein cholesterol (LDL-C), HDL-C, triglyceride, free fatty acid and fecal bifidobacteria. The study was conducted at the Shinagawa Season Terrace Health Care Clinic (Tokyo, Japan) from October 2022 to April 2023 by K.S.O. Corporation (Tokyo, Japan), a contract research organization, and was registered with the University Hospital Medical Information Network Clinical Trials Registry (UMIN-CTR) "http://www.umin.ac.jp/ctr/index.htm (accessed on 26 October 2022)" as UMIN000049328. This article conforms to the Consolidated Standards of Reporting Trials (CONSORT) 2010 guidelines (Supplementary Materials, Table S1).

2.4. Abdominal Fat Area

Abdominal VFA and SFA were measured using CT. Four-slice CT images (120 kVp, 400 mAs tube current, 5.0 mm slice thickness and 420 mm field of view) were acquired at the level of the lumbar 4 vertebra. Abdominal VFA, SFA and TFA (i.e., visceral + subcutaneous) were measured using Fat Scan ver. 4 (East Japan Institute of Technology Co., Ltd., Hitachi, Japan). To avoid unnecessary radiation exposure, CT scans were conducted only once at each measurement point (0, 8 and 12 weeks). The measurement of VFA by CT is reported to be easily affected by the slice site as well as the respiration phase of the subject [45]. Therefore, to investigate the time course changes in VFA accurately, the scanner and principal investigator strictly assessed a series of CT images obtained from the same subjects at each measurement point, treating any inappropriate data as missing values.

2.5. Anthropometric Measures and Body Composition

Weight and height were measured to the nearest 0.1 kg and 0.1 cm, respectively, with the participant standing. BMI was calculated in the standard way: weight (kg) divided by the square of height (m). Waist and hip circumferences were measured to the nearest 0.1 cm in a standing position. Waist circumference was measured around the abdomen at the level of the umbilicus. Hip circumference was measured at the level of maximum extension of the buttocks posteriorly in a horizontal plane.

2.6. Clinical Parameters

Blood pressure, pulse rate and the concentrations of biochemical and hematological parameters in plasma were measured at weeks 0, 4, 8 and 12. The concentrations of urinary parameters were also measured at weeks 0 and 12. Blood samples were drawn from each participant after 10 h of no food or drink except water (fasting) prior to testing. Clinical parameters included hematological tests (white blood cell count, red blood cell count, hemoglobin, hematocrit, platelet count and leukogram), biochemical tests (total protein, albumin, total bilirubin, aspartate aminotransferase, alanine aminotransferase, lactate dehydrogenase, alkaline phosphatase, gamma-glutamyltransferase, urea nitrogen, creatinine, uric acid, sodium, chlorine, potassium, calcium, TC, LDL-C, HDL-C, TG, glucose and HbA1c [NGSP]) and urinalysis (protein, sugar, bilirubin, urinary ketone bodies, occult blood reaction, urobilinogen, pH and specific gravity). All of these tests were performed at LSI Medience Corporation (Tokyo, Japan).

2.7. Fecal Samples

Fecal samples were submitted at weeks 0, 8 and 12. Fecal samples were collected at home between 7 and 2 days before the specified visit. The submitted samples were promptly transported to the Kyoto Institute of Nutrition and Pathology (Kyoto, Japan) by refrigerated transport at temperatures below −15°C.

2.8. Fecal DNA Extraction

Bacterial DNA was extracted from fecal samples, using the ISOSPIN Fecal DNA Kit (Nippon Gene Co., Ltd., Tokyo, Japan), following the procedure of Tourlousse et al. [46]. Specifically, a 0.2 g fecal sample, 700 μL of FE1 buffer and 10 μL of RNase were added to a tube with attached beads. A bead-beating homogenizer (FastPrep-24; MP Biomedicals, Irvine, CA, USA) was used at a rate of 6 m/s for 1 min to crush the cells. The process was repeated three times, during which the sample was kept at room temperature for 5 min. Then, 90 μL of FE2 buffer was added and the samples were centrifuged at 12,000× *g* for 15 min. The supernatant (up to 500 μL) was collected and mixed with FB buffer and isopropanol, each at 0.4× the volume of the supernatant obtained. Finally, the sample was loaded onto a spin column and washed according to the manufacturer's instructions. Purified DNA was eluted with 50 μL of Tris-EDTA buffer (pH 8.0).

2.9. Fecal Bifidobacteria

Bacterial DNA was extracted from 10-fold dilutions of the fecal samples, and the number of gut bifidobacteria was determined by quantitative real-time PCR using bifidobacteria species- and subspecies-specific primers according to a procedure described previously [47]. Total counts of bifidobacteria in the fecal samples are represented as the sum of 10 species (*B. longum* subsp. *longum*, *B. adolescentis*, *B. catenulatum*, *B. pseudocatenulatum*, *B. breve*, *B. bifidum*, *B. longum* subsp. *infantis*, *B. dentium*, *B. angulatum* and *B. animalis* subsp. *lactis*). Endogenous bifidobacteria were regarded as the sum of nine species, without *B. animalis* subsp. *lactis*. The detection limit of each species or subspecies was 2.0×10^5 cells per gram of feces.

2.10. Statistical Analysis

All measurements are expressed as mean, standard deviation (SD) and standard error (SE). Statistical analyses were performed using IBM® SPSS® Statistics 27 (IBM Corp., Armonk, NY, USA) or R® 4.2.1 (R Foundation for Statistical Computing, Vienna, Austria). A p-value < 0.05 was used as the threshold for determining significance. As basic statistics, means, SDs and SEs are expressed to the nearest significant digit and percentages are expressed to one decimal place, with digits adjusted by rounding. Missing data were treated as missing values and no surrogate values were used. Statistical analysis of VFA, SFA, TFA and fecal bifidobacterium was performed with unpaired t-tests, using the Benjamini–Hochberg procedure in order to compare between the active and placebo groups at each examination time. In addition, statistical analyses were performed with paired t-tests, using the Benjamini–Hochberg procedure in order to compare the test results at the start of intake (week 0) with those at 8 and 12 weeks after intake. Fecal bifidobacteria counts were converted to ordinary logarithms before performing statistical analysis. For VFA, SFA and TFA, an intergroup comparison was performed using two-factor repeated-measures analysis of variance (ANOVA) with the actual values. For the other items, comparisons between the active and placebo groups at each examination time were statistically analyzed with an unpaired t-test (two-tailed). In addition, statistical analysis was performed with paired t-tests to compare the test results at the start of the intake with those at 4, 8 and 12 weeks after intake. Regression analysis to correlate bifidobacteria counts with body composition parameters and biomarkers of obesity were performed by applying ANOVA to a mixed linear model, with bifidobacteria count as the objective variable, sex, age, BMI, VFA, SFA, TG, TC, LDL-C, HDL-C, treatment group (active or placebo) and time point (0, 8 and 12 weeks) as explanatory variables and participant ID as a random variable. The lmer function of the R package lmerTest, version 3.1-3, was used for these analyses.

3. Results

3.1. Analysis of the Participant Population

The flow chart of study participation is shown in Figure 1. A total of 473 participants were screened for this study. After screening, 120 participants were eligible: 60 were assigned to the active group and 60 were assigned to the placebo. A significant difference in basophil ratio between the active and placebo groups was observed at the beginning of the study but was deemed acceptable because it was within the reference range. For the other items, there were no differences in the baseline characteristics of the participants' data (Table 2). Dietary consumption for the three days prior to the measurement is summarized in Table 3. It was concluded that the results of dietary consumption did not significantly affect the results of this study. The fat intake of the active group at week 12 was significantly less than that of the placebo group. The difference in mean fat intake between the active and placebo groups was 8.2 g/day or 73.8 kcal/day in terms of calories. Considering the energy intake recommended for the participants in this study [48], this change was only 2.8–3.8% of the daily energy intake. In addition, the difference between groups in the degree of change in VFA, discussed below, was confirmed from week 8; it was determined that this difference between groups, confirmed at week 12, did not have a significant impact on the study. Also, energy and protein intakes at week 12 were significantly reduced compared with week 0 in both the active and placebo groups. Changes in energy and protein intake from baseline in the active and placebo groups were determined not to have affected the study results because there were no significant differences between the groups. Carbohydrate and fiber intakes at week 12 in the placebo group were significantly reduced compared with baseline but this was determined not to have affected the study results because there were no significant differences between groups. By the end of the study, one participant from the active group withdrew due to an illness unrelated to the study that may have affected the results, and one participant from the placebo group withdrew for personal reasons. After the completion of the entire study, one participant from the placebo group was excluded due to a confirmed illness unrelated to the study that may have affected the results. In addition, three participants were excluded because they were found to have consumed drugs or foods during the study period that might have affected the results ($n = 1$ from the active group and $n = 2$ from the placebo group). Thus, a total of 114 patients (58 in the active group and 56 in the placebo group) were included in the analysis. There were no reported harms or unintended effects in either group.

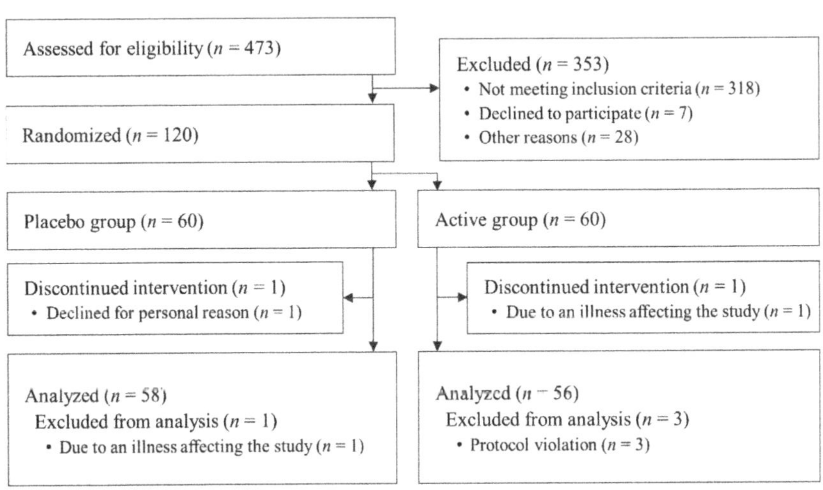

Figure 1. Flowchart of participant selection.

Table 2. Baseline characteristics of the participants (placebo group: n = 60; active group: n = 60).

	Characteristic	Placebo Group	Active Group	p-Value
Clinical findings	Age, years	50.6 (8.0)	50.6 (8.4)	0.973
	Height, cm	165.9 (8.2)	167.8 (8.5)	0.221
	Body weight, kg	73.6 (7.9)	73.8 (8.2)	0.873
	Body mass index, kg/m^2	26.7 (1.5)	26.2 (2.0)	0.129
	Waist circumference, cm	92.5 (5.1)	92.4 (6.2)	0.902
	Hip circumference, cm	98.0 (3.6)	98.2 (4.5)	0.840
	Waist-to-hip ratio	0.9 (0.04)	0.9 (0.04)	0.720
	Visceral fat area, cm^2	124.9 (31.0)	124.0 (31.3)	0.864
	Subcutaneous fat area, cm^2	206.0 (58.1)	207.7 (65.2)	0.881
	Total fat area, cm^2	330.9 (61.6)	331.7 (69.8)	0.953
	Systolic blood pressure, mmHg	127.7 (11.2)	132.4 (15.1)	0.055
	Diastolic blood pressure, mmHg	79.0 (10.6)	82.3 (11.7)	0.108
	Heartbeat, bpm	77.5 (10.5)	78.4 (10.8)	0.644
Laboratory findings	White blood cell count, /μL	6185.0 (1449)	6076.7 (1077)	0.643
	Red blood cell count, ×10^4/μL	495.8 (45.4)	489.0 (43.8)	0.405
	Hemoglobin, g/dL	15.0 (1.3)	14.7 (1.2)	0.214
	Hematocrit, %	47.2 (3.6)	46.6 (3.2)	0.351
	Platelet count, ×10^4/μL	26.7 (4.5)	28.0 (6.0)	0.164
	Neutrophil ratio, %	58.3 (7.2)	57.8 (5.9)	0.675
	Lymphocyte ratio, %	32.2 (6.8)	32.6 (5.7)	0.678
	Monocyte ratio, %	5.6 (1.3)	5.3 (1.0)	0.193
	Eosinophil ratio, %	3.2 (2.0)	3.4 (2.7)	0.652
	Basophil ratio, %	0.7 (0.3)	0.9 (0.3)	0.047
	Total serum protein, g/dL	7.2 (0.4)	7.3 (0.3)	0.237
	Albumin, g/dL	4.4 (0.3)	4.5 (0.3)	0.120
	Aspartate aminotransferase, U/L	24.8 (10.8)	23.5 (6.7)	0.430
	Alanine aminotransferase, U/L	30.4 (21.2)	26.6 (17.1)	0.280
	Lactate dehydrogenase, U/L	183.0 (29.6)	188.9 (32.7)	0.308
	Total bilirubin, mg/dL	0.8 (0.2)	0.9 (0.3)	0.228
	Alkaline phosphatase, U/L	75.7 (19.9)	76.8 (17.3)	0.758
Laboratory findings	γ-Glutamyl transpeptidase, U/L	40.4 (31.3)	44.7 (36.6)	0.485
	Blood urea nitrogen, mg/dL	13.3 (2.8)	13.8 (3.2)	0.356
	Creatinine, mg/dL	0.8 (0.15)	0.8 (0.16)	0.894
	Uric acid, mg/dL	5.8 (1.3)	5.8 (1.3)	0.746
	Sodium (Na), mEq/L	141.3 (1.7)	141.4 (1.4)	0.726
	Chlorine (Cl), mEq/L	104.1 (2.2)	103.9 (1.7)	0.645
	Potassium (K), mEq/L	4.2 (0.3)	4.2 (0.2)	0.145
	Calcium (Ca), mg/dL	9.5 (0.3)	9.5 (0.3)	0.294
	Total cholesterol, mg/dL	218.0 (30.2)	217.8 (31.9)	0.967
	LDL cholesterol, mg/dL	140.0 (27.5)	137.9 (30.0)	0.694
	HDL cholesterol, mg/dL	55.3 (12.4)	58.6 (14.7)	0.185
	Triglycerides, mg/dL	131.6 (65.8)	121.4 (49.7)	0.338
	Glucose, mg/dL	89.8 (9.2)	89.1 (9.6)	0.677
	HbA1c (NGSP), %	5.5 (0.3)	5.5 (0.3)	0.926
	Free fatty acid, mEq/L	0.4 (0.2)	0.5 (0.2)	0.077
	Urine pH	6.2 (0.5)	6.2 (0.6)	0.560
	Urine specific gravity	1.0 (0.007)	1.0 (0.007)	0.491
	Compliance rate of the test sample, % *	99.80 (0.50)	99.88(0.42)	0.349

All data are presented as the mean (standard deviation). Comparisons of value between placebo and active groups were tested by analysis of variance. HDL, high-density lipoprotein; LDL, low-density lipoprotein. * The compliance rate of test sample intake is shown excluding participants who dropped out.

Table 3. Dietary composition during the treatment period.

Parameter		0 Weeks	4 Weeks	8 Weeks	12 Weeks
Energy, kcal	Active	1820.9 (385.4)	1816.5 (428.4)	1792.3 (399.5)	1704.8 (413.2) *
	Placebo	1925.3 (360.3)	1929.8 (411.4)	1880.7 (414.8)	1820.1 (366.6) *
Protein, g	Active	68.7 (15.7)	67.3 (17.8)	66.0 (16.2)	64.9 (17.5) *
	Placebo	72.5 (15.3)	72.1 (15.4)	69.7 (19.0)	68.6 (14.9) *
Fat, g	Active	63.8 (17.9)	61.4 (19.2)	60.7 (20.0)	55.9 (19.1) *#
	Placebo	67.9 (20.0)	66.9 (20.4)	65.3 (20.6)	64.1 (18.7)
Carbohydrate, g	Active	229.7 (53.0)	236.2 (57.4)	233.3 (52.8)	222.6 (54.8)
	Placebo	243.3 (56.2)	247.1 (59.4)	240.4 (53.5)	230.2 (50.6) *
Dietary fiber, g	Active	10.8 (3.3)	10.7 (3.1)	10.6 (3.1)	10.1 (3.2)
	Placebo	10.9 (2.9)	11.0 (3.0)	10.4 (3.0)	9.9 (2.7) *

All data are presented as the mean (standard deviation). * $p < 0.05$ compared with week 0, paired t-test. # $p < 0.05$ compared with placebo, unpaired t-test.

3.2. Abdominal Fat Area

From the viewpoint of the accuracy of the CT data described in Section 2.4, we assessed a series of CT images obtained from the same participants at each measurement point, treating any inappropriate data as missing values. Data from 16 participants (1 in the active group and 15 in the placebo group) were treated as missing values in part or in whole due to overestimation of VFA, caused mainly by compression of the abdominal cavity during inspiration, and 6 participants (all in the active group) were treated as missing values in part or in whole due to underestimation of VFA, caused mainly by the inclusion of an internal organ or gas in the CT scan images. Consequently, data from 102 participants at week 0 (55 in the active group and 47 in the placebo group), data from 100 participants at week 8 (54 in the active group and 46 in the placebo group) and data from 94 participants at week 12 (52 in the active group and 42 in the placebo group) were analyzed. The mean decreases in VFA from baseline to 8 and 12 weeks, respectively, were significantly greater in the active group (-12.5 ± 1.8 cm^2 and -13.6 ± 2.2 cm^2) compared with the placebo group (-3.0 ± 2.0 cm^2 and -2.2 ± 2.2 cm^2). In addition, the mean reduction in TFA from baseline to 8 and 12 weeks, respectively, was significantly greater in the active group (-13.7 ± 2.6 cm^2 and -13.0 ± 3.0 cm^2) compared with the placebo group (-0.7 ± 4.2 cm^2 and 0.3 ± 3.9 cm^2). There were no statistically significant differences in SFA between the two groups and no changes within either group (Figure 2). The actual values of VFA, SFA and TFA are summarized in Table 4; VFA and TFA in the active group at weeks 8 and 12 were significantly reduced compared with baseline. There was a significant group-by-time interaction in VFA and TFA from baseline.

Table 4. Changes in abdominal fat area by CT scan during the treatment period.

Parameter		0 Weeks	8 Weeks	12 Weeks	Time × Group †
Visceral fat area, cm^2	Active	124.0 (4.1)	111.1 (4.2) *	106.8 (3.6) *	<0.0001
	Placebo	119.8 (3.9)	117.5 (4.9)	114.9 (4.0)	
Subcutaneous fat area, cm^2	Active	206.9 (8.3)	207.2 (8.3)	206.7 (8.6)	0.379
	Placebo	211.9 (8.8)	212.2 (8.6)	215.8 (9.4)	
Total fat area, cm^2	Active	331.0 (9.3)	318.4 (9.9) *	313.5 (9.5) *	0.001
	Placebo	331.7 (9.1)	329.7 (10.2)	330.7 (10.3)	

All data are presented as the mean (standard error). * $p < 0.05$ compared with week 0, paired t-test. † p-value represented as a group-by-time interaction effect by two-factor repeated-measures analysis of variance.

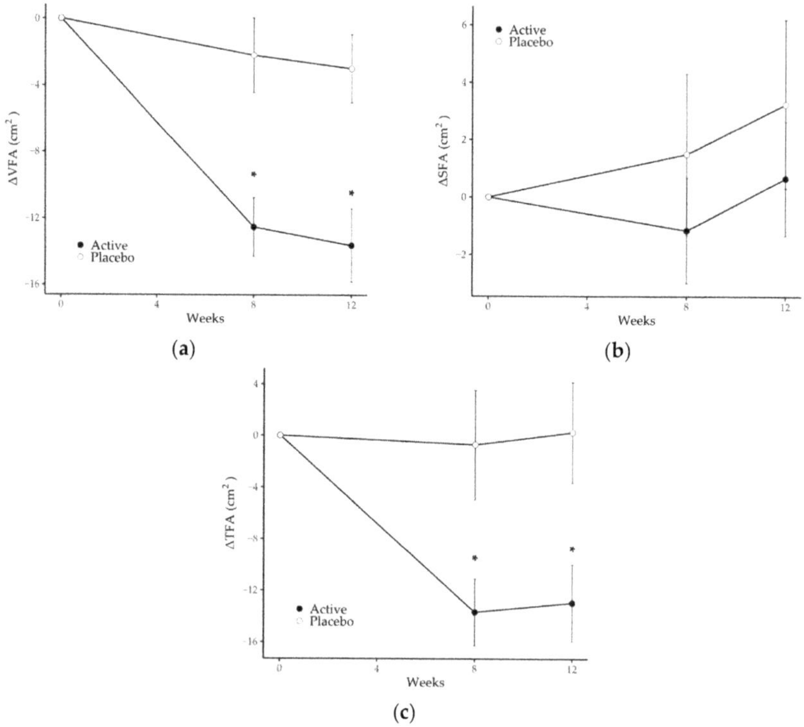

Figure 2. Changes in (**a**) visceral fat area, (**b**) subcutaneous fat area and (**c**) total fat area in the placebo and active groups during the study period. Values are the means, with error bars as standard error. Asterisks (*) indicate a p-value < 0.05 as a result of inter-group difference (the placebo group vs. the active group; unpaired t-test with Benjamini–Hochberg procedure). SFA, subcutaneous fat area; TFA, total fat area; VFA, visceral fat area.

3.3. Anthropometric Parameters

Body weight, BMI and WHR values are summarized in Table 5; there were no statistically significant differences in body weight, BMI or WHR between the two groups. Body weight and BMI in the active group at week 12 were significantly lower compared with baseline. The values for waist and hip circumference that were used to calculate WHR are presented in Table S2 of the Supplementary Materials.

Table 5. Changes in anthropometric parameters during the treatment period.

Parameter		0 Weeks	4 Weeks	8 Weeks	12 Weeks
Body weight, kg	Active	74.9 (1.1)	74.8 (1.1)	74.7 (1.1)	74.6 (1.1) *
	Placebo	74.8 (1.1)	74.8 (1.1)	74.6 (1.1)	74.3 (1.1)
Body mass index, kg/m^2	Active	26.5 (0.3)	26.4 (0.3)	26.4 (0.3)	26.3 (0.3) *
	Placebo	27.0 (0.2)	27.0 (0.2)	26.9 (0.2)	26.9 (0.2)
Waist-to-hip ratio	Active	0.9 (0.005)	0.9 (0.005)	0.9 (0.005)	0.9 (0.005)
	Placebo	0.9 (0.006)	0.9 (0.005)	0.9 (0.005)	0.9 (0.005)

All data are presented as the mean (standard error). * p < 0.05 compared with week 0, paired t-test.

3.4. Biochemical Parameters in Plasma

The values of TC, LDL-C, HDL-C, TG and free fatty acids in plasma are summarized in Table 6. TG in the active group at weeks 4 and 8 was significantly lower compared with the placebo group. TC, LDL-C and HDL-C in the active group at week 8 were significantly lower compared with baseline. Free fatty acids in the active and placebo groups at week 12 were significantly increased compared with baseline.

Table 6. Changes in plasma biochemistry parameters during the treatment period.

Parameter		0 Weeks	4 Weeks	8 Weeks	12 Weeks
Total cholesterol, mg/dL	Active	219.4 (4.2)	217.6 (4.1)	211.6 (4.2) *	217.5 (4.6)
	Placebo	219.2 (4.2)	219.0 (3.7)	216.2 (4.2)	219.4 (4.3)
LDL cholesterol, mg/dL	Active	139.9 (4.2)	136.3 (3.8)	135.3 (3.9) *	136.5 (4.3)
	Placebo	136.3 (3.8)	139.0 (3.4)	138.7 (3.7)	138.1 (4.0)
HDL cholesterol, mg/dL	Active	57.9 (2.1)	59.0 (1.9)	55.8 (1.7) *	56.2 (1.7)
	Placebo	55.5 (1.7)	54.6 (1.8)	54.2 (1.9)	54.5 (1.8)
Triglycerides, mg/dL	Active	113.6 (7.0)	116.2 (6.3) #	113.6 (5.4) #	118.3 (8.8)
	Placebo	130.2 (8.3)	141.7 (11.1)	138.5 (10.1)	143.1 (15.8)
Free fatty acid, mEq/L	Active	0.53 (0.03)	0.50 (0.02)	0.53 (0.02)	0.61 (0.03) *
	Placebo	0.48 (0.03)	0.50 (0.02)	0.48 (0.02)	0.55 (0.03) *

All data are presented as the mean (standard error). HDL, high-density lipoprotein; LDL, low-density lipoprotein. * $p < 0.05$ compared with week 0, paired t-test. # $p < 0.05$ compared with placebo, unpaired t-test.

3.5. Fecal Bifidobacteria

Changes in the number of fecal bifidobacteria are shown in Figure 3 and Table 7. Total bifidobacteria, *B. animalis* subsp. *lactis*, *B. catenulatum* and *B. pseudovatenulatum* counts in the active group at weeks 8 and 12 were significantly higher compared with the placebo group. In addition, total bifidobacteria, *B. animalis* subsp. *lactis*, *B. longum* subsp. *longum*, *B. adolescentis* group, *B. catenulatum* and *B. pseudovatenulatum* in the active group at 8 and 12 weeks were significantly increased compared with baseline.

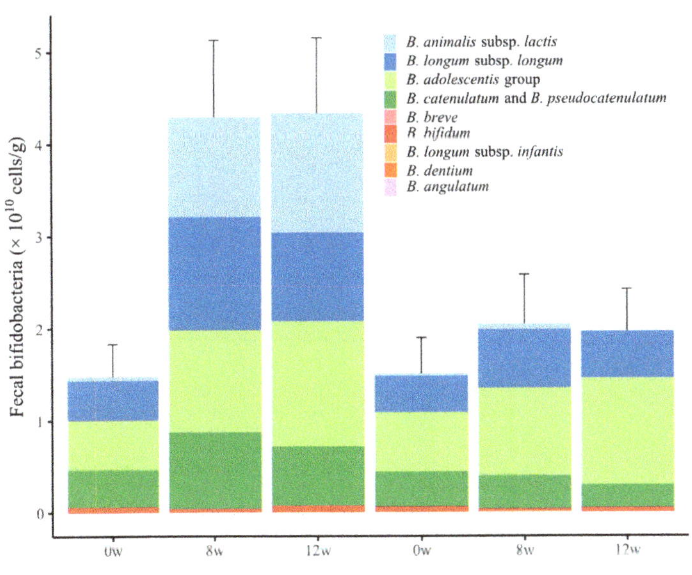

Figure 3. Changes in the number of fecal bifidobacteria in the placebo and active groups during the study period. Values are expressed as the sum of the mean ± standard error values of each species.

Table 7. Changes in the number of fecal bifidobacterial during the treatment period.

Bifidobacteria		0 Weeks	8 Weeks	12 Weeks
Total bifidobacteria	Active	9.62 (0.14)	10.35 (0.08) *#	10.41 (0.07) *#
	Placebo	9.48 (0.17)	9.60 (0.17)	9.61 (0.17)
B. animalis subsp. *lactis*	Active	5.70 (0.15)	9.67 (0.09) *#	9.77 (0.08) *#
	Placebo	5.70 (0.14)	5.72 (0.15)	5.50 (0.10)
B. longum subsp. *longum*	Active	8.62 (0.20)	8.86 (0.23) *	8.96 (0.23) *
	Placebo	8.26 (0.23)	8.35 (0.24)	8.37 (0.23)
B. adolescentis group	Active	7.81 (0.28)	8.14 (0.31) *	8.18 (0.31) *
	Placebo	7.73 (0.29)	8.09 (0.29)	8.19 (0.30)
B. catenulatum and *B. pseudocatenulatum*	Active	8.22 (0.25)	8.40 (0.27) *#	8.37 (0.27) *#
	Placebo	7.89 (0.26)	7.87 (0.26)	7.90 (0.25)
B. breve	Active	5.66 (0.11)	5.77 (0.12)	5.74 (0.12)
	Placebo	5.62 (0.11)	5.62 (0.11)	5.68 (0.12)
B. bifidum	Active	6.02 (0.20)	6.08 (0.20)	6.14 (0.21)
	Placebo	5.89 (0.18)	5.98 (0.19)	5.98 (0.19)
B. longum subsp. *infantis*	Active	5.53 (0.09)	5.59 (0.11)	5.57 (0.10)
	Placebo	5.51 (0.09)	5.51 (0.10)	5.52 (0.09)
B. dentium	Active	5.64 (0.11)	5.66 (0.10)	5.63 (0.10)
	Placebo	5.63 (0.09)	5.71 (0.10)	5.64 (0.09)
B. angulatum	Active	n.d.	n.d.	n.d.
	Placebo	n.d.	n.d.	n.d.
Endogenous bifidobacteria	Active	9.60 (0.14)	9.89 (0.16) *	9.92 (0.16) *
	Placebo	9.45 (0.17)	9.57 (0.17)	9.61 (0.17)

All data are presented as the mean (standard error) of common logarithms of the number of bacteria per 1 g feces. The detection limit of quantitative PCR was 2.0×10^5 cells per gram of feces. n.d., not detected. * $p < 0.05$ compared with placebo, unpaired t-test with Benjamini–Hochberg procedure. # $p < 0.05$ compared with week 0, paired t-test with Benjamini–Hochberg procedure.

3.6. Regression Analysis with Fecal Bifidobacteria and Measured Parameters

Table 8 shows the results of the regression analysis performed to examine the association of total bifidobacteria, endogenous bifidobacteria and *B. animalis* subsp. *lactis* bacteria counts with the participants' characteristics, body compositions and plasma parameters. Regression analysis showed that an increase in the number of total bifidobacteria was significantly associated with a decrease in BMI and VFA, while an increase in the number of *B. animalis* subsp. *lactis* and endogenous bifidobacteria was significantly associated with a decrease in BMI.

Table 8. Association of changes over time in body composition and metabolic parameters with changes in bifidobacteria.

Object Variable	Ratio of Change from Week 0 to 12	Explanatory Variable	Change from Week 0 to 12, %	p-Value
Total bifidobacteria	33.46 (12.51)	BMI	−0.5 (0.2)	0.010
		VFA	−7.2 (1.3)	0.012
Endogenous bifidobacteria	27,875.92 (5977.64)	BMI	−0.5 (0.2)	0.025
B. animalis subsp. *lactis*	8.07 (2.63)	BMI	−0.5 (0.2)	0.025

All data are presented as the mean (standard error). Data were generated by applying analysis of variance to a mixed linear model. BMI, body mass index; VFA, visceral fat area.

4. Discussion

We investigated the effects of consuming a dairy drink containing a synbiotic comprising *Bifidobacterium animalis* subsp. *lactis* GCL2505 and inulin on abdominal fat in overweight adults. The results showed that the consumption of the test beverage resulted in a reduction in abdominal visceral fat and total abdominal fat. In abdominal adipose tissue, visceral and subcutaneous adipose tissue have very different effects on metabolic disorders [49], and several studies have reported that excess VFA, rather than SFA, body weight or BMI, is correlated with metabolic disorders [45,50]. Therefore, abdominal VFA was set as the primary endpoint in this study.

After 12 weeks of consuming a dairy beverage containing GCL2505 and inulin, the reduction in VFA from week 0 to weeks 8 and 12 in the active group was significantly greater than that in the placebo group. Although there was no significant group difference in the reduction in SFA, the reduction in TFA was significantly greater compared with the placebo group, thus confirming the reduction in overall abdominal fat due to ingestion of the synbiotic. In addition, quantification of fecal bifidobacteria showed that total bifidobacteria, *B. animalis* subsp. *lactis*, *B. catenulatum* and *B. pseudocatenulatum* in the active group at weeks 8 and 12 were significantly greater compared with the placebo group. This confirmed an increase in endogenous bifidobacteria due to inulin, as well as an increase in *B. animalis* subsp. *lactis* due to ingestion of GCL2505. In the active group, the total number of endogenous bifidobacteria, *B. longum* subsp. *longum* and *B. adolescentis* was significantly increased at weeks 8 and 12 compared with baseline. These results are in line with a previous study showing that intake of GCL2505 and inulin increases endogenous bifidobacteria (especially *B. longum* subsp. *longum* and *B. adolescentis*) as well as total bifidobacteria counts more compared with GCL2505 alone [29].

At weeks 4 and 8, TG in the active group was significantly lower compared with the placebo group, suggesting an effect of GCL2505 and inulin intake on lipid parameters in plasma. In addition, body weight and BMI at week 12 as well as TC, LDL-C and HDL-C at week 8 in the active group were significantly lower compared with baseline. Changes in body weight and BMI were also reported in a meta-analysis by Koutnikova et al. [17], along with changes in VFA, which showed significant differences between groups. In addition, changes in body weight, BMI, TC and LDL-C were reported in a meta-analysis on synbiotics by Musazadeh et al. [51], along with changes in TG, which showed significant differences between groups. Although further studies are needed because of the absence of differences between groups, it is possible that the changes in these parameters were due to the intake of GCL2505 and inulin. However, the differences between groups in TG at week 12 as well as the differences in TC, LDL-C and HDL-C at week 12 compared with baseline were not significant. In addition, there was a significant increase in free fatty acids at week 12 compared to baseline in both groups. These results might be associated with an increase in the mean changes in TFA from baseline in both groups from weeks 8 to 12. It is hypothesized that intake of GCL2505 and inulin reduced visceral and body fat through a mechanism involving two steps. In the first step, intake of GCL2505 and inulin increases bifidobacteria and production of short-chain fatty acids (SCFAs) in the gut. In this study, total bifidobacteria counts increased significantly in the active group compared with the placebo group. In animal studies, the intake of GCL2505 alone contributed to an increase in the number of fecal bifidobacteria along with a corresponding increase in the concentration of acetic acid in feces and blood [52,53]. In clinical studies, intake of GCL2505 and inulin increased total bifidobacteria counts in feces [29], while intake of inulin alone increased SCFAs such as acetic acid [54] by increasing the number of bifidobacteria in the gut. Thus, in the present study, it was suggested that GCL2505 and inulin in the gut increased acetic acid, one of the SCFAs, by increasing the number of total bifidobacteria. In the second step, the increase in SCFAs in the gut improved glucose tolerance and systemic fatty acid oxidation through their receptor, G protein-coupled receptor 43 (GPR43), leading to a reduction in visceral and body fat. Previous studies in animals showed that increased production of acetic acid improves glucose tolerance, promotes systemic fatty acid oxidation and suppresses body fat accumulation via GPR43 [53]. In addition, intake of inulin led to a reduction in VFA [55]. In clinical studies, it was shown that daily consumption of a test beverage containing a higher amount of GCL2505 (8×10^{10} CFU) compared with standard fermented milk reduced abdominal VFA [14]. Taken together, the findings suggest that combined intake of GCL2505 and inulin may increase the concentration of SCFAs in the gut by increasing the total bifidobacteria count, thereby reducing visceral fat and body fat via GPR43.

The differences in TG between groups at week 8 may be due to the action of SCFAs. It was suggested that SCFAs promote lipid clearance in the liver by downregulating angiopoietin-like protein 4, which inhibits lipoprotein lipase [56]. In fact, the reduction in TG associated with *Lactiplantibacillus plantarum* consumption [10] is thought to be due to SCFA-mediated mechanisms of action. In the present study, VFA at baseline in the active group (124.0 ± 31.3 cm^2) was lower than that reported in a previous study [14] in which GCL2505 alone was ingested eight times (133.4 ± 29.6 cm^2) and the participants had low visceral fat from the start. Nevertheless, the change in VFA (-13.6 ± 2.2 cm^2) was greater than that reported in previous trials (-5.1 ± 1.8 cm^2). In addition, the degree of change was greater than that reported in a meta-analysis [17] (-6.30 cm^2, 95% CI -9.05, -3.56). In addition, as noted above, combined intake of GCL2505 and inulin was shown in clinical studies to increase the total number of intestinal bifidobacteria more compared with consumption of GCL2505 alone [29]. It is possible that the combined intake of GCL2505 and inulin might have reduced VFA and TFA more effectively in the present study by increasing total bifidobacteria.

The relationship between parameters such as VFA and TG (which changed in this study) as well as the number of bifidobacteria were estimated by regression analysis and applying analysis of variance to a mixed effects model. The results suggest that an increase in total bifidobacteria count is significantly associated with a decrease in BMI and VFA, whereas an increase in *B. animalis* subsp. *lactis* and endogenous bifidobacteria is associated only with a decrease in BMI and not with a decrease in VFA. Because the increase in *B. animalis* subsp. *lactis* and endogenous bifidobacteria does not correlate with the decrease in VFA, it is possible that increases in both *B. animalis* subsp. *lactis* and endogenous bifidobacteria contributed to the reduction in VFA that appeared when inulin was ingested in addition to GCL2505. In addition, parameters related to lipids such as TG, TC, LDL-C and HDL-C showed no significant correlation with total bifidobacteria, endogenous bifidobacteria and *B. animalis* subsp. *lactis*. In some lactic acid bacteria, effects on blood lipids were also reported for pathways that were not mediated by SCFAs [57]. It is possible that the changes in the present parameters may have been caused by a pathway that is not mediated by SCFAs derived from bifidobacteria. This study quantified bifidobacteria but did not investigate other intestinal bacteria. Therefore, further research is needed to understand how intestinal microbiota other than bifidobacteria change and affect body composition and blood parameters.

Accumulation of visceral fat induces chronic inflammation. Recent studies have reported that intake of GCL2505 and inulin suppresses chronic inflammation, thereby improving cognitive and vascular function [58,59]. Further research is needed to determine the mechanism by which the intake of GCL2505 and inulin suppresses chronic inflammation by suppressing visceral fat accumulation.

5. Conclusions

In conclusion, the results suggest that the combined intake of GCL2505 and inulin suppresses the accumulation of visceral fat more compared with the intake of GCL2505 alone. Visceral adipose tissue has endocrine functions and it secretes a variety of bioactive substances, including adipocytokines, which affect the risk of developing metabolic abnormalities. In terms of preventing the development of metabolic abnormalities, reducing visceral fat is relatively more important than weight or body fat. It is thought that the combined intake of GCL2505 and inulin, which are functional food components that can be easily applied to daily life, may help to prevent the development of metabolic abnormalities in overweight adults.

Supplementary Materials: The following supporting information can be downloaded at: https://www.mdpi.com/article/10.3390/nu15245025/s1, Table S1: CONSORT 2010 checklist; Table S2: Changes in waist and hip circumference during the treatment period.

Author Contributions: Conceptualization, Y.B.; methodology, Y.B.; validation, Y.B.; formal analysis, Y.B.; investigation, Y.B. and M.K.; writing—original draft preparation, Y.B., N.A. and Y.S.; writing—review and editing, Y.B., N.A. and Y.S.; visualization, Y.B.; supervision, D.T. All authors have read and agreed to the published version of the manuscript.

Funding: This research received no external funding.

Institutional Review Board Statement: The study was conducted in accordance with the Declaration of Helsinki and approved by the Ethical Committee of Kobuna Orthopedics Clinic (approval date: 5 October 2022; approval number: MK-2210-01).

Informed Consent Statement: Informed consent was obtained from all subjects involved in the study.

Data Availability Statement: Datasets generated during the current study and/or analyzed during the current study are available from the responsible author upon reasonable request.

Acknowledgments: We would like to express our sincere appreciation to the team members involved in the production of the test foods for this research.

Conflicts of Interest: Y.B., M.K., N.A. and Y.S. are employees of Ezaki Glico Co., Ltd. The other author, D.T. reports no conflict of interest in this work.

References

1. World Health Organization. Obesity and Overweight. Available online: https://www.who.int/news-room/fact-sheets/detail/obesity-and-overweight (accessed on 19 October 2023).
2. Shuster, A.; Patlas, M.; Pinthus, J.H.; Mourtzakis, M. The clinical importance of visceral adiposity: A critical review of methods for visceral adipose tissue analysis. *Br. J. Radiol.* **2012**, *85*, 1-e25. [CrossRef] [PubMed]
3. Green, M.; Arora, K.; Prakash, S. Microbial medicine: Prebiotic and probiotic functional foods to target obesity and metabolic syndrome. *Int. J. Mol. Sci.* **2020**, *21*, 2890. [CrossRef] [PubMed]
4. Pedroza Matute, S.; Iyavoo, S. Exploring the Gut Microbiota: Lifestyle Choices, Disease Associations, and Personal Genomics. *Front. Nutr.* **2023**, *10*, 1225120. [CrossRef] [PubMed]
5. Hill, C.; Guarner, F.; Reid, G.; Gibson, G.R.; Merenstein, D.J.; Pot, B.; Morelli, L.; Canani, R.B.; Flint, H.J.; Salminen, S.; et al. The International Scientific Association for Probiotics and Prebiotics consensus statement on the Scope and appropriate use of the term probiotic. *Nat. Rev. Gastroenterol. Hepatol.* **2014**, *11*, 506–514. [CrossRef]
6. Kadooka, Y.; Sato, M.; Imaizumi, K.; Ogawa, A.; Ikuyama, K.; Akai, Y.; Okano, M.; Kagoshima, M.; Tsuchida, T. Regulation of abdominal adiposity by probiotics (*Lactobacillus gasseri* SBT2055) in adults with obese tendencies in a randomized controlled trial. *Eur. J. Clin. Nutr.* **2010**, *64*, 636–643. [CrossRef] [PubMed]
7. Ogawa, A.; Kadooka, Y.; Kato, K.; Shirouchi, B.; Sato, M. Lactobacillus gasseri SBT2055 reduces postprandial and fasting serum non-esterified fatty acid levels in Japanese hypertriacylglycerolemic subjects. *Lipids Health Dis.* **2014**, *13*, 36. [CrossRef]
8. Kim, J.; Yun, J.M.; Kim, M.K.; Kwon, O.; Cho, B. Lactobacillus gasseri BNR17 supplementation reduces the visceral fat accumulation and waist circumference in obese adults: A randomized, double-blind, placebo-controlled trial. *J. Med. Food* **2018**, *21*, 454–461. [CrossRef]
9. Rahayu, E.S.; Mariyatun, M.; Putri Manurung, N.E.; Hasan, P.N.; Therdtatha, P.; Mishima, R.; Komalasari, H.; Mahfuzah, N.A.; Pamungkaningtyas, F.H.; Yoga, W.K.; et al. Effect of probiotic Lactobacillus plantarum Dad-13 powder consumption on the gut microbiota and intestinal health of overweight adults. *World J. Gastroenterol.* **2021**, *27*, 107–128. [CrossRef]
10. Sohn, M.; Na, G.Y.; Chu, J.; Joung, H.; Kim, B.-K.; Lim, S. Efficacy and safety of Lactobacillus plantarum K50 on lipids in Koreans with obesity: A randomized, double-blind controlled clinical trial. *Front. Endocrinol.* **2021**, *12*, 790046. [CrossRef]
11. Sohn, M.; Jung, H.; Lee, W.S.; Kim, T.H.; Lim, S. Effect of Lactobacillus plantarum LMT1-48 on body fat in overweight subjects: A randomized, double-blind, placebo-controlled trial. *Diabetes Metab. J.* **2023**, *47*, 92–103. [CrossRef]
12. Minami, J.; Iwabuchi, N.; Tanaka, M.; Yamauchi, K.; Xiao, J.-Z.; Abe, F.; Sakane, N. Effects of Bifidobacterium breve B-3 on body fat reductions in pre-obese adults: A randomized, double-blind, placebo-controlled trial. *Biosci. Microbiota Food Health* **2018**, *37*, 67–75. [CrossRef]
13. Pedret, A.; Valls, R.M.; Calderón-Pérez, L.; Llauradó, E.; Companys, J.; Pla-Pagà, L.; Moragas, A.; Martín-Luján, F.; Ortega, Y.; Giralt, M.; et al. Effects of daily consumption of the probiotic *Bifidobacterium animalis* subsp. lactis CECT 8145 on anthropometric adiposity biomarkers in abdominally obese subjects: A randomized controlled trial. *Int. J. Obes.* **2019**, *43*, 1863–1868. [CrossRef]

14. Takahashi, S.; Anzawa, D.; Takami, K.; Ishizuka, A.; Mawatari, T.; Kamikado, K.; Sugimura, H.; Nishijima, T. Effect of *Bifidobacterium animalis* ssp. lactis GCL2505 on visceral fat accumulation in healthy Japanese adults: A randomized controlled trial. *Biosci. Microbiota Food Health* **2016**, *35*, 163–171. [CrossRef] [PubMed]
15. Sun, J.; Buys, N. Effects of probiotics consumption on lowering lipids and CVD risk factors: A systematic review and meta-analysis of randomized controlled trials. *Ann. Med.* **2015**, *47*, 430–440. [CrossRef] [PubMed]
16. Zhang, Q.; Wu, Y.; Fei, X. Effect of probiotics on body weight and body-mass index: A systematic review and meta-analysis of randomized, controlled trials. *Int. J. Food Sci. Nutr.* **2015**, *67*, 571–580. [CrossRef] [PubMed]
17. Koutnikova, H.; Genser, B.; Monteiro-Sepulveda, M.; Faurie, J.-M.; Rizkalla, S.; Schrezenmeir, J.; Clément, K. Impact of bacterial probiotics on obesity, diabetes and mon-alcoholic fatty liver disease related variables: A systematic review and meta-analysis of randomised controlled trials. *BMJ Open* **2019**, *9*, e017995. [CrossRef] [PubMed]
18. Pineiro, M.; Asp, N.-G.; Reid, G.; Macfarlane, S.; Morelli, L.; Brunser, O.; Tuohy, K. FAO Technical Meeting on Prebiotics. *J. Clin. Gastroenterol.* **2008**, *42 Pt 2* (Suppl. S3), S156–S159. [CrossRef]
19. Beserra, B.T.S.; Fernandes, R.; do Rosario, V.A.; Mocellin, M.C.; Kuntz, M.G.F.; Trindade, E.B.S.M. A systematic review and meta-analysis of the prebiotics and synbiotics effects on glycaemia, insulin concentrations and lipid parameters in adult patients with overweight or obesity. *Clin. Nutr.* **2015**, *34*, 845–858. [CrossRef]
20. Thompson, S.V.; Hannon, B.A.; An, R.; Holscher, H.D. Effects of isolated soluble fiber supplementation on body weight, glycemia, and insulinemia in adults with overweight and obesity: A systematic review and meta-analysis of randomized controlled trials. *Am. J. Clin. Nutr.* **2017**, *106*, 1514–1528. [CrossRef] [PubMed]
21. Huwiler, V.V.; Schönenberger, K.A.; Segesser von Brunegg, A.; Reber, E.; Mühlebach, S.; Stanga, Z.; Balmer, M.L. Prolonged isolated soluble dietary fibre supplementation in overweight and obese patients: A systematic review with meta-analysis of randomised controlled trials. *Nutrients* **2022**, *14*, 2627. [CrossRef]
22. Larsen, I.S.; Choi, B.S.-Y.; Föh, B.; Kristensen, N.N.; Ouellette, A.; Haller, R.F.; Olsen, P.B.; Saulnier, D.; Sina, C.; Jensen, B.A.H.; et al. Experimental diets dictate the metabolic benefits of probiotics in obesity. *Gut Microbes* **2023**, *15*, 2192547. [CrossRef]
23. Wastyk, H.C.; Perelman, D.; Topf, M.; Fragiadakis, G.K.; Robinson, J.L.; Sonnenburg, J.L.; Gardner, C.D.; Sonnenburg, E.D. Randomized controlled trial demonstrates response to a probiotic intervention for metabolic syndrome that may correspond to diet. *Gut Microbes* **2023**, *15*, 2178794. [CrossRef]
24. Ojima, M.N.; Yoshida, K.; Sakanaka, M.; Jiang, L.; Odamaki, T.; Katayama, T. Ecological and molecular perspectives on responders and non-responders to probiotics and prebiotics. *Curr. Opin. Biotechnol.* **2022**, *73*, 108–120. [CrossRef]
25. Kovatcheva-Datchary, P.; Nilsson, A.; Akrami, R.; Lee, Y.S.; De Vadder, F.; Arora, T.; Hallen, A.; Martens, E.; Björck, I.; Bäckhed, F. Dietary fiber-induced improvement in glucose metabolism is associated with increased abundance of prevotella. *Cell Metab.* **2015**, *22*, 971–982. [CrossRef] [PubMed]
26. Swanson, K.S.; Gibson, G.R.; Hutkins, R.; Reimer, R.A.; Reid, G.; Verbeke, K.; Scott, K.P.; Holscher, H.D.; Azad, M.B.; Delzenne, N.M.; et al. The International Scientific Association for Probiotics and Prebiotics (ISAPP) consensus statement on the definition and scope of synbiotics. *Nat. Rev. Gastroenterol. Hepatol.* **2020**, *17*, 687–701. [CrossRef] [PubMed]
27. Childs, C.E.; Röytiö, H.; Alhoniemi, E.; Fekete, A.A.; Forssten, S.D.; Hudjec, N.; Lim, Y.N.; Steger, C.J.; Yaqoob, P.; Tuohy, K.M.; et al. Xylo-oligosaccharides alone or in synbiotic combination with *Bifidobacterium animalis* subsp. lactis induce bifidogenesis and modulate markers of immune function in healthy adults: A double-blind, placebo-controlled, randomised, factorial cross-over study. *Br. J. Nutr.* **2014**, *111*, 1945–1956. [CrossRef] [PubMed]
28. Krumbeck, J.A.; Rasmussen, H.E.; Hutkins, R.W.; Clarke, J.; Shawron, K.; Keshavarzian, A.; Walter, J. Probiotic Bifidobacterium strains and galactooligosaccharides improve intestinal barrier function in obese adults but show no synergism when used together as synbiotics. *Microbiome* **2018**, *6*, 121. [CrossRef] [PubMed]
29. Anzawa, D.; Mawatari, T.; Tanaka, Y.; Yamamoto, M.; Genda, T.; Takahashi, S.; Nishijima, T.; Kamasaka, H.; Suzuki, S.; Kuriki, T. Effects of synbiotics containing *Bifidobacterium animalis* subsp. lactis GCL2505 and inulin on intestinal bifidobacteria: A randomized, placebo-controlled, crossover study. *Food Sci. Nutr.* **2019**, *7*, 1828–1837. [CrossRef] [PubMed]
30. Oh, Y.J.; Kim, H.J.; Kim, T.S.; Yeo, I.H.; Ji, G.E. Effects of Lactobacillus plantarum PMO 08 alone and combined with chia seeds on metabolic syndrome and parameters related to gut health in high-fat diet-induced obese mice. *J. Med. Food* **2019**, *22*, 1199–1207. [CrossRef] [PubMed]
31. Li, Y.; Liu, M.; Liu, H.; Wei, X.; Su, X.; Li, M.; Yuan, J. Oral supplements of combined Bacillus licheniformis Zhengchangsheng® and xylooligosaccharides improve high-fat diet-induced obesity and modulate the gut microbiota in rats. *Biomed. Res. Int.* **2020**, *2020*, 9067821. [CrossRef]
32. Rogha, M.; Esfahani, M.Z.; Zargarzadeh, A.H. The efficacy of a synbiotic containing Bacillus coagulans in treatment of irritable bowel syndrome: A randomized placebo-controlled trial. *Gastroenterol. Hepatol. Bed Bench* **2014**, *7*, 156–163.
33. Skrzydło-Radomańska, B.; Prozorow-Król, B.; Cichoż-Lach, H.; Majsiak, E.; Bierła, J.B.; Kosikowski, W.; Szczerbiński, M.; Gantzel, J.; Cukrowska, B. The effectiveness of synbiotic preparation containing Lactobacillus and Bifidobacterium probiotic strains and short chain fructooligosaccharides in patients with diarrhea predominant irritable bowel syndrome-a randomized double-blind, placebo-controlled study. *Nutrients* **2020**, *12*, 1999.
34. Ipar, N.; Aydogdu, S.D.; Yildirim, G.K.; Inal, M.; Gies, I.; Vandenplas, Y.; Dinleyici, E.C. Effects of synbiotic on anthropometry, lipid profile and oxidative stress in obese children. *Benef. Microbes* **2015**, *6*, 775–782. [CrossRef]

35. Mohammadi-Sartang, M.; Bellissimo, N.; Totosy de Zepetnek, J.O.; Brett, N.R.; Mazloomi, S.M.; Fararouie, M.; Bedeltavana, A.; Famouri, M.; Mazloom, Z. The effect of daily fortified yogurt consumption on weight loss in adults with metabolic syndrome: A 10-week randomized controlled trial. *Nutr. Metab. Cardiovasc. Dis.* **2018**, *28*, 565–574. [CrossRef]
36. Eslamparast, T.; Poustchi, H.; Zamani, F.; Sharafkhah, M.; Malekzadeh, R.; Hekmatdoost, A. Synbiotic supplementation in nonalcoholic fatty liver disease: A randomized, double-blind, placebo-controlled pilot study. *Am. J. Clin. Nutr.* **2014**, *99*, 535–542. [CrossRef] [PubMed]
37. Lalitsuradej, E.; Sirilun, S.; Sittiprapaporn, P.; Sivamaruthi, B.S.; Pintha, K.; Tantipaiboonwong, P.; Khongtan, S.; Fukngoen, P.; Peerajan, S.; Chaiyasut, C. The effects of synbiotics administration on stress-related parameters in Thai subjects-a preliminary study. *Foods* **2022**, *11*, 759. [CrossRef] [PubMed]
38. Valle, M.C.P.R.; Vieira, I.A.; Fino, L.C.; Gallina, D.A.; Esteves, A.M.; da Cunha, D.T.; Cabral, L.; Benatti, F.B.; Marostica Junior, M.R.; Batista, Â.G.; et al. Immune status, well-being and gut microbiota in military supplemented with synbiotic ice cream and submitted to field training: A randomised clinical trial. *Br. J. Nutr.* **2021**, *126*, 1794–1808. [CrossRef] [PubMed]
39. Chiu, Y.-H.; Lin, S.-L.; Tsai, J.-J.; Lin, M.-Y. Probiotic actions on diseases: Implications for therapeutic treatments. *Food Funct.* **2014**, *5*, 625–634. [CrossRef] [PubMed]
40. Taipale, T.J.; Pienihäkkinen, K.; Isolauri, E.; Jokela, J.T.; Söderling, E.M. *Bifidobacterium animalis* subsp. lactis BB-12 in reducing the risk of infections in early childhood. *Pediatr. Res.* **2016**, *79*, 65–69. [CrossRef] [PubMed]
41. Ishizuka, A.; Tomizuka, K.; Aoki, R.; Nishijima, T.; Saito, Y.; Inoue, R.; Ushida, K.; Mawatari, T.; Ikeda, T. Effects of administration of *Bifidobacterium animalis* subsp. lactis GCL2505 on defecation frequency and bifidobacterial microbiota composition in humans. *J. Biosci. Bioeng.* **2012**, *113*, 587–591. [CrossRef]
42. Aoki, R.; Tsuchida, S.; Arai, Y.; Ohno, K.; Nishijima, T.; Mawatari, T.; Mikami, Y.; Ushida, K. Effect of *Bifidobacterium animalis* subsp. lactis GCL2505 on the physiological function of intestine in a rat model. *Food Sci. Nutr.* **2016**, *4*, 782–790. [CrossRef] [PubMed]
43. Lim, S.; Moon, J.H.; Shin, C.M.; Jeong, D.; Kim, B. Effect of Lactobacillus sakei, a probiotic derived from kimchi, on body fat in Koreans with obesity: A randomized controlled study. *Endocrinol. Metab.* **2020**, *35*, 425–434. [CrossRef] [PubMed]
44. Cho, Y.G.; Yang, Y.J.; Yoon, Y.S.; Lee, E.S.; Lee, J.H.; Jeong, Y.; Kang, C.H. Effect of MED-02 containing two probiotic strains, Limosilactobacillus fermentum MG4231 and MG4244, on body fat reduction in overweight or obese subjects: A randomized, multicenter, double-blind, placebo-controlled study. *Nutrients* **2022**, *14*, 3583. [CrossRef]
45. Ryo, M.; Kishida, K.; Nakamura, T.; Yoshizumi, T.; Funahashi, T.; Shimomura, I. Clinical significance of visceral adiposity assessed by computed tomography: A Japanese perspective. *World J. Radiol.* **2014**, *6*, 409–416. [CrossRef] [PubMed]
46. Tourlousse, D.M.; Narita, K.; Miura, T.; Sakamoto, M.; Ohashi, A.; Shiina, K.; Matsuda, M.; Miura, D.; Shimamura, M.; Ohyama, Y.; et al. Validation and standardization of DNA extraction and library construction methods for metagenomics-based human fecal microbiome measurements. *Microbiome* **2021**, *9*, 95. [CrossRef]
47. Tanaka, Y.; Takami, K.; Nishijima, T.; Aoki, R.; Mawatari, T.; Ikeda, T. Short- and long-term dynamics in the intestinal microbiota following ingestion of *Bifidobacterium animalis* subsp. lactis GCL2505. *Biosci. Microbiota Food Health* **2015**, *34*, 77–85. [CrossRef]
48. Ministry of Health, Labour and Welfare. *Dietary Reference Intakes for Japanese (2020)*; Ministry of Health, Labour and Welfare: Tokyo, Japan, 2019. Available online: https://www.mhlw.go.jp/content/001151422.pdf (accessed on 26 October 2023).
49. Ibrahim, M.M. Subcutaneous and visceral adipose tissue: Structural and functional differences. *Obes. Rev.* **2010**, *11*, 11–18. [CrossRef]
50. Chaston, T.B.; Dixon, J.B. Factors associated with percent change in visceral versus subcutaneous abdominal fat during weight loss: Findings from a systematic review. *Int. J. Obes.* **2008**, *32*, 619–628. [CrossRef]
51. Musazadeh, V.; Mohammadi Anilou, M.; Vajdi, M.; Karimi, A.; Sedgh Ahrabi, S.; Dehghan, P. Effects of synbiotics supplementation on anthropometric and lipid profile parameters: Finding from an umbrella meta-analysis. *Front. Nutr.* **2023**, *10*, 1121541. [CrossRef]
52. Aoki, R.; Kamikado, K.; Suda, W.; Takii, H.; Mikami, Y.; Suganuma, N.; Hattori, M.; Koga, Y. A proliferative probiotic Bifidobacterium strain in the gut ameliorates progression of metabolic disorders via microbiota modulation and acetate elevation. *Sci. Rep.* **2017**, *7*, 43522. [CrossRef]
53. Horiuchi, H.; Kamikado, K.; Aoki, R.; Suganuma, N.; Nishijima, T.; Nakatani, A.; Kimura, I. *Bifidobacterium animalis* subsp. lactis GCL2505 modulates host energy metabolism via the short-chain fatty acid receptor GPR43. *Sci. Rep.* **2020**, *10*, 4158. [CrossRef] [PubMed]
54. Birkeland, E.; Gharagozlian, S.; Birkeland, K.I.; Valeur, J.; Måge, I.; Rud, I.; Aas, A.-M. Prebiotic effect of inulin-type fructans on faecal microbiota and short-chain fatty acids in type 2 diabetes: A randomised controlled Trial. *Eur. J. Nutr.* **2020**, *59*, 3325–3338. [CrossRef] [PubMed]
55. Nakajima, H.; Nakanishi, N.; Miyoshi, T.; Okamura, T.; Hashimoto, Y.; Senmaru, T.; Majima, S.; Ushigome, E.; Asano, M.; Yamaguchi, M.; et al. Inulin reduces visceral adipose tissue mass and improves glucose tolerance through altering gut metabolites. *Nutr. Metab.* **2022**, *19*, 50. [CrossRef] [PubMed]
56. He, M.; Shi, B. Gut microbiota as a potential target of metabolic syndrome: The role of probiotics and prebiotics. *Cell Biosci.* **2017**, *7*, 54. [CrossRef]
57. Ogawa, A.; Kobayashi, T.; Sakai, F.; Kadooka, Y.; Kawasaki, Y. Lactobacillus gasseri SBT2055 suppresses fatty acid release through enlargement of fat emulsion size in vitro and promotes fecal fat excretion in healthy Japanese subjects. *Lipids Health Dis.* **2015**, *14*, 20. [CrossRef]

58. Azuma, N.; Mawatari, T.; Saito, Y.; Tsukamoto, M.; Sampei, M.; Iwama, Y. Effect of continuous ingestion of bifidobacteria and dietary fiber on improvement in cognitive function: A randomized, double-blind, placebo-controlled trial. *Nutrients* **2023**, *15*, 4175. [CrossRef]
59. Azuma, N.; Saito, Y.; Nishijima, T.; Aoki, R.; Nishihira, J. Effect of daily ingestion of Bifidobacterium and dietary fiber on vascular endothelial function: A randomized, double-blind, placebo-controlled, parallel-group comparison study. *Biosci. Biotechnol. Biochem.* **2023**, zbad148. [CrossRef]

nutrients

MDPI

Article

Effect of Continuous Ingestion of Bifidobacteria and Dietary Fiber on Improvement in Cognitive Function: A Randomized, Double-Blind, Placebo-Controlled Trial

Naoki Azuma [1,*], Takashi Mawatari [1], Yasuo Saito [1], Masashi Tsukamoto [1], Masatoshi Sampei [1] and Yoshitaka Iwama [2]

1 R&D Laboratory, Ezaki Glico Co., Ltd., 4-6-5, Utajima, Nishiyodogawa-Ku, Osaka 555-8502, Japan; takashi.mawatari@glico.com (T.M.); yasuo.saito@glico.com (Y.S.); masashi.tsukamoto@glico.com (M.T.); masatoshi.sampei@glico.com (M.S.)

2 Nihonbashi Cardiology Clinic, Kyodo Bldg. #201, 13-4 Nihonbashi Kodenmacho, Chuo-Ku, Tokyo 103-0001, Japan; yiwama@well-sleep.jp

* Correspondence: naoki.azuma@glico.com; Tel.: +81-70-4008-4615

Abstract: *Bifidobacterium animalis* subsp. *lactis* GCL2505 has been shown to have some positive effects on health, including improved defecation frequency and reduced visceral fat. These effects are thought to be due to GCL2505′s unique ability to reach the intestine in a viable form and proliferate after a single intake. This leads to an increased number of intestinal bifidobacteria. This randomized, double-blind, placebo-controlled, parallel-group study was conducted to confirm that intake of GCL2505 and inulin (a prebiotic) improve cognitive function ($n = 80$). Participants consumed test drinks containing 1×10^{10} colony-forming units of GCL2505 per 100 g and 2.0 g of inulin per 100 g for 12 weeks. The change in cognitive function assessment scores was set as the primary endpoint. There were significant improvements in scores in the neurocognitive index domain, which is an assessment of overall cognitive function, in addition to overall attention, cognitive flexibility, and executive function domains. The intervention significantly increased the number of fecal bifidobacteria and affected the levels of several inflammatory markers. These results suggest that intake of GCL2505 and inulin improves cognitive function by improving the intestinal environment and alleviating inflammation.

Keywords: *Bifidobacterium animalis* subsp. *lactis*; probiotics; visceral fat; metabolic syndrome; gut microbacteria; anti-inflammatory; mild cognitive impairment; aging

1. Introduction

It is estimated that more than 55 million people have dementia worldwide, and nearly 10 million new cases of the disease occur each year. Dementia is now the seventh leading cause of death and one of the leading causes of disability and dependency among the elderly. The global economic cost of dementia has reached USD 1.3 trillion, approximately half of which is attributed to care provided by informal caregivers (e.g., family members and close friends), who provide an average of 5 h of care and supervision per day [1]. Furthermore, it is estimated that the number of people with dementia will continue to rise, reaching 78 million by 2030 and 139 million by 2050 [2].

Alzheimer's disease (AD), a disease of progressive cognitive decline, is the most common form of dementia, with a prevalence of more than 60%, and represents a serious threat to public health [1]. Although the pathogenesis of AD remains to be elucidated, a number of therapeutic approaches have been developed based on the "amyloid cascade hypothesis", that neurofibrillary tangles are caused by the increase and accumulation of amyloid-β (Aβ) in the brain, followed by abnormal phosphorylation of tau protein [3,4]. AD is a chronic disease of the brain that progresses over decades, with Aβ accumulation in the brain beginning decades before clinical symptoms appear [5]. Once AD develops, it is difficult to improve its symptoms, and currently available AD drugs can only slow the progression of the disease; thus, there is no fundamental cure. For this reason, it is desirable to prevent the onset of AD. Mild cognitive impairment (MCI) is a pre-dementia condition characterized as a cognitive state between normal cognitive aging and dementia and is associated with an increased risk of developing AD [6]. Clinical studies have shown that patients with MCI progress to AD at a rate of 10% to 15% per year [7]. Meanwhile, it has been reported that it is possible to regress from MCI to a cognitively normal state [8,9]. Therefore, implementing appropriate intervention at the MCI stage is critical to reduce the number of patients with dementia. In addition, AD is not only caused by genetic factors and aging but also by environmental factors such as lifestyle habits, including sleep and diet. Accordingly, it is desirable for substances that contribute to the prevention of AD to be included in daily-use foods and used on a routine basis [10].

Various studies have investigated the triggers of AD, clarifying the relationship between obesity and cognitive function as well as the mechanisms of obesity-induced cognitive decline. An observational study reported that cognitive function was lower in participants with more visceral fat [11]. Another study reported that the insulin resistance that developed with obesity promoted the accumulation of Aβ and the formation of neurofibrillary tangles [12]. The association between inflammation and cognitive function has also received attention. Acute and chronic systemic inflammation associated with increased levels of tumor necrosis factor (TNF)-α, a typical inflammation-inducing cytokine, are associated with increased cognitive decline in AD [13]. Thus, management of environmental factors is now considered important for the prevention of AD, and lifestyle interventions aimed at preventing the onset and progression of AD as well as preventive measures in daily life with functional foods are necessary as one approach to solving the problem [14]. In addition, food ingredients such as chlorogenic acid [15], propolis extract [16], and astaxanthin [17] have been reported to improve cognitive function, while probiotics are expected to play a major role in future dementia countermeasures, based on the results of meta-analyses showing that probiotics improve cognitive function in patients with MCI [18]. The cognitive improvement effect of the probiotic bacterial strain *Lactobacillus rhamnosus* GG might be attributable to the improvement of signaling markers [19]. Meanwhile, intake of *Bifidobacterium breve* A1 might contribute to the modulation of brain immune response through the production of short-chain fatty acids (SCFAs), thereby contributing to the improvement of cognitive function [20,21]. It has also been reported that intake of *Lactiplantibacillus plantarum* OLL2712 protects against memory decline in the elderly due to its high IL-10 induction activity in immune cells [22].

Bifidobacterium animalis subsp. *lactis* GCL2505, a probiotic strain originally isolated from the feces of healthy adults [23–25], has been shown to reduce visceral fat [26]. Horiuchi et al. reported that GCL2505 affects host metabolic homeostasis (e.g., enhanced glucose tolerance, suppressed body fat accumulation) in a GPR43-dependent manner, due to enhanced SCFA production in the gut [27]. In clinical trials, daily consumption of fermented milk containing GCL2505 was shown to reduce abdominal visceral fat mass [28]. Furthermore, clinical studies have shown that GCL2505, when taken in combination with inulin [29], a typical prebiotic material, increases the total number of bifidobacteria in the gut more than GCL2505 alone [30].

In our study based on previous results, it was newly found that the combined intake of GCL2505 and inulin may reduce the risk of cognitive decline via visceral fat reduction. We also speculate that the anti-inflammatory effect of acetic acid produced by GCL2505 in the gut might improve cognitive function. SCFAs are known to exhibit anti-inflammatory effects by modulating immune cell chemotaxis as well as the release of reactive oxygen species (ROS) and cytokines [31]. However, it has not yet been demonstrated that the visceral-fat-reducing and anti-inflammatory effects of probiotics contribute directly to improving cognitive function. Therefore, in the present study, we conducted a randomized, double-blind, placebo-controlled, parallel-group study to test the hypothesis that the combined intake of GCL2505 and inulin improve cognitive function.

2. Materials and Methods

2.1. Participants

Participants were Japanese men and women between the ages of 50 and 80 years at the time of consent, who satisfied the inclusion criteria, did not satisfy any of the exclusion criteria, and were deemed eligible to participate by the study investigator. In this study, participants had to be selected from healthy people, not sick people, because the effects of food consumption—bifidobacteria and inulin—had to be confirmed. Young adults with documented cognitive decline were excluded from the study because they were more likely to have AD or other illnesses. Participants had to be drawn from healthy individuals with mild cognitive decline due to aging. The inclusion criteria were as follows: (1) score of 24 or higher on the Mini Mental State Examination—Japanese (MMSE-J), 17 or higher on the Japanese version of the Montreal Cognitive Assessment (MoCa-J), and 5 or less on the Geriatric Depression Scale—short version—Japanese (GDS-S-J); (2) subjective symptoms of memory loss or reported by close relatives or acquaintances to have other symptoms of memory loss; and (3) fully informed of the purpose and content of the study, deemed to have the capacity to consent, and volunteered of their own accord to participate in the study based on a thorough understanding of the purpose and content of the study, and provided written informed consent to participate in the study. Exclusion criteria were as follows: (1) current or past history of mental disorders (including depressive symptoms), cerebrovascular diseases, and sleep disorders; (2) serious liver, kidney, heart, respiratory, endocrine, or metabolic diseases; (3) smoker; (4) regular alcohol user (consuming $\geq$ 60 g pure alcohol equivalent daily) with an extremely irregular diet; (5) unable to follow the restrictions on foods or supplements that affect the intestinal environment during the study period; (6) use of antibiotics within 1 month prior to the start of the study; (7) have undergone digestive surgery (excluding appendicitis); (8) have experienced allergies to any of the study food ingredients; (9) currently taking medications that may affect cognitive function (e.g., antipsychotic, anxiolytic, antidepressant, antiparkinsonian, antisomatics, antiepileptic, and anticoagulant medications); (10) visual or hearing impairment that may interfere with cognitive function tests; (11) routinely consume foods or supplements that may affect cognitive function; (12) current or former drug or alcohol dependence; (13) participation in research involving the ingestion of other foods or the use of pharmaceuticals, the application of cosmetics or pharmaceuticals, or participation in other research while participating in this study; and (14) deemed ineligible by the principal investigator.

2.2. Test Foods

The test products were a dairy drink (active drink) containing inulin (Orafti GR; BENEO GmbH, Mannheim, Germany) and GCL2505 or placebo. The active drink contained 1×10^{10} colony-forming units of GCL2505 and 2.0 g of inulin per 100 g. The placebo was prepared using the same ingredients as the active drink, with the addition of food-grade acetic acid and lactic acid to adjust the flavor and pH; the basic ingredients were skim milk powder, fructose, dextrose, sucrose, yeast extract, acidifier, stabilizer, and flavoring. The nutritional details of the test products are shown in Table 1.

Table 1. Nutritional details of the test drinks.

	Placebo	Active
Energy, kcal/100 g	48.0	52.0
Moisture, g/100 g	86.9	84.9
Protein, g/100 g	2.8	2.8
Fat, g/100 g	0.1	0.1

Table 1. *Cont.*

	Placebo	Active
Carbohydrate, g/100 g	9.1	11.2
Ash, g/100 g	1.1	1.1

The active drink contained 2.0 g of inulin and 1.0×10^{10} colony-forming units of GCL2505.

2.3. Experimental Design

This was a randomized, placebo-controlled, double-blind, parallel-group study. Participants were stratified by age at screening and sex, and Cognitrax (short), MMSE-J, and MoCa-J scores served as stratification factors for randomizing in block sizes of four by computer-generated randomization to two groups (1:1). The controller (allocation manager) assigned the two groups to the test drink intake group and the control food intake group. For sample size, the final target number of subjects was set at 80, referring to previous reports on probiotic-induced cognitive function [21,22]. Participants in the active and placebo groups consumed 100 g of dairy beverage once daily for 12 weeks. Both the participants and observers were blinded to the group allocation for the duration of the study. Double blinding was accomplished by labeling the test drink with only an identification number. The change in Cognitrax (long) scores between weeks 0 and 12 was set as the primary endpoint. The secondary endpoints were Cognitrax (long) scores between weeks 0 and 8, fecal bifidobacteria, SF-36v2® scores, blood inflammation markers (Olink Target 96 Inflammation), blood high-sensitivity C-reactive protein (*hs*-CRP), blood interleukin (IL)-1β, and serum brain-derived neurotrophic factor (BDNF). The study was conducted at Nihonbashi Cardiology Clinic (Tokyo, Japan) from September to December 2022 by K.S.O. Corporation (Tokyo, Japan), a contract research organization, and was registered with the University Hospital Medical Information Network Clinical Trials Registry (UMIN-CTR) "http://www.umin.ac.jp/ctr/index.htm (accessed on 15 July 2022)" as UMIN000048386. This article conforms to the Consolidated Standards of Reporting Trials (CONSORT) 2010 guidelines (Supplementary Materials, Table S1).

2.4. Cognitrax Test

Participants' cognitive function was measured using Cognitrax, a computer-based battery of cognitive function tests that was developed as a Japanese version of CNS Vital Signs [32]. Based on a previous study [33], the cognitive function tests were administered in the following order: verbal memory test, visual memory test, finger tapping test, symbol digit coding test, Stroop test, shift attention test, continuous performance test, perception of emotion test, nonverbal reasoning test, and four-part continuous performance test.

2.5. Quality of Life Test

The SF-36v2®, a widely used quality of life rating scale, consists of the following eight scales: "physical functioning", "role physical", "bodily pain", "general health perceptions", "vitality", "social functioning", "role emotional", and "mental health". The score for each scale was estimated based on national norms (norm-based scoring) and calculated as a standard score (mean, 50) [34].

2.6. Mental Health Status

The Japanese version of the World Health Organization Five Well-Being Index (WHO-5-J) survey was conducted at weeks 0, 8, and 12 [35] to confirm that participants had no abnormal mental health status during the study period.

2.7. Biochemical Parameters

Blood pressure, pulse rate, and body weight were measured at weeks 0, 8, and 12. The concentrations of biochemical parameters were also measured at weeks 0 and 12. Blood samples were drawn from each participant after 4 h of no food or drink except water (fasting) prior to testing. Biochemical parameters included hematological tests (white blood cell count, red blood cell count, hemoglobin, hematocrit, mean corpuscular volume, mean corpuscular hemoglobin, mean corpuscular hemoglobin concentration, platelet count, leukogram), biochemical tests (total protein, albumin, total bilirubin, aspartate aminotransferase, alanine aminotransferase, lactate dehydrogenase (IFCC), alkaline phosphatase (IFCC), gamma-glutamyltransferase, urea nitrogen, creatinine, uric acid, sodium, chlorine, potassium, calcium, total cholesterol, low-density lipoprotein cholesterol, high-density lipoprotein cholesterol, triglycerides, phospholipids, glucose, HbA1c (NGSP), insulin), and urinalysis (protein, sugar, bilirubin, urinary ketone bodies, occult blood reaction, urobilinogen, pH, and specific gravity). In addition, serum BDNF was quantified. All of these tests were performed at LSI Medience Corporation (Tokyo, Japan).

2.8. Inflammatory Protein Profile

Blood inflammation markers in frozen serum were determined using Olink® Target 96 Inflammation Panels (Olink Proteomics AB, Uppsala, Sweden) with proximity expansion technology, a high-throughput multiplex proteomic immunoassay [36]. The panel contains 92 immune-related proteins, mostly cytokines and chemokines. The assay involves epitope-specific binding and hybridization of a set of paired oligonucleotide antibody probes, followed by amplification using quantitative PCR, normalized on a log base 2 to Olink's own relative abundance units (normalized protein expression values). Quality control was performed on samples (using spiked internal controls) and external controls for each sample plate. This inspection was performed by Pharma Foods Corporation (Tokyo, Japan). In addition, blood *hs*-CRP (LSI Medience Corporation) and blood IL-1β (Filgen, Inc., Aichi, Japan) were quantified.

2.9. Fecal Samples

Fecal samples were submitted at weeks 0 and 12. Fecal samples were collected at home between 7 and 2 days before the specified visit. The submitted samples were promptly transported to the laboratory by refrigerated transport.

2.10. Fecal DNA Extraction

Bacterial DNA was extracted from fecal samples using the ISOSPIN Fecal DNA Kit (Nippon Gene Co., Ltd., Tokyo, Japan), following the procedure of Tourlousse et al. [37]. Specifically, a sample (here, 0.2 g fecal sample), 700 μL of FE1 buffer, and 10 μL of RNase were added to a tube with attached beads. A bead-beating homogenizer (FastPrep-24; MP Biomedicals, Irvine, CA) was used at a rate of 6 m/s for 1 min to crush the cells. The process was repeated three times, during which the sample was kept at room temperature for 5 min. Then, 90 μL of FE2 buffer was added and the samples were centrifuged at 12,000× *g* for 15 min. The supernatant (up to 500 μL) was collected and mixed with FB buffer and isopropanol, each at 0.4× the volume of the supernatant obtained. Finally, the sample was loaded onto a spin column and washed according to the manufacturer's instructions. Purified DNA was eluted with 50 μL of Tris-EDTA buffer (pH 8.0).

2.11. Fecal Bifidobacteria

Following Tanaka et al. [38], real-time polymerase chain reaction (PCR) was performed using genus-specific primers capable of detecting *Bifidobacterium* spp., including GCL2505. The primer sequences were as follows: *Bifidobacterium* spp. sense primer, 5′-GATTCTGGCTCAGGATGAACGC-3′; *Bifidobacterium* spp. antisense primer, 5′-CTGATAGGACGCGACCCCAT-3′. Each PCR reaction mixture consisted of 20 pmol of each primer in a total volume of 10 μL; 5 μL of SYBR® premix Ex taq (Takara Bio, Shiga, Japan); and 1 μL of DNA solution [38]. This inspection was performed by the Kyoto Institute of Nutrition and Pathology (Kyoto, Japan).

2.12. Statistical Analysis

All measurements were expressed as mean and standard deviation (SD). All statistical analyses were performed using IBM® SPSS® Statistics 27 (IBM Corp., Armonk, NY, USA). A p-value < 0.05 was used as the threshold for determining significance. As basic statistics, means and SDs are expressed to the nearest significant digit, percentages are expressed to one decimal place, and finally digits were adjusted by rounding. Missing data were treated as missing values and no surrogate values were used; Cognitrax (Long) statistical analyses were performed with paired t-tests in order to compare the test results at the start date of the intake (week 0) with those at 8 and 12 weeks after intake. For the other items, Dunnett's test (two-tailed) was used to compare the test results at the start of intake (week 0) with those at 8 and 12 weeks after the start of intake, and the Wilcoxon signed rank test was used for qualitative items. Comparisons between the active and placebo groups at each examination time were statistically analyzed with an unpaired t-test (two-tailed), and a Wilcoxon's rank-sum test was used to compare the qualitative endpoints.

3. Results

3.1. Analysis of the Participant Population

The participant selection process is shown in Figure 1. A total of 255 participants were screened for this study. After screening, 80 participants were eligible: 40 were assigned to the active group and 40 were assigned to the placebo. A significant difference in alkaline phosphatase levels between the active and placebo groups was observed at the beginning of the study but was deemed acceptable because it was within the reference range. For the other items, there were no differences in the baseline characteristics of the participants' data (Table 2). By the end of the study, one participant from the active group withdrew for personal reasons. After the completion of the entire study, one participant from the placebo group was dropped due to an extremely irregular lifestyle. One participant from the active group was dropped due to a confirmed illness unrelated to the study that may have affected the results. In addition, nine participants were also excluded because they were found to have consumed drugs or foods during the study period that might have affected the results (n = 5 from the active group and n = 4 from the placebo group). Another participant in the active group was dropped due to partial missing primary-endpoint data. Thus, a total of 67 patients (32 in the active group and 35 in the placebo group) were included in the analysis. Moreover, four participants with reduced WHO-5-J scores were also excluded for Cognitrax analysis (n = 1 from the active group and n = 3 from the placebo group) because they had a score of $\leq$50, which is used as the cut-off for assigning a 'screening diagnosis' of depression in the global version of WHO-5 [39]; reduced WHO-5-J scores might indicate the possibility of earlier depression caused by isolation due to the COVID-19 pandemic, and thus Cognitrax tests may not have been performed properly in these participants. There were no reported harms or unintended effects in each group.

Table 2. Baseline characteristics of the participants (placebo group: n = 40; active group: n = 40).

	Placebo Group	Active Group	p-Value
Age, years	62.7 (6.9)	64.6 (7.1)	0.229
MMSE-J	28.0 (1.6)	28.0 (1.3)	0.878
Corrected MOCA-J	22.9 (1.8)	22.9 (2.1)	0.955
GDS-S-J	2.1 (1.6)	1.8 (1.7)	0.372
Height, cm	161.5 (9.0)	161.8 (8.0)	0.907
Body weight, kg	59.0 (11.3)	59.5 (12.1)	0.842
Body mass index, kg/m^2	22.4 (2.8)	22.6 (3.1)	0.861
Systolic blood pressure, mmHg	129.5 (17.0)	128.3 (16.3)	0.733
Diastolic blood pressure, mmHg	77.8 (12.8)	75.7 (10.7)	0.439
Heartbeat, bpm	70.6 (10.7)	71.1 (12.0)	0.830
White blood cell count, /μL	5817.5 (1731.7)	5685.0 (1543.8)	0.719
Red blood cell count, $\times 10^4$/μL	446 (40.9)	445.5 (41.1)	0.957

Table 2. *Cont.*

	Placebo Group	Active Group	*p*-Value
Hemoglobin, g/dL	13.7 (1.3)	13.7 (1.1)	0.978
Hematocrit, %	43.3 (3.7)	43.4 (3.3)	0.914
Platelet count, $\times 10^4/\mu$L	24.6 (5.1)	23.5 (5.1)	0.354
Mean corpuscular volume, fL	97.3 (4.3)	97.6 (3.8)	0.764
Mean corpuscular hemoglobin, pg	30.9 (1.8)	30.9 (1.3)	0.966
Mean corpuscular hemoglobin concentration, %	31.7 (1.1)	31.7 (0.9)	0.711
Neutrophil ratio, %	54.9 (7.7)	55.4 (8.8)	0.777
Lymphocyte ratio, %	35.1 (6.7)	34.2 (7.7)	0.573
Monocyte ratio, %	6.3 (1.8)	6.0 (1.5)	0.501
Eosinophil ratio, %	2.9 (2.0)	3.6 (2.5)	0.173
Basophil ratio, %	0.8 (0.3)	0.8 (0.3)	0.461
Total serum protein, g/dL	7.1 (0.4)	7.1 (0.4)	0.358
Albumin, g/dL	4.4 (0.2)	4.4 (0.2)	0.718
Aspartate aminotransferase, U/L	22.0 (6.2)	22.2 (9.6)	0.923
Alanine aminotransferase, U/L	18.1 (9.0)	17.5 (9.1)	0.777
Lactate dehydrogenase, U/L	185.1 (23.8)	187.5 (23.5)	0.655
Total bilirubin, mg/dL	1.0 (0.4)	0.9 (0.3)	0.175
Alkaline phosphatase, U/L	59.8 (14.6)	66.1 (13.0)	0.045
γ-Glutamyl transpeptidase, U/L	26.3 (13.7)	30.4 (32.7)	0.467
Blood urea nitrogen, mg/dL	14.8 (2.5)	14.4 (3.4)	0.594
Creatinine, mg/dL	0.8 (0.2)	0.8 (0.1)	0.788
Uric acid, mg/dL	5.6 (1.6)	5.1 (1.3)	0.181
Sodium (Na), mEq/L	141.5 (1.7)	141.2 (1.9)	0.392
Chlorine (Cl), mEq/L	104.3 (2.0)	104.1 (2.1)	0.664
Potassium (K), mEq/L	4.3 (0.4)	4.3 (0.3)	0.667
Calcium (Ca), mg/dL	9.3 (0.3)	9.3 (0.3)	0.966
Total cholesterol, mg/dL	216.3 (35.0)	212.7 (36.0)	0.649
LDL cholesterol, mg/dL	125.0 (30.8)	118.9 (25.6)	0.338
HDL cholesterol, mg/dL	71.3 (17.0)	73.1 (24.1)	0.705
Triglycerides, mg/dL	98.8 (49.1)	103.0 (52.6)	0.711
Phospholipid, mg/dL	233.9 (31.5)	233.9 (37.9)	0.995
Glucose, mg/dL	88.6 (8.4)	87.5 (9.8)	0.601
HbA1c (NGSP), %	5.5 (0.3)	5.5 (0.3)	0.778
Urine pH	6.2 (0.7)	6.2 (0.8)	0.759

All data are presented as mean (standard deviation). Comparisons between the placebo and active groups were tested by analysis of variance.

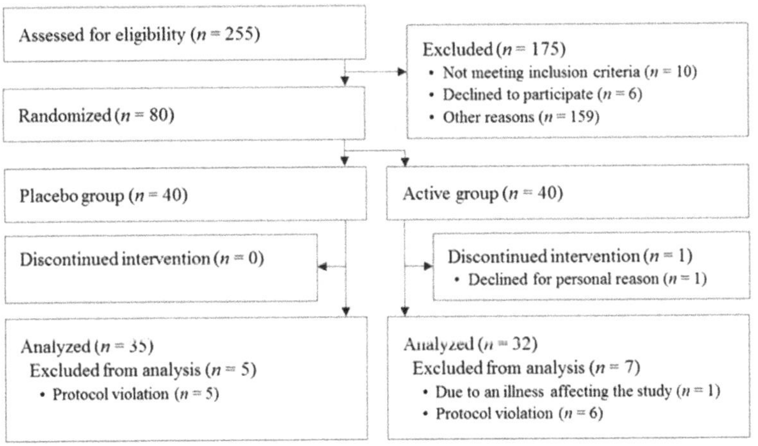

Figure 1. Flowchart of participant selection.

3.2. Cognitrax Test

The change in neurocognitive index domain score from week 0 to week 12 in the active group (5.5 ± 7.1) was greater than that in the placebo group (2.3 ± 4.0), and there was a statistically significant difference between them (p = 0.027 by the unpaired t-test). Furthermore, the changes in complex attention domain score (8.3 ± 11.8 vs. 3.2 ± 6.9, p = 0.041 by the unpaired t-test), cognitive flexibility domain score (9.8 ± 11.5 vs. 4.8 ± 6.7, p = 0.038 by the unpaired t-test), and executive function domain score (9.5 ± 11.8 vs. 4.5 ± 6.9, p = 0.044 by the unpaired t-test) in the active group from week 0 to 12 were significantly higher than those in the placebo group (Table 3). The results of the Cognitrax task to calculate scores for each domain are presented in Supplementary Materials, Table S2.

Table 3. Post-intervention changes in each cognitive function parameter (placebo group: n = 32; active group: n = 31).

		Week 0	Week 8		Week 12	
		Mean (SD)	Mean (SD)	p-Value	Mean (SD)	p-Value
Neurocognitive index	Placebo	103.5 (5.7)	104.8 (4.9)	0.090	105.7 (5.9)	0.003
	Active	101.6 (6.8)	104.5 (8.2)	0.045	107.2 (5.1)	<0.001
Δ Neurocognitive index	Placebo		1.4 (4.4)	0.345	2.3 (4.0)	0.027
	Active		2.9 (7.6)		5.5 (7.1)	
Composite memory	Placebo	105.3 (13.6)	102.7 (15.9)	0.264	102.6 (12.3)	0.231
	Active	104.4 (14.8)	107.6 (13.6)	0.147	107.2 (13.3)	0.297
Δ Composite memory	Placebo		−2.7 (13.2)	0.070	−2.8 (12.7)	0.114
	Active		3.2 (12.1)		2.8 (14.7)	
Verbal memory	Placebo	105.3 (13.6)	106.1 (15.0)	0.746	105.1 (13.2)	0.926
	Active	104.6 (14.3)	107.9 (13.6)	0.129	108.5 (14.6)	0.142
Δ Verbal memory	Placebo		0.8 (14.1)	0.452	−0.2 (13.3)	0.241
	Active		3.3 (11.7)		3.9 (14.5)	
Visual memory	Placebo	103.7 (14.8)	98.3 (15.7)	0.035	99.2 (14.8)	0.124
	Active	103.3 (12.8)	105.1 (11.8)	0.495	103.7 (11.0)	0.861
Δ Visual memory	Placebo		−5.4 (14.0)	0.047	−4.5 (16.1)	0.202
	Active		1.8 (14.3)		3.9 (14.5)	
Psychomotor speed	Placebo	106.3 (8.8)	108.5 (9.3)	0.045	108.3 (10.1)	0.073
	Active	106.2 (8.3)	108.9 (8.4)	0.016	108.6 (7.6)	0.054
Δ Psychomotor speed	Placebo		2.1 (5.7)	0.664	1.9 (5.8)	0.720
	Active		2.8 (6.1)		2.5 (6.9)	
Reaction time	Placebo	94.7 (11.3)	94.8 (11.8)	0.906	98.8 (11.1)	<0.001
	Active	96.6 (9.1)	98.8 (10.6)	0.112	100.8 (9.5)	0.003
Δ Reaction time	Placebo		0.1 (6.0)	0.230	4.2 (4.8)	0.981
	Active		2.2 (7.3)		4.2 (7.1)	
Complex attention	Placebo	108.3 (8.9)	111.9 (6.2)	0.028	111.5 (6.4)	0.015
	Active	104.0 (11.2)	105.2 (23.4)	0.762	112.3 (4.7)	<0.001
Δ Complex attention	Placebo		3.6 (8.8)	0.580	3.2 (6.9)	0.041
	Active		1.2 (22.4)		8.3 (11.8)	
Cognitive flexibility	Placebo	102.8 (9.3)	106.1 (7.3)	0.004	107.5 (8.4)	<0.001
	Active	96.7 (10.9)	102.2 (9.8)	0.014	106.5 (6.4)	<0.001
Δ Cognitive flexibility	Placebo		3.3 (6.0)	0.359	4.8 (6.7)	0.038
	Active		5.5 (11.6)		9.8 (11.5)	
Processing speed	Placebo	114.4 (9.3)	115.9 (9.9)	0.312	116.8 (10.2)	0.100
	Active	114 (9.3)	117.2 (11.1)	0.058	118.4 (9.3)	0.004
Δ Processing speed	Placebo		1.4 (7.9)	0.403	2.3 (7.7)	0.291
	Active		3.3 (9.2)		4.4 (7.8)	
Executive function	Placebo	102.8 (9.3)	105.3 (7.5)	0.016	107.2 (9)	0.001
	Active	96.8 (11.1)	102.0 (9.9)	0.022	106.3 (6.4)	<0.001
Δ Executive function	Placebo		2.5 (5.5)	0.265	4.5 (6.9)	0.044
	Active		5.2 (11.9)		9.5 (11.8)	
Social acuity	Placebo	86.2 (14.7)	92.0 (14.5)	0.052	94.0 (18.0)	0.013
	Active	90.0 (17.6)	94.5 (17.2)	0.213	98.1 (14.9)	0.018

Table 3. *Cont.*

		Week 0	Week 8		Week 12	
		Mean (SD)	Mean (SD)	*p*-Value	Mean (SD)	*p*-Value
Δ Social acuity	Placebo		5.8 (16.3)	0.790	7.8 (16.8)	0.937
	Active		4.6 (20.1)		8.2 (18.1)	
Reasoning	Placebo	94.7 (16.5)	91.3 (18.3)	0.264	95.0 (17.5)	0.929
	Active	92.8 (15.9)	96.6 (14.5)	0.166	93.9 (14.2)	0.706
Δ Reasoning	Placebo		−3.4 (16.9)	0.078	0.3 (15.8)	0.837
	Active		3.8 (14.9)		1.1 (15.6)	
Working memory	Placebo	105.7 (12.1)	105.0 (14.1)	0.779	105.6 (11.0)	0.952
	Active	103.5 (13.7)	109.2 (9.3)	0.010	107.7 (10.6)	0.095
Δ Working memory	Placebo		−0.7 (14.3)	0.056	−0.1 (11.6)	0.180
	Active		5.7 (11.6)		4.2 (13.5)	
Sustained attention	Placebo	108.3 (10.0)	108.8 (9.0)	0.767	109.0 (10.3)	0.737
	Active	106.7 (9.9)	110.6 (9.1)	0.020	111.7 (7.2)	0.013
Δ Sustained attention	Placebo		0.6 (10.6)	0.184	0.7 (11.5)	0.126
	Active		3.9 (8.8)		5.0 (10.6)	
Simple attention	Placebo	102.5 (11.6)	103.0 (14.1)	0.892	103.8 (9.8)	0.528
	Active	105.6 (7.3)	81.4 (136.6)	0.337	105.7 (6.8)	0.948
Δ Simple attention	Placebo		0.5 (19.3)	0.320	1.3 (11.3)	0.683
	Active		−24.2 (137.7)		0.1 (10.9)	
Motor speed	Placebo	99.7 (10.7)	101.4 (10.0)	0.101	100.5 (11.4)	0.523
	Active	99.2 (11.7)	100.8 (10.4)	0.129	99.8 (10.2)	0.673
Δ Motor speed	Placebo		1.8 (5.9)	0.926	0.8 (7.1)	0.901
	Active		1.6 (5.8)		0.6 (7.6)	

All data are presented as mean (SD) of Cognitrax scores. Data at week 8 and week 12 were compared with those at week 0 using the paired *t*-test. Comparisons between the placebo and active groups were calculated as changes from week 0, indicated by the "Δ" symbol, and tested using the unpaired *t*-test.

3.3. Quality of Life Test

Changes in quality of life during the study period were assessed by the SF-36v2®. No statistically significant differences were found between the active and placebo groups in terms of change in score (Table 4).

Table 4. Post-intervention changes in each quality-of-life parameter (placebo group: n = 35; active group: n = 32).

		Week 0	Week 8		Week 12	
		Mean (SD)	Mean (SD)	*p*-Value	Mean (SD)	*p*-Value
Physical functioning	Placebo	52.2 (5.4)	51.7 (6.0)	0.682	52.8 (4.8)	0.633
	Active	53.2 (5.0)	53.6 (4.0)	0.660	53.7 (5.1)	0.556
Δ Physical functioning	Placebo		−0.6 (5.4)	0.115	0.6 (4.0)	0.473
	Active		0.4 (3.0)		0.5 (3.1)	
Role physical	Placebo	52.8 (5.6)	53.6 (5.1)	0.534	53.6 (5.5)	0.471
	Active	53.3 (6.8)	53.7 (4.7)	0.873	54 (4.4)	0.710
Δ Role physical	Placebo		0.8 (4.3)	0.895	0.9 (6.0)	0.777
	Active		0.4 (7.2)		0.7 (6.0)	
Bodily pain	Placebo	49.9 (8.8)	48.6 (7.9)	0.472	49.3 (10.1)	0.825
	Active	50.9 (10.3)	50.9 (9.2)	1.000	51.8 (9.4)	0.648
Δ Bodily pain	Placebo		−1.4 (6.8)	0.273	−0.7 (9.6)	0.296
	Active		0.0 (6.2)		0.9 (7.7)	
General health perceptions	Placebo	58.2 (6.9)	57.2 (6.9)	0.256	58.2 (6.8)	1.000
	Active	57.1 (6.8)	57.5 (7.1)	0.846	57.2 (8.0)	0.981
Δ General health perceptions	Placebo		−1.0 (4.3)	0.877	0.0 (4.7)	0.575
	Active		0.4 (3.9)		0.1 (5.2)	
Vitality	Placebo	55.5 (7.0)	55.8 (7.7)	0.930	56.0 (6.2)	0.854
	Active	56.8 (6.7)	57.2 (6.2)	0.862	57.0 (6.4)	0.963

Table 4. *Cont.*

		Week 0	Week 8		Week 12	
		Mean (SD)	Mean (SD)	*p*-Value	Mean (SD)	*p*-Value
Δ Vitality	Placebo		0.3 (6.6)	0.443	0.5 (6.5)	0.535
	Active		0.4 (5.0)		0.2 (5.6)	
Social functioning	Placebo	54.2 (6.6)	55.3 (6.3)	0.423	55.0 (6.3)	0.632
	Active	53.5 (7.0)	55.1 (5.7)	0.183	55.2 (5.2)	0.128
Δ Social functioning	Placebo		1.1 (5.1)	0.877	0.8 (7.0)	0.850
	Active		1.6 (6.7)		1.8 (5.6)	
Role emotional	Placebo	52.8 (6.3)	54.9 (4.5)	0.041	54.6 (4.3)	0.093
	Active	54.4 (5.1)	53.8 (4.8)	0.779	54.9 (4.2)	0.783
Δ Role emotional	Placebo		2.1 (6.1)	0.332	1.8 (6.3)	0.720
	Active		−0.6 (7.2)		0.6 (3.2)	
Mental health	Placebo	56.9 (4.4)	57.6 (5.5)	0.663	57.6 (6.9)	0.718
	Active	57.1 (6.7)	57.1 (6.5)	0.993	57.5 (5.4)	0.775
Δ Mental health	Placebo		0.7 (6.1)	0.732	0.7 (5.2)	0.982
	Active		0.1 (5.1)		0.5 (4.4)	

All data are presented as mean (SD) of SF-36v2® scores. Data at week 8 and week 12 were compared with those at week 0 using Dunnett's test. Comparisons between the placebo and active groups were calculated as changes from week 0, indicated by the "Δ" symbol, and tested using the unpaired *t*-test.

3.4. Fecal *Bifidobacteria*

Quantification of the bifidobacteria in the feces (Figure 2A) revealed that the total number of bifidobacteria in the active group at week 12 increased significantly compared to week 0, while that of the placebo group did not change much. And the total number of bifidobacteria was significantly increased in the active group (7.71 ± 0.56 log cells/g feces) compared with the placebo group (7.31 ± 0.90 log cells/g feces) at week 12 (p = 0.031 using the unpaired *t*-test).

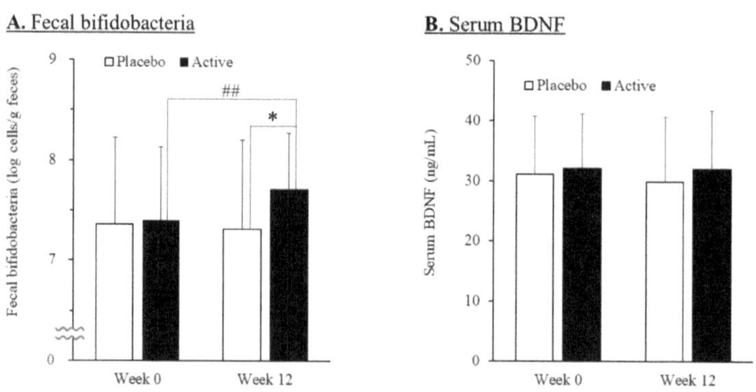

Figure 2. Changes in fecal bifidobacteria (**A**) and in serum BDNF (**B**) in the placebo (n = 35) and active (n = 32) groups during the study period. Values are means, with error bars as standard deviation. Double number signs (##) indicate p-value < 0.01 as a result of intra-group difference (week 0 vs. week 12; paired *t*-test). Asterisk (*) indicates p-value < 0.05 as a result of inter-group difference (the placebo group vs. the active group; unpaired *t*-test).

3.5. Serum BDNF

Serum BDNF levels were quantified. The results showed no statistically significant difference between the active and placebo groups during the study period (Figure 2B).

3.6. Blood Inflammation Markers

The expression levels of 92 inflammatory markers were examined at weeks 0 and 12 using Olink® Target 96 Inflammation Panels; 75 inflammatory markers that were determined to be quantifiable were analyzed (Supplementary Materials, Table S3). The results showed that there were statistically significant differences between the change in scores from week 0 to week 12 for leukemia inhibitory factor receptor, sulfotransferase 1A1, C-C motif hemokine (CCL)23, and TNF (ligand) superfamily member 12 (TWEAK) in the active and the placebo groups (using the unpaired t-test) (Table 5). In addition, there were no statistically significant differences between the change in scores from week 0 to week 12 of adenosine deaminase, osteoprotegerin, eotaxin, glial cell line-derived neurotrophic factor, fractalkine, interleukin-8 (IL-8), CCL28, IL-18, IL-10, CCL19, and T-cell surface glycoprotein CD5 in the active and placebo groups, but the trends were considered significant because the p-value was <0.1 (Table 5). Blood *hs*-CRP and blood IL-1β were also measured at weeks 0 and 12, but neither showed a statistically significant difference between the active and placebo groups (Supplementary Materials, Table S4).

Table 5. Post-intervention changes in each cognitive function parameter (placebo group: n = 32; active group: n = 31).

Parameter		Week 0	Week 12		Change	
		Mean (SD)	Mean (SD)	p-Value	Mean (SD)	p-Value
LIF-R, NPX	Placebo	3.36 (0.27)	3.39 (0.23)	0.459	0.02 (0.19)	0.012
	Active	3.49 (0.30)	3.40 (0.20)	0.008	−0.10 (0.19)	
ST1A1, NPX	Placebo	2.19 (1.16)	1.75 (0.83)	0.111	−0.36 (1.10)	0.020
	Active	1.82 (0.82)	2.10 (1.00)	0.079	0.33 (0.86)	
CCL23, NPX	Placebo	10.86 (0.35)	10.88 (0.25)	0.718	0.02 (0.25)	0.021
	Active	11.09 (0.46)	10.90 (0.40)	0.012	−0.15 (0.32)	
TWEAK, NPX	Placebo	9.10 (0.34)	9.08 (0.33)	0.658	−0.01 (0.18)	0.033
	Active	9.22 (0.26)	9.10 (0.20)	0.002	−0.12 (0.19)	
ADA, NPX	Placebo	5.39 (0.41)	5.46 (0.39)	0.068	0.08 (0.23)	0.052
	Active	5.53 (0.37)	5.50 (0.30)	0.345	−0.04 (0.24)	
OPG, NPX	Placebo	10.49 (0.4)	10.58 (0.41)	0.134	0.09 (0.34)	0.052
	Active	10.60 (0.30)	10.60 (0.20)	0.188	−0.04 (0.17)	
CCL11, NPX	Placebo	8.66 (0.43)	8.76 (0.36)	0.029	0.09 (0.24)	0.054
	Active	8.72 (0.29)	8.70 (0.20)	0.649	0.02 (0.21)	
GDNF, NPX	Placebo	1.98 (0.47)	2.19 (0.44)	0.003	0.20 (0.33)	0.055
	Active	2.15 (0.33)	2.10 (0.30)	0.447	0.04 (0.28)	
CX3CL1, NPX	Placebo	3.66 (0.48)	3.82 (0.51)	0.059	0.17 (0.48)	0.058
	Active	3.86 (0.41)	3.80 (0.40)	0.520	−0.04 (0.34)	
IL-8, NPX	Placebo	5.97 (0.41)	6.22 (0.49)	0.012	0.24 (0.53)	0.063
	Active	6.15 (0.75)	6.20 (0.70)	0.733	0.02 (0.37)	
CCL28, NPX	Placebo	2.62 (0.50)	2.64 (0.49)	0.711	0.02 (0.27)	0.063
	Active	2.73 (0.46)	2.60 (0.50)	0.009	−0.09 (0.18)	
IL-18, NPX	Placebo	9.10 (0.67)	9.28 (0.68)	0.002	0.18 (0.30)	0.064
	Active	9.14 (0.63)	9.20 (0.50)	0.912	0.01 (0.41)	
IL-10, NPX	Placebo	3.22 (0.52)	3.44 (0.78)	0.035	0.22 (0.56)	0.064
	Active	3.47 (0.44)	3.50 (0.40)	0.975	0.00 (0.28)	
CCL19, NPX	Placebo	8.73 (0.77)	8.88 (0.90)	0.163	0.15 (0.60)	0.081
	Active	8.72 (0.49)	8.70 (0.60)	0.278	−0.06 (0.31)	
CD5, NPX	Placebo	5.92 (0.40)	5.91 (0.37)	0.864	−0.01 (0.19)	0.092
	Active	6.02 (0.39)	5.90 (0.30)	0.017	−0.09 (0.20)	

All data were obtained using Olink® Target 96 Inflammation Panels and are presented as the mean (SD) of log base 2-normalized protein expression values (NPX). Data at week 12 were compared with those at week 0 using the paired t-test. Comparisons between the placebo and active groups were evaluated by calculating the change in measurements at week 0 and 12 in both groups using unpaired t-tests. LIF-R, leukemia inhibitory factor receptor; ST1A1, sulfotransferase 1A1; CCL23, C-C motif hemokine 23; TWEAK, tumor necrosis factor (ligand) superfamily, member 12; ADA, adenosine deaminase; OPG, osteoprotegerin; CCL11, eotaxin; GDNF, glial cell line-derived neurotrophic factor; CX3CL1, fractalkine; IL-8, interleukin 8; CCL28, C-C motif chemokine 28; IL-18, interleukin 18; IL-10, interleukin 10; CCL19, C-C motif chemokine 19; CD5, T-cell surface glycoprotein CD5.

4. Discussion

We investigated the effects of consuming a dairy beverage containing *Bifidobacterium animalis* subsp. *lactis* GCL2505 and inulin on cognitive function in healthy adults. The results showed that the test drink had a positive effect on cognitive function.

Cognitive function was assessed using Cognitrax, a computerized battery of neurocognitive tests developed for routine clinical screening applications. Cognitrax has been reported to have very similar characteristics to traditional psychological tests. It can measure a wide range of cognitive functions and is suitable for accurate assessment of scored cognitive functions because of its high sensitivity to discriminating between individuals with MCI and healthy individuals [32]. A similar test, CNS Vital Signs, is reported to adequately discriminate between healthy individuals, patients with MCI, and patients with dementia [40]. Cognitrax has also been used to assess cognitive function in clinical trials [15–17]. Therefore, Cognitrax was considered a suitable tool for testing cognitive function in this study.

The change in neurocognitive index domain scores from week 0 to week 12 was significantly higher in the active group than in the placebo group. The Neurocognitive Index domain score is calculated by averaging the scores from the total memory, cognitive functioning speed, reaction time, total attention, and cognitive flexibility domains, and is, therefore, used to assess a person's overall neurocognitive status. Furthermore, intake of GCL2505 and inulin was associated with statistically significant improvements in scores in the complex attention, cognitive flexibility, and executive function domains. Total attention refers to the ability to process things accurately while maintaining attention. Cognitive flexibility refers to the ability to understand and process changes in instructions. Executive function refers to the ability to make decisions based on an understanding of background rules and concepts. Thus, it is considered that intake of GCL2505 and inulin may contribute to improved activities of daily living by improving these functions.

It is hypothesized that intake of GCL2505 and inulin improves cognitive function through a mechanism involving the following three steps. Step 1: Intake of GCL2505 and inulin improves the intestinal environment. In this study, the number of fecal bifidobacteria was significantly increased in the active group compared with the placebo group. It was reported that intake of GCL2505 increases the number of bifidobacteria in the feces and the concentration of acetic acid in the feces and blood [23,25–27]. Clinical studies have shown the effects of inulin [41] as well as the enhanced effect of GCL2505 in combination with inulin in terms of increased total bifidobacteria count in the intestine [30]. Thus, participants in the active group who consumed a test drink containing GCL2505 and inulin in the present study had an improved intestinal environment via increasing the number of bifidobacteria in the gut. This change may possibly have increased the level of acetic acid, an SCFA, in the gut of the active group. Step 2: Improvement in the intestinal environment leads to alleviation of inflammation. Increased SCFA levels in the gut alleviate inflammation in the body. It was reported that administration of dietary fiber or acetic acid to mice reduced blood levels of IL-1, a known inflammatory cytokine [42]. SCFAs are also known to have positive effects on inflammation by reducing visceral fat and improving glucose metabolism. Experiments with mice showed that GCL2505 played a role in reducing visceral fat area by increasing the number of bifidobacteria in the intestine, as evidenced by the increased numbers in the feces as well as the higher acetic acid levels in feces and blood [26]. Daily consumption of yogurt containing GCL2505 has been shown in clinical studies to reduce visceral fat mass in the abdomen of humans [28]. Inulin consumption has also been proven to reduce visceral fat area [43], and Lauridsen et al. found that obese individuals have decreased expression of IL-10, which contributes to reducing inflammation, and increased expression of nitric oxide synthase 2, which triggers inflammation, in the brain [44]. A correlation between high-fat diet intake and hypothalamic inflammatory status has been reported [45,46], and Mao et al. reported that administration of a high-fat diet in combination with (-)-epigallocatechin gallate in mice suppressed the increased expression of IL-6, TNF-α, and IL-1β in the hypothalamus by inhibiting body weight gain [47]. Step

3: Alleviating inflammation improves cognitive function. It has recently been shown that inflammatory conditions and cognitive function are closely linked. Studies on patients with AD have reported a correlation between attenuated cognitive function and acute and chronic systemic inflammation associated with increased TNF-α levels [13], and that reduced levels of NLRP3 (nucleotide-binding oligomerization domain-like receptor family, pyrin domain-containing 3) inflammasome-derived inflammatory cytokines alleviate the progression of AD pathology [48]. TNF-α is a typical pro-inflammatory cytokine [49], and Habbas et al. suggested a link between increased TNF-α levels in the brain and cognitive impairment [50]. Increasing the number of SCFA-producing bacteria by fecal transplantation in rats resulted in increased SCFA levels in the gut and a reduction in cognitive decline [51]. The hypothesis thus far indicates that the effects of GCL2505 and inulin on cognitive function may be realized by increasing intestinal SCFA levels and alleviating inflammatory conditions.

In this study, the changes in expression of 75 inflammation markers from week 0 to 12 were analyzed by principal component analysis and tested by permutational multivariate analysis of variance, but there were no statistically significant differences between the active and placebo groups ($p = 0.109$) (Supplementary Materials, Figure S1). However, some reductions in pro-inflammatory cytokine levels were observed. For example, TWEAK is expressed in animals with chronic intestinal inflammation [52] and induces secretion of the pro-inflammatory cytokine IL-8 [53]. CCL23 is also secreted by neutrophils stimulated by lipopolysaccharides and TNF-α [54]. These results suggest that the chronic inflammatory state of the intestinal tract may be somehow affected by GCL2505 and inulin and that the inhibitory effect of *Bifidobacterium breve* A1 on brain atrophy may be related to the suppression of inflammation [55]. The effect of *Lactiplantibacillus plantarum* OLL2712, which has been shown to strongly induce IL-10 production and have an effect on chronic inflammation, was confirmed to improve cognitive function, suggesting that the suppressive effect of OLL2712 on gut and nerve inflammation may be the reason for the improvement [22]. Intake of *Bifidobacterium longum* BB68S led to a decrease in the numbers of the inflammation-inducing bacteria *Solobacterium* and *Oribacterium*, indicating that improved cognitive function might be due to reduced inflammation [56]. Akbari et al. reported that administration of multiple probiotics had a favorable effect on *hs*-CRP as well as MMSE scores, malondialdehyde, insulin metabolic markers, and triglyceride levels in patients with AD [57]. An intervention involving the intake of *Bifidobacterium bifidum* BGN4 and *Bifidobacterium longum* BORI decreased intestinal levels of inflammatory bacteria and increased BDNF levels, suggesting that these bifidobacteria might have an anti-inflammatory effect [58]. The effects of probiotics on cognitive function have also been verified by multiple clinical trials. Many of these effects are presumed to be due to the alleviation of inflammation, but in many cases, the results from indirectly assessing inflammatory conditions have been used. In contrast, in the present study, multiple inflammation markers were measured for the first time in a study in which probiotics were ingested, thereby directly demonstrating the relationship between inflammation and the effect of probiotics on cognitive function. Although further research is needed to understand the inflammatory state, it is believed that the effects of GCL2505 and inulin on cognitive function are strongly related to the inflammatory state in the gut.

The results of the SF-36v2® health-related quality of life assessment in the present study did not reveal any differences between the active and placebo groups. All scores at week 0 were above 50 except for the bodily pain scale. It is possible that the participants in this study did not originally have low SF-36v2® scores, so the improvement effect was not apparent. The selection criteria "subjective or other symptoms of forgetfulness" established in this study suggested that the person's quality of life was not significantly impaired. Also, we did not confirm the effect of GCL2505 and inulin on BDNF, which is a neurotrophic factor that is essential for synaptogenesis, plasticity, and neuroimmune responses and plays an important role in learning, memory formation, and affective disorders [59,60]. Although BDNF levels have been reported to be associated with gut microbiota [61,62], meta-analyses have shown no correlation between probiotic intake and BDNF levels. [63]. Given the inconsistent relationship between BDNF and the gut microbiota observed here, further research is warranted.

Just to be sure, no serious health problems have been reported as a result of the consumption of foods containing GCL2505. In addition, foods using *Bifidobacterium animalis* subsp. *lactis* have been sold all over the world, and no health hazard has been reported to be caused by these bifidobacteria. So the safety of GCL2505 is assured, but in some cases, intake of GCL2505 may cause slightly increased farting or softer stools in some people.

5. Conclusions

This is the first randomized controlled trial to demonstrate the efficacy of GCL2505 and inulin in improving memory function in the elderly. Elderly patients with early memory loss who consumed GCL2505 and inulin for 12 weeks showed significant improvements in scores in the neurocognitive index domain, which is an assessment of overall cognitive function, in addition to the complex attention, cognitive flexibility, and executive function domains, and the number of bifidobacteria in feces increased significantly. Because there is currently no effective pharmacological therapy to prevent the onset and progression of cognitive decline in the pre-dementia stage, the findings of this study suggest that continuous intake of GCL2505 and inulin may be an effective approach to protect memory function in the elderly.

Supplementary Materials: The following supporting information can be downloaded at: https://www.mdpi.com/article/10.3390/nu15194175/s1, Table S1: CONSORT 2010 checklist of information to include when reporting a randomized trial. Table S2: Post-intervention changes in Cognitrax scores (placebo group: $n = 32$; active group: $n = 31$), Table S3: Post-intervention changes in each inflammatory marker (all data), Table S4: Post-intervention changes in IL-1β and *hs*-CRP scores, Figure S1: Principal coordinate analysis with *t*-distribution ellipses between the active and the placebo groups.

Author Contributions: Conceptualization, N.A., T.M. and Y.S.; methodology, T.M. and Y.S.; validation, T.M. and Y.S.; formal analysis, T.M. and M.S.; investigation, N.A., T.M. and M.T.; writing—original draft preparation, N.A. and Y.S.; writing—review and editing, T.M., Y.S., M.T. and Y.I.; visualization, N.A., M.T. and M.S.; supervision, Y.I.; funding acquisition, Y.S. All authors have read and agreed to the published version of the manuscript.

Funding: This research received no external funding.

Institutional Review Board Statement: The study was conducted in accordance with the Declaration of Helsinki and approved by the Ethical Committee of Kobuna Orthopedics Clinic (approval date: 12 March 2022; approval number: MK-2205-01).

Informed Consent Statement: Informed consent was obtained from all participants involved in the study.

Data Availability Statement: Datasets generated during the current study and/or analyzed during the current study are available from the responsible author upon reasonable request.

Conflicts of Interest: N.A., T.M., Y.S., M.T., and M.S. are employees of Ezaki Glico Co., Ltd. The other author, Y.I., reports no conflict of interest in this work.

References

1. World Health Organization. Dementia. Available online: https://www.who.int/news-room/fact-sheets/detail/dementia (accessed on 15 March 2023).
2. World Health Organization. *Global Status Report on the Public Health Response to Dementia*; World Health Organization: Geneva, Switzerland, 2021.
3. Hosaka, A.; Araki, W.; Oda, A.; Tomidokoro, Y.; Tamaoka, A. Statins reduce amyloid β-peptide production by modulating amyloid precursor protein maturation and phosphorylation through a cholesterol-independent mechanism in cultured neurons. *Neurochem. Res.* **2013**, *3*, 589–600. [CrossRef] [PubMed]

4. Soeda, Y.; Yoshikawa, M.; Almeida, O.F.X.; Sumioka, A.; Maeda, S.; Osada, H.; Kondoh, Y.; Saito, A.; Miyasaka, T.; Kimura, T.; et al. Toxic tau oligomer formation blocked by capping of cysteine residues with 1,2-dihydroxybenzene groups. *Nat. Commun.* **2015**, *6*, 10216. [CrossRef] [PubMed]
5. Jack, C.R., Jr.; Knopman, D.S.; Jagust, W.J.; Shaw, L.M.; Aisen, P.S.; Weiner, M.W.; Petersen, R.C.; Trojanowski, J.Q. Hypothetical model of dynamic biomarkers of the Alzheimer's pathological cascade. *Lancet Neurol.* **2010**, *9*, 119–128. [CrossRef] [PubMed]
6. Petersen, R.C.; Smith, G.E.; Waring, S.C.; Ivnik, R.J.; Tangalos, E.G.; Kokmen, E. Mild cognitive impairment: Clinical characterization and outcome. *Arch. Neurol.* **1999**, *56*, 303–308. [CrossRef]
7. Plassman, B.L.; Langa, K.M.; Fisher, G.G.; Heeringa, S.G.; Weir, D.R.; Ofstedal, M.B.; Burke, J.R.; Hurd, M.D.; Potter, G.G.; Rodgers, W.L.; et al. Prevalence of cognitive impairment without dementia in the United States. *Ann. Intern. Med.* **2008**, *148*, 427–434. [CrossRef]
8. Ganguli, M.; Jia, Y.; Hughes, T.F.; Snitz, B.E.; Chang, C.-C.H.; Berman, S.B.; Sullivan, K.J.; Kamboh, M.I. Mild cognitive impairment that does not progress to dementia: A population-based study. *J. Am. Geriatr. Soc.* **2019**, *67*, 232–238. [CrossRef]
9. Shimada, H.; Makizako, H.; Park, H.; Doi, T.; Lee, S. Validity of the national center for geriatrics and gerontology-functional assessment tool and mini-mental state examination for detecting the incidence of dementia in older Japanese adults. *Geriatr. Gerontol. Int.* **2017**, *17*, 2383–2388. [CrossRef]
10. Yusufov, M.; Weyandt, L.L.; Piryatinsky, I. Alzheimer's disease and diet: A systematic review. *Int. J. Neurosci.* **2017**, *127*, 161–175. [CrossRef]
11. Ozato, N.; Saitou, S.; Yamaguchi, T.; Katashima, M.; Misawa, M.; Jung, S.; Mori, K.; Kawada, H.; Katsuragi, Y.; Mikami, T.; et al. Association between visceral fat and brain structural changes or cognitive function. *Brain Sci.* **2021**, *11*, 1036. [CrossRef]
12. Toppala, S.; Ekblad, L.L.; Lötjönen, J.; Helin, S.; Hurme, S.; Johansson, J.; Jula, A.; Karrasch, M.; Koikkalainen, J.; Laine, H.; et al. Midlife insulin resistance as a predictor for late-life cognitive function and cerebrovascular lesions. *J. Alzheimers Dis.* **2019**, *72*, 215–228. [CrossRef]
13. Holmes, C.; Cunningham, C.; Zotova, E.; Woolford, J.; Dean, C.; Kerr, S.; Culliford, D.; Perry, V.H. Systemic inflammation and disease progression in Alzheimer disease. *Neurology* **2009**, *73*, 768–774. [CrossRef] [PubMed]
14. Pistollato, F.; Cano, S.S.; Elio, I.; Vergara, M.M.; Giampieri, F.; Battino, M. Role of gut microbiota and nutrients in amyloid formation and pathogenesis of Alzheimer disease. *Nutr. Rev.* **2016**, *74*, 624–634. [CrossRef] [PubMed]
15. Saitou, K.; Ochiai, R.; Kozuma, K.; Sato, H.; Koikeda, T.; Osaki, N.; Katsuragi, Y. Effect of chlorogenic acids on cognitive function: A randomized, double-blind, placebo-controlled trial. *Nutrients* **2018**, *10*, 1337. [CrossRef] [PubMed]
16. Asama, T.; Hiraoka, T.; Ohkuma, A.; Okumura, N.; Yamaki, A.; Urakami, K. Cognitive improvement and safety assessment of a dietary supplement containing propolis extract in elderly Japanese: A placebo-controlled, randomized, parallel-group, double-blind human clinical study. *Evid. Based Complement. Alternat. Med.* **2021**, *2021*, 6664217. [CrossRef] [PubMed]
17. Sekikawa, T.; Kizawa, Y.; Li, Y.; Takara, T. Cognitive function improvement with astaxanthin and tocotrienol intake: A randomized, double-blind, placebo-controlled study. *J. Clin. Biochem. Nutr.* **2020**, *67*, 307–316. [CrossRef] [PubMed]
18. Zhu, G.; Zhao, J.; Zhang, H.; Chen, W.; Wang, G. Probiotics for mild cognitive impairment and Alzheimer's disease: A systematic review and meta-analysis. *Foods* **2021**, *10*, 1672. [CrossRef]
19. Sanborn, V.; Azcarate-Peril, M.A.; Updegraff, J.; Manderino, L.; Gunstad, J. Randomized clinical trial examining the impact of *Lactobacillus rhamnosus* GG probiotic supplementation on cognitive functioning in middle aged and older adults. *Neuropsychiatr Dis. Treat.* **2020**, *16*, 2765–2777. [CrossRef]
20. Kobayashi, Y.; Sugahara, H.; Shimada, K.; Mitsuyama, E.; Kuhara, T.; Yasuoka, A.; Kondo, T.; Abe, K.; Xiao, J.-Z. Therapeutic potential of *Bifidobacterium breve* strain A1 for preventing cognitive impairment in Alzheimer's disease. *Sci. Rep.* **2017**, *7*, 13510. [CrossRef]
21. Xiao, J.; Katsumata, N.; Bernier, F.; Ohno, K.; Yamauchi, Y.; Odamaki, T.; Yoshikawa, K.; Ito, K.; Kaneko, T. Probiotic *Bifidobacterium breve* in improving cognitive functions of older adults with suspected mild cognitive impairment: A randomized, double-blind, placebo-controlled trial. *J. Alzheimers Dis.* **2020**, *77*, 139–147. [CrossRef]
22. Sakurai, K.; Toshimitsu, T.; Okada, E.; Anzai, S.; Shiraishi, I.; Inamura, N.; Kobayashi, S.; Sashihara, T.; Hisatsune, T. Effects of *Lactiplantibacillus plantarum* OLL2712 on memory function in older adults with declining memory: A randomized placebo-controlled trial. *Nutrients* **2022**, *14*, 4300. [CrossRef]
23. Ishizuka, A.; Tomizuka, K.; Aoki, R.; Nishijima, T.; Saito, Y.; Inoue, R.; Ushida, K.; Mawatari, T.; Ikeda, T. Effects of administration of *Bifidobacterium animalis* subsp. *lactis* GCL2505 on defecation frequency and bifidobacterial microbiota composition in humans. *J. Biosci. Bioeng.* **2012**, *113*, 587–591. [PubMed]
24. Tanaka, Y.; Takami, K.; Nishijima, T.; Aoki, R.; Mawatari, T.; Ikeda, T. Short- and long-term dynamics in the intestinal microbiota following ingestion of *Bifidobacterium animalis* subsp. *lactis* GCL2505. *Biosci. Microbiota Food Health* **2015**, *34*, 77–85. [CrossRef] [PubMed]
25. Aoki, R.; Tsuchida, S.; Arai, Y.; Ohno, K.; Nishijima, T.; Mawatari, T.; Mikami, Y.; Ushida, K. Effect of *Bifidobacterium animalis* subsp. *lactis* GCL2505 on the physiological function of intestine in a rat model. *Food Sci. Nutr.* **2016**, *4*, 782–790.
26. Aoki, R.; Kamikado, K.; Suda, W.; Takii, H.; Mikami, Y.; Suganuma, N.; Hattori, M.; Koga, Y. A proliferative probiotic *Bifidobacterium* strain in the gut ameliorates progression of metabolic disorders via microbiota modulation and acetate elevation. *Sci. Rep.* **2017**, *7*, 43522. [CrossRef]

27. Horiuchi, H.; Kamikado, K.; Aoki, R.; Suganuma, N.; Nishijima, T.; Nakatani, A.; Kimura, I. *Bifidobacterium animalis* subsp. *lactis* GCL2505 modulates host energy metabolism via the short-chain fatty acid receptor GPR43. *Sci. Rep.* **2020**, *10*, 4158.
28. Takahashi, S.; Anzawa, D.; Takami, K.; Ishizuka, A.; Mawatari, T.; Kamikado, K.; Sugimura, H.; Nishijima, T. Effect of *Bifidobacterium animalis* ssp. lactis GCL2505 on visceral fat accumulation in healthy Japanese adults: A randomized controlled trial. *Biosci. Microbiota Food Health* **2016**, *35*, 163–171. [CrossRef]
29. Mensink, M.A.; Frijlink, H.W.; van der Voort Maarschalk, K.; Hinrichs, W.L.J. Inulin, a flexible oligosaccharide I: Review of its physicochemical characteristics. *Carbohydr. Polym.* **2015**, *130*, 405–419. [CrossRef]
30. Anzawa, D.; Mawatari, T.; Tanaka, Y.; Yamamoto, M.; Genda, T.; Takahashi, S.; Nishijima, T.; Kamasaka, H.; Suzuki, S.; Kuriki, T. Effects of synbiotics containing *Bifidobacterium animalis* subsp. *lactis* GCL2505 and inulin on intestinal bifidobacteria: A randomized, placebo-controlled, crossover study. *Food Sci. Nutr.* **2019**, *7*, 1828–1837.
31. Wu, W.; Sun, M.; Chen, F.; Cao, A.T.; Liu, H.; Zhao, Y.; Huang, X.; Xiao, Y.; Yao, S.; Zhao, Q.; et al. Microbiota metabolite short-chain fatty acid acetate promotes intestinal IgA response to microbiota which is mediated by GPR43. *Mucosal Immunol.* **2017**, *10*, 946–956. [CrossRef]
32. Gualtieri, C.T.; Johnson, L.G. Reliability and validity of a computerized neurocognitive test battery, CNS Vital Signs. *Arch. Clin. Neuropsychol.* **2006**, *21*, 623–643. [CrossRef]
33. Baba, Y.; Inagaki, S.; Nakagawa, S.; Kaneko, T.; Kobayashi, M.; Takihara, T. Effect of daily intake of green tea catechins on cognitive function in middle-aged and older subjects: A randomized, placebo-controlled study. *Molecules* **2020**, *25*, 4265. [CrossRef] [PubMed]
34. Ware, J.E., Jr.; Sherbourne, C.D. The MOS 36-item short-form health survey (SF-36). I. Conceptual framework and item selection. *Med. Care* **1992**, *30*, 473–483. [CrossRef] [PubMed]
35. Awata, S.; Bech, P.; Yoshida, S.; Hirai, M.; Suzuki, S.; Yamashita, M.; Ohara, A.; Hinokio, Y.; Matsuoka, H.; Oka, Y. Reliability and validity of the Japanese version of the World Health Organization-Five Well-Being Index in the context of detecting depression in diabetic patients. *Psychiatry Clin. Neurosci.* **2007**, *61*, 112–119. [CrossRef] [PubMed]
36. Assarsson, E.; Lundberg, M.; Holmquist, G.; Björkesten, J.; Thorsen, S.B.; Ekman, D.; Eriksson, A.; Dickens, E.R.; Ohlsson, S.; Edfeldt, G.; et al. Homogenous 96-plex PEA immunoassay exhibiting high sensitivity, specificity, and excellent scalability. *PLoS ONE* **2014**, *9*, e95192. [CrossRef] [PubMed]
37. Tourlousse, D.M.; Narita, K.; Miura, T.; Sakamoto, M.; Ohashi, A.; Shiina, K.; Matsuda, M.; Miura, D.; Shimamura, M.; Ohyama, Y.; et al. Validation and standardization of DNA extraction and library construction methods for metagenomics-based human fecal microbiome measurements. *Microbiome* **2021**, *9*, 95. [CrossRef]
38. Tanaka, K.; Nakamura, Y.; Terahara, M.; Yanagi, T.; Nakahara, S.; Furukawa, O.; Tsutsui, H.; Inoue, R.; Tsukahara, T.; Koshida, S. Poor bifidobacterial colonization is associated with late provision of colostrum and improved with probiotic supplementation in low birth weight infants. *Nutrients* **2019**, *11*, 839. [CrossRef]
39. Topp, C.H.; Østergaard, S.D.; Søndergaard, S.; Bech, P. The WHO-5 Well-Being Index: A systematic review of the literature. *Psychother. Psychosom.* **2015**, *84*, 167–176. [CrossRef]
40. Gualtieri, C.T.; Johnson, L.G. Neurocognitive testing supports a broader concept of mild cognitive impairment. *Am. J. Alzheimers Dis. Other Demen.* **2005**, *20*, 359–366. [CrossRef]
41. Birkeland, E.; Gharagozlian, S.; Birkeland, K.I.; Valeur, J.; Måge, I.; Rud, I.; Aas, A.-M. Prebiotic effect of inulin-type fructans on faecal microbiota and short-chain fatty acids in type 2 diabetes: A randomised controlled trial. *Eur. J. Nutr.* **2020**, *59*, 3325–3338. [CrossRef]
42. Marques, F.Z.; Nelson, E.; Chu, P.-Y.; Horlock, D.; Fiedler, A.; Ziemann, M.; Tan, J.K.; Kuruppu, S.; Rajapakse, N.W.; El-Osta, A.; et al. High-fiber diet and acetate supplementation change the gut microbiota and prevent the development of hypertension and heart failure in hypertensive mice. *Circulation* **2017**, *135*, 964–977. [CrossRef]
43. Nakajima, H.; Nakanishi, N.; Miyoshi, T.; Okamura, T.; Hashimoto, Y.; Senmaru, T.; Majima, S.; Ushigome, E.; Asano, M.; Yamaguchi, M.; et al. Inulin reduces visceral adipose tissue mass and improves glucose tolerance through altering gut metabolites. *Nutr. Metab.* **2022**, *19*, 50. [CrossRef] [PubMed]
44. Lauridsen, J.K.; Olesen, R.H.; Vendelbo, J.; Hyde, T.M.; Kleinman, J.E.; Bibby, B.M.; Brock, B.; Rungby, J.; Larsen, A. High BMI levels associate with reduced mRNA expression of IL10 and increased mRNA expression of iNOS (NOS2) in human frontal cortex. *Transl. Psychiatry* **2017**, *7*, e1044. [CrossRef] [PubMed]
45. Waise, T.M.Z.; Toshinai, K.; Naznin, F.; NamKoong, C.; Moin, A.S.M.; Sakoda, H.; Nakazato, M. One-day high-fat diet induces inflammation in the nodose ganglion and hypothalamus of mice. *Biochem. Biophys. Res. Commun.* **2015**, *464*, 1157–1162. [CrossRef] [PubMed]
46. Thaler, J.P.; Yi, C.-X.; Schur, E.A.; Guyenet, S.J.; Hwang, B.H.; Dietrich, M.O.; Zhao, X.; Sarruf, D.A.; Izgur, V.; Maravilla, K.R.; et al. Obesity is associated with hypothalamic injury in rodents and humans. *J. Clin. Investig.* **2012**, *122*, 153–162. [CrossRef]
47. Mao, L.; Hochstetter, D.; Yao, L.; Zhao, Y.; Zhou, J.; Wang, Y.; Xu, P. Green tea polyphenol (-)-epigallocatechin gallate (EGCG) attenuates neuroinflammation in palmitic acid-stimulated BV-2 microglia and high-fat diet-induced obese mice. *Int. J. Mol. Sci.* **2019**, *20*, 5081. [CrossRef]
48. Heneka, M.T.; Kummer, M.P.; Stutz, A.; Delekate, A.; Schwartz, S.; Vieira-Saecker, A.; Griep, A.; Axt, D.; Remus, A.; Tzeng, T.-C.; et al. NLRP3 is activated in Alzheimer's disease and contributes to pathology in APP/PS1 mice. *Nature* **2013**, *493*, 674–678. [CrossRef]

49. Tzanavari, T.; Giannogonas, P.; Karalis, K.P. TNF-alpha and obesity. *Curr. Dir. Autoimmun.* **2010**, *11*, 145–156.
50. Habbas, S.; Santello, M.; Becker, D.; Stubbe, H.; Zappia, G.; Liaudet, N.; Klaus, F.R.; Kollias, G.; Fontana, A.; Pryce, C.R.; et al. neuroinflammatory TNFα impairs memory via astrocyte signaling. *Cell* **2015**, *163*, 1730–1741. [CrossRef]
51. Xiao, W.; Su, J.; Gao, X.; Yang, H.; Weng, R.; Ni, W.; Gu, Y. The microbiota-gut-brain axis participates in chronic cerebral hypoperfusion by disrupting the metabolism of short-chain fatty acids. *Microbiome* **2022**, *10*, 62.
52. Dohi, T.; Burkly, L.C. The TWEAK/Fn14 pathway as an aggravating and perpetuating factor in inflammatory diseases: Focus on inflammatory bowel diseases. *J. Leukoc. Biol.* **2012**, *92*, 265–279. [CrossRef]
53. Lee, S.J.; Kim, J.; Ko, J.; Lee, E.J.; Koh, H.J.; Yoon, J.S. Tumor necrosis factor-like weak inducer of apoptosis induces inflammation in Graves' orbital fibroblasts. *PLoS ONE* **2018**, *13*, e0209583. [CrossRef] [PubMed]
54. Arruda-Silva, F.; Bianchetto-Aguilera, F.; Gasperini, S.; Polletti, S.; Cosentino, E.; Tamassia, N.; Cassatella, M.A. Human neutrophils produce CCL23 in response to various TLR-agonists and TNFα. *Front. Cell. Infect. Microbiol.* **2017**, *7*, 176. [CrossRef] [PubMed]
55. Asaoka, D.; Xiao, J.; Takeda, T.; Yanagisawa, N.; Yamazaki, T.; Matsubara, Y.; Sugiyama, H.; Endo, N.; Higa, M.; Kasanuki, K.; et al. Effect of probiotic *Bifidobacterium breve* in improving cognitive function and preventing brain atrophy in older patients with suspected mild cognitive impairment: Results of a 24-week randomized, double-blind, placebo-controlled trial. *J. Alzheimers Dis.* **2022**, *88*, 75–95. [CrossRef] [PubMed]
56. Shi, S.; Zhang, Q.; Sang, Y.; Ge, S.; Wang, Q.; Wang, R.; He, J. Probiotic *Bifidobacterium longum* BB68S improves cognitive functions in healthy older adults: A randomized, double-blind, placebo-controlled trial. *Nutrients* **2022**, *15*, 51. [CrossRef] [PubMed]
57. Akbari, E.; Asemi, Z.; Kakhaki, R.D.; Bahmani, F.; Kouchaki, E.; Tamtaji, O.R.; Hamidi, G.A.; Salami, M. Effect of probiotic supplementation on cognitive function and metabolic status in Alzheimer's disease: A randomized, double-blind and controlled trial. *Front. Aging Neurosci.* **2016**, *8*, 256. [CrossRef]
58. Kim, C.-S.; Cha, L.; Sim, M.; Jung, S.; Chun, W.Y.; Baik, H.W.; Shin, D.-M. Probiotic supplementation improves cognitive function and mood with changes in gut microbiota in community-dwelling older adults: A randomized, double-blind, placebo-controlled, multicenter trial. *J. Gerontol. A Biol. Sci. Med. Sci.* **2021**, *76*, 32–40. [CrossRef]
59. Melzack, R. The McGill Pain Questionnaire: Major properties and scoring methods. *Pain* **1975**, *1*, 277–299. [CrossRef]
60. Schulz, K.F.; Altman, D.G.; Moher, D. CONSORT 2010 Statement: Updated guidelines for reporting parallel group randomised trials. *BMC Med.* **2010**, *8*, 18. [CrossRef]
61. Bercik, P.; Denou, E.; Collins, J.; Jackson, W.; Lu, J.; Jury, J.; Deng, Y.; Blennerhassett, P.; Macri, J.; McCoy, K.D.; et al. The intestinal microbiota affect central levels of brain-derived neurotropic factor and behavior in mice. *Gastroenterology* **2011**, *141*, 599–609. [CrossRef]
62. Rahimlou, M.; Hosseini, S.A.; Majdinasab, N.; Haghighizadeh, M.H.; Husain, D. Effects of long-term administration of multi-strain probiotic on circulating levels of BDNF, NGF, IL-6 and mental health in patients with multiple sclerosis: A randomized, double-blind, placebo-controlled trial. *Nutr. Neurosci.* **2022**, *25*, 411–422. [CrossRef]
63. Dehghani, F.; Abdollahi, S.; Shidfar, F.; Clark, C.C.T.; Soltani, S. Probiotics supplementation and brain-derived neurotrophic factor (BDNF): A systematic review and meta-analysis of randomized controlled trials. *Nutr. Neurosci.* **2023**, *26*, 942–952. [CrossRef] [PubMed]

nutrients

MDPI

Article

Glucosinolates Extracts from *Brassica juncea* Ameliorate HFD-Induced Non-Alcoholic Steatohepatitis

Ming-Jen Sheu [1,†], Mei-Chen Yeh [2,†], Ming-Chang Tsai [3,4,5], Chi-Chih Wang [3,4,5], Yen-Ling Chang [3], Chau-Jong Wang [6,7,*] and Hui-Pei Huang [7,8,*]

1. Division of Hepatogastroenterology, Department of Internal Medicine, Chi Mei Medical Center, No. 901, Zhonghua Rd. Yongkang Dist., Tainan City 71004, Taiwan; hmj@mail.chimei.org.tw
2. Division of Metabolism and Endocrinology, Department of Internal Medicine, Chi Mei Medical Center, Tainan 71004, Taiwan; jamie.wien@gmail.com
3. Institute of Medicine, Chung Shan Medical University, Taichung 402, Taiwan; tsaimc1110@gmail.com (M.-C.T.); bananaudwang@gmail.com (C.-C.W.); hhpei0422@gmail.com (Y.-L.C.)
4. School of Medicine, Chung Shan Medical University, Taichung 402, Taiwan
5. Division of Gastroenterology and Hepatology, Department of Internal Medicine, Chung Shan Medical University Hospital, Taichung 402, Taiwan
6. Department of Health Industry Technology Management, Chung Shan Medical University, Taichung 402, Taiwan
7. Department of Medical Research, Chung Shan Medical University Hospital, Taichung 402, Taiwan
8. Department of Biochemistry, School of Medicine, Chung Shan Medical University, Taichung 40242, Taiwan

* Correspondence: wcj@csmu.edu.tw (C.-J.W.); hhpei@csmu.edu.tw (H.-P.H.); Tel.: +886-4-24730022 (ext. 11670) (C.-J.W.); +886-4-24730022 (ext. 26032) (H.-P.H.)

† These authors contributed equally to this work.

Abstract: Non-alcoholic fatty liver disease (NAFLD) is mainly characterized by excessive fat accumulation in the liver. It spans a spectrum of diseases from hepatic steatosis to non-alcoholic steatohepatitis (NASH), fibrosis, cirrhosis, and hepatocellular carcinoma (HCC). *Brassica juncea* is rich in glucosinolates and has been proven to possess many potential pharmacological properties, including hypoglycemic, anti-oxidation, anti-inflammatory, and anti-carcinogenic activities. This study aims to investigate whether whole-plant *Brassica juncea* (WBJ) and its glucosinolates extracts (BGE) have hepatoprotective effects against a high-fat diet (HFD)-induced NAFLD and further explore the mechanism underlying this process in vivo and in vitro. WBJ treatment significantly reduced body fat, dyslipidemia, hepatic steatosis, liver injury, and inflammation; WBJ treatment also reversed the antioxidant enzyme activity to attenuate oxidative stress in HFD-fed rat liver. Moreover, WBJ and BGE enhanced the activation of AMPK to reduce SREBPs, fatty acid synthase, and HMG-CoA reductase but increased the expression of CPT-I and PPARα to improve hepatic steatosis. In addition, WBJ and BGE could ameliorate NAFLD by inhibiting TNF-α and NF-κB. Based on the above results, this study demonstrates that WBJ and BGE ameliorate HFD-induced hepatic steatosis and liver injury. Therefore, these treatments could represent an unprecedented hope toward improved strategies for NAFLD.

Keywords: non-alcoholic fatty liver disease (NAFLD); *Brassica juncea*; glucosinolates; high-fat diet (HFD); AMPK

1. Introduction

Worldwide, the incidence of chronic liver disease is increasing year by year, resulting in liver-related morbidity and increased mortality [1]. Non-alcoholic fatty liver disease (NAFLD) is the most common causes of chronic liver disease. Many studies have clearly shown that NAFLD is a multi-organ disease that is related to type 2 diabetes mellitus (T2DM), cardiovascular diseases (CVD), and chronic kidney disease (CKD) [2]. Methods to effectively ameliorate NAFLD have become a global issue that should not be underestimated. NAFLD is related to lipotoxicity, which is caused by the accumulation of toxic metabolites derived from triglycerides in the liver, pancreas, and muscles, leading to a cascade of inflammation and insulin resistance [3]. The pathogenic mechanism of NAFLD is still unclear. So far, the two-hit theory proposed by Day et al. is the most widely accepted hypothesis [4]. The first hit is brought about by obesity and insulin resistance (IR). IR can promote the synthesis of triglycerides and reduce fatty acid β-oxidation in mitochondria, which accelerates the accumulation of excessive triglycerides and free fatty acids in the liver, resulting in hepatic steatosis [5]. The second hit is related to oxidative stress and abnormal mitochondrial function. Severe steatosis leads to lipid toxicity, induces the increase in the CYP2E1 and CYP4A activities of cytochrome P-450, produces oxidative free radicals, and releases inflammatory cytokines (e.g., TNF-α, IL-6, etc.), which eventually causes more serious complications such as steatohepatitis, liver fibrosis, liver cirrhosis, or liver cancer [6].

Research has demonstrated that, if dysregulated, AMPK and its related pathways are associated with metabolic diseases such as cancer, inflammatory diseases, obesity, and diabetes. AMPK activates the upstream kinases of 3-hydroxy-3-methylglutaryl-CoA reductase (HMG CoAR) and acetyl-CoA carboxylase (ACC), which, in turn, phosphorylate HMG CoAR and ACC, therefore inhibiting cholesterol biosynthesis and fatty acid synthesis/lipogenesis [7]. In addition, AMPK inhibits glycolysis to control hepatic glucose production by downregulating phosphoenolpyruvate carboxykinase and glucose 6-phosphatase [8]. Therefore, restoring or activating the expression of the AMPK pathway could potentially be therapeutic in NAFLD.

Peroxisome proliferator-activated receptor α (PPARα) is a ligand-activated transcription factor that is essential for fatty acid metabolism [9]. Many studies have shown that fatty acids are lipophilic molecules, which can be used as ligands to activate PPARα. Activated PPARα triggers several genes, including lipoprotein lipase genes that allow lipoprotein particles to release fatty acids, which conversely control the rates of fat catabolism, lipogenesis, and ketone body synthesis [10]. Therefore, PPARα is expressed in many organs with active fatty acid oxidation such as the liver, heart, and skeletal muscle [11]. The influences of PPARα on both acute and chronic inflammatory processes have been identified in in vitro and in vivo studies. PPARα are essential to the prevention of NAFLD, and extra-hepatocyte PPARα activity contributes to whole-body lipid homeostasis. Moreover, PPARα also improves metabolic syndromes, including obesity, atherosclerosis, and NASH, by inhibiting chronic inflammation [12]. For instance, PPARα negatively regulates hepatic inflammatory responses induced by an improper diet by repressing the signaling pathways controlled by the AP-1 and NF-κB transcription factors [13].

Brassica juncea (mustard) contains ascorbic acid, carotenoids, fiber, polyphenols, glucosinolates, and other ingredients, which have the effects of improving metabolic syndromes such as obesity, dyslipidemia, and diabetes [14]. Polyphenols and glucosinolates are considered to be the main functional components for disease prevention. Many studies have confirmed that glucosinolates have the effects of anti-inflammation and anti-oxidation, increasing the activity of detoxifying enzymes and inhibiting the mitosis of cancer cells, thereby preventing the occurrence of cancer [15]. Glucosinolates and their hydrolysates can activate phase II detoxification enzymes and antioxidant enzymes, as well as eliminate reactive oxygen species and lipid peroxidation by regulating the Nrf2/ARE signaling pathway [16]. However, the potential effects of glucosinolates and the mechanism of their action against a high-fat diet (HFD)-induced NAFLD remain unclear. In this study, the efficacy and mechanism of *Brassica juncea* and its functional component, glucosinolates, were analyzed in improving NAFLD in vivo and in vitro.

2. Materials and Methods

2.1. Preparation of Whole-Plant Brassica juncea (WBJ) and Glucosinolates Extracts from Brassica juncea (BGE)

Whole plants of *Brassica juncea* (WBJ) were freeze-dried using a rotary evaporator. The glucosinolates were extracted from 2 g WBJ with 20 mL 70% aqueous methanol solution in a water bath at 75 °C for 15 min. The extract was centrifuged at 3000 rpm for 5 min, and the supernatant was collected. All of the supernatant was pipetted onto a column containing a DEAE Sephadex A-25 anion exchange resin (GE Healthcare, Uppsala, Sweden). Glucosinolates were eluted from the column with 0.5 mol/L potassium sulphate. The eluent was vacuum freeze-dried to yield a powder as the glucosinolates extract of *Brassica juncea* (BGE). All BGE were dissolved with sterile ddH_2O and filtered through a 0.22 μm syringe filter and stored at −20 °C until analysis with high-performance liquid chromatography (HPLC) or until performance of the cell experiment.

2.2. HPLC Analysis

The glucosinolate composition in *Brassica Juncea* was determined using HPLC according to a previous study [17]. The HPLC system (Hitachi, Danbury, CT, USA) consisted of a pump (L-6200A), an ultraviolet detector (L-4250), and a Hitachi D-7000 HPLC system manager program. Then, 10 μL BGE was injected into the GL Sciences Inertsil ODS-2 column (5 μm particle size, 4.6 × 250 mm). Mobile-phase solvent A (100% distilled water) and solvent B (20% acetonitrile in water) were used for the elution of compounds at a flow rate of 1.5 mL/min. The absorbance spectrum was detected at a wavelength of 229 nm. Sinigrin (SIN), progoitrin (PRO), and gluconapin (NAP) were used as standards.

2.3. Animal Models and Treatment

Male Wistar rats (200 ± 10 g) were purchased from the BioLASCO Taiwan Co., Ltd. (Taipei, Taiwan). The rats were housed and used according to the guidelines of the Taiwan regulations for animal care. All animal experimental protocols were reviewed and approved by the Institutional Animal Care and Use Committee (IACUC) of the Chung Shan Medical University, Taichung, Taiwan (Approval Number 1770). The rats were housed in standard laboratory conditions (22 ± 2 °C, 60 ± 5% relative humidity, and under a 12 h light/dark cycle). Food and water were provided ad libitum to help rats adapt to the new environment for at least 1 week before the experimental procedures. In total, 60 rats were randomly divided into 5 groups (n = 12 per group): control; HFD; HFD + 0.5% WBJ; HFD + 1.0% WBJ; HFD + 2.0% WBJ (Figure 1). The feed formulation for the rats is presented in Table 1. All rats except the control group were fed with the HFD for 4 weeks prior to WBJ administration to trigger non-alcoholic steatohepatitis. After 4 weeks, the rats were fed an HFD and given different concentrations of WBJ (0.5%, 1.0%, and 2.0%) simultaneously for 8 weeks. During this period, all rats were weighed every week. At the end of the experiment, all animals were euthanized, and the liver, blood, and visceral fat were subsequently harvested for further analysis.

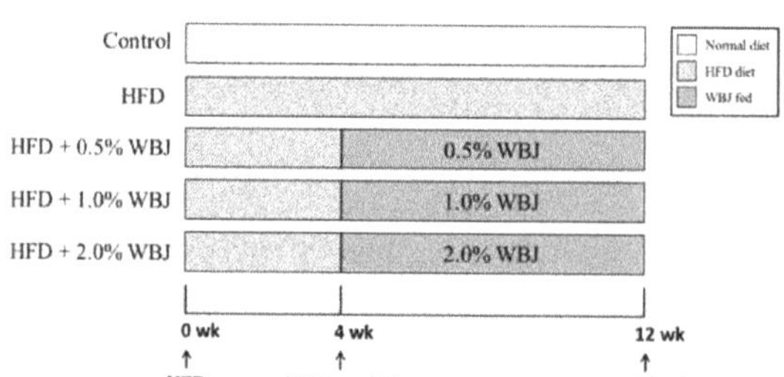

Figure 1. Schematic representation of WBJ treatment in Wistar rats.

Table 1. Feed formulation of Wistar rats.

Ingredients (g/kg Dietary Weight)	Control	HFD			
		-	LD	MD	HD
Casein	260	260	260	260	260
Corn sstarch	500	150	150	150	150
Sucrose	90	90	90	90	90
Corn oil	50				
Beef tallow		400	400	400	400
Cellulose	50	50	50	50	50
Mineral mixture [a]	40	40	40	40	40
Vitamin mixture [a]	10	10	10	10	10
WBJ			5	10	20

[a], Mineral and vitamin mixtures (AIN-76) were purchased from Oriental Yeast (Tokyo, Japan). Control, normal diet; HFD, high-fat diet; LD, low dose (contained 0.5% WBJ); MD, middle dose (contained 1.0% WBJ); HD, high dose (contained 2.0 % WBJ). WBJ, whole-plant *Brassica juncea*.

2.4. Blood Sample Analysis

Samples were harvested with cardiac puncture and immediately collected into BD vacutainer TM heparin blood collection tubes. The blood samples were centrifuged at 3000 rpm for 5 min at 4 °C. The serum was separated and stored at −20 °C for further analysis. The levels of total cholesterol, triglyceride, aspartate transaminase (AST), and alanine transaminase (ALT) in the serum were measured with clinical chemistry reagent kits (Randox Laboratories, Ltd., Antrim, UK). The free fatty acid (FFA) level in the serum was determined with a quantification kit (BioVision, Inc., Mountain View, CA, USA) according to the manufacturer's instructions.

2.5. Hepatic Histologic Analysis

Liver tissues were excised and fixed in 10% (*v/v*) neutral buffered formalin overnight. Then, the specimens were paraffin embedded. The paraffin blocks were sectioned at 5 μm and stained with a hematoxylin and eosin (H&E) reagent. Frozen sections were stained with Oil red O. The sections were observed at 200× or 400× magnification under optical microscopy.

2.6. Determination of Total Triglyceride and Cholesterol Contents in the Liver

The rat liver samples (0.5 g) were homogenized with chloroform/methanol (1:2, *v*/*v*) at 4 °C for 3 min. Then, chloroform and distilled water (1:1, *v*/*v*) were added to the homogenate and vortexed for 1 min. The homogenate was centrifuged at 3000 rpm for 12 min, and the lipid extract in the lowest layer was dissolved with 200 μL isopropanol after it was completely air dried. The triglyceride and cholesterol levels in the hepatic samples were measured through commercial kits (Human, Wiesbaden, Germany) according to the manufacturer's protocol.

2.7. Antioxidant Enzyme Activity Assays

The rat liver samples (0.5 g) were mechanically disrupted with 5 mL phosphate–EDTA (0.1 M KH_2PO_4 + 0.1 mM EDTA, pH 7.0) buffer with a homogenizer at 4 °C for 3 min. The homogenate was centrifuged at 3000 rpm for 30 min. After that, the supernatant was recentrifuged at 12,000 rpm for 5 min. The cleared hepatic homogenate was collected and stored at −20 °C. The activity of SOD was detected according to a previous method [18]. Then, 50 μL hepatic homogenate was mixed with 100 μL Tris–cacodylic acid buffer (50 mM, pH 8.2) and ultrapure water added to 980 μL. Lastly, 20 μL 0.2 mM pyrogallol was blended in the mixed solution. The autoxidation rate of pyrogallol was immediately assayed every 30 s for 3 min at 420 nm with a spectrophotometer and calculated to analyze the superoxide anion radical scavenging activity of SOD. The glutathione (GSH) activity was measured as follows: 0.5 mL hepatic homogenate was mixed with 4.5 mL phosphate–EDTA buffer (0.1 M Na_2HPO_4, 5 mM EDTA, pH 8.0), and the mixture was diluted 10 times in phosphate–EDTA buffer. After diluting, 0.1 mL o-phthalaldehyde (5 mg/mL, dissolved in ethanol) was mixed with the dilute solution.

The final reaction was performed for 15 min at room temperature and protected from light. The fluorescence intensity was detected at an emission spectrum of 420 nm and an excitation spectrum of 350 nm through a fluorophotometer. GSH-Rd activity was assayed as described previously [19]. Then, 10 μL hepatic homogenate was mixed with 90 μL potassium phosphate buffer solution (20 mM, pH 7.0), followed by incubation in 900 μL of the reaction solution (1.1 mM $MgCl_2$•$6H_2O$, 5.0 mM GSSG and 0.1 mM NADPH in potassium phosphate buffer (100 mM, pH7.0)). The reduction rate of NADPH was immediately monitored every 30 s for 5 min at a wavelength of 340 nm with a spectrophotometer.

2.8. Cell Culture and Treatment

The human hepatocellular carcinoma cell line HepG2 was purchased as a commercially available product (Food Industry Research and Development Institute, Hsinchu, Taiwan). HepG2 cells were cultured in high-glucose Dulbecco's Modified Eagle Medium (DMEM) that was supplemented with 10% fetal bovine serum (FBS), 1.5 g/L sodium bicarbonate, 4 mM L-glutamine, 1 mM sodium pyruvate, 1 mM non-essential amino acids solution (NEAA), and 1 mM PSA. The cultured cells were grown at 37 °C in a humidified incubator with 5% CO_2. To establish the in vitro model of cellular fat accumulation, HepG2 cells at 70% confluence were stimulated with 0.3 mM of a long-chain oleic acid (OA)/1% BSA complex and treated with various concentrations of BGE (50 μM SIN, 50 μM PRO, 50 μM NAP, and 2.0–6.0 mg/mL BGE) for 24 h.

2.9. Cell Viability and Cytotoxicity Assays

Cell viability was calculated through the reduction of MTT (3-(4,5-dimethylthiazol-2-yl)-2,5-diphenyltetrazolium bromide). Cells were frequently cultured for 24 h, followed by treatment with 0.3 mM OA/1% BSA and various concentrations of BGE for 24 h. Thereafter, the medium was removed, and the MTT reagent (0.5 mg/mL in culture medium) was subsequently added to each well for 4 h. The reduced purple-blue MTT formazan crystals produced by viable cells were solubilized with 1 mL isopropanol, and the solution was centrifuged at 1200 rpm for 5 min. After centrifuging, the supernatant was collected and measured at 563 nm with an ELISA reader.

2.10. Nile Red Stain

Cellular lipid accumulation was assessed with Nile red staining. HepG2 cells were treated with 0.3 mM OA/1% BSA and various concentrations of BGE (50 μM SIN, 50 μM PRO, 50 μM NAP, and 2.0 mg/mL BGE) for 24 h. Adherent cells were separated with trypsin for 5 min and then centrifuged at 1000 rpm for 10 min. The supernatant was removed and washed twice with PBS. Cells stained with the Nile red solution were removed from light exposure for 30 min at room temperature. The cellular lipid contents were analyzed with flow cytometry (BD Biosciences, Franklin Lakes, NJ, USA).

2.11. Preparation of Cells and Hepatic Proteins

The protein extracts from cultured cells or liver tissues were lysed and extracted with a RIPA buffer (150 mM NaCl, 0.5% deoxycholic acid, 50 mM Tris-Base, 1% NP-40, 1% SDS, 10 μg/mL PMSF, and 10 μg/mL leupeptin, pH 7.5) containing 1% protease inhibitor (17 μg/mL leupeptin and 10 μg/mL sodium orthovanadate) and phosphatase inhibitor. The cell lysates or liver tissues were homogenized with the RIPA buffer on ice for 3 min. Then, all protein extracts were centrifuged at 12,000 rpm for 20 min at 4 °C. The protein concentration was measured with Coomassie blue (Kenlor Industries, Inc., Santa Ana, CA, USA) and BSA (bovine serum albumin) was used as a standard. The absorbance was detected at 595 nm with a spectrophotometer.

2.12. Western Blot Analysis

Protein samples were separated with a 6–15% SDS-PAGE and transferred onto nitrocellulose membranes. Membranes were blocked with commercial BlockPRO Protein-Free blocking buffer (Visual Protein) for 1 h at room temperature. Blots were washed with TBST (Tris-buffered saline with Tween 20, pH7.6) 3 times (10 min each time) and hybridized with various primary antibodies specific for p-AMPK, AMPK, SREBP-1, FAS, SREBP-2, HMG-CoR, CPT1, TNF-α, or NF-κB overnight at 4 °C. The primary antibodies were removed, and the membranes were washed with TBST followed by incubation with anti-mouse or anti-rabbit horseradish peroxidase HRP-conjugated secondary antibodies for 1 h at room temperature. The secondary antibodies were removed, and the membranes were washed with TBST. The blotted membrane was activated with enhanced chemiluminescence (ECL) and imaged with a LAS-4000 Super CCD Remote Control Science Imaging System (Fuji, Tokyo, Japan). Protein quantity was determined with densitometry using Fujifilm MultiGauge, version 3.0, software.

2.13. Statistical Analysis

Statistical analyses were performed using SigmaPlot 12.5 software (12.5.0.38). The statistical significance ($p < 0.05$) among all of the different groups was determined using the student's t test or one-way ANOVA. Results are expressed as the mean value ± SD. All results are representative of at least three independent experiments.

3. Results

3.1. Daily Intake of BGE in Rats

The major glucosinolates in BGE were sinigrin (SIN, 97.33%) and gluconapin (NAP, 2.67%) (Figure 2A–C). To simulate the pattern of human ingestion of *Brassica juncea*, the rats were directly fed whole-plant *Brassica juncea* powder (WBJ) to investigate the efficacy of *Brassica juncea* in NAFLD. According to the HPLC analysis, the daily food intake of rats of 0.5% WBJ, 1.0% WBJ, and 2.0% WBJ contained 2.314 g, 4.628 g, and 9.256 g of sinigrin and 63.7 μg, 127.4 μg, and 254.8 μg of gluconapin, respectively (Figure 2D).

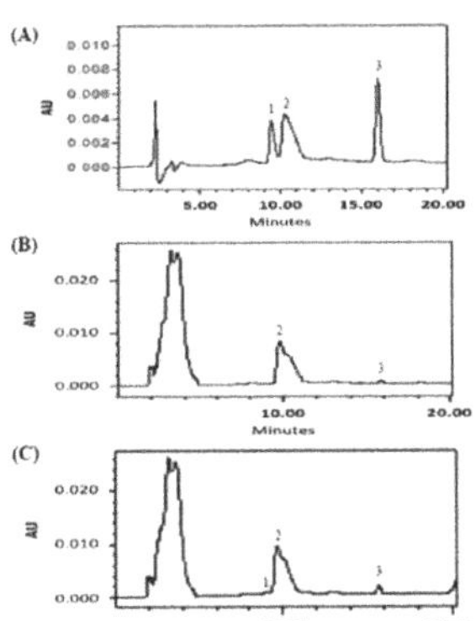

(D)

Peak No.	RT (min)	Compounds name	BGE (μg/mg)	Daily intake of BGE in animal (μg/0.5% diet) [a]	Daily intake of BGE in animal (μg/1.0% diet) [b]	Daily intake of BGE in animal (μg/2.0% diet) [c]
1	9.45	Progoitrin	N.D.	-	-	-
2	10.26	Sinigrin	35.60±1.06	2314±1.06	4628±1.06	9256±1.06
3	15.9	Gluconapin	0.98±0.08	63.7±0.08	127.4±0.08	254.8±0.08

Figure 2. The HPLC chromatogram of BGE. (**A**) HPLC chromatogram of three kinds of glucosinolates standards. Peak: 1, sinigrin (SIN); 2, progoitrin (PRO); 3, gluconapin (NAP). (**B**) HPLC chromatogram of glucosinolates from BGE (100 mg/mL, 10 μL). (**C**) HPLC chromatogram profiles of mixture BGE and 2 μg/mL glucosinolate standards. (**D**) Glucosinolate compounds identified in BGE. [a] The average daily intake in the rats was 13 g/day and BGE was 0.5% of the diet. [b] The average daily intake in the rats was 13 g/day and BGE was 1.0% of the diet. [c] The average daily intake in the rats was 13 g/day and BGE was 2.0% of the diet.

3.2. Effect of WBJ Treatment on Serum Lipid Parameters

In order to reflect a human dietary pattern, the WBJ was freeze-dried and fed to obese rats induced by the HFD. After treatment with an HFD in rats, the serum triglyceride (TG) and free fatty acids (FFA) in the HFD group were higher than the control group, whereas WBJ decreased the TG level induced by the HFD. Low-density lipoprotein cholesterol (LDL-C) and high-density lipoprotein cholesterol (HDL-C) are clinically regarded as the key factors for evaluating cardiovascular disease [20]. This study further analyzed the contents of LDL-C and HDL-C in rat serum to explore whether WBJ could reduce blood lipids and reduce the incidence of cardiovascular disease. Table 2 shows that WBJ could decrease the total cholesterol content even though the HFD did not induce the increase in cholesterol in rats. The LDL-C level of the HFD group was higher than the control group. However, the value of the LDL-C after treatment with WBJ was lower than the HFD group. Meanwhile, after induction of the high-fat diet, the HDL-C in the HFD group was decreased. The value of HDL-C tended to rise in the WBJ treatment groups. A similar result was also observed for the LDL-C/HDL-C ratio.

Table 2. Effects of WBJ on the serum biochemical parameters and hepatic lipids in HFD-fed rats.

	Control	HFD	HFD + 0.5% WBJ	HFD + 1.0% WBJ	HFD + 2.0%WBJ
Total cholesterol (mg/dL)	70.9 ± 3.57 [a]	71.3 ± 9.25	51.9 ± 6.19 [c]	55.2 ± 7.09 [c]	59.2 ± 7.23 [c]
Total triglyceride (mg/dL)	34.7 ± 7.24	121.4 ± 26.98 [b]	64.4 ± 8.77 [c]	74.2 ± 16.47 [c]	62.1 ± 15.09 [c]
FFA (mmol/L)	0.73 ± 0.10	1.12 ± 0.20 [b]	0.77 ± 0.16 [c]	0.95 ± 0.07 [c]	0.85 ± 0.10 [c]
LDL-C (mg/dL)	11.4 ± 3.27	18 ± 3.20 [b]	15.8 ± 1.75	13.9 ± 2.13 [c]	12 ± 2.11 [c]
HDL-C (mg/dL)	49.3 ± 5.96	30.8 ± 4.78 [b]	36.1 ± 6.31 [c]	37.4 ± 7.72 [c]	43.1 ± 6.61 [c]
LDL-C/HDL-C ratio	0.23 ± 0.05	0.60 ± 0.15	0.45 ± 0.09 [c]	0.38 ± 0.08 [c]	0.28 ± 0.06 [c]
Glucose (mg/dL)	155.7 ± 52.95	267.7 ± 51.48 [b]	235 ± 71.04	268.6 ± 65.61	248 ± 38.34
AST (U/L)	121.7 ± 10.45	140.2 ± 27.32	145 ± 26.44	133.8 ± 17.81	127.9 ± 11.76
ALT (U/L)	33.2 ± 3.91	59.4 ± 9.16 [b]	49.3 ± 5.81 [c]	44.2 ± 7.45 [c]	41.6 ± 8.42 [c]
BUN (mg/dL)	20.64 ± 1.81	12.92 ± 1.37 [b]	9.74 ± 0.92 [c]	11.4 ± 0.96 [c]	10.54 ± 0.39 [c]
UA (mg/dL)	4.01 ± 1.25	4.58 ± 1.05	4.75 ± 1.03	5.13 ± 1.34	4.93 ± 1.22
Creatinine (mg/dL)	0.61 ± 0.05	0.68 ± 0.04 [b]	0.62 ± 0.07 [c]	0.66 ± 0.05	0.63 ± 0.04
Liver-triglyceride (mg/dL)	253.74 ± 30.21	328.05 ± 26.48 [b]	309.23 ± 58.54	265.17 ± 54.82 [c]	194.04 ± 58.57 [c]
Liver cholesterol (mg/dL)	38.12 ± 12.43	136.38 ± 21.00 [b]	113.12 ± 31.21	99.96 ± 27.93 [c]	96.69 ± 39.85 [c]

[a] Each value is expressed as the mean ± SD (n = 10/ group). Results were statistically analyzed with a one-way ANOVA. [b] $p < 0.05$ compared to the control group. [c] $p < 0.05$ compared to the HFD group.

The liver function in rat serum was also analyzed to further observe whether WBJ could reduce liver injury. The results showed that the GOT and GPT of the HFD group were higher than in the control group. After feeding the rats WBJ, the values decreased with the increase in WBJ concentration, which indicated that WBJ could alleviate the liver injury induced by the high-fat diet. However, WBJ had a less significant effect on improving blood sugar levels and renal function in this animal model (Table 2).

3.3. Effect of WBJ on Fat Distribution in Rats

The peripheral fat around the kidneys, intestinal interstitium, and accessory testicles were collected and weighed when the mice were euthanized. The results show that the fat around these organs in the group fed with the HFD increased significantly compared with the control group. The fat around the organs decreased when treated with WBJ, especially in the 2.0% WBJ group, indicating that WBJ could effectively reduce body fat production (Table 3). Similarly, the levels of TG and CHO in the feces of the HFD group were higher than in the control group, whereas the levels of TG and CHO in the feces increased with the increase in WBJ concentration compared to the HFD group (Figure 3). The above results reveal the effect of WBJ on the inhibition of lipid accumulation.

Table 3. WBJ reduced organ peripheral fat weight in HFD-fed rats.

Tissue Weights (mg)	Control	HFD	HFD + 0.5% WBJ	HFD + 1.0% WBJ	HFD + 2.0% WBJ
Kidney fat	21.63 ± 25.56 [a]	7766.00 ± 1408.37 [b]	5978.75 ± 1844.83 [c]	5622.25 ± 1346.33 [c]	5310.25 ± 1976.94 [c]
Intestinal fat	331.63 ± 230.70	8306.75 ± 2210.45 [b]	6125.63 ± 1394.05 [c]	6037.50 ± 994.79 [c]	5783.75 ± 577.86 [c]
Gonad fat	326.63 ± 116.79	5936.75 ± 874.64 [b]	4827.13 ± 1891.91	4766.75 ± 1046.01	4426.00 ± 1184.69 [c]

[a] Each value is expressed as the mean ± SD (n = 12/group). Results were statistically analyzed with one-way ANOVA. [b] $p < 0.05$ compared to the control group. [c] $p < 0.05$ compared to the HFD group.

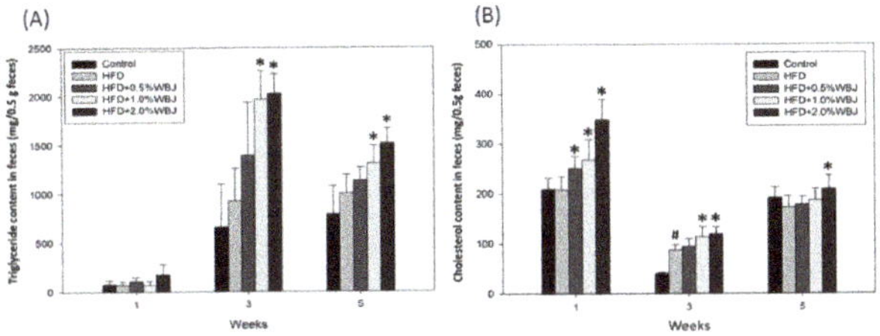

Figure 3. Effects of WBJ on lipid content in feces from HFD-fed rats. Feces levels of TG (**A**) and CHO (**B**) were measured at weeks 1, 3, and 5. Each value is expressed as the mean ± SD (n = 12/group). Results were statistically analyzed with one-way ANOVA. # $p < 0.05$ compared to the control group. * $p < 0.05$ compared to the HFD group.

3.4. WBJ Reduced Hepatic Steatosis Induced by an HFD

The liver weight/body weight ratio was used to evaluate whether WBJ ameliorated NAFLD induced by the high-fat diet. Compared with the HFD group, the ratio of liver weight to body weight was slightly decreased after feeding with WBJ. Notably, the ratio of liver weight to body weight in the 2.0% WBJ group was significantly lower than that in the HFD group (Figure 4A,B). Accordingly, WBJ could slow down hepatomegaly caused by a high-fat diet. Next, we observed the fat accumulation in rat livers with H&E staining and Oil red O staining. The results showed that there was no hepatic steatosis in the control group. However, many lipid droplets were found in the livers of the HFD group. The results indicate that WBJ could effectively reduce hepatic steatosis, especially in the 2.0% WBJ treatment group (Figure 4C).

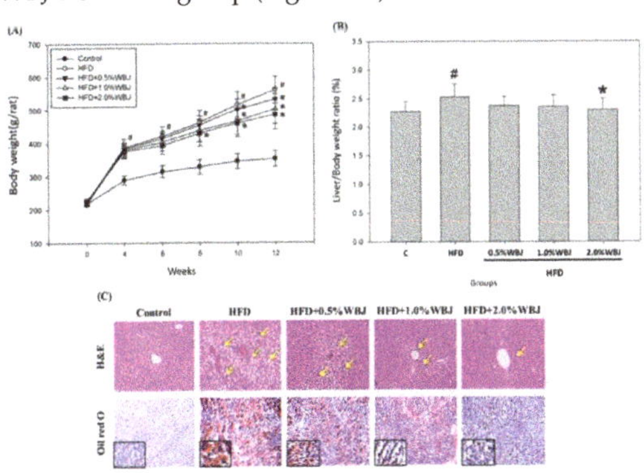

Figure 4. Effects of WBJ on the liver weight and histopathological examination of the liver tissue in HFD-fed rats. (**A**) Body weight change versus time in rats fed various diets. (**B**) Liver/body weight ratio. (**C**) H&E and Oil red O staining images of liver sections. All values are the mean ± SD (n = 12). Control, rats fed normal chow; HFD, rats fed with 40% beef tallow; HFD + 0.5% WBJ, rats fed an HFD with 0.5% WBJ; HFD + 1.0% WBJ, rats fed an HFD with 1.0% WBJ; HFD + 2.0% WBJ, rats were fed an HFD with 2.0% WBJ. Results were statistically analyzed with a one-way ANOVA. # $p < 0.05$ compared to the control group. * $p < 0.05$ compared to the HFD group. The sections were photographed at 200× or 400× magnification. The yellow arrow indicated the lipid droplet.

3.5. Mechanisms of WBJ Reducing Hepatic Steatosis Induced by an HFD

Our previous studies demonstrated that *Mulberry* leaf polyphenol extracts and *Solanum* polyphenol extracts reduced the expression of triglyceride synthesis (SREBP-1, FAS) and cholesterol synthesis (SREBP-2, HMGoCR), and increased the expression of the fatty acid β-oxidation protein CPT-1 by activating the AMPK pathway [21–23]. This study further observed the expression of lipid metabolism-related proteins with a western blot assay. The results showed that the activity of p-AMPK and the level of CPT-1 in the HFD group were lower than in the control group, and the expression of SREBP-1, FAS, SREBP-2, and HMGCoR was increased significantly. However, after treatment with WBJ, p-AMPK activity significantly increased compared to the HFD group. WBJ inhibited the expression of SREBP-1, FAS, SREBP-2, and HMGCoR induced by the HFD, and promoted CPT-1 expression (Figure 5A–D). Inflammation plays a crucial role in the development of NAFLD into non-alcoholic steatohepatitis (NASH). This study further analyzed the protein expression of inflammatory factors. The results showed that WBJ could reduce the activities of TNF-α and NF-κB in the liver induced by the HFD and had a significant effect at a low concentration of 0.5% WBJ (Figure 5E). In summary, it might be concluded that WBJ can produce lipid-lowering and anti-inflammatory effects by inhibiting the synthesis pathways of lipids and the expression of inflammatory factors, thereby reducing the occurrence of NAFLD.

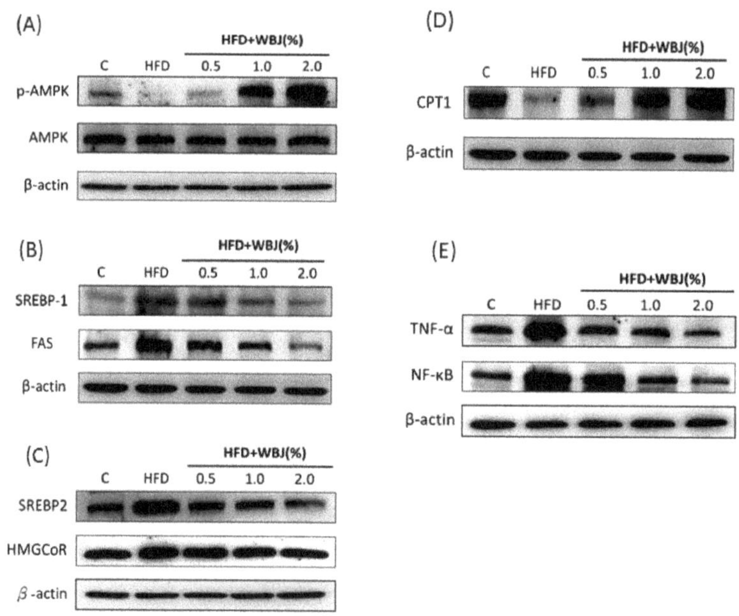

Figure 5. WBJ treatment reduced lipid synthesis-related protein and inflammatory protein expression in the liver of HFD-fed rats. Animals were fed a control diet, HFD, or HFD pair feeding with a 0.5%, 1.0%, or 2.0% WBJ diet (n = 12). Protein extracts from rat liver were measured with western blotting to detect (**A**) p-AMPK, AMPK; (**B**) SREBP-1, FAS; (**C**) SREBP-2, HMGCoR; (**D**) CPT1; and (**E**) TNF-α, NF-κb. β-actin was used as a loading control. Data are presented as the mean ± SD from three independent experiments and were statistically analyzed with a one-way ANOVA.

3.6. WBJ Enhanced Antioxidative Enzyme Activities in Rat Liver

Severe liver lipid accumulation may promote the generation of oxidative stress, leading to the deterioration into NAFLD [24]. We further explored the effect of WBJ on reducing the liver oxidative stress induced by an HFD. After feeding rats with an HFD for 12 weeks, the antioxidant enzyme activities of glutathione (GSH), glutathione reductase (GSH-Rd), and superoxide dismutase (SOD) in the liver showed a significant decrease compared to the control group. However, after treatment with WBJ, the liver antioxidant enzyme activity rose compared to the HFD group in a dose-dependent manner. Therefore, it could be concluded that WBJ can diminish NAFLD induced by an HFD by increasing the activity of antioxidant enzymes in the liver of rats (Figure 6).

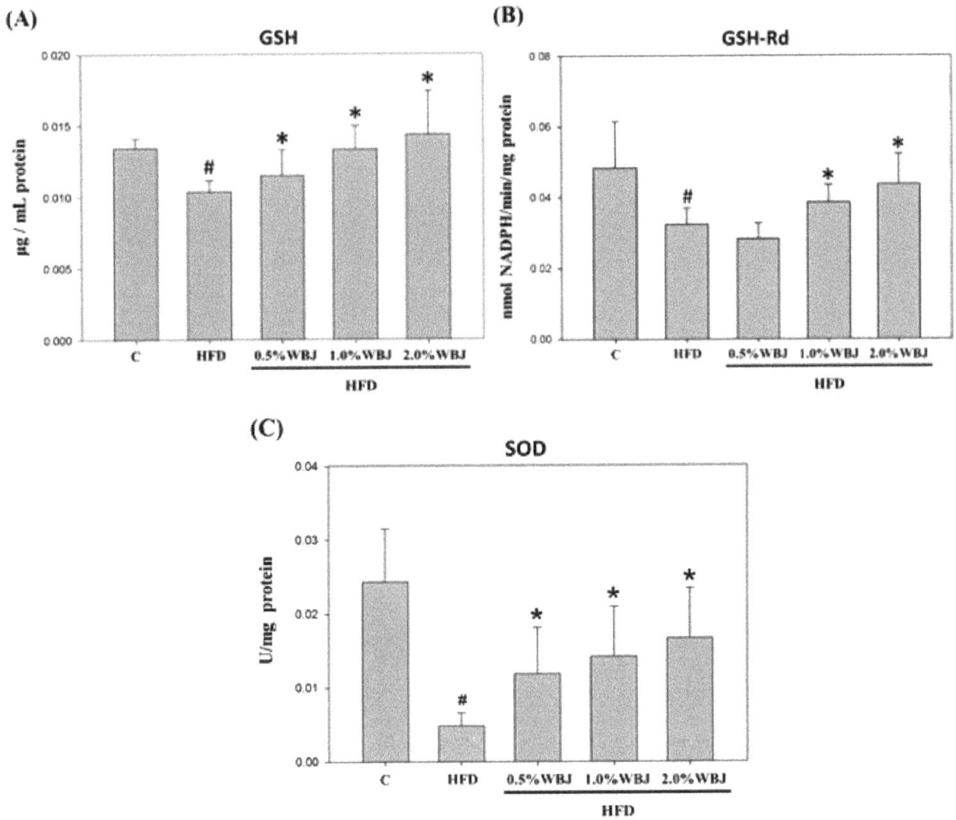

Figure 6. Effects of WBJ on the antioxidant enzyme activities in the liver of HFD-fed rats. The levels of GSH (**A**), GSH-Rd (**B**), and SOD (**C**) in the liver tissue of rats were measured as described in the Materials and Methods. All values are presented as the mean ± SD (n = 12). Results were statistically analyzed with a one-way ANOVA. # $p < 0.05$ compared to the control group. * $p < 0.05$ compared to the HFD group.

3.7. *The Cytotoxicity Effects of Brassica juncea Glucosinolates on HepG2 Cells*

Glucosinolates are a group of phytochemicals that have been shown to be beneficial against a wide range of chronic liver diseases [25]. HepG2 cells were treated with 50 μM SIN, 50 μM NAP, 2.0 mg/mL BGE, 4.0 mg/mL BGE, and 6.0 mg/ml BGE for 24 h, and the cell viability was tested with an MTT assay. The survival rate of cells in the 4.0 mg/mL BGE and 6.0 mg/mL BGE groups showed a significant decrease compared to the control group. Therefore, the dosage that did not significantly cause the death of HepG2 cells (50 μM SIN, NAP, and 2.0 mg/mL BGE) was selected for follow-up experiments (Figure 7A).

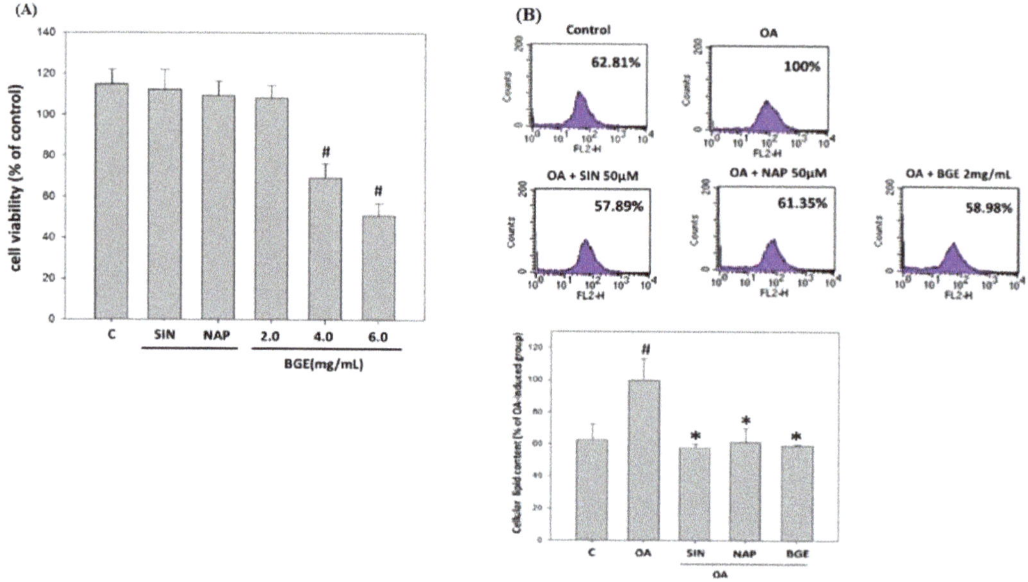

Figure 7. Effects of *Brassica juncea* glucosinolates on lipid accumulation in HepG2 cells. HepG2 cells were incubated with different glucosinolates (50 μM SIN, NAP, and 2.0, 4.0, 6.0 mg/mL BGE) at 37 °C for 24 h. (**A**) Cell viability was analyzed with an MTT assay. (**B**) HepG2 cells were stained with Nile red and analyzed using flow cytometry. The data are presented as the mean ± SD for three replicates per treatment. # $p < 0.05$ compared to the control group. * $p < 0.05$ compared to the HFD group.

3.8. *Brassica juncea* Glucosinolates Inhibited the Lipid Accumulation in HepG2 Cells

To evaluate the effect of glucosinolates on lipid accumulation in hepatocytes, HepG2 cells were exposed to oleic acid (OA); then, 50 μM SIN, 50 μM NAP, and 2.0 mg/mL BGE were added at the same time, for 24 h. The lipids in the cells were stained with Nile red and analyzed quantitatively with flow cytometry. The results showed that the lipid content in HepG2 cells in the SIN, NAP, and BGE groups was lower than that in the induction group, and reducing SIN by about 40% had the most significant effect (Figure 7B). Therefore, SIN and BGE were used in the subsequent experiment.

3.9. The Mechanisms of *Brassica juncea* Glucosinolates Reversing NAFLD

Previous studies have demonstrated that the activation of AMPK reduced the expression of triglyceride synthesis proteins, such as SREBP-1 and FAS, activated the expression of fatty acid β-oxidation proteins, such as CPT-1 and PPARα, and then inhibited the occurrence of NAFLD [26]. Our results showed that OA induced excessive lipid accumulation and inhibited the activity of p-AMPK in HepG2 cells, whereas SIN and BGE restored the p-AMPK activity (Figure 8A). Moreover, SIN and BGE decreased the levels of SREBP1 and FAS in HepG2 cells induced by OA (Figure 8B). Meanwhile, SIN and BGE accelerated the lipid β-oxidation via increasing CPT-1 and PPARα expression reduced by OA (Figure 8C). Inflammation plays a crucial role in the progression from NAFLD to more severe non-alcoholic steatohepatitis (NASH) [27]. Figure 8D shows that SIN inhibited TNF-α and NF-κB expression induced by OA in HepG2 cells, whereas BGE inhibited the inflammation of HepG2 cells in a dose-dependent manner. In summary, it can be concluded that *Brassica juncea* glucosinolates SIN and BGE can regulate lipid synthesis and inflammatory factors by activating AMPK, and these glucosinolates can also produce lipid-lowering and anti-inflammatory effects.

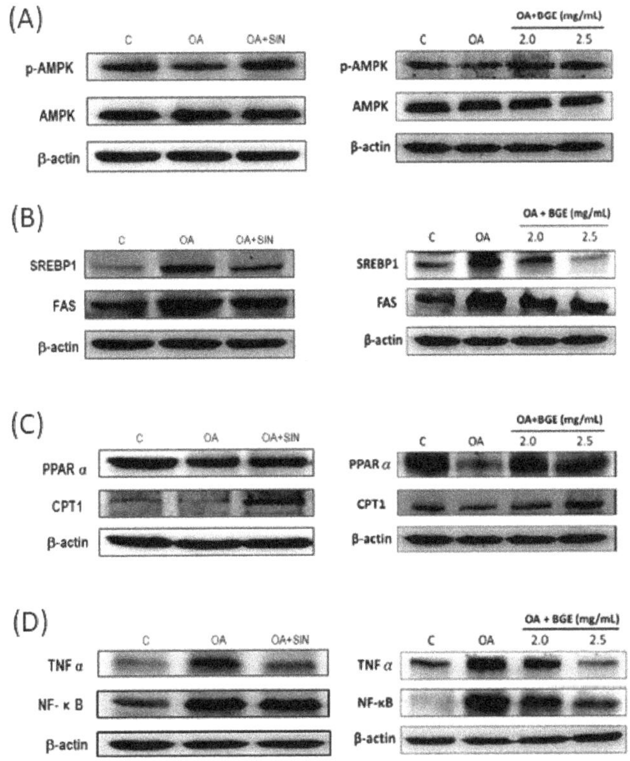

Figure 8. *Brassica juncea* glucosinolates reduced lipid synthesis-related protein and inflammatory protein expression in OA-induced HepG2 cells. HepG2 cells were treated with 0.3 mM OA and 50 μM SIN or different concentrations of BGE (2.0, 2.5 mg/mL BGE) at 37 °C for 24 h. Protein extracts from HepG2 cells were measured with western blotting to detect (**A**) p-AMPK, AMPK; (**B**) SREBP-1, FAS; (**C**) PPARα, CPT1; and (**D**) TNF-α, NF-κb. β-actin was used as a loading control. Data are presented as the mean ± SD from three independent experiments and statistically analyzed with a student's *t*-test.

4. Discussion

The pathogenesis of NAFLD is closely related to obesity, and the World Health Organization (WHO) regards obesity as the most important public health issue in the 21st century. Studies have shown that obesity induces oxidative stress and lipid peroxidation in the body. Lipid accumulation, oxidative stress, and inflammation have been widely regarded as the key factors for simple steatosis, leading to NASH [4,28,29]. Unfortunately, there are still many limitations to the medical treatment of NAFLD induced by a high-fat diet. So far, diet control and exercise are the main suggestions to treat NAFLD because there are currently no FDA-approved medications for this disease. Therefore, if the supplement of natural products can help people prevent or ameliorate NAFLD, it will improve the health quality of the whole society. Our laboratory previously demonstrated that the functional components from *Mulberry* leaf, *Nelumbo nucifera* leaf, and *Hibiscus sabdariffa* can effectively reduce body fat and improve alcoholic and non-alcoholic fatty liver disease [30–32]. The hypoglycemic, neuroprotective, antibacterial, and anticancer effects of *Brassica juncea* have been proven previously [33]. Here, our results showed that WBJ reduced body weight, body fat, and hepatomegaly in rats induced by an HFD. In both in vivo and in vitro experiments, WBJ and BGE upregulated p-AMPK expression and exhibited antioxidant enzyme activities, thereby helping to reduce lipid synthesis and promote fatty acid β-oxidation, as well as reducing the production of inflammatory factors.

In the serum lipid parameters, we observed that WBJ had a stronger effect on improving HDL values than reducing LDL values following treatment with WBJ (Table 2). This might be the reason why WBJ decreased the CHO level induced by an HFD in rats in a dose-dependent manner. Moreover, the levels of serum CHO, TG, or FFA, whether given 0.5% WBJ, 1.0% WBJ, or 2.0% WBJ, were significantly lower than those in the HFD group. Therefore, WBJ indeed had a lipid-lowering effect, and it had a significant effect at a low dose (0.5% WBJ) (Table 2).

Brassica juncea is rich in dietary fibers (3.2 g dietary fiber/100 g), which can promote gastrointestinal peristalsis, the digestion and absorption of nutrients, and regulate the interactions among intestinal microorganisms [34,35]. In addition, several studies have demonstrated that dietary fiber can elevate the excretion of lipids, reduce the production of cytokines in white adipose tissue, and inhibit systemic inflammation [36,37]. This study confirmed that WBJ promoted lipid excretion in feces. However, during the experiment, we found that there was an occult blood reaction in rat feces from the sixth week after an HFD, which may have been caused by the long-term intake of the HFD [38]. Fecal TG and CHO analyses were therefore limited in the later stage of the experiment. Nevertheless, we still found that WBJ promoted lipid excretion in the third and fifth weeks after the HFD treatment (Figure 3). The imbalance of gut microbiota causes abnormal lipid metabolism in the adipose tissue and liver, as well as immune disorders, which are closely related to metabolic syndromes such as obesity, fatty liver, diabetes, and cardiovascular disease [39, 40]. In the future, we will evaluate the effect of functional components of WBJ on increasing beneficial intestinal bacteria and reducing bad bacteria, and we will further explore the role of *Brassica juncea* in disease prevention.

Increasingly, studies have indicated that the AMPK signaling pathway plays a vital role in improving lipid metabolism disorders [41]. LKB1 (liver kinase B1, also known as serine/threonine kinase 11) phosphorylates AMPK$\alpha^{Thr\text{-}172}$, thus activating AMPK. Following this, activated AMPK (p-AMPK) inhibits the synthesis pathways of ATP consumption (such as glycogenesis, glycogen synthesis, fatty acid, and cholesterol synthesis) and promotes catabolism (such as fatty acid oxidation, glycolysis) to produce adenosine triphosphate (ATP) [7]. In addition to the LKB1 signaling pathway, there are other AMPK upstream kinases (AMPKK) that can activate AMPK, such as CaMKK and TAK1, to maintain energy balance [26,42]. Our results also showed that WBJ and BGE activated AMPK, inhibited the pathway of SREBP1/FAS, SREBP2/HMGCoR, reduced the synthesis of TG and CHO, and promoted the expression of CPT1 to increase the β-oxidation of fatty acids, thereby reducing the accumulation of lipids in the liver. In the future, the effect of *Brassica juncea* on the expression of LKB1, CaMKK, and TAK1 proteins will be evaluated to further explore the possible mechanism of mustard in improving NAFLD.

Nowadays, for many health foods, it is known that excessive supplementation is not beneficial for health. Indeed, high doses of glucosinolates may have side effects on the body. After feeding rats with 50 μmol/kg sinigrin or its degradation product, allyl isothiocyanate (25 μmol/kg and 50 μmol/kg), for 4 h, rats showed mild hyperglycemia. After continuously feeding sinigrin and allyl isothiocyanate for two weeks, the rats exhibited hyperinsulinemia, hyperlipidemia, and other systemic metabolic disorders [43]. However, the detailed mechanism of metabolic abnormalities caused by excessive glucosinolates is still not fully understood, which is also one of the issues that we need to clarify in the future.

Cruciferae plants are vegetables that are often eaten in our daily life, such as broccoli, cabbage, Chinese cabbage, and mustard. Glucosinolates, a unique component of Cruciferae, have been proved to be a multi-target natural substance that can effectively lower body fat. Glucosinolates also have detoxifying, anti-inflammatory, antioxidant, and anticancer effects [44]. However, rich secondary metabolites, such as glucosinolates, lead to the spicy and bitter taste of some cruciferous plants. The bitterness of cruciferous plants is often associated with consumer rejection and poor taste. If functional components, such as BGE, can be extracted and applied to the development of hepatoprotective nutrient foods, it is believed that its economic value can effectively be improved.

Author Contributions: Conceptualization, M.-J.S. and M.-C.Y.; Methodology, C.-J.W. and H.-P.H.; Software, M.-C.T. and C.-C.W.; Validation, C.-J.W.; Formal analysis, H.-P.H.; Resources, M.-J.S., M.-C.Y., M.-C.T., C.-C.W. and C.-J.W.; Data curation, Y.-L.C.; Writing—original draft, H.-P.H.; Writing—review & editing, H.-P.H.; Project administration, M.-J.S. and M.-C.Y. All authors have read and agreed to the published version of the manuscript.

Funding: This research was funded by Chung Shan Medical University (CSMU-CMMC-107-05) and Chi Mei Medical Center (CMCSMU10702).

Institutional Review Board Statement: The study was conducted in accordance with the Institutional Animal Care and Use Committee (IACUC) of the Chung Shan Medical University, Taichung, Taiwan (Approval Number 1770).

Informed Consent Statement: Not applicable.

Data Availability Statement: No new data were created or analyzed in this study. Data sharing is not applicable to this article.

Acknowledgments: This investigation was supported by the Chung Shan Medical University (CSMU-CMMC-107-05) and Chi Mei Medical Center (CMCSMU10702).

Conflicts of Interest: The authors declare that they have no competing interest.

References

1. Younossi, Z.M.; Koenig, A.B.; Abdelatif, D.; Fazel, Y.; Henry, L.; Wymer, M. Global epidemiology of nonalcoholic fatty liver disease-Meta-analytic assessment of prevalence, incidence, and outcomes. *Hepatology* **2016**, *64*, 73–84. [CrossRef] [PubMed]
2. Byrne, C.D.; Targher, G. NAFLD: A multisystem disease. *J. Hepatol.* **2015**, *62* (Suppl. S1), S47–S64. [CrossRef]
3. Targher, G.; Marchesini, G.; Byrne, C.D. Risk of type 2 diabetes in patients with non-alcoholic fatty liver disease: Causal association or epiphenomenon? *Diabetes Metab.* **2016**, *42*, 142–156. [CrossRef]
4. Day, C.P.; James, O.F. Steatohepatitis: A tale of two "hits"? *Gastroenterology* **1998**, *114*, 842–845. [CrossRef] [PubMed]
5. Lomonaco, R.; Ortiz-Lopez, C.; Orsak, B.; Webb, A.; Hardies, J.; Darland, C.; Finch, J.; Gastaldelli, A.; Harrison, S.; Tio, F.; et al. Effect of adipose tissue insulin resistance on metabolic parameters and liver histology in obese patients with nonalcoholic fatty liver disease. *Hepatol. Baltim. Md.* **2012**, *55*, 1389–1397. [CrossRef] [PubMed]
6. Farrell, G.C.; van Rooyen, D.; Gan, L.; Chitturi, S. NASH is an Inflammatory Disorder: Pathogenic, Prognostic and Therapeutic Implications. *Gut Liver* **2012**, *6*, 149–171. [CrossRef]
7. Mihaylova, M.M.; Shaw, R.J. The AMPK signalling pathway coordinates cell growth, autophagy and metabolism. *Nat. Cell Biol.* **2011**, *13*, 1016–1023. [CrossRef]
8. Kim, T.; Davis, J.; Zhang, A.J.; He, X.; Mathews, S.T. Curcumin activates AMPK and suppresses gluconeogenic gene expression in hepatoma cells. *Biochem. Biophys. Res. Commun.* **2009**, *388*, 377–382. [CrossRef]
9. Inagaki, T.; Dutchak, P.; Zhao, G.; Ding, X.; Gautron, L.; Parameswara, V.; Li, Y.; Goetz, R.; Mohammadi, M.; Esser, V.; et al. Endocrine regulation of the fasting response by PPARalpha-mediated induction of fibroblast growth factor 21. *Cell Metab.* **2007**, *5*, 415–425. [CrossRef]
10. Dixon, E.D.; Nardo, A.D.; Claudel, T.; Trauner, M. The Role of Lipid Sensing Nuclear Receptors (PPARs and LXR) and Metabolic Lipases in Obesity, Diabetes and NAFLD. *Genes* **2021**, *12*, 645. [CrossRef]
11. Francque, S.; Verrijken, A.; Caron, S.; Prawitt, J.; Paumelle, R.; Derudas, B.; Lefebvre, P.; Taskinen, M.R.; Van Hul, W.; Mertens, I.; et al. PPARalpha gene expression correlates with severity and histological treatment response in patients with non-alcoholic steatohepatitis. *J. Hepatol.* **2015**, *63*, 164–173. [CrossRef] [PubMed]
12. Fuentes, E.; Guzman-Jofre, L.; Moore-Carrasco, R.; Palomo, I. Role of PPARs in inflammatory processes associated with metabolic syndrome (Review). *Mol. Med. Rep.* **2013**, *8*, 1611–1616. [CrossRef] [PubMed]
13. Pawlak, M.; Bauge, E.; Bourguet, W.; De Bosscher, K.; Lalloyer, F.; Tailleux, A.; Lebherz, C.; Lefebvre, P.; Staels, B. The transrepressive activity of peroxisome proliferator-activated receptor alpha is necessary and sufficient to prevent liver fibrosis in mice. *Hepatology* **2014**, *60*, 1593–1606. [CrossRef] [PubMed]
14. Raiola, A.; Errico, A.; Petruk, G.; Monti, D.M.; Barone, A.; Rigano, M.M. Bioactive Compounds in Brassicaceae Vegetables with a Role in the Prevention of Chronic Diseases. *Molecules* **2017**, *23*, 15. [CrossRef]
15. Arumugam, A.; Abdull Razis, A.F. Apoptosis as a Mechanism of the Cancer Chemopreventive Activity of Glucosinolates: A Review. *Asian Pac. J. Cancer Prev.* **2018**, *19*, 1439–1448. [CrossRef]

16. Krajka-Kuzniak, V.; Paluszczak, J.; Szaefer, H.; Baer-Dubowska, W. The activation of the Nrf2/ARE pathway in HepG2 hepatoma cells by phytochemicals and subsequent modulation of phase II and antioxidant enzyme expression. *J. Physiol. Biochem.* **2015**, *71*, 227–238. [CrossRef] [PubMed]
17. Bhandari, S.R.; Jo, J.S.; Lee, J.G. Comparison of Glucosinolate Profiles in Different Tissues of Nine Brassica Crops. *Molecules* **2015**, *20*, 15827–15841. [CrossRef]
18. Marklund, S.; Marklund, G. Involvement of the superoxide anion radical in the autoxidation of pyrogallol and a convenient assay for superoxide dismutase. *Eur. J. Biochem.* **1974**, *47*, 469–474. [CrossRef]
19. Bellomo, G.; Mirabelli, F.; DiMonte, D.; Richelmi, P.; Thor, H.; Orrenius, C.; Orrenius, S. Formation and reduction of glutathione-protein mixed disulfides during oxidative stress. A study with isolated hepatocytes and menadione (2-methyl-1,4-naphthoquinone). *Biochem. Pharmacol.* **1987**, *36*, 1313–1320. [CrossRef]
20. Hao, W.; Friedman, A. The LDL-HDL profile determines the risk of atherosclerosis: A mathematical model. *PLoS ONE* **2014**, *9*, e90497. [CrossRef]
21. Yang, T.Y.; Yu, M.H.; Wu, Y.L.; Hong, C.C.; Chen, C.S.; Chan, K.C.; Wang, C.J. Mulberry Leaf (*Morus alba* L.) Extracts and Its Chlorogenic Acid Isomer Component Improve Glucolipotoxicity-Induced Hepatic Lipid Accumulation via Downregulating miR-34a and Decreased Inflammation. *Nutrients* **2022**, *14*, 4808. [CrossRef] [PubMed]
22. Peng, C.H.; Cheng, J.J.; Yu, M.H.; Chung, D.J.; Huang, C.N.; Wang, C.J. Solanum nigrum polyphenols reduce body weight and body fat by affecting adipocyte and lipid metabolism. *Food Funct.* **2020**, *11*, 483–492. [CrossRef] [PubMed]
23. Chang, J.J.; Chung, D.J.; Lee, Y.J.; Wen, B.H.; Jao, H.Y.; Wang, C.J. Solanum nigrum Polyphenol Extracts Inhibit Hepatic Inflammation, Oxidative Stress, and Lipogenesis in High-Fat-Diet-Treated Mice. *J. Agric. Food Chem.* **2017**, *65*, 9255–9265. [CrossRef] [PubMed]
24. Ma, Y.; Lee, G.; Heo, S.Y.; Roh, Y.S. Oxidative Stress Is a Key Modulator in the Development of Nonalcoholic Fatty Liver Disease. *Antioxidants* **2021**, *11*, 91. [CrossRef]
25. Mohammed, E.D.; Abdel-Naim, A.B.; Kangpeng, J.; Jiang, R.; Wei, J.; Sun, B. The mother relationship between insulin resistance and non-alcoholic steatohepatitis: Glucosinolates hydrolysis products as a promising insulin resistance-modulator and fatty liver-preventer. *Life Sci.* **2021**, *264*, 118615. [CrossRef]
26. Wang, Q.; Liu, S.; Zhai, A.; Zhang, B.; Tian, G. AMPK-Mediated Regulation of Lipid Metabolism by Phosphorylation. *Biol. Pharm. Bull.* **2018**, *41*, 985–993. [CrossRef]
27. Cotter, T.G.; Rinella, M. Nonalcoholic Fatty Liver Disease 2020: The State of the Disease. *Gastroenterology* **2020**, *158*, 1851–1864. [CrossRef]
28. Gomes, A.L.; Teijeiro, A.; Buren, S.; Tummala, K.S.; Yilmaz, M.; Waisman, A.; Theurillat, J.P.; Perna, C.; Djouder, N. Metabolic Inflammation-Associated IL-17A Causes Non-alcoholic Steatohepatitis and Hepatocellular Carcinoma. *Cancer Cell* **2016**, *30*, 161–175. [CrossRef]
29. Zhao, L.; Zhong, S.; Qu, H.; Xie, Y.; Cao, Z.; Li, Q.; Yang, P.; Varghese, Z.; Moorhead, J.F.; Chen, Y.; et al. Chronic inflammation aggravates metabolic disorders of hepatic fatty acids in high-fat diet-induced obese mice. *Sci. Rep.* **2015**, *5*, 10222. [CrossRef]
30. Lee, Y.J.; Hsu, J.D.; Lin, W.L.; Kao, S.H.; Wang, C.J. Upregulation of caveolin-1 by mulberry leaf extract and its major components, chlorogenic acid derivatives, attenuates alcoholic steatohepatitis via inhibition of oxidative stress. *Food Funct.* **2017**, *8*, 397–405. [CrossRef]
31. Tang, C.C.; Lin, W.L.; Lee, Y.J.; Tang, Y.C.; Wang, C.J. Polyphenol-rich extract of Nelumbo nucifera leaves inhibits alcohol-induced steatohepatitis via reducing hepatic lipid accumulation and anti-inflammation in C57BL/6J mice. *Food Funct.* **2014**, *5*, 678–687. [CrossRef] [PubMed]
32. Kao, E.S.; Yang, M.Y.; Hung, C.H.; Huang, C.N.; Wang, C.J. Polyphenolic extract from Hibiscus sabdariffa reduces body fat by inhibiting hepatic lipogenesis and preadipocyte adipogenesis. *Food Funct.* **2016**, *7*, 171–182. [CrossRef]
33. Kumar, V.; Thakur, A.K.; Dev Barothia, N.; Chatterjee, S.S. Therapeutic potentials of Brassica juncea: An overview. *CellMed* **2011**, *1*, 2.1–2.16. [CrossRef]
34. Moyano, G.; Sáyago-Ayerdi, S.G.; Largo, C.; Caz, V.; Santamaria, M.; Tabernero, M. Potential use of dietary fibre from Hibiscus sabdariffa and Agave tequilana in obesity management. *J. Funct. Foods* **2016**, *21*, 1–9. [CrossRef]
35. Tabernero, M.; Venema, K.; Maathuis, A.J.; Saura-Calixto, F.D. Metabolite production during in vitro colonic fermentation of dietary fiber: Analysis and comparison of two European diets. *J. Agric. Food Chem.* **2011**, *59*, 8968–8975. [CrossRef]
36. Al-Lahham, S.; Roelofsen, H.; Rezaee, F.; Weening, D.; Hoek, A.; Vonk, R.; Venema, K. Propionic acid affects immune status and metabolism in adipose tissue from overweight subjects. *Eur. J. Clin. Investig.* **2012**, *42*, 357–364. [CrossRef]
37. Mirmiran, P.; Bahadoran, Z.; Khalili Moghadam, S.; Zadeh Vakili, A.; Azizi, F. A Prospective Study of Different Types of Dietary Fiber and Risk of Cardiovascular Disease: Tehran Lipid and Glucose Study. *Nutrients* **2016**, *8*, 686. [CrossRef] [PubMed]
38. Chen, J.G.; Cai, J.; Wu, H.L.; Xu, H.; Zhang, Y.X.; Chen, C.; Wang, Q.; Xu, J.; Yuan, X.L. Colorectal cancer screening: Comparison of transferrin and immuno fecal occult blood test. *World J. Gastroenterol.* **2012**, *18*, 2682–2688. [CrossRef]
39. Tremaroli, V.; Backhed, F. Functional interactions between the gut microbiota and host metabolism. *Nature* **2012**, *489*, 242–249. [CrossRef] [PubMed]
40. Henao-Mejia, J.; Elinav, E.; Jin, C.; Hao, L.; Mehal, W.Z.; Strowig, T.; Thaiss, C.A.; Kau, A.L.; Eisenbarth, S.C.; Jurczak, M.J.; et al. Inflammasome-mediated dysbiosis regulates progression of NAFLD and obesity. *Nature* **2012**, *482*, 179–185. [CrossRef]

41. Fang, C.; Pan, J.; Qu, N.; Lei, Y.; Han, J.; Zhang, J.; Han, D. The AMPK pathway in fatty liver disease. *Front. Physiol.* **2022**, *13*, 970292. [CrossRef] [PubMed]
42. Hurley, R.L.; Anderson, K.A.; Franzone, J.M.; Kemp, B.E.; Means, A.R.; Witters, L.A. The Ca^{2+}/calmodulin-dependent protein kinase kinases are AMP-activated protein kinase kinases. *J. Biol. Chem.* **2005**, *280*, 29060–29066. [CrossRef] [PubMed]
43. Okulicz, M. Multidirectional time-dependent effect of sinigrin and allyl isothiocyanate on metabolic parameters in rats. *Plant Foods Hum. Nutr.* **2010**, *65*, 217–224. [CrossRef] [PubMed]
44. Paul, S.; Geng, C.A.; Yang, T.H.; Yang, Y.P.; Chen, J.J. Phytochemical and Health-Beneficial Progress of Turnip (*Brassica rapa*). *J. Food Sci.* **2019**, *84*, 19–30. [CrossRef] [PubMed]

nutrients

Article

Association between Nonfood Pre- or Probiotic Use and Cognitive Function: Results from NHANES 2011–2014

Jingyi Chen [1,2,†], Nian Yang [1,2,†], Yilei Peng [2], Honghao Zhou [1,*] and Qing Li [2,3,*]

1 Institute of Precision Medicine, The First Affiliated Hospital of Shantou University Medical College, Shantou 515041, China; 19jychen@stu.edu.cn (J.C.); 20nyang@stu.edu.cn (N.Y.)
2 Department of Clinical Pharmacology, Xiangya Hospital, Central South University, Changsha 410008, China; 228101062@csu.edu.cn
3 Institute of Clinical Pharmacology, Central South University, Hunan Key Laboratory of Pharmacogenetics, Changsha 410078, China
* Correspondence: 801004@csu.edu.cn (H.Z.); liqing9251026@csu.edu.cn (Q.L.); Tel.: +86-731-8480-5380 (Q.L.)
† These authors contributed equally to this work.

Abstract: In this study, we collected data from the National Health and Nutrition Examination Survey (NHANES) for the years 2011–2014. Multiple linear regression and logistic regression were used to analyse the association between nonfood pro- or prebiotic use and cognitive function among elderly Americans. To estimate the potential unobserved results, propensity score matching (PSM) was used to analyse the causal effect. Nonfood pro- or prebiotic use was analysed through the Dietary Supplement Use 30-Day Study. Cognitive function was evaluated by the Digit Symbol Substitution Test (DSST), the Animal Fluency Test (AFT), the Consortium to Establish a Registry for Alzheimer's Disease (CERAD), and a composite Z-score calculated by summing the Z-scores of three tests. Male participants who used nonfood pro- or prebiotics tended to have higher comprehensive cognitive function (sum.z) with a β-coefficient of 0.64 (95% CI: 0.08–1.19). Probiotics or prebiotics may be a protective factor against cognitive impairment in males, with an odds ratio of 0.08 (95% CI: 0.02–0.29). Furthermore, the average treatment effect for the treated (ATT) with nonfood pro- or prebiotics (0.555) on sum.z in males was statistically significant ($p < 0.05$). Our research revealed that nonfood pre- or probiotic use was an effective method to improve cognitive function in elderly men from the USA.

Keywords: probiotic; prebiotic; cognitive function; NHANES

1. Introduction

Aging can have negative effects on cognitive skills, including learning and memory. Cognitive health has emerged as a critical public health concern, especially for the elderly. The elderly are expected to account for more than one-fifth of the world's population by 2050 [1]. In the United States, the population aged 65 and above is expected to nearly double from 52 million in 2018 to 95 million in 2060. This increase will lead to a higher proportion of this age group in the total population, rising from 16% to 23% [2]. Therefore, the number of Americans suffering from age-related cognitive decline is expected to increase [3]. This decline may be caused by a combination of genetic and environmental factors, as well as physiological, psychological, social, lifestyle, and dietary considerations [4].

There are multiple causes for cognitive impairment, but one potential factor is that the decrease of microbial diversity in elder individuals results in disruption of intestinal barrier permeability [5]. Recent studies have revealed that gut microbiota can affect brain function and behaviour. The gut–brain axis not only maintains the muscular, sensory, and secretory pathways in the gastrointestinal tract but also affects brain growth, function, and behaviour [6]. The gut microbiota plays a circular role in microglia functions, neuronal shape, and blood–brain barrier integrity [7]. Compared with young people, elderly people have a lower abundance of beneficial microbiota, specifically *Bifidobacterium* and *Lactobacillus* [8]. Probiotics are food ingredients or supplements that contain living microbes, while prebiotics are composed of non-digestive substrates which selectively stimulate the growth of beneficial microbes. The intake of prebiotics can elevate the levels of beneficial gut microbiota in older adults [9]. In addition, probiotic supplements can suppress the NF-κB signalling pathway mediated by TLR4 and RIG-I, as well as the inflammatory response, thereby improving cognitive function in aged SAMP8 mice [10]. Therefore, the intake of probiotics or prebiotics may have positive effects on human health. Moreover, the consumption of pre- or probiotics is high in the United States, with the highest intake found among older adults, reaching 8.8% [11]. Based on these findings, there is a growing interest in using nonfood pre- or probiotics as medicine to regulate the gut microbiota and return to a more physiological state.

Although some systematic reviews on the effects of pre- or probiotics on cognitive outcomes have been performed, no consistent conclusions have been drawn [12,13], which are insufficient to provide definitive evidence that the use of pre- or probiotics has effects on cognitive function. Therefore, we aimed to conduct a well-controlled and population-based study to better understand the role of nonfood pre- or probiotics in cognitive function. In this study, we aimed to investigate the association between nonfood pre- or probiotic use and cognitive function in older adults through analysing the data from the National Health and Nutrition Examination Survey (NHANES) for the years 2011–2014.

2. Materials and Methods

2.1. Population under Investigation

NHANES is a study adopting a multistage sampling approach to assess health conditions and lifestyle alterations in the United States. These data are collected through personal interviews, physical assessments, biological specimen collection, and field investigations involving representative samples from the national population. We collected data on 19,931 individuals from the NHANES (2011–2014). Then, we excluded the interviewees under 60 years old (n = 16,299) and those with missing data on BMI, smoking, drinking, hypertension, stroke, diabetes mellitus (DM) and cardiovascular disease (CVD), and stroke (n = 745). Additionally, participants who lacked pre- or probiotic dietary supplement information (n = 923) and those who did not receive cognitive function tests or failed to complete four cognitive tests were also excluded (n = 176). Finally, only 1788 participants were included in our analysis, as shown in Figure 1, which describes the whole screening procedure.

2.2. Assessment of Nonfood Pre- or Probiotic Use

We analysed the Dietary Supplement Use 30-Day Study before the interview date to determine whether the sample used nonfood pre- or probiotics. Detailed nonfood pre- or probiotic information can be found in Table S1 [11].

2.3. Cognitive Functioning Evaluation

The NHANES cognitive functioning test was conducted at the Mobile Exam Center (MEC), which consisted of the CERAD word learning test (CERAD), Animal Fluency Test (AFT) and Digit Symbol Substitution Test (DSST) (https://wwwn.cdc.gov/Nchs/Nhanes/2013-2014/CFQ_H.htm (accessed on March 2017)). We used Z-score to standardise the scores of CERAD, AFT, and DSST. The sum of the three standardised scores is recorded as 'sum.z'.

2.4. Covariates

The NHANES collects information on demographic, socioeconomic, and health-related issues. We adopted some of them as covariates, including age (60–70 years and ≥70 years), gender (male and female), ethnicity (Mexican-American, non-Hispanic white, non-Hispanic black, and others), educational level (less than high school, high school or higher) [14], ratio of family income to poverty (PIR) (<1.3, ≥1.3–3.5, and >3.5), body mass index (BMI) (normal: <25 kg/m^2, overweight: 25 to <30 kg/m^2, obesity: ≥30 kg/m^2) [15], drinking (never, former, current) [16] and smoking status (never, former, current) [17]. In addition, disease history (hypertension, stroke, DM, and CVD) was included as covariates.

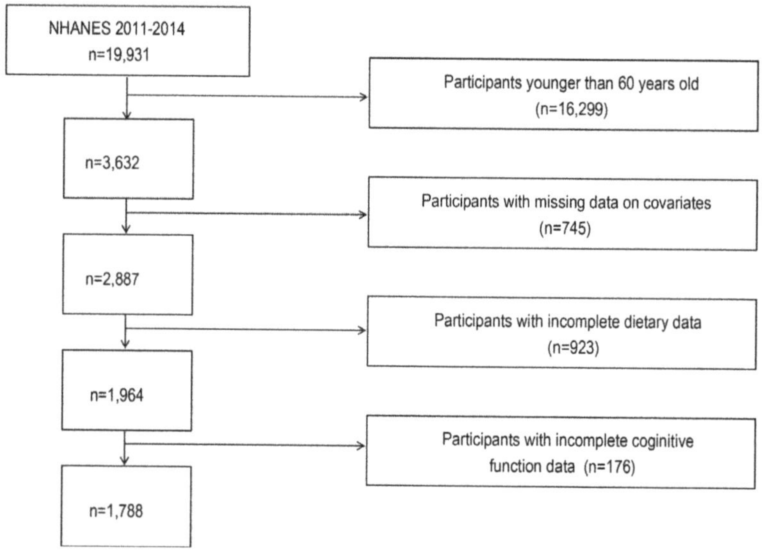

Figure 1. The flow diagram of the sample screening procedure.

2.5. Statistical Analysis

We chose 'wtmec2yr' for 2011–2014 and calculated these weights using the following formula:

$$wt = 1/2 * WTDR2D. \quad (1)$$

Continuous data were represented by mean and standard deviation (SD), while categorical variables were denoted by sample size and weighted percentage (%). The participants were divided into two groups: one group included individuals who used nonfood pre- or probiotics, and the other group comprised those who did not consume either. To investigate the relationship between cognitive function and nonfood pre- or probiotic use, we conducted a linear regression analysis with defined risk factors of cognitive function as covariates based on previous studies [18]. Model 1 did not make any adjustments. Model 2 adjusted for age, gender, race, educational level, PIR, BMI, alcohol drinking status, and smoking status. Model 3 further adjusted for hypertension, stroke, DM, and CVD in addition to the factors from Model 2. We also stratified our analysis by gender, age, ethnicity, and BMI to evaluate their impact on cognitive function. Furthermore, we analysed the interaction effects of these three factors (age, ethnicity, and BMI) with nonfood pre- or probiotic use. The p value < 0.05 was statistically significant.

To better understand the effects of nonfood pre- or probiotic use on cognitive impairment, we performed a logistic regression analysis. Since there is no clear diagnostic measure for cognitive impairment, we established the cutoff point for the two age groups based on previous studies to define cognitive impairment [19]. The lowest quartile of sum.z was used as the threshold, with −0.812 for 60–69 years old and −2.311 for ≥70 years old. All analyses were performed using R (4.2.2) software.

Although regression models have been used to investigate the relationship between nonfood pre- or probiotic use and cognitive function, cross-sectional observational research is methodologically challenged by the limitation of causal inference. In order to address this limitation when randomized data are unattainable, we adopted propensity score matching

(PSM) as a suitable alternative for estimating effects. Using binary random variables 'Di' to represent whether nonfood pre- or probiotics were used or not and 'Yi' to measure comprehensive cognitive function (sum.z), we established a simple linear regression model:

$$Yi = \alpha + \beta Di + \mu i, \tag{2}$$

The value of i = 1 represents 'nonfood pre- or probiotic use', while the value of i = 0 represents 'no nonfood pre- or probiotic use'. Kernel matching was adopted to eliminate bias, and the average treatment effect for the treated (ATT), representing the β-coefficient, was denoted as the anticipated difference in outcomes between the nonfood pre- or probiotic use group and the group with no nonfood pre- or probiotic use [20]. The data analysis was performed using Stata (15.1) software.

3. Results

3.1. Descriptive Statistics

This analysis included 1788 elderly individuals aged 60 or above. In this study cohort, the average age of the two groups was 69.05 for those who used nonfood pre- or probiotics and 69.36 for those who did not. Participants who used nonfood pre- or probiotics were more likely to be white. The baseline characteristics of the population are displayed in Table 1. Three different cognitive tests (z.CERD, z.AFT, and z.DSST) and their sum (sum.z) were compared between the participants who used nonfood pre- or probiotics and those who did not. The *p*-values indicated statistical significance between the two groups for each test. The sum.z, which represents the comprehensive cognitive function, showed an improvement of 0.68 when comparing the participants who used nonfood pre- or probiotics with those who did not.

Table 1. Characteristics of all participants according to nonfood pre- or probiotic use.

Characteristic	Nonfood Pre- or Probiotic Use	No Nonfood Pre- or Probiotic Use	*p*
	135 (7.56%)	1653 (92.4%)	
Age	69.05 (67.78, 70.32)	69.36 (68.92, 69.80)	0.65
Age_subgroup			0.72
60–69	67 (57.00)	838 (54.86)	
≥70	68 (43.00)	815 (45.14)	
Gender			0.56
Female	68 (54.94)	940 (57.55)	
Male	67 (45.06)	713 (42.45)	
Ethnicity			0.01
White	87 (89.63)	892 (82.67)	
Black	26 (4.81)	345 (6.95)	
Mexican	8 (1.85)	120 (2.63)	
Other	14 (3.71)	296 (7.75)	
Education			0.1
Less than high school	12 (6.51)	347 (13.31)	
High school or higher	123 (93.49)	1306 (86.69)	
PIR			0.16
<1.3	21 (9.34)	448 (15.62)	
1.3–3.5	53 (37.32)	632 (37.96)	
>3.5	61 (53.34)	573 (46.42)	
BMI			0.32
<25	40 (32.50)	444 (25.91)	
25–29.9	45 (37.06)	588 (36.82)	
≥30	50 (30.44)	621 (37.27)	
Smoker			0.75
Never	68 (52.96)	855 (51.63)	
Former	57 (40.19)	617 (39.00)	
Current	10 (6.84)	181 (9.37)	

Table 1. *Cont.*

Characteristic	Nonfood Pre- or Probiotic Use	No Nonfood Pre- or Probiotic Use	p
	135 (7.56%)	1653 (92.4%)	
Alcohol			0.37
Current	82 (69.56)	928 (64.13)	
Former	38 (21.66)	458 (22.57)	
Never	15 (8.78)	267 (13.30)	
Hypertension			0.12
No	45 (41.40)	470 (32.25)	
Yes	90 (58.60)	1183 (67.75)	
Stroke			0.28
No	123 (90.64)	1534 (93.70)	
Yes	12 (9.36)	119 (6.30)	
DM			0.43
No	85 (66.90)	958 (63.43)	
Yes	50 (33.10)	695 (36.57)	
CVD			0.57
No	103 (76.22)	1283 (78.78)	
Yes	32 (23.78)	370 (21.22)	
z.CERD	0.27 (0.03, 0.51)	0.14 (0.05, 0.24)	0.3
z.AFT	0.51 (0.26, 0.75)	0.26 (0.18, 0.33)	0.05
z.DSST	0.61 (0.45, 0.76)	0.31 (0.25, 0.38)	0.002
sum.z	1.39 (0.88, 1.89)	0.71 (0.53, 0.90)	0.02

PIR: ratio of family income to poverty; BMI: body mass index; DM: diabetes mellitus; CVD: cardiovascular disease.

Furthermore, we compared cognitive function by gender among the participants who did not use nonfood pre- or probiotics. The results revealed that females had worse comprehensive cognitive function than males (0.54 vs. 0.91) in the group that did not use nonfood pre- or probiotics (Table S2).

3.2. Modulation of Cognitive Function Score According to Nonfood Pre- or Probiotic Use

To explore the relationship between cognitive function and nonfood pre- or probiotic use, we performed a multiple linear regression analysis with a crude model and two multivariable-adjusted models (Table 2). In the crude model (Model 1), the better performance of cognitive function was significantly associated with the use of nonfood pre- or probiotics. The β-coefficient for z.DSST and comprehensive cognitive function (sum.z) were 0.29 (95% CI: 0.11–0.47) and 0.67 (95% CI: 0.14–1.21), respectively. However, after adjusting for age, sex, race, BMI, smoking, alcohol use, education, and poverty (Model 2) and further adjusting for diseases including hypertension, stroke, DM, and CVD (Model 3), no significant relationship was found. Furthermore, we carried out a subgroup analysis and found that males performed better cognitive functions. In Model 1, the β-coefficient of z.DSST was 0.41 (95% CI: 0.11,0.72) and the β-coefficient of sum.z was 1.09 (95% CI: 0.30,1.88). After adjusting for covariates, the association between sum.z and nonfood pre- or probiotic use remained significant, with the value of 0.73 (95% CI: 0.19 to 1.27) in Model 2 and 0.64 (95% CI: 0.13 to 1.24) in Model 3, although the coefficients were attenuated (Table 2). However, no significant association was found in females.

3.3. Interaction Effects

To learn more about the interaction between nonfood pre- or probiotic use and factors such as age, ethnicity, and BMI, we conducted interaction analyses on these variables. The results indicated that there was no interaction between these variables in the model when other covariates were considered. However, participants who used nonfood pre- or probiotics showed better z.AFT, z.DSST, and sum.z in the population of BMI < 25, with β-coefficients of 0.31(95% CI: 0.06, 0.56), 0.41 (95% CI: 0.12, 0.70) and 0.88 (95% CI: 0.17,

1.59), respectively (Figure 2). Moreover, white individuals who used pre -or probiotics showed better z.DSST, with the β-coefficient of 0.19 (95% CI: 0.01, 0.37).

Table 2. Regression coefficients and 95% confidence intervals in nonfood pre- or probiotic use group when compared with no nonfood pre- or probiotic use group.

	z.AFT	z.CEART	z.DSST	Sum.z
All participants (n = 1788)				
Model 1	0.25 (0.00, 0.50)	0.13 (−0.12, 0.38)	0.29 (0.11, 0.47) **	0.67 (0.14, 1.21) *
Model 2	0.14 (−0.08, 0.36)	0.06 (−0.17, 0.29)	0.15 (−0.01, 0.32)	0.35 (−0.08, 0.78)
Model 3	0.13 (−0.09, 0.36)	0.07 (−0.17, 0.30)	0.16 (−0.01, 0.33)	0.36 (−0.09, 0.80)
Male (n = 780)				
Model 1	0.4 (−0.02, 0.83)	0.27 (−0.05, 0.60)	0.41 (0.11, 0.72) *	1.09 (0.30, 1.88) *
Model 2	0.29 (−0.07, 0.65)	0.19 (−0.07, 0.45)	0.25 (0.00, 0.50)	0.73 (0.19, 1.27) *
Model 3	0.25 (−0.11, 0.61)	0.18 (−0.09, 0.45)	0.26 (−0.01, 0.52)	0.69 (0.13, 1.24) *
Female (n = 1008)				
Model 1	0.12(−0.15, 0.39)	0.03 (−0.31, 0.37)	0.2 (−0.01, 0.41)	0.35 (−0.35, 1.05)
Model 2	0.01 (−0.25, 0.27)	−0.05 (−0.36, 0.25)	0.06 (−0.13, 0.25)	0.01 (−0.57, 0.60)
Model 3	0.03 (−0.24, 0.31)	−0.03 (−0.35, 0.29)	0.08 (−0.12, 0.29)	0.09 (−0.54, 0.72)

Model 1: no adjustment; Model 2: adjusted for age, gender, ethnicity, educational level, PIR, BMI, smoke, and drink; Model 3: adjusted for all the factors in Model 2 plus hypertension, stroke, DM, and CVD. PIR: ratio of family income to poverty; BMI: body mass index; CVD: cardiovascular disease. * $p < 0.05$, ** $p < 0.01$

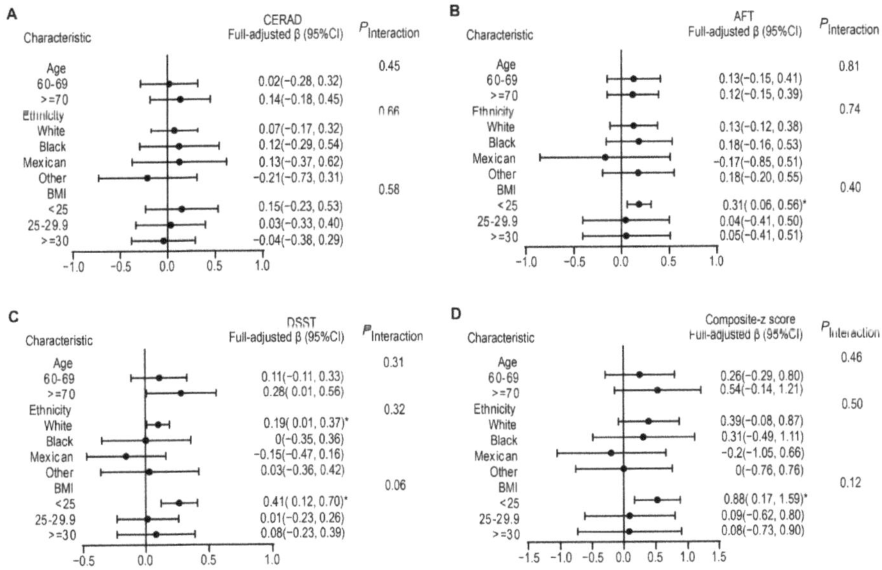

Figure 2. Association between nonfood pre- or probiotic use and different cognitive function scores stratified by age, ethnicity, and BMI. Adjusted for age, gender, ethnicity, BMI, drink, smoke, education, PIR, hypertension, stroke, DM, and CVD. The strata variable was not included in the model when stratifying by itself. (**A**) CERAD; (**B**) AFT; (**C**) DSST; (**D**) Composite-z score (sum.z). * $p < 0.05$.

3.4. Cognitive Impairment and Nonfood Pre- or Probiotic Use

To further explore the association between nonfood pre- or probiotic use and cognitive impairment, the sum.z score was divided into quartiles based on age subgroups, with the lowest quartile considering cognitive impairment. The basic characteristics of the two groups are displayed in Table 3. The results showed that participants with cognitive impairment used almost none of the nonfood pre- or probiotics. Subsequently, a logistic regression analysis was conducted to examine the association between nonfood pre- or probiotic use and cognitive impairment in males and females. The results suggested that nonfood pre- or probiotic use was a protective factor for cognition impairment in males, with significant odds ratios in both the unadjusted model and the two adjusted models ($p < 0.001$), while no significant difference was observed in women in the adjusted models (Table 4).

Table 3. Characteristics of all participants in cognitive impairment and non-cognitive impairment groups.

Characteristic	Cognitive Impairment	Non-Cognitive Impairment	*p*
	447 (25.00)	1341 (75.00)	
Age	71.51 (70.70, 72.33)	68.94 (68.46, 69.42)	<0.0001
Age_subgroup			0.003
60–69	226 (46.49)	679 (56.59)	
≥70	221 (53.51)	662 (43.41)	
Gender			0.57
Female	223 (55.54)	785 (57.64)	
Male	224 (44.46)	556 (42.36)	
Ethnicity			<0.0001
White	157 (65.82)	822 (86.43)	
Black	143 (16.20)	228 (5.06)	
Mexican	48 (6.15)	80 (1.92)	
Other	99 (11.83)	211 (6.59)	
Education			<0.0001
Less than high school	180 (29.57)	179 (9.67)	
High school or higher	267 (70.43)	1162 (90.33)	
PIR			<0.0001
<1.3	205 (36.65)	264 (11.18)	
1.3–3.5	151 (40.26)	534 (37.48)	
>3.5	91 (23.09)	543 (51.34)	
BMI			0.59
<25	122 (29.01)	362 (26.05)	
25–29.9	157 (35.00)	476 (37.17)	
≥30	168 (35.98)	503 (36.78)	
Smoker			0.02
Never	229 (52.43)	694 (51.63)	
Former	151 (34.63)	523 (39.91)	
Current	67 (12.94)	124 (8.46)	
Alcohol			<0.0001
Current	183 (43.27)	827 (68.45)	
Former	163 (32.51)	333 (20.69)	
Never	101 (24.22)	181 (10.86)	
Hypertension			<0.0001
No	101 (19.10)	414 (35.58)	
Yes	346 (80.90)	927 (64.42)	
Stroke			0.003
No	388 (86.10)	1269 (94.74)	
Yes	59 (13.90)	72 (5.26)	
DM			0.01
No	230 (54.74)	813 (65.35)	
Yes	217 (45.26)	528 (34.65)	
CVD			0.01
No	310 (68.38)	1076 (80.37)	
Yes	137 (31.62)	265 (19.63)	
Group			<0.001
No pre- or probiotic use	432 (97.29)	1221 (89.89)	
Pre- or probiotic use	15 (2.71)	120 (10.11)	

PIR: ratio of family income to poverty; BMI: body mass index; CVD: cardiovascular disease.

Table 4. Odds ratio (95% confidence intervals) of nonfood pre- or probiotic use in cognitive impairment in males and females.

	Model 1		Model 2		Model 3	
	OR (95% CI)	*p*	OR (95% CI)	*p*	OR (95% CI)	*p*
Male	0.06 (0.02, 0.19)	<0.0001	0.08 (0.02, 0.25)	<0.001	0.08 (0.02, 0.27)	<0.001
Female	0.38 (0.17, 0.83)	0.02	0.52 (0.21, 1.26)	0.14	0.50 (0.20, 1.24)	0.13

Model 1: no adjustment; Model 2: adjusted for age, gender, ethnicity, educational level, PIR, BMI, smoke, and drink; Model 3: adjusted for all the factors in Model 2 plus hypertension, stroke, diabetes, and CVD; PIR: ratio of family income to poverty; BMI: body mass index; DM: diabetes mellitus; CVD: cardiovascular disease.

3.5. Balance Test and PSM Results

To estimate the causal effect of nonfood pre- or probiotic use on the sum.z score, we used propensity score matching (PSM) to avoid selectivity bias caused by potential outcomes. As previous analyses have indicated that there is gender bimodality in the impact of nonfood pre- or probiotic use on cognitive function, we stratified our samples by gender and conducted PSM on males and females separately. Before kernel matching, we redefined several covariates, including age, BMI, PIR, education level, smoking status, alcohol consumption, hypertension, stroke, DM, and CVD, as shown in Table S3. Tables 5 and 6 show the balance tests for males and females. The results showed that after matching, all covariates between the nonfood pre- or probiotic use group and the no nonfood pre- or probiotic use group were almost balanced ($p < 0.05$), which meant that sample equilibrium was achieved to some extent. Table 7 shows the ATT for males and females. For males, the result was significantly different ($p < 0.05$) between the treated group and the control group, with a difference of 0.555 and a standard error (SE) of 0.282. However, for females, the difference was not statistically significant ($p > 0.05$), with a difference of 0.235 and a standard error of 0.266. From this, it is suggested that nonfood pre- or probiotic use is an effective method to improve cognitive function in elderly men.

Table 5. Balance test of covariable in males.

Variable	Sample	Mean Value		Standard Bias (%)	Bias Reduction (%)	T	*p*
		Nonfood Pre- or Probiotic Use	No Nonfood Pre- or Probiotic Use				
Age	U	70.015	70.289	−4.1		−0.32	0.752
	M	70.015	70.212	−2.9	28.2	−0.17	0.867
Ethnicity	U	0.731	0.889	−14.5		−1.09	0.277
	M	0.731	0.792	−5.6	61.7	−0.32	0.746
PIR	U	1.478	1.149	45.3		3.32	0.001
	M	1.478	1.377	13.9	69.4	0.84	0.401
Education	U	0.94	0.781	47.1		3.1	0.002
	M	0.94	0.927	4	91.5	0.31	0.756
BMI	U	1.03	1.077	−6.1		−0.48	0.628
	M	1.03	1.059	−3.8	37.6	−0.22	0.827
Smoking	U	0.642	0.776	−21.1		−1.58	0.114
	M	0.642	0.7	−9.3	56.2	−0.54	0.589
Alcohol	U	1.642	1.539	16.8		1.28	0.201
	M	1.642	1.594	7.7	54.1	0.45	0.654
Hypertension	U	0.612	0.697	−17.9		−1.44	0.15
	M	0.612	0.663	−10.7	40.4	−0.61	0.545
Stroke	U	0.06	0.067	−3.1		−0.24	0.811
	M	0.06	0.061	−0.6	80.5	−0.04	0.971
CVD	U	0.224	0.264	−9.2		−0.71	0.478
	M	0.224	0.242	−4.2	54.3	−0.25	0.805
DM	U	0.358	0.467	−22.2		−1.71	0.088
	M	0.358	0.406	−9.7	56.1	−0.57	0.573

PIR: ratio of family income to poverty; BMI: body mass index; DM: diabetes mellitus; CVD: cardiovascular disease; M: matched; U: Unmatched.

Table 6. Balance test of covariable in females.

Variable	Sample	Mean Value		Standard Bias (%)	Bias Reduction (%)	T	*p*
		Nonfood Pre- or Probiotic Use	No Nonfood Pre- or Probiotic Use				
Age	U	70.235	69.766	7.1		0.55	0.584
	M	70.235	70.013	3.4	52.6	0.19	0.846
Ethnicity	U	0.515	0.893	−35.7		−2.64	0.008
	M	0.515	0.67	−14.7	59	−0.91	0.362
PIR	U	1.118	1.02	12.8		1	0.316
	M	1.118	1.072	6	53.5	0.35	0.728
Education	U	0.882	0.797	23.4		1.71	0.087
	M	0.882	0.846	10	57.2	0.62	0.537
BMI	U	1.118	1.13	−1.5		−0.12	0.906
	M	1.118	1.133	−1.9	−28.3	−0.11	0.913
Smoking	U	0.5	0.453	7.1		0.57	0.569
	M	0.5	0.473	4.1	42.4	0.24	0.812
Alcohol	U	1.353	1.295	7.5		−0.16	0.869
	M	1.353	1.325	3.5	52.9	−0.05	0.962
Hypertension	U	0.721	0.73	−2.1		1.25	0.212
	M	0.721	0.724	−0.8	59.8	0.83	0.408
Stroke	U	0.118	0.076	14.2		1.13	0.26
	M	0.118	0.075	14.3	−0.3	0.72	0.474
CVD	U	0.25	0.194	13.5		−0.05	0.964
	M	0.25	0.198	12.4	8.3	0.04	0.971
DM	U	0.382	0.385	−0.6		0.57	0.566
	M	0.382	0.379	0.6	−12.4	0.21	0.838

PIR: ratio of family income to poverty; BMI: body mass index; DM: diabetes mellitus; CVD: cardiovascular disease; M: matched; U: Unmatched.

Table 7. PSM result in males and females.

	Nonfood Pre- or Probiotic Use	No Nonfood Pre- or Probiotic Use	Difference (ATT)	SE	T	*p*
Male	0.54	−0.015	0.555	0.282	1.97	<0.05
Female	0.604	0.37	0.235	0.266	0.88	>0.05

ATT: the average treatment effect for the nonfood pre- or probiotic use.

4. Discussion

This cross-sectional analysis aimed to evaluate the association between nonfood pre- or probiotic use and cognitive function in older adults. We found that nonfood pre- or probiotic use is significantly positively correlated with comprehensive composite cognitive function, particularly among males, both before and after adjusting for demographic and potential confounding factors. Moreover, obesity significantly altered the association between nonfood pre- or probiotic use and cognitive function. Our results suggest that alterations in the gut microbiota may contribute to the prevention of cognitive impairment in older adults. In previous studies, the association between nonfood pre- or probiotic use and cognitive function has been inconsistent. Several meta-analyses have shown that probiotic treatment improves cognitive impairment [21–24]. However, some studies do not support the positive effects of probiotics, prebiotics, and fermented foods on cognitive function in elder populations [12,25,26]. Inconsistent results may be due to different sociodemographic characteristics or small sample sizes. In our analysis, we found that nonfood pre- or probiotic use improved comprehensive cognitive function. This study is a national survey with a large sample size, providing reliable evidence. Furthermore, we conducted subgroup analyses by examining demographic characteristics (gender, age, ethnicity, and BMI). Males and individuals with a BMI < 25 are more likely to benefit from nonfood pre- or probiotic use. It is well-known that elder women with higher oestrogen levels are at higher risk for Alzheimer's disease. The observed gender dimorphism in cognitive function may be due to the metabolism of sex hormones regulated by the gut microbiota [27]. Therefore, preventive strategies for cognitive impairment in women may need to consider other intervention measures. Some evidence also suggests that obesity can impair cognitive function [28,29], leading us to speculate that obesity may interfere with the effectiveness of probiotics or prebiotics. Obesity is a metabolic disorder and is associated with gut microbiota dysbiosis, which may affect the function of prebiotics and probiotics in the gut. Therefore, it may be necessary to consider the relationship between obesity, gut microbiota, and cognitive function when designing interventions for improving cognitive function strategies with prebiotics or probiotics. In the future, research focusing on the effects of specific probiotic and prebiotic types on individuals with different BMI levels may develop more personalized intervention strategies to meet the needs of different populations, thereby improving the success rate and applicability of these approaches. In addition, it is important to explore how obesity interferes with the effectiveness of prebiotics or probiotics.

To further understand the effects of nonfood pre- or probiotic use on cognitive function, after conducting correlational analyses (multiple linear regression and multiple logistic regression), we also adopted the kernel matching method for propensity scores to infer causal relationships between nonfood pre- or probiotic use and cognitive function. PSM is primarily used to address confounding biases in estimating causal effects in observational studies. By matching individuals with similar propensity scores in the experimental and control groups, the reliability of causal inference is improved by simulating a randomized controlled trial. ATT was calculated to assess the effect of nonfood pre- or probiotic use on cognitive function. Our results indicated (ATT = 0.555, $p < 0.05$) that, after considering confounding factors, nonfood pre- or probiotic use had a significant positive effect on cognitive function in males. However, it should be noted that this study was based on observational data. Although propensity score matching (PSM) was used to control for potential confounding factors, there may still be some unconsidered confounders, which may continue to bias results. Therefore, it just implies that nonfood pre- or probiotic use is a potential causal relationship and could be an effective strategy for improving cognitive function in elderly males.

This study has several advantages. First, it avoids the heterogeneity of small samples by using a large, nationally representative sample. Second, cognitive function was assessed by using three cognitive scores (DSST, AFT, and CEARD) and a composite cognitive score (sum.z). Multiple linear regression was used to analyse the effect of nonfood pro- or prebiotic use on cognitive function, and then logistic regression was used to further confirm that nonfood pro- or prebiotic use was a protective factor for cognitive impairment. In the regression analyses, covariates were also adjusted to eliminate potential confounding effects. Third, this study was an observational study, so causal inference was methodologically challenging. To address this limitation, and to perform causal inference when random data are not available, we used kernel matching to eliminate biases and propensity score matching (PSM) in the form of ATT to interpret potential causal effects. However, there are some limitations to this study. First, dietary data were collected in the NHANES questionnaire, which may introduce information bias. For instance, data from the Dietary Supplement Use 30-Day Study may be subject to recall bias. Second, in this study, we only analysed whether the participants were using nonfood pro- or prebiotics. Since it is difficult to standardize the amount or unit of pre- or probiotics in the diet, we did not quantify their intake, so we were unable to explore the effect of different types or amounts of pre- or prebiotics on cognitive function. Moreover, the data from the Dietary Supplement Use 30-Day Study only recorded participants' intake, not the duration of use of pre- or prebiotic supplements. Therefore, it is difficult to distinguish between short-term and long-term differences. Despite some limitations, the obtained results have been encouraging and motivating for further research in this direction. Further long-term prospective research is necessary.

5. Conclusions

In conclusion, by adopting a nationally representative cohort of elder people from the USA, we found that the positively association between nonfood pre- or probiotic use and cognitive function is stronger in males, the population of BMI < 25 and white individuals. Our study emphasized that nonfood pre- or probiotic use is an effective method to improve cognitive function in elderly men.

Supplementary Materials: The following supporting information can be downloaded at: https://www.mdpi.com/article/10.3390/nu15153408/s1. Table S1: The classification of probiotics and prebiotics; Table S2: Charateristics of participants who did not use pre- or probiotic; Table S3: Definition various variables for PSM.

Author Contributions: Investigation, data curation, writing—original draft preparation, J.C. and N.Y.; conceptualization, writing—review and editing, Q.L. and H.Z.; validation, visualization, Y.P. All authors have read and agreed to the published version of the manuscript.

Funding: This research was funded by the National Natural Science Foundation (NNSF) of China (81973401).

Institutional Review Board Statement: Not applicable.

Informed Consent Statement: Not applicable.

Data Availability Statement: Not applicable.

Conflicts of Interest: The authors declare no conflict of interest.

References

1. Rehman, T. Role of the gut microbiota in age-related chronic inflammation. *Endocr. Metab. Immune Disord. Drug Targets* **2012**, *12*, 361–367. [CrossRef] [PubMed]
2. Zhu, Y.; Tan, Z.; Li, S.; Zhu, F.; Qin, C.; Zhang, Q.; Yang, Y. Serum creatinine to cystatin C ratio and cognitive function among middle-aged and older adults in China. *Front. Aging Neurosci.* **2022**, *14*, 919430. [CrossRef]
3. Hebert, L.E.; Weuve, J.; Scherr, P.A.; Evans, D.A. Alzheimer disease in the United States (2010–2050) estimated using the 2010 census. *Neurology* **2013**, *80*, 1778–1783. [CrossRef] [PubMed]
4. Iranpour, S.; Saadati, H.M.; Koohi, F.; Sabour, S. Association between caffeine intake and cognitive function in adults; effect modification by sex: Data from National Health and Nutrition Examination Survey (NHANES) 2013–2014. *Clin. Nutr.* **2020**, *39*, 2158–2168. [CrossRef]
5. Brenner, S.R. Hypothesis: Intestinal barrier permeability may contribute to cognitive dysfunction and dementia. *Age Ageing* **2010**, *39*, 278. [CrossRef] [PubMed]
6. Serra, M.C.; Nocera, J.R.; Kelleher, J.L.; Addison, O. Prebiotic Intake in Older Adults: Effects on Brain Function and Behavior. *Curr. Nutr. Rep.* **2019**, *8*, 66–73. [CrossRef] [PubMed]
7. Wiley, N.C.; Dinan, T.G.; Ross, R.P.; Stanton, C.; Clarke, G.; Cryan, J.F. The microbiota-gut-brain axis as a key regulator of neural function and the stress response: Implications for human and animal health. *J. Anim. Sci.* **2017**, *95*, 3225–3246. [PubMed]
8. Woodmansey, E.J.; McMurdo, M.E.; Macfarlane, G.T.; Macfarlane, S. Comparison of compositions and metabolic activities of fecal microbiotas in young adults and in antibiotic-treated and non-antibiotic-treated elderly subjects. *Appl. Environ. Microbiol.* **2004**, *70*, 6113–6122. [CrossRef]
9. Białecka-Dębek, A.; Granda, D.; Szmidt, M.K.; Zielińska, D. Gut Microbiota, Probiotic Interventions, and Cognitive Function in the Elderly: A Review of Current Knowledge. *Nutrients* **2021**, *13*, 2514. [CrossRef]
10. Yang, X.; Yu, D.; Xue, L.; Li, H.; Du, J. Probiotics modulate the microbiota-gut-brain axis and improve memory deficits in aged SAMP8 mice. *Acta Pharm. Sinica B* **2020**, *10*, 475–487. [CrossRef]
11. O'Connor, L.E.; Gahche, J.J.; Herrick, K.A.; Davis, C.D.; Potischman, N.; Vargas, A.J. Nonfood Prebiotic, Probiotic, and Synbiotic Use Has Increased in US Adults and Children From 1999 to 2018. *Gastroenterology* **2021**, *161*, 476–486.e3. [CrossRef]
12. Marx, W.; Scholey, A.; Firth, J.; D'Cunha, N.M.; Lane, M.; Hockey, M.; Ashton, M.M.; Cryan, J.F.; O'Neil, A.; Naumovski, N.; et al. Prebiotics, probiotics, fermented foods and cognitive outcomes: A meta-analysis of randomized controlled trials. *Neurosci. Biobehav. Rev.* **2020**, *118*, 472–484. [CrossRef] [PubMed]
13. Baldi, S.; Mundula, T.; Nannini, G.; Amedei, A. Microbiota shaping—The effects of probiotics, prebiotics, and fecal microbiota transplant on cognitive functions: A systematic review. *World J. Gastroenterol.* **2021**, *27*, 6715–6732. [CrossRef] [PubMed]
14. Lovden, M.; Fratiglioni, L.; Glymour, M.M.; Lindenberger, U.; Tucker-Drob, E.M. Education and Cognitive Functioning Across the Life Span. *Psychol. Sci. Public Interes.* **2020**, *21*, 6–41. [CrossRef]
15. Marioni, R.E.; Yang, J.; Dykiert, D.; Mottus, R.; Campbell, A.; Group, C.C.W.; Davies, G.; Hayward, C.; Porteous, D.J.; Visscher, P.M.; et al. Assessing the genetic overlap between BMI and cognitive function. *Mol. Psychiatry* **2016**, *21*, 1477–1482. [CrossRef]

16. Topiwala, A.; Ebmeier, K.P. Effects of drinking on late-life brain and cognition. *Evid. Based Ment. Health* **2018**, *21*, 12–15. [CrossRef] [PubMed]
17. Swan, G.E.; Lessov-Schlaggar, C.N. The effects of tobacco smoke and nicotine on cognition and the brain. *Neuropsychol. Rev.* **2007**, *17*, 259–273. [CrossRef] [PubMed]
18. Fan, Y.; Zhang, Y.; Li, J.; Liu, Y.; Chang, H.; Jiang, Y.; Tuo, X.; Zhou, L.; Yu, Y. Association between healthy eating index-2015 and various cognitive domains in US adults aged 60 years or older: The National Health and Nutrition Examination Survey (NHANES) 2011–2014. *BMC Public Health* **2021**, *21*, 1862. [CrossRef]
19. Shi, Y.; Wang, H.; Zhu, Z.; Ye, Q.; Lin, F.; Cai, G. Association between exposure to phenols and parabens and cognitive function in older adults in the United States: A cross-sectional study. *Sci. Total Environ.* **2023**, *858*, 160129. [CrossRef]
20. Austin, P.C. An Introduction to Propensity Score Methods for Reducing the Effects of Confounding in Observational Studies. *Multivariate Behav. Res.* **2011**, *46*, 399–424. [CrossRef]
21. Lv, T.; Ye, M.; Luo, F.; Hu, B.; Wang, A.; Chen, J. Probiotics treatment improves cognitive impairment in patients and animals: A systematic review and meta-analysis. *Neurosci. Biobehav. Rev.* **2021**, *120*, 159–172. [CrossRef]
22. Den, H.; Dong, X.; Chen, M.; Zou, Z. Efficacy of probiotics on cognition, and biomarkers of inflammation and oxidative stress in adults with Alzheimer's disease or mild cognitive impairment—A meta-analysis of randomized controlled trials. *Aging* **2020**, *12*, 4010–4039. [CrossRef] [PubMed]
23. Ruiz-Gonzalez, C.; Roman, P.; Rueda-Ruzafa, L.; Rodriguez-Arrastia, M.; Cardona, D. Effects of probiotics supplementation on dementia and cognitive impairment: A systematic review and meta-analysis of preclinical and clinical studies. *Prog. Neuropsychopharmacol. Biol. Psychiatry* **2021**, *108*, 110189. [CrossRef] [PubMed]
24. Liu, C.; Guo, X.; Chang, X. Intestinal Flora Balance Therapy Based on Probiotic Support Improves Cognitive Function and Symptoms in Patients with Alzheimer's Disease: A Systematic Review and Meta-analysis. *Biomed. Res. Int.* **2022**, *2022*, 4806163. [CrossRef] [PubMed]
25. Tahmasbi, F.; Mirghafourvand, M.; Shamekh, A.; Mahmoodpoor, A.; Sanaie, S. Effects of probiotic supplementation on cognitive function in elderly: A systematic review and Meta-analysis. *Aging Ment. Health* **2022**, *26*, 1778–1786. [CrossRef] [PubMed]
26. Krüger, J.F.; Hillesheim, E.; Pereira, A.C.S.N.; Camargo, C.Q.; Rabito, E.I. Probiotics for dementia: A systematic review and meta-analysis of randomized controlled trials. *Nutr. Rev.* **2021**, *79*, 160–170. [CrossRef]
27. Thion, M.S.; Low, D.; Silvin, A.; Chen, J.; Grisel, P.; Schulte-Schrepping, J.; Blecher, R.; Ulas, T.; Squarzoni, P.; Hoeffel, G.; et al. Microbiome Influences Prenatal and Adult Microglia in a Sex-Specific Manner. *Cell* **2018**, *172*, 500–516.e16. [CrossRef]
28. Fitzpatrick, S.; Gilbert, S.; Serpell, L. Systematic review: Are overweight and obese individuals impaired on behavioural tasks of executive functioning? *Neuropsychol. Rev.* **2013**, *23*, 138–156. [CrossRef]
29. Prickett, C.; Brennan, L.; Stolwyk, R. Examining the relationship between obesity and cognitive function: A systematic literature review. *Obes. Res. Clin. Pract.* **2015**, *9*, 93–113. [CrossRef]

nutrients

MDPI

Article

The Usefulness of Resistant Maltodextrin and Chitosan Oligosaccharide in Management of Gut Leakage and Microbiota in Chronic Kidney Disease

Weerapat Anegkamol [1], Panumas Kamkang [1], Sittiphong Hunthai [1], Maroot Kaewwongse [2], Mana Taweevisit [3], Natthaya Chuaypen [1,4], Pakkapon Rattanachaisit [1,5] and Thasinas Dissayabutra [1,*]

1 Metabolic Disease in Gastrointestinal and Urinary System Research Unit, Department of Biochemistry, Faculty of Medicine, Chulalongkorn University, Bangkok 10330, Thailand; 6174767730@student.chula.ac.th (W.A.); 6370041030@student.chula.ac.th (P.K.); 6370056530@student.chula.ac.th (S.H.); natthaya.c@chula.ac.th (N.C.); pakkapon.r@chula.ac.th (P.R.)
2 Division of Physiology, School of Medical Sciences, University of Phayao, Phayao 56000, Thailand; maroot.ka@up.ac.th
3 Department of Pathology, Faculty of Medicine, Chulalongkorn University, Bangkok 10330, Thailand; dr.mana4@gmail.com
4 Center of Excellence in Hepatitis and Liver Cancer, Department of Biochemistry, Faculty of Medicine, Chulalongkorn University, Bangkok 10330, Thailand
5 Department of Physiology, Faculty of Medicine, Chulalongkorn University, Bangkok 10330, Thailand
* Correspondence: thasinas@chula.md

Abstract: Microbiota-dysbiosis-induced gut leakage is a pathophysiologic change in chronic kidney disease (CKD), leading to the production of several uremic toxins and their absorption into the bloodstream to worsen the renal complications. We evaluate the benefits of resistant maltodextrin (RMD) and chitosan oligosaccharide (COS) supplements in cell culture and CKD-induced rats. The RMD exerted a significant anti-inflammatory effect in vitro and intestinal occludin and zonula occluden-1 up-regulation in CKD rats compared with inulin and COS. While all prebiotics slightly improved gut dysbiosis, RMD remarkably promoted the relative abundance and the combined abundance of *Lactobacillus*, *Bifidobacteria*, *Akkermansia*, and *Roseburia* in CKD rats. Supplements of RMD should be advantageous in the treatment of gut leakage and microbiota dysbiosis in CKD.

Keywords: gut microbiota; prebiotics; chronic kidney disease; chitosan oligosaccharide; maltodextrin; gut leakage; tight junction

1. Introduction

Chronic kidney disease (CKD) is a global health crisis. Patients with CKD suffer from the retention of serum uremic toxins that subsequently instigate systemic complications. Uremic toxins are mainly produced in the intestine by the gut microbiota and absorbed into the bloodstream. In CKD patients, harmful gut microbiota overgrowth leads to the production of uremic toxins, inflammation, and an increase in gut permeability [1,2]. The retention of serum uremic toxins such as urea, ammonia, indoxyl sulfate, P-cresol, and trimethylamine-N-oxide causes fatal cardiovascular and metabolic complications and subsequently worsens the renal function.

Several approaches to suppress uremic toxin synthesis and absorption have been recently introduced to alleviate CKD progression or induce renal recovery. Among these, dietary supplementation with prebiotics and probiotics was one of the most promising methods [3–6]. Prebiotics serve several beneficial functions in gut ecology. They act as a food source to promote the growth of commensal gut microbiota. Additionally, they are fermented and metabolized into short-chain fatty acids (SCFA), which provide more favorable functions, including anti-inflammation, suppression of pathogenic bacteria, immunomodulation, mucin production, maintenance of a normal gut barrier, and increased colonic mineral absorption [7]. Inulin, or fructo-oligosaccharide (FOS), and galacto-oligosaccharide (GOS) are common prebiotics used in various products and were evidently beneficial in correcting gut dysbiosis and lowering serum uremic toxins [8–10].

Recent studies disclosed the usefulness of other prebiotics in modifying gut microbiota and modulating gut and other systemic diseases. Chitosan, as a non-plant derived prebiotic, was used to treat bowel inflammation and metabolic syndrome [11,12], and maltodextrin, a starch commonly used as a placebo, was lately reported to contain prebiotic properties [13]. In the present study, we focused on the effects of chitosan oligosaccharide (COS) and resistant maltodextrin (RMD) supplementation on the gut barrier integrity in cell culture and cisplatin-induced CKD rats. We anticipated that COS and RMD supplementation may improve the gut barrier and alleviate gut dysbiosis found in CKD.

2. Materials and Methods

2.1. *Tested Chemicals*

The resistant maltodextrin (RMD) and inulin used in this study were purchased from Banpong Novitat Co., Ltd. (Bangkok, Thailand). RMD is an oligosaccharide with random 1–2, 1–3, 1–4, and 1–6 α and β glycosidic linkages, making it partially resistant to human digestive enzymes. The glycemic index of RMD was evaluated to be approximately 59, and it has a 2000 Da molecular weight. Inulin is a chain of fructose molecules linked together, with a glucose molecule at one end. The molecular weight of inulin typically ranges from approximately 500 to 4000 daltons.

Chitosan oligosaccharide (COS) is an oligomer composed of β-(1-4)-linked d-glucosamine units. It is derived from the deacetylation and hydrolysis of chitin. The molecular weight of COS is less than 10,000 daltons, and each molecule consists of approximately 55 monomers. A small molecule COS used in this study was synthesized and provided by Assist. Prof. Dr. Rath Pitchyangura from the Department of Biochemistry, Faculty of Science, Chulalongkorn University, with a molecular weight of around 5–9 kDa.

2.2. *In Vitro Study*

2.2.1. MTT Survivability Test

Caco-2, human colon adenocarcinoma cells were used in this study. The cells were cultures in sterilized DMEM (Dulbecco's modified Eagle's medium, containing 4 mM L-glutamine, 4500 mg/L glucose, 1 mM sodium pyruvate, and 1500 mg/L sodium bicarbonate) supplemented with 10% *v/v* FBS (fetal bovine serum), 1% *v/v* penicillin, and 1% *v/v* streptomycin and incubated at 5% CO_2 and 37 °C. The cells were seeded in a 96-well plate with 15,000 cells per plate and incubated for 24 h. The cells were washed with PBS (phosphate-buffered saline) and new DMEM was added to the test substances for 24 h. The test substances included inulin, RMD, and COS at 10, 50, 100, 500, and 1000 μg/mL. PBS, 0.05% acetic acid, and 0.3% H_2O_2 were used as positive control, vehicle, and negative control for viable cells, respectively. After 24 h, 3-(4,5-dimethylthiazol-2-yl)-2,5-diphenyltetrazolium bromide (MTT) was added to the cells. A multidetection microplate reader was used to detect the absorbance at 570 nm on the culture plate. The survivability of cells exposed to different concentrations of the test substances were compared to the positive control.

2.2.2. Transepithelial Electrical Resistance (TEER) Assay

Caco-2 cells of 500,000 cells were seeded into transwell (ThinCertTM, 0.4 μm pore size, 1.131 cm^2 culture surface) and incubated for 21 days with high glucose DMEM supplemented with 10% FBS, 1% penicillin, and 1% streptomycin. Then, the cells were washed by PBS and the medium was replaced by FBS-free media. The cells' TEER values were measured using an epithelial voltohmmeter (EVOM2) to obtain TEER values at 0 h. The cells in the transwell were treated with PBS, as a control, and the tested substances, including 100 μg/mL of inulin, COS, and RMD. A concentration of 60 ng/mL of TGF-β, growth factor, was used to see the TEER value in integrity-enhanced epithelial tissue. The study was carried out in 2 settings: normal and inflammation. In the inflammation setting, the cells were challenged by adding 50 ng/mL TNF-α, pro-inflammatory cytokine, in both the apical and basolateral sides to induce cell inflammation that disturbs the epithelial integrity of the cells 1 h before adding the test substances. The cells from each treatment were incubated for 24 h and then their TEER values were remeasured to obtain TEER values at 24 h. Fold changes were calculated from the ratio between TEER values at 24 h and TEER values at 0 h. The fold change in TEER values from each treatment was compared to the fold change in TEER from the control to calculate the %TEER to control.

2.3. *Animal Study*

2.3.1. Animal Preparation, CKD Induction, and Sample Collection

The rats were raised in the Animal Center at the Faculty of Medicine, Chulalongkorn University, in a 35 cm × 75 cm × 18 cm cage, at 22 °C, under a 12/12 light and dark cycle, 40–60% relative humidity, with ad libitum food and drink. Four-week-old Wistar rats were purchased from Nomura Siam Co., Ltd. and acclimatized for two weeks before the initiation of the experiment. The rats were randomly assigned to five groups: control, CKD, inulin, COS, and RMD, each consisting of 6 animals. To induce kidney damage, the rats in the CKD, inulin, COS, and RMD groups received an intraperitoneal injection of cisplatin at a dosage of 10 mg/kg of body weight. In contrast, the control group rats were injected with 1 mL of normal saline. Following the injection, all rats were monitored for a period of five weeks to allow the acute renal injury to subside.

During this period, the rats were provided with normal chow (CP 802), comprised of carbohydrate 52%, fat 19.77%, and protein 28.23%, and drinking water containing 1% *w*/*w* phosphate (as phosphoric acid), along with the designated treatments for the experiment. Additionally, the rats were orally administered 54 mg of inulin per kg, 16 mg of COS per kg, or 54 mg of RMD per kg of rat. An approximately 0.5 mL blood sample was obtained from each rat's tail vein using heparinized 1.5 mL microcentrifuge tubes. The blood was then centrifuged at 2500 G for 10 min to separate the plasma, which was subsequently stored at −80 °C. The rats' fresh feces were collected and preserved in DNA/RNA Shield®® reagent (ZYMO Research, Irvine, CA, USA) at −80 °C.

Blood samples were collected from all rats before the administration of cisplatin, at the start of the treatments, and on the day of sacrifice. At the end of the twelve-week study period, the rats were euthanized using long exposure to CO_2. On the day of sacrifice, blood samples were collected via cardiac puncture. The kidneys were longitudinally split, with half of the specimens placed in 4% paraformaldehyde and stored at 4 °C, while the other half was collected in RNAlater™ and stored at −80 °C. The jejunum section of the intestine was cut into 1 cm pieces and stored at −80 °C. Fecal samples were also collected and stored in DNA/RNA Shield®® reagent (ZYMO Research) at −80 °C. The remaining sections of the intestine and right femur bones were collected and preserved in paraformaldehyde at 4 °C.

2.3.2. Gene Expression of Tight Junction Protein

RNA was extracted from the jejunum sections of the intestinal tissue, and quantitative reverse transcription-polymerase chain reaction (RT-qPCR) was performed. RT-qPCR utilized primers specific to GAPDH (a housekeeping gene), claudin-1, occludin, and zonula occludens-1 (ZO-1). The goal was to measure the relative expression of each tight junction protein in response to different treatments.

To determine the fold changes of each tight junction protein relative to the control group, the expression levels were calculated and normalized to the housekeeping gene. This analysis allowed for an assessment of the impact of each treatment on the expression of these crucial tight junction proteins.

2.3.3. Serum Creatinine, Calcium, Phosphate, and PTH Profiling

The concentrations of creatinine, calcium, and phosphate in the samples were analyzed using the automated Alinity ci system at the Department of Laboratory Medicine, Faculty of Medicine, Chulalongkorn University. The serum parathyroid hormone (PTH) concentration from each treatment was determined using an enzyme-linked immunosorbent assay (ELISA) of a parathyroid hormone ELISA kit in 96-well plates purchased from Wuhan Fine Biotech Co., Ltd. (Wuhan, China). The optical densities were measured at a wavelength of 450 nm using multidetection microplate readers. In addition, the estimated creatinine clearance (eClCr) (Supplementary Figure S1 and Table S1) was calculated using the web-based ACLARA (https://idal.uv.es/aclara/, accessed on 13 July 2023) [14].

2.3.4. Histopathological Evaluation

The tissue samples were subjected to permanent slide-making processes at the Department of Pathology, Faculty of Medicine, Chulalongkorn University. Sections of jejunum tissues were stained using the hematoxylin and eosin (H&E) staining technique to visualize the ZO-1 protein. Immunofluorescent staining was performed using a rabbit polyclonal antibody specific to ZO-1. The evaluation of the stained tissues was conducted by a pathologist affiliated with the Department of Pathology, Faculty of Medicine, Chulalongkorn University.

2.3.5. Intestinal Microbiota Analysis

The DNA extraction process from fecal samples involved using the ZymoBIOMICS™ DNA Miniprep Kit according to the manufacturer's instructions provided by Zymo Research Corp and supplied by S.M.Chemical Supplies Co., Ltd., Bangkok, Thailand. The resulting DNA was then assessed for concentration and purity using a DeNovix™ UV-vis spectrophotometer (Purchased from Bio-Active Co., Ltd., Bangkok, Thailand) and stored at −20 °C until further analysis.

To analyze the intestinal microbiome, the 16S/ITS Microbiome Profiling Service offered by Modgut Genomic Service at King Mongkut's University of Technology Thonburi (Thailand), was utilized. The V3–V4 region of the 16S rRNA gene, serving as the target sequence, was amplified. The richness of microbial taxonomic groups in the samples was determined by the number of groups, while the evenness of their distribution indicated the evenness of the groups. For data visualization, an analysis of the alpha diversity, which includes the Shannon index and Pielou's evenness, was conducted to summarize the ecological communities of the gut microbiota in terms of richness and evenness. The top relative abundance taxa were determined by calculating the average abundance of each taxon in each group divided by the total abundance of that taxon. The relative abundance of the bacteria was further analyzed and correlated with physiological parameters.

2.4. Statistical Analysis

The statistical analysis in the current study was conducted using SPSS version 22.0. Continuous data were assessed using the Student *t*-test for comparing two independent groups, while the ANOVA with post hoc Bonferroni test was used for comparisons involving more than two groups. Non-parametric variables were analyzed using the Mann–Whitney U test. In an animal experiment, the Student *t*-test was employed to compare two independent groups, and the Kruskal–Wallis test was used to evaluate differences in means for non-parametric data. To examine correlations, Pearson correlation was employed. Significance was determined at $p < 0.05$. Figures, diagrams, and graphs were created using GraphPad Prism 9 (Windows 64-bit) v9.5.1.733.

2.5. Ethical Consideration

The research adhered to the principles outlined in the Helsinki Declaration and followed the guidelines of good clinical practice when involving human participants, who provided fecal samples. The experimental procedures involving animals were conducted in accordance with the protocols of the Institutional Animal Care and Use Committee (IACUC). Ethical approval for the study was obtained from the Institutional Review Board of the Faculty of Medicine, Chulalongkorn University (IRB number 914/64). The protocols for animal experiments were approved by the Chulalongkorn University Animal Care and Use Committee (CU-ACUC protocol number 004/2563).

3. Results

3.1. In Vitro Study

In this study, we conducted an MTT survivability test on Caco-2 cells to assess the potential cytotoxicity of inulin, chitosan oligosaccharide (COS), and resistant maltodextrin (RMD). Our results demonstrated a significant decrease in cell viability at a concentration of 1000 µg/mL for inulin and COS, and at 500 µg/mL for RMD (Figure 1A). These findings indicated the presence of a toxic dose of the tested treatments on Caco-2 cells. Based on these results, we selected the concentration of 100 µg/mL for inulin, COS, and RMD to investigate the impact on transepithelial electrical resistance (TEER) across the Caco-2 cell monolayer.

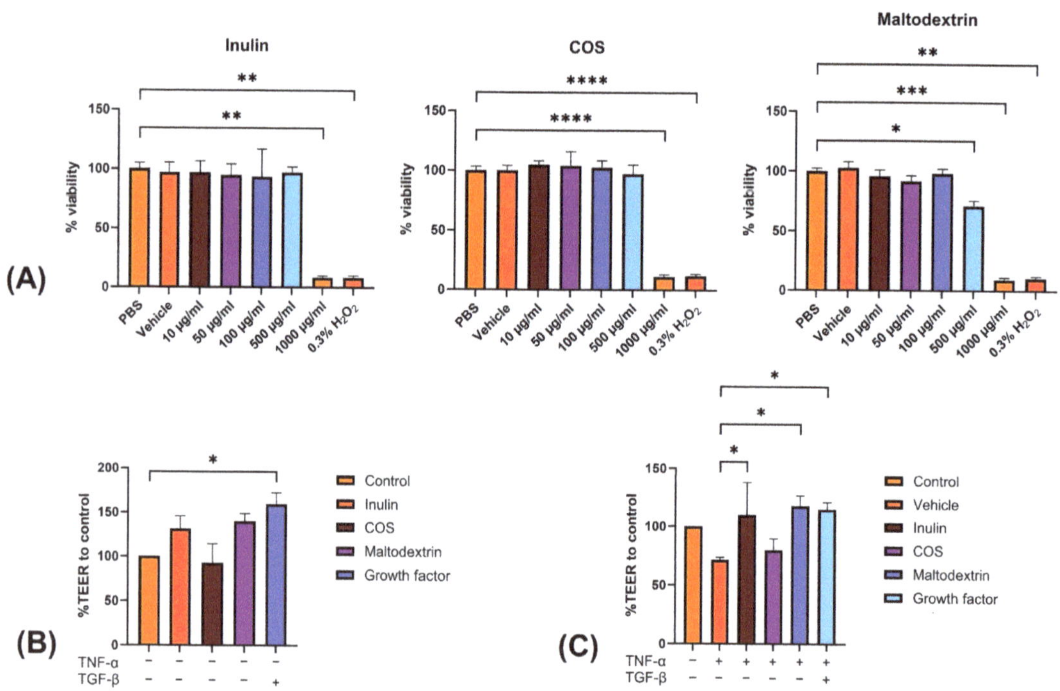

Figure 1. Effects of prebiotics on Caco-2 cells' viability and TEER after 24 h of treatment. (**A**) MTT cell, (**B**) %TEER under normal conditions, (**C**) %TEER in an inflammation-induced condition. * $p < 0.05$, ** $p < 0.005$, *** $p < 0.001$, and **** $p < 0.0001$ as compared to PBS.

To further investigate the effects of prebiotic treatments on intestinal barrier integrity, we employed the TEER assay. The TEER values of the Caco-2 cell monolayers were compared between each prebiotic treatment and the control group. As a positive control for TEER enhancement, TGF-β was included. In the normal setting, none of the prebiotic treatments exhibited a significant increase in the %TEER compared to the control group, as shown in Figure 1B. However, under an inflammation-induced setting using TNF-α induction, both inulin and maltodextrin demonstrated a substantial increase in TEER compared to the vehicle-treated group, as shown in Figure 1C.

3.2. Animal Studies

All rats with CKD exhibited significantly higher serum creatinine levels compared to the control group, indicating the induction of CKD through cisplatin administration. The estimated creatinine clearance of CKD rats was about 41.1% lower than the control (3.80 ± 0.42 vs. 2.24 ± 0.97 mL/min in control and CKD groups, respectively). However, no significant differences were observed between prebiotic-treated and untreated rats in terms of serum creatinine levels (Figure 2A), suggesting that the prebiotic treatments did not have a significant impact on renal function.

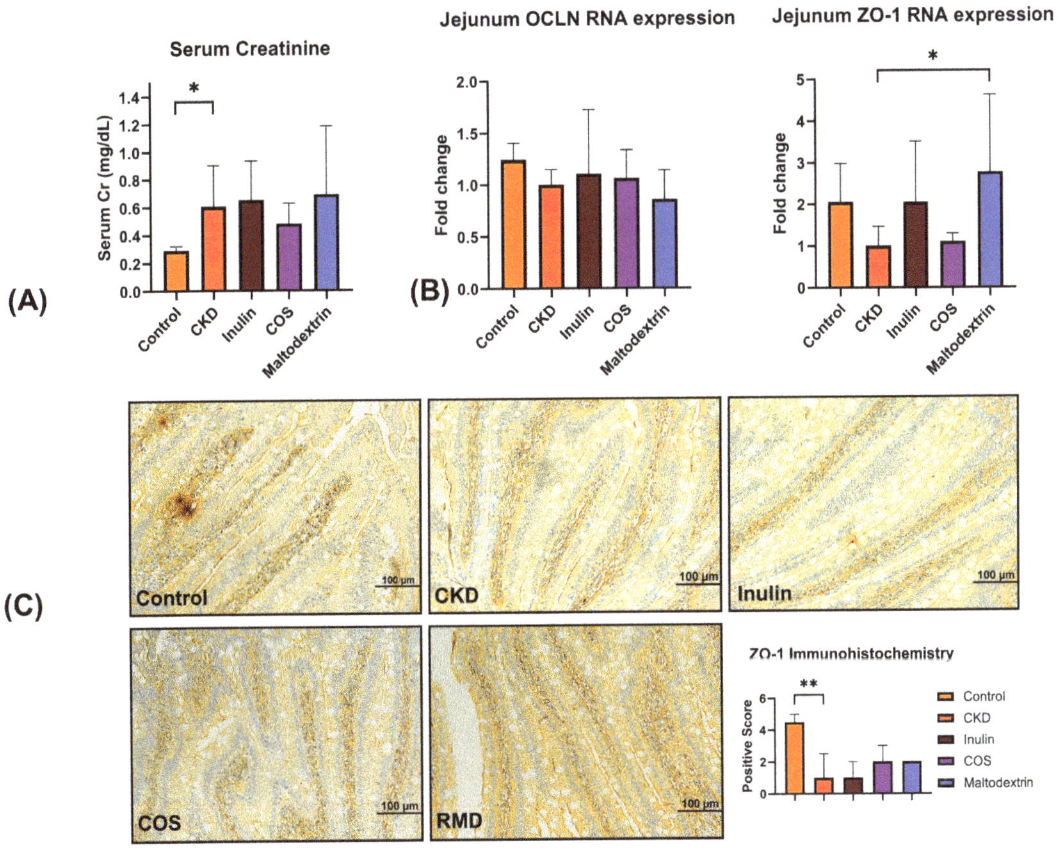

Figure 2. (**A**) Serum creatinine. (**B**) The RNA expression levels of occludin (OCLN) and zonula occluden-1 (ZO-1) in the jejunum. (**C**) Immunohistochemistry against ZO-1 protein in intestinal epithelium of rat jejunum. * $p < 0.05$, ** $p < 0.005$.

The RNA expression levels of occludin (OCLN) and zonula occluden-1 (ZO-1) in the jejunum of rats were analyzed. We found no significant differences in the RNA expression of OCLN among any of the groups. However, treatment with RMD resulted in a promotion of jejunal ZO-1 expression in rats with CKD, as shown in Figure 2B. This suggests that RMD may play a role in enhancing the expression of ZO-1, a critical tight junction protein involved in gut barrier integrity.

Immunohistochemistry studies also revealed that COS and RMD slightly rescued ZO-1 protein expression in the jejunum of rats with CKD, further supporting the potential protective effects of RMD on gut barrier integrity (Figure 2C).

3.3. Gut Microbiome Study

3.3.1. Fecal Microbiota Diversity

The Shannon index and Pielou's evenness were evaluated to assess the diversity and evenness of bacterial communities at week 12 of the experiment. However, no significant differences were observed in the Shannon index and Pielou's evenness among the different groups (Figure 3A).

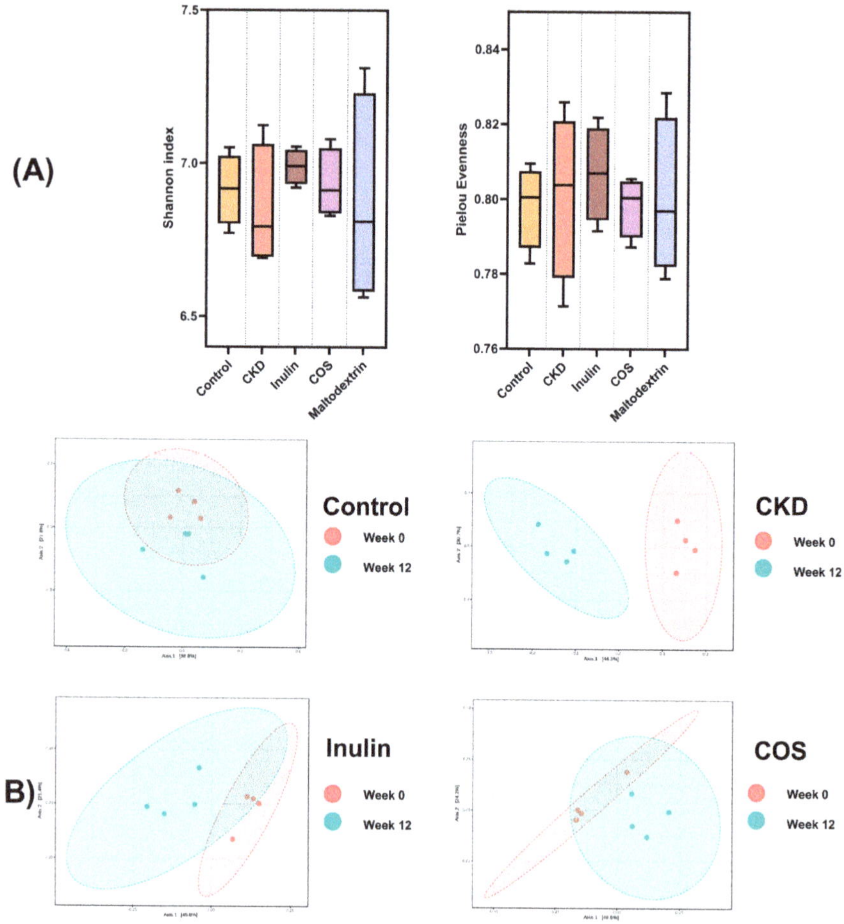

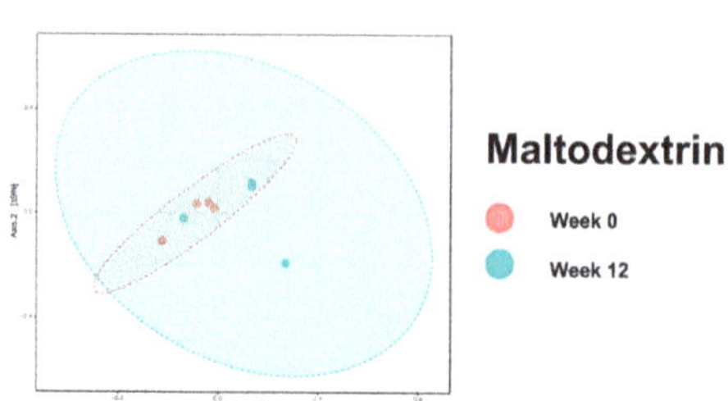

Group	F-value	R-squared	p-value
Control	1.6337	0.21401	0.106
CKD	4.5464	0.43108	0.039*
Inulin	4.1316	0.40779	0.039*
COS	3.6561	0.37863	0.039*
RMD	1.9612	0.24634	0.12

Figure 3. Species diversity of fecal microbiota: (**A**) alpha-diversity indices. The reported values are median ± IQR (n = 4). A significant difference among all the groups was not detected. (**B**) Beta diversity (Bray–Curtis dissimilarity) PCoA index. Results are means for n = 4; * $p < 0.05$.

In contrast, when examining the control group, beta diversity analysis to compare the gut microbiota alteration between pre- and post-study in each treatment group using principal coordinate analysis (PCoA) of Bray–Curtis dissimilarity demonstrated no significant difference in diversity between week 0 and week 12. This suggests that the overall composition of bacterial communities in the control group remained relatively stable over the course of the experiment. The CKD group exhibited a significantly lower diversity at the 12th week, while inulin supplementation led to a relatively lesser but still significantly lower diversity. This indicates a notable change in the bacterial community composition over time in these groups. On the other hand, the rats treated with RMD did not show a significant change in beta diversity, similar to the control group, suggesting that RMD treatment maintained a relatively stable and consistent bacterial community composition throughout the 12-week period. Interestingly, the COS group displayed a significant increase in diversity, suggesting a notable shift in the composition of bacterial communities caused by the COS treatment.

3.3.2. Relative Abundances and Correlation

Regarding the relative abundance of bacterial phyla in the fecal microbiota, comparisons were made between each treatment group and the CKD group (Figure 4A). At week 0, in the control group, *Cyanobacteria* were found to be higher in abundance compared to the CKD group, while *Proteobacteria* were lower. By week 12, significant differences in relative abundance emerged. In the control group, *Verrucomicrobiota* and *Campylobacterota* were higher in abundance compared to the CKD group. Similarly, in the RMD group, a higher relative abundance of *Verrucomicrobiota* compared to CKD was observed. These findings indicate that both the control and RMD groups exhibited higher levels of *Verrucomicrobiota* compared to the CKD group at the end of the 12-week period. These results indicated the potential influence of RMD on the relative abundance of the specific bacterial phylum *Verrucomicrobiota*.

The analysis of relative abundances at the genus level revealed slight differences in several bacterial genera across each treatment group; however, no significant differences were observed. While the relative abundances of various genera showed some variability, these differences did not reach statistical significance. Of particular interest were the beneficial bacteria, including *Lactobacillus*, *Bifidobacterium*, *Roseburia*, and *Akkermansia*. The relative abundances of these genera were aggregated for each treatment group. Remarkably, a significant increase was observed in the RMD treatment group compared to the other groups, as shown in Figure 4B. This finding suggests that the administration of RMD had a notable impact on the relative abundances of these beneficial bacterial genera. The significant increase in the relative abundances of the aforementioned genera in the RMD treatment group indicates the potential of RMD in promoting the growth and proliferation of these beneficial bacteria. These findings support the notion that RMD may exert beneficial effects on the gut microbiota composition by selectively enriching beneficial bacterial populations.

To explore the relationships between the fecal microbiota and various physiological parameters, including serum creatinine levels, ZO-1 RNA expression, and occludin RNA expression, Pearson correlation coefficients (r-values) were calculated. These correlation coefficients provide insights into the strength and direction of associations between variables. A heatmap was generated based on the correlation coefficients to visualize the relationships between the variables. The heatmap allows for a comprehensive overview of the correlation patterns among the measured parameters (Figure 4C).

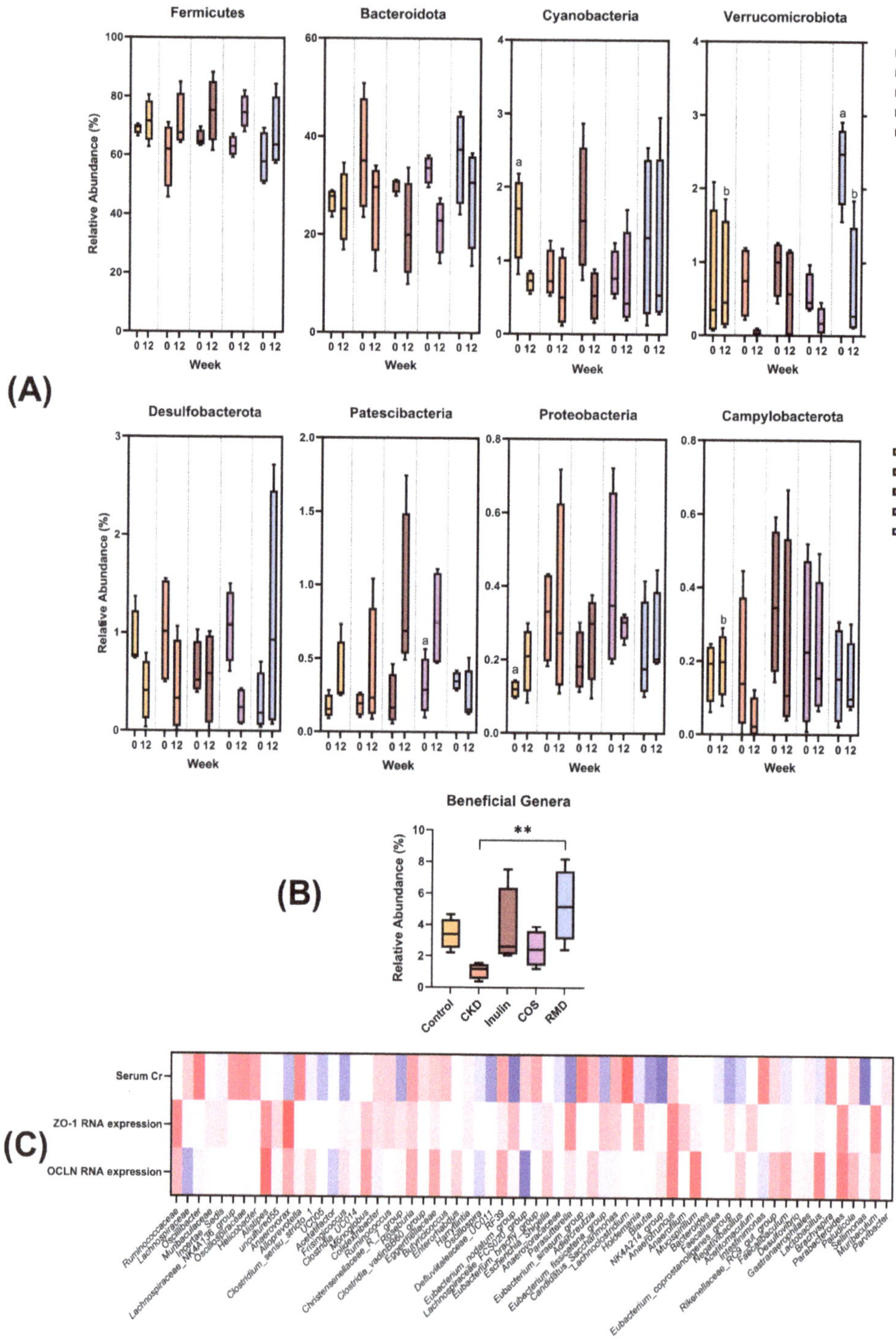

Figure 4. (**A**) Relative abundance of fecal microbiota at phylum level. (a) represents significant difference compared to CKD group at week 0, (b) represents significant difference compared to CKD

at week 12. $p < 0.05$. (**B**) Relative abundance of 4 beneficial genera, *Lactobacillus*, *Bifidobacterium*, *Roseburia*, and *Akkermansia*, ** $p < 0.05$. (**C**) Heatmap of correlation between fecal microbiota, serum creatinine, ZO-1 RNA expression, and occludin RNA expression. r-values were calculated using Pearson correlation test. Red color indicates negative correlation, blue color indicates positive correlation.

4. Discussion

Prebiotic is a term used for a group of nutrients that are degraded and function as a food source by gut microbiota [15]. Several types of prebiotics are currently available. The most common prebiotics used in food products are fructans or inulin and fructo-oligosaccharide (FOS), which evidently promoted lactic acid bacteria growth, and galacto-oligosaccharide (GOS) that boosted *Bifidobacteria* and *Lactobacilli* [16]. Other prebiotics such as lactulose, arabinoxylans, xylo-oligosacharide (XOS), resistant starch, and other carbohydrate oligosaccharides are sometimes used. Currently, most scientists believe that prebiotics only work as the bacterial food source to maintain a balanced gut microbiota and restore gut microbiota homeostasis or gut eubiosis [17].

The present study focused on utilizing prebiotics to alleviate gut leakage. As an initial step, we tested candidate prebiotics to assess their ability to enhance tight junction protein expression and increase transepithelial electrical resistance, particularly in an inflammatory state, using in vitro models. Overall, our investigation demonstrated that in vitro studies could partially reflect the in vivo results. Prebiotics that showed stronger effects on Caco-2 cells tended to exert favorable effects in animal studies.

We used inulin as a standard prebiotic and found that even though inulin could restore transepithelial electrical resistance of inflammation-induced Caco-2 cells, in the in vivo study, inulin was ineffective in rescuing tight junction expression in the intestine. Lately, several groups of researchers utilized inulin as a prebiotic, or a component in synbiotics, to treat renal injury, and most of these studies found that prebiotics and/or synbiotics containing inulin improved serum creatinine, inflammatory cytokines, uremic toxins, glucose, and lipid profiles, but did not mitigate the pathological change in affected tissues, such as cardiac, kidney, or intestinal integrity [18–20]. Melekoglu, et al. reported that inulin supplementation to CKD rats could reduce serum creatinine, p-cresyl sulfate, and IL-6 but had no effect on serum indoxyl sulfate and colonic claudin-1 and occludin protein expression [21]. Regarding this, inulin is not the most appropriate prebiotic to treat kidney disease patients, and new prebiotics should be investigated. In the present study, we explored COS due to the previous report of microbiota-independent prevention of intestinal epithelial inflammation in vitro [22], and RMD, which has been commonly used as a placebo in numerous prebiotics studies, but also presents some prebiotic properties.

Chitosan is a polymer of randomly distributed acetylated and deacetylated forms of D-glucosamine which is derived from chitin found in the outer skeleton of shellfish. Chitosan has a lot of benefits as a biomedical polymer for tissue engineering, artificial organ synthesis, and wound healing [23]. Intake of chitosan may reduce intestinal fat absorption and was assumed to be advantageous in the treatment of hypertension, hypercholesterolemia, and obesity. The degradation of chitosan yields COS, which has lower viscosity and high intestinal epithelial absorption [24]. COS contains an anti-bacterial activity against certain microorganisms such as *Escherichia coli*, *Pseudomonas aeruginosa*, *Acinetobacter baumannii*, *Staphylococcus aureus*, etc. [25,26].

In the aspect of prebiotics, COS promotes Bifidobacteria, Lactobacilli, Prevotella, Rosuburia, and Faecalibacterium prausnitzii growth, while suppressing the number of Firmicutes, Streptococci, Bacteroides fragilis, Clostridium spp., and E. coli [27,28]. In certain conditions, COS improves gut dysbiosis and intestinal epithelial leakage [12]. Additionally, COS was reported to be beneficial in metabolic syndrome, diabetes mellitus and fatty liver disease since it can inhibit hepatic fat accumulation, reduce adipogenesis, and stimulate white fat cell browning and promote glucose homeostasis in diabetic rats [11,29–31]. Recent studies revealed the beneficial effects of chitosan nanoparticles and COS in ulcerative colitis, autoimmune encephalitis, autoimmune arthritis, lupus nephritis, and autoimmune hepatitis [32–34]; however, there have been a limited number of studies about COS in renal diseases. Chitosan nanoparticle-encapsulated drugs combined with metformin reduced creatinine, proteinuria, and downregulated TNF-α, IL-6, and IL-1β in type 2 DM rats [35,36]. Zhang H., et al., revealed that COS ameliorated proteinuria and the expression of kidney injury markers and may reverse pathologic change in diabetic rats' kidneys [37]. Recent research revealed that COS attenuates oxidative damage, renal fibrosis, and renal cyst growth [38,39]. Accordingly, we expected that supplementation with COS could be beneficial in chronic renal disease. However, our in vitro study showed that COS could not enhance the integrity of colonic epithelial cells in normal or inflammatory states and giving COS supplements to CKD rats could not restore the intestinal tight junction protein expression and the glomerular function. The gut microbiome study demonstrated that COS slightly improved the number of Firmicutes and Patescibacteria, while reducing Bacteroidota, Vercurromicrobiota, and Desulfobacterota, similar to inulin feeding, and supposedly partially normalized gut dysbiosis. We infer that COS may not be an ideal prebiotic to treat gut leakage in CKD.

It should be noted that a dosage of 8 mg/kg of COS was used in this experiment, while 20 mg/kg of COS, given to animals in a previous study, was claimed to have an anti-inflammatory effect. Regarding this, we supposed that a higher dose or alternative preparation of COS may be advantageous in CKD. Further investigation is required to validate this hypothesis.

RMD is a polymer of D-glucose derived from the enzymatical process of starch, especially from plants. RMD has been used as the placebo control in many studies focusing on prebiotic efficacy. However, indigestible, fermentable, resistant maltodextrin was proven to have prebiotic functions, as consumption of RMD increased gut microbiota richness, particularly the *Bifidobacteria* count, and short chain fatty acid (SCFA) production [40–42]. Recent reports revealed that when compared to FOS, RDM intake was more efficient in the production of acetate, butyrate, propionate, and total SCFA, as well as lower in trimethylamine synthesis [13,43]. However, evidence was lacking about the impact of RMD on disease prevention, including intestinal epithelium and kidney health.

Our study showed that RMD treatment efficiently increased monolayer Caco-2 cell integrity in normal and inflammatory-induced conditions and promoted intestinal ZO-1 expression in CKD rats. These effects are stronger than inulin at the same dosage. The RMD supplement substantially boosted the relative abundance of gut microbiota in CKD rats, and partly suppressed *Verrucomicrobiota* and 'bad bacteria' *Campyrobacterota* numbers. In addition, we found that the combination of 'good bacteria' *Lactobacillus*, *Bifidobacteria*, *Akkermansia*, and *Roseburia* increased significantly in RMD-fed rats compared to their CKD counterparts. This group of bacteria is responsible for SCFA synthesis and mucous membrane protection. These results indicated that RMD supplementation was better than dose-dependent inulin and COS in the modulation of gut dysbiosis.

Gut bacteria such as *Lactobacillus*, *Bifidobacteria*, *Akkermansia*, and *Roseburia* are commonly recognized as beneficial microbiota. These bacteria have been found to enhance metabolism, reduce inflammation, improve intestinal barrier function, and maintain microbiota homeostasis. Their higher abundance is associated with positive outcomes in the treatment of diabetes mellitus, obesity, and inflammatory bowel disease (IBD) [44–46]. In CKD patients, lower levels of *Lactobacillus*, *Bifidobacteria*, *Akkermansia*, and *Roseburia* have been strongly linked to poorer glomerular function, malnutrition, and fatal complications [10,46,47]. Therefore, promoting the growth of these beneficial bacteria through prebiotics such as RMD is considered advantageous for CKD therapy.

Based on the correlation study between the relative abundance of gut microbiota and serum creatinine or tight junction protein expression (Figure 4C), significant correlations were observed between specific components of the gut microbiota and serum creatinine levels. Notably, *Oscillibacter* and *Lachnoclostridium* showed a negative correlation, while *Sellimonas, Eubacterium nodatum,* and NK 4A214 exhibited a positive correlation with serum creatinine. Serum creatinine is commonly utilized as a marker of kidney function, particularly glomerular filtration rate. Normally, creatinine production is independent of gut microbiota activity, as it primarily originates from muscle metabolism and glomerular filtration. However, recent research has indicated that the gut microbiota can influence various aspects of human health, including metabolism and inflammation, which may indirectly impact kidney function [48]. It is important to note that correlation does not imply causation and may be linked to factors such as the animal's activities, hydration levels, and underlying health conditions that can influence creatinine levels independently of the gut microbiota.

Moreover, we demonstrated the deleterious effects of *Alistipes* and *Anaerovorax* on the expression of intestinal tight junction protein expression. Previous studies reviewed the increase in *Alistipes* abundance in bowel inflammation related to the downregulation of intestinal claudin-1, ZO-1, and occludin expression [49,50], while *Anaerovorax* was associated with low tight junction protein expression in heat stress [51]. Regarding renal diseases, these bacteria were reported to be elevated in CKD patients [52,53]. We speculated that the abundance of *Alistipes* and *Anaerovorax* could reflect the severity of gut leakage and uremic toxin absorption in CKD patients.

5. Conclusions

We proposed the resistant type of maltodextrin as a desirable prebiotic supplement in CKD patients to improve gut integrity, lower uremic toxin absorption, and modulate beneficial gut microbiota growth. RMD is generally safe and has no contraindication for hypertension, diabetes mellitus, or renal insufficiency. We also expected that COS may be useful in the management of CKD, but the optimal preparation and dosage should be validated. Our future plan includes utilizing COS and RMD as components of synbiotics to alleviate complications of CKD in animal models.

Supplementary Materials: The following supporting information can be downloaded at: https://www.mdpi.com/article/10.3390/nu15153363/s1, Figure S1: Estimated creatinine clearance calculated from serum creatinine by ACRALA (https://idal.uv.es/aclara/); Table S1: Estimated creatinine clearance calculated from serum creatinine by ACRALA (https://idal.uv.es/aclara/).

Author Contributions: Conceptualization, T.D.; methodology, T.D. and M.K.; validation, W.A., P.K. and T.D.; formal analysis, W.A., N.C. and T.D.; investigation, W.A., M.K., P.K. and S.H.; resources, T.D.; data curation, W.A., M.T. and M.K.; writing—original draft preparation, W.A.; writing—review and editing, T.D.; visualization, W.A.; supervision, N.C. and T.D.; project administration, T.D.; funding acquisition, T.D. All authors have read and agreed to the published version of the manuscript.

Funding: This research was funded by the Rachadapiseksompode Fund, Chulalongkorn University (RA-MF-33/64 and GA65/33), the Fundamental Fund, Chulalongkorn University (HEA663000050), the Thailand Research Fund (TRF) (RSA280077), and the Program Management Unit for Human Resources and Institutional Development, Research, and Innovation (PMU-B) (B36G660010).

Institutional Review Board Statement: The research adhered to the principles outlined in the Helsinki Declaration and followed the guidelines of good clinical practice when involving human participants, who provided fecal samples. The experimental procedures involving animals were conducted in accordance with the protocols of the Institutional Animal Care and Use Committee (IACUC). Ethical approval for the study was obtained from the Institutional Review Board of the Faculty of Medicine, Chulalongkorn University (IRB number 914/64). The protocols for animal experiments were approved by the Chulalongkorn University Animal Care and Use Committee (CU-ACUC protocol number 004/2563).

Informed Consent Statement: Not applicable.

Data Availability Statement: The data presented in this study are openly available at https://doi.org/10.6084/m9.figshare.23618073 (accessed on 2 July 2023).

Acknowledgments: We would like to thank Pisit Tungkijvanich from the Department of Biochemistry, Somying Tumwasorn and Asada Leelahavanichkul from the Department of Microbiology, and Lecturer Kroonpong Iampenkhae from the Department of Pathology, Faculty of Medicine, Chulalongkorn University. Rath Pichyangura from the Department of Biochemistry, Faculty of Science, Chulalongkorm University, Chainan Suksmith from Banpong Novitat Co. Ltd., and Pinidphon Prombutara from Mod Gut Co. Ltd., for your kind suggestions and assistance. We were also grateful for the cooperation from the staff of the Department of Microbiology and the Center of Excellence in Hepatitis and Liver Cancer, Department of Biochemistry, Faculty of Medicine, Chulalongkorn University, as well as Tanyaporn Keratibumrungpong and Tawatchai Chumponsuk from the Metabolic Disease in Gastrointestinal and Urinary System Research Unit.

Conflicts of Interest: The authors declare no conflict of interest. The funders had no role in the design of the study; in the collection, analyses, or interpretation of data; in the writing of the manuscript; or in the decision to publish the results.

References

1. Voroneanu, L.; Burlacu, A.; Brinza, C.; Covic, A.; Balan, G.G.; Nistor, I.; Popa, C.; Hogas, S.; Covic, A. Gut Microbiota in Chronic Kidney Disease: From Composition to Modulation towards Better Outcomes—A Systematic Review. *J. Clin. Med.* **2023**, *12*, 1948. [CrossRef] [PubMed]
2. Rysz, J.; Franczyk, B.; Ławiński, J.; Olszewski, R.; Ciałkowska-Rysz, A.; Gluba-Brzózka, A. The Impact of CKD on Uremic Toxins and Gut Microbiota. *Toxins* **2021**, *13*, 252. [CrossRef] [PubMed]
3. Lydia, A.; Indra, T.A.; Rizka, A.; Abdullah, M. The effects of synbiotics on indoxyl sulphate level, constipation, and quality of life associated with constipation in chronic haemodialysis patients: A randomized controlled trial. *BMC Nephrol.* **2022**, *23*, 259. [CrossRef]
4. Ebrahim, Z.; Proost, S.; Tito, R.Y.; Raes, J.; Glorieux, G.; Moosa, M.R.; Blaauw, R. The Effect of ß-Glucan Prebiotic on Kidney Function, Uremic Toxins and Gut Microbiome in Stage 3 to 5 Chronic Kidney Disease (CKD) Predialysis Participants: A Randomized Controlled Trial. *Nutrients* **2022**, *14*, 805. [CrossRef]
5. Shamloo, M.; Mollard, R.; Wang, H.; Kingra, K.; Tangri, N.; MacKay, D. A randomized double-blind cross-over trial to study the effects of resistant starch prebiotic in chronic kidney disease (ReSPECKD). *Trials* **2022**, *23*, 72. [CrossRef] [PubMed]
6. McFarlane, C.; Krishnasamy, R.; Stanton, T.; Savill, E.; Snelson, M.; Mihala, G.; Kelly, J.T.; Morrison, M.; Johnson, D.W.; Campbell, K.L. Synbiotics Easing Renal Failure by Improving Gut Microbiology II (SYNERGY II): A Feasibility Randomized Controlled Trial. *Nutrients* **2021**, *13*, 4481. [CrossRef]
7. Salvatore, S.; Battigaglia, M.S.; Murone, E.; Dozio, E.; Pensabene, L.; Agosti, M. Dietary Fibers in Healthy Children and in Pediatric Gastrointestinal Disorders: A Practical Guide. *Nutrients* **2023**, *15*, 2208. [CrossRef]
8. Ribeiro, M.C.; Levi, Y.; Moraschini, V.; Messora, M.R.; Furlaneto, F.A.C. Effects of Prebiotic Therapy on Gastrointestinal Microbiome of Individuals with Different Inflammatory Conditions: A Systematic Review of Randomized Controlled Trials. Probiotics Antimicrob. *Proteins* **2023**, 1–23. [CrossRef]
9. Beker, B.M.; Colombo, I.; Gonzalez-Torres, H.; Musso, C.G. Decreasing microbiota-derived uremic toxins to improve CKD outcomes. *Clin. Kidney J.* **2022**, *15*, 2214–2219. [CrossRef]
10. Tain, Y.-L.; Hsu, C.-N. Role of the Gut Microbiota in Children with Kidney Disease. *Children* **2023**, *10*, 269. [CrossRef]
11. Wang, Y.; Liu, S.; Tang, D.; Dong, R.; Feng, Q. Chitosan Oligosaccharide Ameliorates Metabolic Syndrome Induced by Overnutrition via Altering Intestinal Microbiota. *Front. Nutr.* **2021**, *8*, 743492. [CrossRef] [PubMed]
12. He, N.; Wang, S.; Lv, Z.; Zhao, W.; Li, S. Low molecular weight chitosan oligosaccharides (LMW-COSs) prevent obesity-related metabolic abnormalities in association with the modification of gut microbiota in high-fat diet (HFD)-fed mice. *Food Funct.* **2020**, *11*, 9947–9959. [CrossRef] [PubMed]
13. Cheon, S.; Kim, G.; Bae, J.-H.; Lee, D.H.; Seong, H.; Kim, D.H.; Han, J.-S.; Lim, S.-Y.; Han, N.S. Comparative analysis of prebiotic effects of four oligosaccharides using in vitro gut model: Digestibility, microbiome, and metabolome changes. *FEMS Microbiol. Ecol.* **2023**, *99*, fiad002. [CrossRef] [PubMed]
14. Pellicer-Valero, Ó.J.; Massaro, G.A.; Casanova, A.G.; Paniagua-Sancho, M.; Fuentes-Calvo, I.; Harvat, M.; Martín-Guerrero, J.D.; Martínez-Salgado, C.; López-Hernández, F.J. Neural Network-Based Calculator for Rat Glomerular Filtration Rate. *Biomedicines* **2022**, *10*, 610. [CrossRef] [PubMed]
15. Davani-Davari, D.; Negahdaripour, M.; Karimzadeh, I.; Seifan, M.; Mohkam, M.; Masoumi, S.; Berenjian, A.; Ghasemi, Y. Prebiotics: Definition, Types, Sources, Mechanisms, and Clinical Applications. *Foods* **2019**, *8*, 92. [CrossRef]
16. Scott, K.P.; Martin, J.C.; Duncan, S.H.; Flint, H.J. Prebiotic stimulation of human colonic butyrate-producing bacteria and bifidobacteria, in vitro. *FEMS Microbiol. Ecol.* **2013**, *87*, 30–40. [CrossRef]
17. Quigley, E.M. Prebiotics and Probiotics in Digestive Health. *Clin. Gastroenterol. Hepatol.* **2018**, *17*, 333–344. [CrossRef]
18. Jerez-Morales, A.; Merino, J.S.; Díaz-Castillo, S.T.; Smith, C.T.; Fuentealba, J.; Bernasconi, H.; Echeverría, G.; García-Cancino, A. The Administration of the Synbiotic *Lactobacillus bulgaricus* 6c3 Strain, Inulin and Fructooligosaccharide Decreases the Concentrations of Indoxyl Sulfate and Kidney Damage in a Rat Model. *Toxins* **2021**, *13*, 192. [CrossRef]

19. Lai, S.; Mazzaferro, S.; Muscaritoli, M.; Mastroluca, D.; Testorio, M.; Perrotta, A.; Esposito, Y.; Carta, M.; Campagna, L.; Di Grado, M.; et al. Prebiotic Therapy with Inulin Associated with Low Protein Diet in Chronic Kidney Disease Patients: Evaluation of Nutritional, Cardiovascular and Psychocognitive Parameters. *Toxins* **2020**, *12*, 381. [CrossRef]
20. Lai, S.; Molfino, A.; Testorio, M.; Perrotta, A.M.; Currado, A.; Pintus, G.; Pietrucci, D.; Unida, V.; La Rocca, D.; Biocca, S.; et al. Effect of Low-Protein Diet and Inulin on Microbiota and Clinical Parameters in Patients with Chronic Kidney Disease. *Nutrients* **2019**, *11*, 3006. [CrossRef]
21. Melekoglu, E.; Cetinkaya, M.A.; Kepekci-Tekkeli, S.E.; Kul, O.; Samur, G. Effects of prebiotic oligofructose-enriched inulin on gut-derived uremic toxins and disease progression in rats with adenine-induced chronic kidney disease. *PLoS ONE* **2021**, *16*, e0258145. [CrossRef] [PubMed]
22. Yousef, M.; Pichyangkura, R.; Soodvilai, S.; Chatsudthipong, V.; Muanprasat, C. Chitosan oligosaccharide as potential therapy of inflammatory bowel disease: Therapeutic efficacy and possible mechanisms of action. *Pharmacol. Res.* **2012**, *66*, 66–79. [CrossRef] [PubMed]
23. Jiménez-Gómez, C.P.; Cecilia, J.A. Chitosan: A Natural Biopolymer with a Wide and Varied Range of Applications. *Molecules* **2020**, *25*, 3981. [CrossRef] [PubMed]
24. Milkova, V. Electrosteric stabilization of oil/water emulsions by adsorption of chitosan oligosaccharides—An electrokinetic study. *Carbohydr. Polym.* **2021**, *265*, 118072. [CrossRef]
25. Yu, D.; Feng, J.; You, H.; Zhou, S.; Bai, Y.; He, J.; Cao, H.; Che, Q.; Guo, J.; Su, Z. The Microstructure, Antibacterial and Antitumor Activities of Chitosan Oligosaccharides and Derivatives. *Mar. Drugs* **2022**, *20*, 69. [CrossRef]
26. Saito, H.; Sakakibara, Y.; Sakata, A.; Kurashige, R.; Murakami, D.; Kageshima, H.; Saito, A.; Miyazaki, Y. Antibacterial activity of lysozyme-chitosan oligosaccharide conjugates (LYZOX) against Pseudomonas aeruginosa, Acinetobacter baumannii and Methicillin-resistant Staphylococcus aureus. *PLoS ONE* **2019**, *14*, e0217504. [CrossRef]
27. Lee, H.-W.; Park, Y.-S.; Jung, J.-S.; Shin, W.-S. Chitosan oligosaccharides, dp 2–8, have prebiotic effect on the *Bifidobacterium bifidium* and *Lactobacillus* sp. *Anaerobe* **2002**, *8*, 319–324. [CrossRef]
28. Guan, Z.; Feng, Q. Chitosan and Chitooligosaccharide: The Promising Non-Plant-Derived Prebiotics with Multiple Biological Activities. *Int. J. Mol. Sci.* **2022**, *23*, 6761. [CrossRef]
29. Tao, W.; Wang, G.; Wei, J. The Role of Chitosan Oligosaccharide in Metabolic Syndrome: A Review of Possible Mechanisms. *Mar. Drugs* **2021**, *19*, 501. [CrossRef]
30. Pan, H.; Fu, C.; Huang, L.; Jiang, Y.; Deng, X.; Guo, J.; Su, Z. Anti-Obesity Effect of Chitosan Oligosaccharide Capsules (COSCs) in Obese Rats by Ameliorating Leptin Resistance and Adipogenesis. *Mar. Drugs* **2018**, *16*, 198. [CrossRef]
31. Liu, S.-H.; Chen, R.-Y.; Chiang, M.-T. Effects and Mechanisms of Chitosan and ChitosanOligosaccharide on Hepatic Lipogenesis and Lipid Peroxidation, Adipose Lipolysis, and Intestinal Lipid Absorption in Rats with High-Fat Diet-Induced Obesity. *Int. J. Mol. Sci.* **2021**, *22*, 1139. [CrossRef]
32. Sepasi, T.; Ghadiri, T.; Ebrahimi-Kalan, A.; Bani, F.; Talebi, M.; Rahbarghazi, R.; Khodakarimi, S.; Beyrampour-Basmenj, H.; Seidi, K.; Abbaspour-Ravasjani, S.; et al. CDX-modified chitosan nanoparticles remarkably reduce therapeutic dose of fingolimod in the EAE model of mice. *Int. J. Pharm.* **2023**, *636*, 122815. [CrossRef] [PubMed]
33. Kim, C.S.; Mathew, A.P.; Vasukutty, A.; Uthaman, S.; Joo, S.Y.; Bae, E.H.; Ma, S.K.; Park, I.-K.; Kim, S.W. Glycol chitosan-based tacrolimus-loaded nanomicelle therapy ameliorates lupus nephritis. *J. Nanobiotechnology* **2021**, *19*, 109. [CrossRef] [PubMed]
34. Bai, K.; Hong, B.; He, J.; Huang, W. Antioxidant Capacity and Hepatoprotective Role of Chitosan-Stabilized Selenium Nanoparticles in Concanavalin A-Induced Liver Injury in Mice. *Nutrients* **2020**, *12*, 857. [CrossRef] [PubMed]
35. Khater, S.I.; Mohamed, A.A.-R.; Arisha, A.H.; Ebraheim, L.L.; El-Mandrawy, S.A.; Nassan, M.A.; Mohammed, A.T.; Abdo, S.A. Stabilized-chitosan selenium nanoparticles efficiently reduce renal tissue injury and regulate the expression pattern of aldose reductase in the diabetic-nephropathy rat model. *Life Sci.* **2021**, *279*, 119674. [CrossRef]
36. Chen, X.; Dai, W.; Li, H.; Yan, Z.; Liu, Z.; He, L. Targeted drug delivery strategy: A bridge to the therapy of diabetic kidney disease. *Drug Deliv.* **2022**, *30*, 2160518. [CrossRef]
37. Zhang, H.; Zhang, S.; Wang, L.; Liu, X.; Wu, Y. Chitooligosaccharide guanidine inhibits high glucose-induced activation of DAG/PKC pathway by regulating expression of GLUT2 in type 2 diabetic nephropathy rats. *J. Funct. Foods* **2018**, *41*, 41–47. [CrossRef]
38. Wu, J.; Xu, Y.; Geng, Z.; Zhou, J.; Xiong, Q.; Xu, Z.; Li, H.; Han, Y. Chitosan oligosaccharide alleviates renal fibrosis through reducing oxidative stress damage and regulating TGF-β1/Smads pathway. *Sci. Rep.* **2022**, *12*, 19160. [CrossRef]
39. Pathomthongtaweechai, N.; Soodvilai, S.; Pichyangkura, R.; Muanprasat, C. Novel Potential Application of Chitosan Oligosaccharide for Attenuation of Renal Cyst Growth in the Treatment of Polycystic Kidney Disease. *Molecules* **2020**, *25*, 5589. [CrossRef]
40. Fastinger, N.D.; Karr-Lilienthal, L.K.; Spears, J.K.; Swanson, K.S.; Zinn, K.E.; Nava, G.M.; Ohkuma, K.; Kanahori, S.; Gordon, D.T.; Fahey, G.C., Jr. A Novel Resistant Maltodextrin Alters Gastrointestinal Tolerance Factors, Fecal Characteristics, and Fecal Microbiota in Healthy Adult Humans. *J. Am. Coll. Nutr.* **2008**, *27*, 356–366. [CrossRef]
41. Burns, A.M.; Solch, R.J.; Dennis-Wall, J.C.; Ukhanova, M.; Nieves, C., Jr.; Mai, V.; Christman, M.C.; Gordon, D.T.; Langkamp-Henken, B. In healthy adults, resistant maltodextrin produces a greater change in fecal bifidobacteria counts and increases stool wet weight: A double-blind, randomized, controlled crossover study. *Nutr. Res.* **2018**, *60*, 33–42. [CrossRef]

42. Baer, D.J.; Stote, K.S.; Henderson, T.; Paul, D.R.; Okuma, K.; Tagami, H.; Kanahori, S.; Gordon, D.T.; Rumpler, W.V.; Ukhanova, M.; et al. The Metabolizable Energy of Dietary Resistant Maltodextrin Is Variable and Alters Fecal Microbiota Composition in Adult Men. *J. Nutr.* **2014**, *144*, 1023–1029. [CrossRef]
43. Nishimoto, Y.; Mizuguchi, Y.; Mori, Y.; Ito, M.; Miyazato, S.; Kishimoto, Y.; Yamada, T.; Fukuda, S. Resistant Maltodextrin Intake Reduces Virulent Metabolites in the Gut Environment: A Randomized Control Study in a Japanese Cohort. *Front. Microbiol.* **2022**, *13*, 644146. [CrossRef] [PubMed]
44. Li, J.; Yang, G.; Zhang, Q.; Liu, Z.; Jiang, X.; Xin, Y. Function of Akkermansia muciniphila in type 2 diabetes and related diseases. *Front. Microbiol.* **2023**, *14*, 1172400. [CrossRef]
45. Memon, H.; Abdulla, F.; Reljic, T.; Alnuaimi, S.; Serdarevic, F.; Asimi, Z.V.; Kumar, A.; Semiz, S. Effects of combined treatment of probiotics and metformin in management of type 2 diabetes: A systematic review and meta-analysis. Diabetes Res. *Clin. Pract.* **2023**, *202*, 110806. [CrossRef] [PubMed]
46. Wu, R.; Xiong, R.; Li, Y.; Chen, J.; Yan, R. Gut microbiome, metabolome, host immunity associated with inflammatory bowel disease and intervention of fecal microbiota transplantation. *J. Autoimmun.* **2023**, 103062, *in press*. [CrossRef] [PubMed]
47. Martín-Del-Campo, F.; Avesani, C.M.; Stenvinkel, P.; Lindholm, B.; Cueto-Manzano, A.M.; Cortés-Sanabria, L. Gut microbiota disturbances and protein-energy wasting in chronic kidney disease: A narrative review. *J. Nephrol.* **2023**, *36*, 873–883. [CrossRef]
48. Andrianova, N.V.; Popkov, V.A.; Klimenko, N.S.; Tyakht, A.V.; Baydakova, G.V.; Frolova, O.Y.; Zorova, L.D.; Pevzner, I.B.; Zorov, D.B.; Plotnikov, E.Y. Microbiome-Metabolome Signature of Acute Kidney Injury. *Metabolites* **2020**, *10*, 142. [CrossRef]
49. Li, H.; Wang, Q.; Chen, P.; Zhou, C.; Zhang, X.; Chen, L. Ursodeoxycholic Acid Treatment Restores Gut Microbiota and Alleviates Liver Inflammation in Non-Alcoholic Steatohepatitic Mouse Model. *Front. Pharmacol.* **2021**, *12*, 788558. [CrossRef]
50. Li, D.-P.; Cui, M.; Tan, F.; Liu, X.-Y.; Yao, P. High Red Meat Intake Exacerbates Dextran Sulfate-Induced Colitis by Altering Gut Microbiota in Mice. *Front. Nutr.* **2021**, *8*, 646819. [CrossRef]
51. Liu, W.-C.; Pan, Z.-Y.; Zhao, Y.; Guo, Y.; Qiu, S.-J.; Balasubramanian, B.; Jha, R. Effects of Heat Stress on Production Performance, Redox Status, Intestinal Morphology and Barrier-Related Gene Expression, Cecal Microbiome, and Metabolome in Indigenous Broiler Chickens. *Front. Physiol.* **2022**, *13*, 761. [CrossRef] [PubMed]
52. Mo, Y.; Sun, H.; Zhang, L.; Geng, W.; Wang, L.; Zou, C.; Wu, Y.; Ji, C.; Liu, X.; Lu, Z. Microbiome-Metabolomics Analysis Reveals the Protection Mechanism of α-Ketoacid on Adenine-Induced Chronic Kidney Disease in Rats. *Front. Pharmacol.* **2021**, *12*, 657827. [CrossRef] [PubMed]
53. Liu, F.; Xu, X.; Chao, L.; Chen, K.; Shao, A.; Sun, D.; Hong, Y.; Hu, R.; Jiang, P.; Zhang, N.; et al. Alteration of the Gut Microbiome in Chronic Kidney Disease Patients and Its Association With Serum Free Immunoglobulin Light Chains. *Front. Immunol.* **2021**, *12*, 609700. [CrossRef] [PubMed]

nutrients

MDPI

Review

From Structure to Function: How Prebiotic Diversity Shapes Gut Integrity and Immune Balance

Lucas de Freitas Pedrosa [1,2], Paul de Vos [2] and João Paulo Fabi [1,3,4,5,*]

1 Department of Food Science and Experimental Nutrition, School of Pharmaceutical Sciences, University of São Paulo, São Paulo 05508-000, SP, Brazil; lfpedrosa@usp.br
2 Department of Pathology and Medical Biology, University Medical Center Groningen, University of Groningen, 9713 GZ Groningen, The Netherlands; p.de.vos@umcg.nl
3 Food and Nutrition Research Center (NAPAN), University of São Paulo, São Paulo 05508-000, SP, Brazil
4 Food Research Center (FoRC), CEPID-FAPESP (Research, Innovation, and Dissemination Centers), São Paulo 05508-080, SP, Brazil
5 Food Research Center (FoRC), CEPIX-USP, University of São Paulo, São Paulo 05508-000, SP, Brazil
* Correspondence: jpfabi@usp.br

Abstract: The microbiota stability, diversity, and composition are pillars for an efficient and beneficial symbiotic relationship between its host and itself. Microbial dysbiosis, a condition where a homeostatic bacterial community is disturbed by acute or chronic events, is a predisposition for many diseases, including local and systemic inflammation that leads to metabolic syndrome, diabetes, and some types of cancers. Classical dysbiosis occurs in the large intestine. During this period, pathogenic strains can multiply, taking advantage of the compromised environment. This overgrowth triggers an exaggerated inflammatory response from the human immune system due to the weakened integrity of the intestinal barrier. Such inflammation can also directly influence higher polyp formation and/or tumorigenesis. Prebiotics can be instrumental in preventing or correcting dysbiosis. Prebiotics are molecules capable of serving as substrates for fermentation processes by gut microorganisms. This can promote returning the intestinal environment to homeostasis. Effective prebiotics are generally specific oligo- and polysaccharides, such as FOS or inulin. However, the direct effects of prebiotics on intestinal epithelial and immune cells must also be taken into consideration. This interaction happens with diverse prebiotic nondigestible carbohydrates, and they can inhibit or decrease the inflammatory response. The present work aims to elucidate and describe the different types of prebiotics, their influence, and their functionalities for health, primarily focusing on their ability to reduce and control inflammation in the intestinal epithelial barrier, gut, and systemic environments.

Keywords: microbiota; prebiotics; inflammation; polysaccharides; dysbiosis

1. Introduction

The correlations among dietary components, gut microbiota modulation, and the resulting health-related effects are a constant topic of interest throughout the scientific community. Prebiotics are one type of dietary component that is notably explored. Prebiotics might stimulate the activity or growth of certain bacterial strains, specifically fermented by local microbiota, which can establish or strengthen a symbiotic relationship between microbiota and the host [1].

Prebiotics support intestinal health by maintaining, repairing, and straightening the gut barrier. This barrier is a complex and multifaceted structure comprised of a protective mucus layer, a tightly regulated epithelial cell layer, and specialized proteins known as tight junctions, which act as gatekeepers for the gut lining. Together, these elements form a robust defense and gatekeeper system that prevents the entry of harmful substances and allows beneficial nutrients and molecules to pass through.

When the gut barrier is weakened, whether due to dysbiosis, antibiotic usage, or dietary insufficiencies, prebiotics can help restore its integrity. By selectively nourishing beneficial gut bacteria, prebiotics help these organisms thrive, promoting the production of short-chain fatty acids (SCFAs) such as butyrate. These SCFAs directly enforce epithelial health, strengthen tight junctions, and enhance mucus production. Also, direct effects on epithelial and mucus integrity have been reported for prebiotics [2–4].

Additionally, regular intake of prebiotics can contribute to the long-term resilience of the gut barrier, making it less susceptible to disruptions from dietary changes or external stressors. Through this ongoing support, prebiotics give rise to a balanced microbial environment and the gut's essential protective functions, ensuring a healthier, more resilient intestinal system over time [5,6].

Gut barrier disruption may lead to substantial intestinal tissue inflammation due to an unnatural over-interaction between bacterial-derived molecules and the host immune system apparatus. While a typical gut environment has a basal stimulation of immune cells, resulting in a ready-to-act immune system, high inflammation can lead to problems such as metabolic disorders, higher toxicity (due to increased permeability), and colon carcinogenesis [5].

In addition to prebiotics, probiotics might also support health. Probiotics are defined as beneficial microorganisms consumed to preserve and promote gut health. According to Patra et al. (2022), through a robust systematic network and meta-analysis, probiotics or probiotic-derived bacteriocins could interact directly with immune enzymes involved in colorectal cancer (CRC) pathogenesis, such as COX-2, and even modulate nod-like receptor protein-3 (NLRP3) or NF-kB pathways, reducing CRC-associated inflammation [7].

The aim of this review is to elucidate, based on the literature, if those crucial interactions with immune and metabolic receptors also happen with prebiotics. While some prebiotics have been extensively studied in preclinical models, this review focuses on updating the current understanding of both emerging and conventional prebiotics. Specifically, it explores their role in enhancing gut barrier integrity, reducing inflammation, and mitigating the risk of colon cancer. This review aims to connect findings from preclinical research with evidence from clinical trials and human studies. By doing so, it provides a thorough analysis of the effectiveness of prebiotics and their potential applications for enhancing human health.

2. Bibliometric Analysis Methodology

For the present analysis, the Scopus database was utilized mainly due to its excellent and intuitive interface for replicating such criteria described here in the future. Two sets of criteria were used. The first focused on tracking the level of interest and the publication rates of reviews and articles on prebiotics over time. The following three sets of keywords were used in this first approach, merged by the Boolean operator "AND":

(1) "prebiotics" AND "inflammation";
(2) "prebiotics" AND "colon" AND "cancer";
(3) "prebiotics" AND "TLR".

The time of publication was set to start in 1997, the first year reporting any publication for all three search settings. The language was limited to English for its position as the universal academic language. Finally, the subject areas chosen to restrict the search were focused on those more inclined to interfere with health-related matters regarding prebiotics, including "Medicine"; "Biochemistry, Genetics, and Molecular Biology"; "Immunology and Microbiology"; "Pharmacology, Toxicology, and Pharmaceutics"; and "Chemistry".

The purpose of Figure 1 is to show the comparisons between different search terms and the number of reviews (indicated by letters followed by the number 2 and represented with dotted lines) and original articles (indicated by letters followed by the number 1 and represented with solid lines) for each term. As visualized in Figure 1, "inflammation" is a search term that appears more often than the other search strings in any given year. This also comes from a terminology point of view since inflammation is a significant umbrella term that, aside from being studied by itself, supports several physiological and pathological processes involving—or not—human health issues. Nevertheless, from this, we can expect that such terminology often resonates more in the literature due to its relevance and complexity and as a "primary pillar" topic. Terminology C, "prebiotics and TLR" is a more straightforward, smaller concept within the larger "inflammation topic" and has kept quite a small and constant uprising, especially from 2020 onwards. It is assumed that this topic may have gained attention through its connection with other critical subjects. These include immunotherapy research focused on immune receptors, the interaction of specific prebiotics with these receptors, and a growing interest in tight junction proteins, gut barrier stability, and related pathologies. Additionally, terminology B, "prebiotics and colon cancer", although potentially associated with these areas, has a distinct biomedical background. This unique background supports methodological independence, encompassing research into chromosomal instabilities, genetic disorders, potential surgical treatments, and other related factors. Even then, specific trends are similar for all the terms used, at least when using the methodology shown here, with particular attention to a plateau of publications observed from 2022 to 2024. This presentation offers an overview tailored to both researchers and lay readers. It highlights the emerging publication trends over recent years, identifies areas with limited or underexplored literature, and pinpoints sharp increases in interest around specific keywords. Additionally, it aids in constructing VOSviewer map inputs and refining subsequent searches. For example, interest in the interaction between TLRs) and prebiotics has shown a gradual but modest increase. This trend lacks the significant surge observed between 2016 and 2021 for the broader keyword "inflammation". In the last four years, the topic has reached a plateau and has even been surpassed by reviews focusing on the same subject.

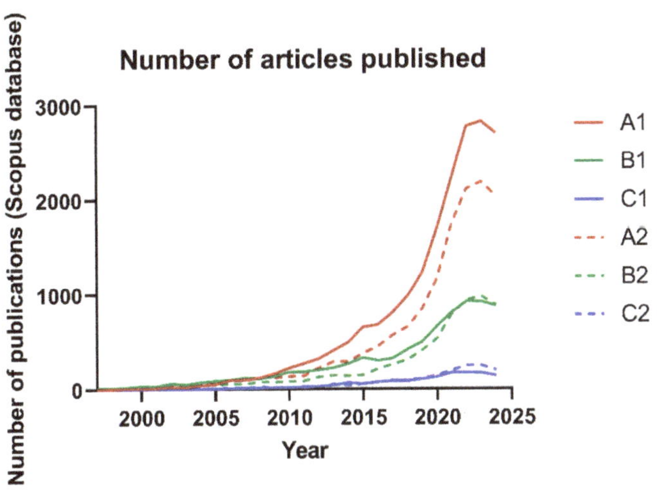

Figure 1. Number of articles published in the Scopus database starting 1997 until now. Number 2 (dotted lines) represents the number of "review articles" under a determined set of terms, while number 1 (solid lines) shows the number of "original articles". A: "prebiotics" AND "inflammation"; B: "prebiotics" AND "colon" AND "cancer"; C: "prebiotics" AND "TLR".

The second search approach was similar to the one described above, except for the publication date being set between 2021 and 2025, to select only for very recent literature for the VOSViewer software analysis. The Scopus database search document data were exported as .csv files using all the citation, bibliographical, and abstract/keyword data. Using the VOSViewer software, version 1.6.18, a map was created based on bibliographical data. With the Scopus file extension selected, a co-occurrence of all keywords with complete counting was performed, and the simulations were chosen based on a minimum occurrence of 100 times for all the terms searched. Visualization was standard, with attraction and repulsion values set as the default.

Such relevance of the data, briefly discussed above, can be observed from the difference in densification among Figures 2–4. Figure 2 shows the result from the search "A" (prebiotics and inflammation), which shows an intensely dense formation of four keyword clusters. Each cluster is well-defined, and boundaries can be drawn.

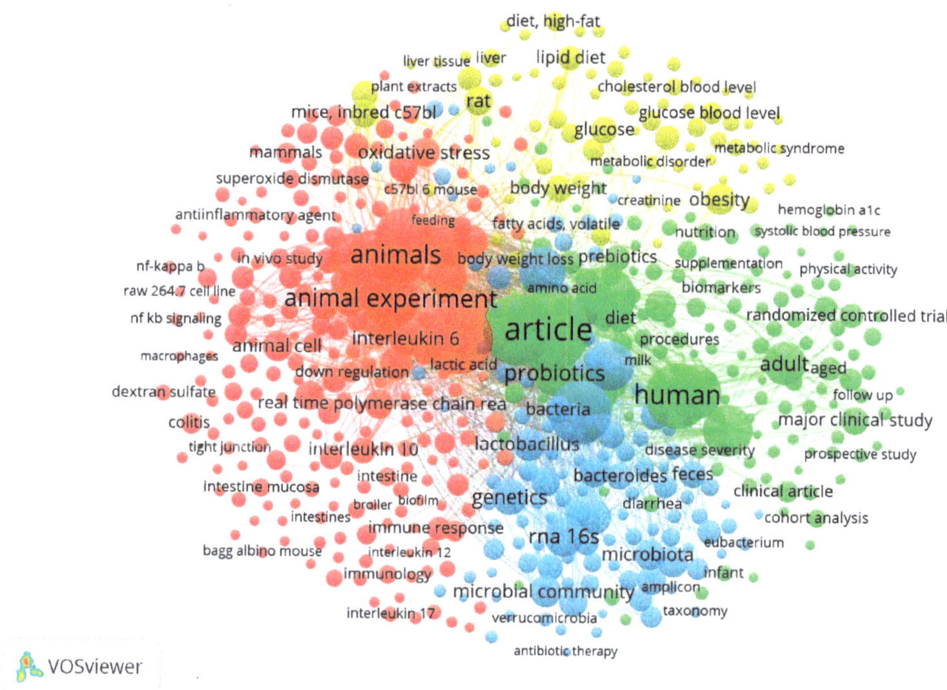

Figure 2. VOSViewer map from the search combination of the terms "prebiotics" and "inflammation" (search A).

The red cluster is focused on animal and cell experimentation data and methodology, ranging from animal model types, such as "albino mouse", "inbred c57", and "raw 264.7 cell line", which refer not only to animal and cell types but also to how the study is set as an intervention/condition, as shown by terms such as "colitis" and "dextran sulfate".

The green cluster holds terms related to human clinical studies, with both methodologies such as "cohort analysis", "randomized clinical trial", and "prospective study", as well as population characteristics, such as "adult", "aged", "supplementation", and "diet".

The blue cluster is mainly focused on microbiome-related terms, which was highly expected to be relevant. Some of the terms include "*Bacteroides*", "*Verrucomicrobia*", and "*Lactobacillus*".

Finally, the yellow cluster is the smallest one and, interestingly, highly related to metabolic syndromes, with terms such as "liver tissue", "glucose", "cholesterol", blood levels, "metabolic disorder", among others.

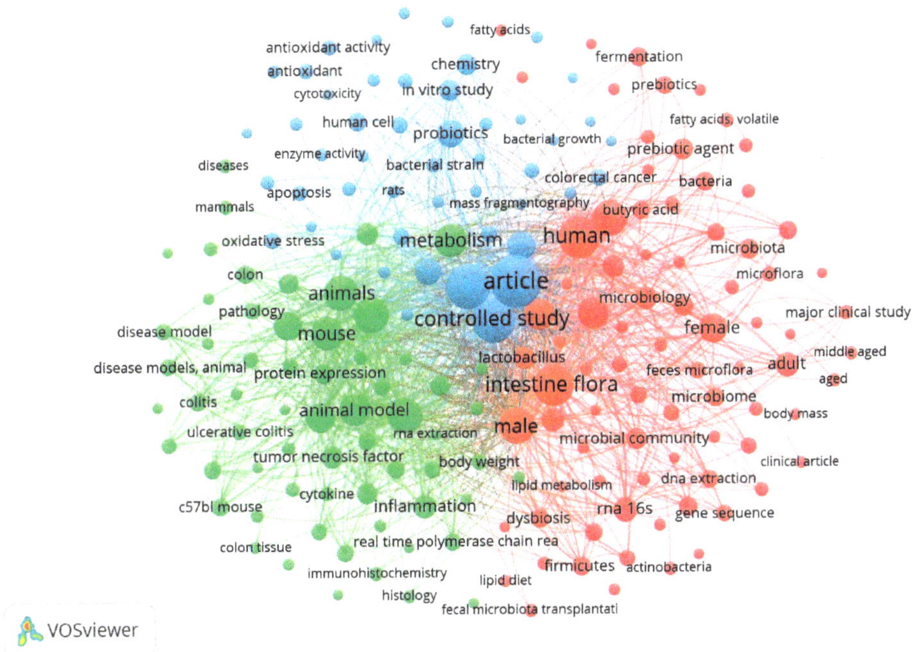

Figure 3. VOSViewer map from the search combination of the terms "prebiotics" and "colon" and "cancer" (search B).

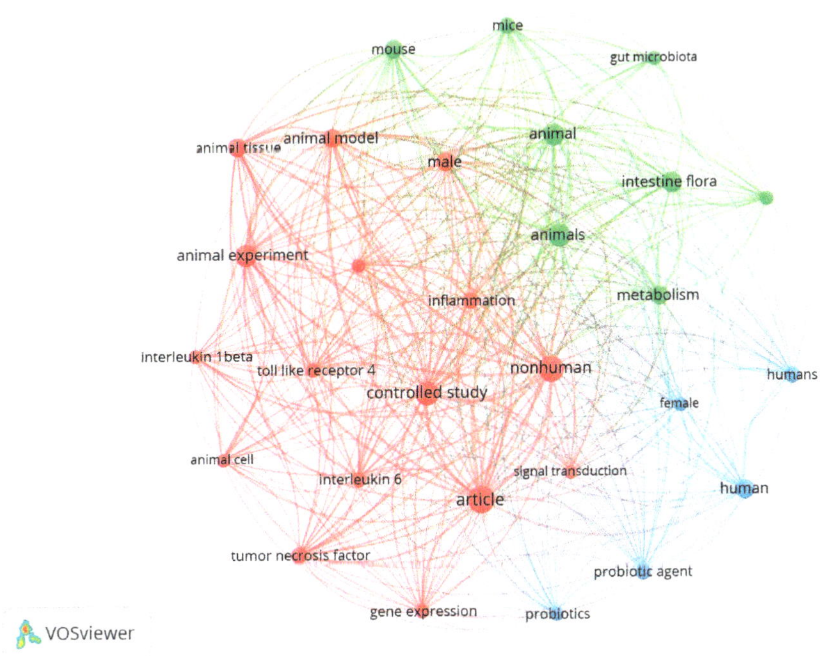

Figure 4. VOSViewer map from the search combination of the terms "prebiotics" and "TLR" (search C).

Figure 2 shows the proximity between the microbiome-related and metabolic disorder terms, intermediated by inflammation-related keywords. Based on that, additional research with the same parameters as described for the maps but changing the search terms to "prebiotics AND microbiome AND metabolic disorder", as well as excluding two reoccurring "contamination" terms through the usage of the "AND NOT" Boolean operator (probiotics and review), resulted in 407 original article documents in the last five years in the Scopus database, giving us an average of about 81 original articles exploring such correlations per year. Examples of metabolic disorders identified were glucose tolerance, cardiomyopathies, insulin resistance, and hyperlipidemia.

Regarding search "B" (prebiotics and colon cancer), three less dense clusters reflecting the lower amount of published data available are highlighted in Figure 3.

The Green cluster is related to the animal modeling and analytical methodology used, with terms such as "ulcerative colitis" (a risk factor for colorectal cancer—CRC), "immunohistochemistry", "histology", and "RNA extraction".

The red cluster merges terms related to human clinical studies ("middle-aged", "major clinical study", and "female"), the microbiome ("prebiotics", "lactobacillus", "intestine flora", and "dysbiosis"), and finally microbiome-related methodologies ("RNA 16s", "fecal microbiota transplantation", and "DNA extraction".

The blue cluster is mainly related to in vitro methodologies such as "mass fragmentography", "apoptosis", "enzyme activity", and "antioxidant activity".

We applied the above exclusion criteria to check the literature convergence between microbiome, colorectal cancer, and prebiotics. We obtained 258 original article papers over the past five years (averaging 52 papers/year). Similar to what is observed in Figure 3, the search results were intrinsically connected to inflammation and fermentation. In this search, a common approach to colorectal cancer evaluation in animal models, "aberrant crypt foci", was also identified.

Figure 4 shows the scarcity of terms with similar levels of relevance/occurrence when using the combination of terms "prebiotics" and "TLR". Nevertheless, a particular pattern is maintained due to their natural separations between animal experimentation and human studies (blue cluster). But this time, the red and green clusters were, albeit separated, highly related and similar, with terms shared between them such as "animal experiment", "mice", "animals", among others. Complementary terms related to the microbiome, such as "intestine flora", were under the green cluster. Meanwhile, terms more related to inflammation, such as "toll-like receptor 4", "interleukin-6", and "tumor necrosis factor", were under the red cluster.

Despite the scarcity, one final search was performed to identify the proximity relationships among the microbiome-, TLR-, and prebiotic-related terms. Here, the findings were reduced, with 62 original articles published over the last five years, which was an average of about 12 articles per year. Again, converging with the terms found in Figure 4, many articles explored TNF-α expression/secretion (red cluster). Although not present in the figure, butyrate production was also relevant in the search, alongside high-fat diet-induced obesity, epithelium integrity, and immune function.

The following sections were written based on the articles in the above bibliometric analyses.

3. Prebiotics Definition, Types, and Effects

While the first definition of prebiotics can be tracked to as early as three decades ago, more updated versions have been distributed throughout the literature. The most accepted updated definition is the one from ISAPP (International Scientific Association for Probiotics and Prebiotics), which describes prebiotics as "substrates that are selectively utilized by host microorganisms conferring a health benefit" [8]. The World Gastroenterology Organization Global Guidelines specify the following two extra points: (1) selective fermentation, not general "utilization"; and (2) resulting in compositional or activity changes of the gastrointestinal microbiota [9].

The most usual prebiotics are β-glucans, pectins, fructo- and galactooligosaccharides (FOSs and GOSs), inulins, lactulose, and human milk oligosaccharides (HMOs). Other fermentable polysaccharides can also be characterized as prebiotics [9,10]. Polyphenols, resistant starch, and conjugated linoleic acids are also being studied as potential candidates for being described as prebiotics, further expanding this realm [8,11]. Being non-digestible is a requirement since it must be available for use by the beneficial bacteria in the host intestinal tract. Such molecules naturally occur in healthy diets, either present in whole grain foods, vegetables, and fruits, but they can also be supplemented [10]; some of the classic beneficial effects of this compound class are shown in Figure 5.

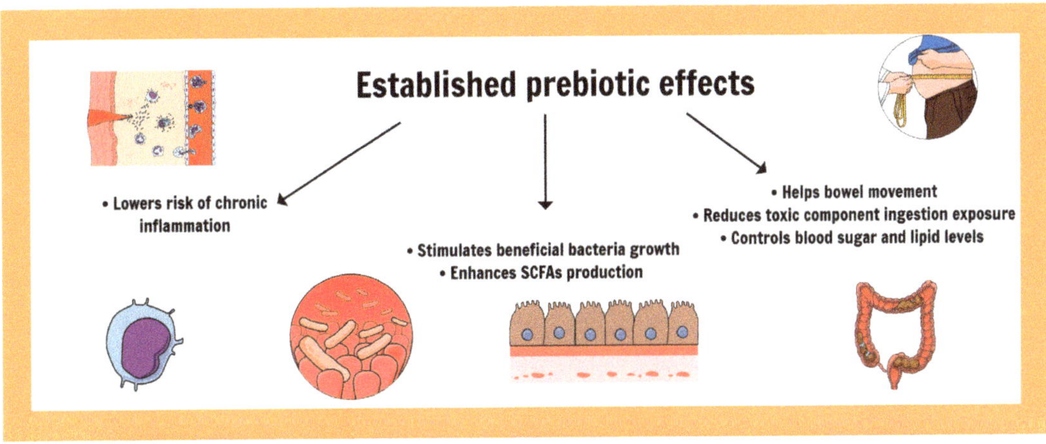

Figure 5. Established prebiotic effects for intestinal and systemic health.

The chemical structure of a substance is a key factor in determining whether it qualifies as a prebiotic. More complex structures present challenges because they are harder to standardize, making it difficult to predict their effectiveness when used for prevention or treatment. The term "prebiotics" is not limited to carbohydrate-based molecules. In fact, many dietary fibers can exert prebiotic effects, depending on their fermentability. Simpler sugars, such as fructans and glucans, are already recognized as prebiotics. However, more complex structures, like heteropolysaccharides (for example, pectins), can also qualify as prebiotics. For those, some factors such as high molecular variability can be influenced by the extraction method, source, molecular size, and the availability of glycosidic bonds necessary for bacterial breakdown, changing their prebiotic behavior [12].

3.1. Gut Barrier Function

The gut barrier is a complex, multi-layered system crucial in maintaining gut homeostasis and protecting the body from external injuries. It comprises a physical barrier of the gut microbiota, a mucus layer, epithelial cells, and immune cells, working together to prevent bacterial adhesion and regulate other processes [13,14]. The gut microbiota keeps the host healthy by competing with pathogenic and symbiotic organisms. They compete for space and resources, primarily derived from diet, to grow and expand their populations at the expense of pathogens [15]. In the mucus layer, antimicrobial products and secretory IgA are released, and its thickness can be regulated via the proteolytic activity ratio of bacteria and the secretory capability of goblet cells [16].

As one of the mentioned parts of the gut barrier, epithelial cells form an intestinal epithelial barrier (IEB). This IEB serves as a critical interface between the external environment and the body's internal milieu, regulating the passage of nutrients and preventing the entry of harmful antigens and microorganisms [17]. Tight junctions (TJs) within the IEB are essential to controlling paracellular permeability and maintaining barrier integrity. When

TJs are dysfunctional, they lead to increased gut permeability, generating the so-called "leaky gut syndrome" [17,18].

Inflammation is closely connected to the function and stability of the gut barrier. For instance, an imbalance in the gut microbiota can disrupt the epithelial barrier by breaking down the mucus layer or causing cellular stress through exposure to the underlying epithelial cell layer. Conversely, certain beneficial strains can produce SCFAs from fermentable substances, and they provide a rapid energy source for epithelial cells, which can help strengthen both the cellular and mucus layers of the gut barrier [19].

The local immune system is also influenced by antigens derived from the microbiota. Regulatory T cells (Tregs) with a tissue repair phenotype, known as ST2+, make up a significant portion of the Foxp3+ Treg population in the gut's lamina propria. ST2 interacts and responds to IL-33, an alarmin, and amphiregulin (AREG), a growth factor. Overall, Tregs help reduce the intensity of proinflammatory responses to common microbial molecules. For example, Tregs use aryl hydrocarbon receptors (AhRs) to increase basal resistance and minimize tissue damage, promoting a more controlled immune response [20,21].

3.2. *Inflammation and Carcinogenesis*

Colon cancer is one of the most prevalent and lethal types of human cancer, according to the Globocan statistics [22]. The interplays between environmental interferences and genetic alterations heavily induce its pathogenesis. For instance, chromosomal instability (CIN) is a hallmark of most colon tumors and significantly contributes to tumor progression by increasing genetic abnormalities. Errors during mitosis and issues with spindle checkpoint activity often facilitate CIN, resulting in aneuploidy and the loss of heterozygosity. CIN can be triggered by chromosomal segregation molecules, such as aurora kinase B (AURKB), which is activated by BOP1. BOP1 is stabilized by the overexpression of the long noncoding RNA CCAT2. Additionally, epigenetic factors may influence this process. Certain microRNAs (miRNAs) can regulate epithelial–mesenchymal transition, a mechanism linked to the onset of colorectal cancer (CRC) [23].

The tumor microenvironment comprises immune cells, stromal cells, and the intestinal microbiome, all playing a role in CRC pathogenesis. For example, MAPK-activated protein kinase 2 (MK2) in macrophages has been linked to a pivotal contribution to colon tumorigenesis under chronic inflammation. Tumor angiogenesis (vascularization) is one of the consequences of the MK2 activity [24,25]. Chronic inflammation is a significant risk factor for the development of colorectal cancer (CRC). Patients with chronic inflammatory bowel diseases, such as ulcerative colitis and Crohn's disease, face an elevated risk of developing colitis-associated cancer (CAC). The persistent inflammatory state in the colon is strongly associated with tumor development, and the risk of CAC increases with the duration of inflammation [26,27].

Some critical immune signaling pathways are implicated in the pathogenesis of inflammation-associated colon cancer. The NF-kB pathway, for example, is crucial for regulating immune responses and inflammation. Still, its activation promotes gene expression in cell proliferation and survival, which can contribute to tumorigenesis. Prostaglandins and cyclooxygenases have also been linked to the development of CRC. Moreover, cell survival and strength can be promoted by the STAT3 signaling pathway, which can be activated via IL-6, a proinflammatory cytokine [26].

Cytokines are essential mediators of the immune response and can influence colorectal cancer (CRC) development. TNF-α and IL-1β are key inflammatory cytokines that promote CRC progression by stimulating interactions between various cell types in the gut. Maintaining a balance between pro- and anti-inflammatory cytokines is crucial for preserving tissue homeostasis and preventing malignant transformations [28,29].

As briefly mentioned before, the gut microbiota significantly influences the inflammatory environment in the colon, and alterations in the diversity and function of the microbiome are associated with changes in the immune response. The toll-like receptors (TLRs) on immune cells respond to microbial components, promoting tolerance or activating inflammatory signaling pathways, leading to CRC. Consequently, the microbiota and its metabolites are part of a "microbiome-inflammation" axis [27,30,31].

3.3. Prebiotics and Inflammation: A Clinical View

Targeting inflammatory pathways and cytokines is a therapeutic strategy to prevent or treat inflammation-associated CRC. The use of monoclonal antibodies, tyrosine kinase inhibitors (e.g., cetuximab), and nucleic acid drugs (e.g., siRNAs and antisense oligonucleotides) as modulators of inflammatory responses is already a trend in clinical exploration for not only CRC but other types of cancers as well [32–36].

However, in this review, we propose to evaluate whether prebiotics can significantly impact inflammation and CRC as well. Postbiotics, which are a combination of probiotics and prebiotics, such as butyrate (an SCFA), can modulate specific immune cells. For example, Kang et al. (2023) found out through a big-cohort study that butyrate boosted anti- programmed cell death protein 1 (PD-1) efficacy by inducing functional CD8+ T cells. PD-1 is considered an immune system "brake" protein, limiting immune response throughout CRC development [37].

Polyphenols are also classified as potential prebiotic candidates. Molinari and colleagues (2021) pointed out that human clinical trials analyzing such treatments are rather scarce, primarily focusing on inflammatory biomarkers, as opposed to animal studies, which the authors deeply dive into [38]. Nevertheless, Macis et al. (2023) tested both curcumin and anthocyanin commercial formulations in a phase II presurgical trial in patients with adenomatous polyps. Still, they did not achieve statistical significance related to the inflammatory biomarkers evaluated (IL-6, IL-10, and TNF- α) [39]. Moorthy et al. (2021) reported in their systematic review, specifically taking animal studies into account, that almost all studies sampled with polyphenolic extracts and all those utilizing pure polyphenols resulted in improvements in the metabolic and gut microbiota profiles. Gut microbiota genus level abundances were correlated to the symptom-alleviating trend after high-fat diet-induced obesity, while alpha diversity was not altered consistently throughout the gathered literature [40].

Another type of emerging prebiotic is resistant starch (RS), which is often a product of the starch retrogradation process but can also be found naturally in some grains, beans, and green fruits. In a placebo-controlled clinical trial, a patented resistant potato starch (Solnul™) reduced diarrhea and constipation effects compared to the control and significantly increased *Akkermansia* and *Bifidobacterium* levels [41]. Within in vitro digestion and fermentation simulations, mixed bacterial cultures exhibited significant growth compared to the control. Notably, the growth of *Lactobacillus* and *Bifidobacterium* mono-cultures was particularly enhanced by RS [42]. Technologies to extract more RS from different sources continue to be explored. Das et al. (2022) reported an enzyme-mediated biotransformation that doubled the RS yield. The RS-rich banana flour generated had a better prebiotic and symbiotic potential, enhancing the growth of the *Bacillus coagulans*, *Lactobacillus rhamnosus*, and *Saccharomyces boulardii* strains [43]. Notably, those prebiotics can often be employed as tools for other purposes, such as carrying compounds that are sensitive to the upper gastric chemical environment, as happens with RS nanoparticles. Wang et al. (2022) found improvements in *Lactobacillus plantarum* growth, its S-layer protein stability, as well as higher butyric acid levels after the addition of RS nanoparticles in a 0.5% concentration [44].

Conventional prebiotics may benefit inflammatory-derived health conditions, such as psoriasis, which are not typically associated with gut health. Buhas et al. (2023) found that after 12 weeks of consuming probiotics and prebiotics, an intervention group of 42 subjects showed significant improvements in the quality of life scores related to dermatological issues. Regarding inflammatory markers, TNF-α significantly decreased in the intervention group, while IL-10, an anti-inflammatory cytokine, significantly increased. Although IL-6 and IL-17a showed numerical changes, these were not statistically significant [45].

The symbiotic administration of *Bifidobacterium animalis* and fructo-oligosaccharides (FOS) for 30 days in middle-aged subjects impacted systemic inflammation, where IL-6,

IL-8, IL-17a, and IFNγ were significantly reduced in the plasma samples. Such effects were not observed in the placebo group, and the authors stated that it was probably gut permeability-independent results due to the lack of change in fecal albumin or plasma intestinal fatty acid-binding protein (iFABP) presence [46].

In an elderly population, the symbiotic effect of *Bifidubacterium animalis* subsp. *lactis* and inulin that focused on cognitive function improvements also achieved the relevant changes regarding inflammation biomarkers. A significant increase in the leukemia inhibitory factor receptor (LIR), CCL-23, and TNF ligand superfamily member 12, as well as a decrease in sulfotransferase 1A1, was detected [47]. CCL-23, for example, is a chemotactic cytokine for resting T cells, monocytes, and neutrophils. This could signify a proinflammatory treatment profile, contrary to what is often expected.

De Giani et al. (2022) compared the effects of administering prebiotics alone versus a symbiotic treatment that included two strains of *Lactobacillus*, one strain of *Bifidobacterium*, and two types of fructans (DP 3-5 and DP10 inulin). In elderly participants, they observed that those in the symbiotic group showed increased levels of fecal calprotectin and β-defensin 2 after 28 days (T28), which remained elevated at a 28-day follow-up (T56). In contrast, the prebiotic group showed an increase in β-defensin 2 at T28, but this effect did not persist to T56. Fecal calprotectin levels in the prebiotic group remained unchanged [48]. Moludi et al. (2021) also reported more significant benefits in groups receiving a combination of probiotics and inulin (symbiotic) than inulin alone or a placebo. While the inulin-only group showed a modest increase in the anti-inflammatory cytokine IL-10, reduced levels of lipopolysaccharide (LPS; indicating reduced microbial translocation), and decreased TNF-α, the symbiotic treatment produced much more potent effects. Specifically, the symbiotic group had increases of 0.37 and 10.10 ng/mL for IL-10, reductions of -1.86 and -22.02 EU/L for LPS, and reductions of -5.73 and -25.05 ng/mL for TNF-α in the inulin-only and symbiotic groups, respectively [49].

In a study by Neyrinck et al. (2021) involving obese patients, fecal calprotectin levels were reduced in the intervention group despite using only inulin, without any probiotics. This finding suggests that inulin alone may help reduce gut inflammation. However, the treatment did not alter the composition of SCFAs. Interestingly, rumenic acid, a conjugated linoleic acid with immunomodulatory potential, was increased in the fecal content and was correlated with an increased abundance of *Bifidobacterium* [46]. In another study involving obese children, inulin administration for up to six months did not impact fecal calprotectin levels or cytokines like IL-1β and IL-6 [50]. The variability in outcomes across these studies may have resulted from significant population differences, highlighting the need for further comparable research to confirm these effects.

A meta-analysis of randomized controlled trials investigated the effects of prebiotics, probiotics, and symbiotics on patients with non-alcoholic fatty liver disease (NAFLD). Pan et al. (2024) found that prebiotics alone can reduce inflammation by lowering TNF- α, IL-6, and C-reactive protein (CRP) levels [51]. Apart from being related to metabolic and liver health, they can also increase local inflammation in the large intestine, compromising gut health.

3.4. Prebiotics and TLRs: Direct Modulation Matters?

Several preclinical studies have elucidated another aspect of dietary prebiotics, especially the ones of carbohydrate origin, which is the role of the direct modulation of immune receptors and other molecules. Toll-like receptors (TLRs), for example, are vital triggers for innate immune responses and are necessary for microbial tolerance in the intestine, but they have also been implicated in other biological scenarios. What if the inflammatory marker outcomes observed and discussed in the previous topic could also have a role in this direct modulation and are not only driven by microbiota regulation? A summary of the observed outcomes is shown in Figure 6.

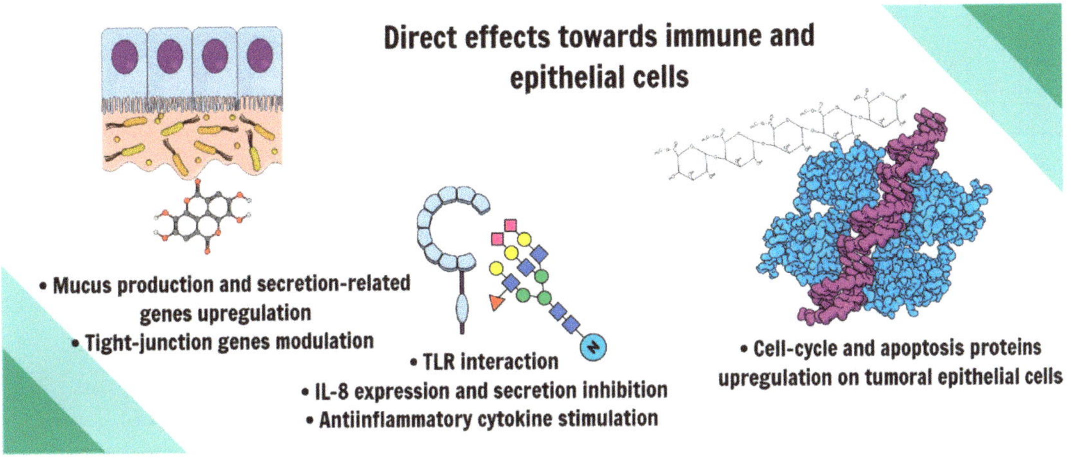

Figure 6. Direct effects observed of prebiotics in both immune and epithelial cells.

A robust cross-sectional study evaluating over 14,000 participants found (after adjustments towards age, gender, race, and other biological parameters) inverse associations of dietary fiber (including some prebiotics) intake and systemic inflammation index, systemic inflammation response index, neutrophil-to-lymphocyte ratio, and other immune-related metrics [52]. The authors argued for both sides; although the microbiota changes caused by prebiotics are established mechanisms, the significant contribution of the direct effects from dietary fiber in the immune system is a particularly interesting trending topic.

Imiquimod is a pharmacological agonist of TLR7, and there is a psoriasis-like induction model of rats with the topical application of imiquimod. A diet containing inulin administered to the animals improved skin thickness, erythema, scaling, and other visual aspects of inflammation while decreasing the number of inflammatory-infiltrating cells. The authors, however, focused on microbiota modulation and concluded that propionate production, at least partly, plays a role in attenuating inflammation. However, the inhibition of epidermal hyperproliferation was only achieved through the high-inulin diet and not with oral propionate ingestion, influenced by keratin-16 expression [53]. The mechanisms, nevertheless, still need to be more conclusive.

By using inulin as a treatment approach, the integrity of the blood–brain barrier (BBB) and inflammation after chronic stress were restored and managed by the inclusion of this prebiotic. TJ protein expression in the gut barrier, such as claudin-5, VE-cadherin, occludin, claudin-1, and ZO-1, were higher in the inulin group than in the condition-positive group. Aside from the structural help, TLR4 abundance and NF-kB were higher in the condition-positive group, leading to a higher release of iNOS, COX-2, and TNF- α. A reversion of this scenario was observed using inulin treatment. Once more, inulin enhanced the SCFA levels in the feces and serum, corroborating data from other studies. Therefore, modulation of the TLR4/MyD88/NF-kB pathway was pointed out as the most influenced by inulin treatment in chronically stressed mice [54].

Metabolic syndrome-related inflammation (on systemic adipose tissue) was also controlled in rats that ingested 10% prebiotic inulin. Adipocyte area decreased, alongside lower mRNA expression of IL-6 and TLR4 transmembrane protein, which led to a better inflammatory profile of the epididymal adipose tissue of the animals [55]. Different fructans, such as inulin-like, levans, and graminans, also performed as immune modulators in several in vitro experiments. Akkerman et al. (2024) found that fructose-based levan exopolysaccharides had a molecular-weight-dependent activation on TLR2 and 4 dependently on the MyD88 adapter molecule. Inhibition of TLR5 and 8 was also achieved, showing both pro- and anti-inflammatory potentials [56]. Meanwhile, in another study, graminan-type fructans (GTFs) activated TLR3, 7, and 9, and inhibited TLR2 and 4. Coincidently, inulin-like fructans (ITFs) activated TLR2 and 4 but did not exert inhibition of any particular TLR. While stimulating dendritic cells, GTFs and long-chain ITFs reduced proinflammatory cytokine production, such as TNF-α and MCP-1 [57]. Activation needs to be considered since some types of patients, for example, those with inflammatory bowel disease, may not be able to tolerate fructan fibers that are not totally fermented, potentially leading to impairment of the condition [58].

The most commonly studied prebiotics are inulin, FOS, and GOS. Their relatively simple structures facilitate regulatory approval and standardization for population health. When considering other types of dietary fibers, the structural differences between molecules play a significant role in interactions with TLRs. For example, Lagos et al. (2024) found that type B-resistant starch crystals led to higher TLR2-dependent NF-κB activation in both in vitro and in silico experiments. TLR4 activation was also observed exclusively with these B-type crystal structures, while no activation was detected for TLR5. In monocytes, this activation triggered the release of IL-8, TNF-α, and IL-1RA. Structural factors such as polydispersity, crystallinity, and chain length are components that may be adjusted to achieve targeted activation of specific receptors [59].

Pectins are also candidates for direct immunomodulation, especially regarding TLR interaction. One of the main structural components for pectins in this regard has been the degree of methylation, which is the proportion of methylated galacturonic acid (Gal*p*A) units throughout the backbone, and the degree of blockiness, which is the interval ratio of methylations in the pectic chain. TLR2/1 dimerization was better inhibited through blocks of non-methylated Gal*p*A units intercalated with methylations [60]. Gasaly et al. (2024) identified that both low and high DM lemon pectins inhibited the IL-8 secretion induced by TLR2/1, with low DM pectin being more efficient. Also, low DM pectin was still the most potent anti-inflammatory and suitable for TJ protein gene expression modulation [61]. The inhibition was more pronounced for both samples in the Caco-2 cell line and less in the DLD-1 cell line. While both lemon pectin samples differed between high and low DM, both had a blockwise distribution of those methyl groups, which can form anchor points when binding to some TLR pockets [61]. Pam3csk4 (TLR2/1 agonist) enhanced claudin-1 and 3 in both cell lines while enhancing claudin-2 only in the DLD-1 cell line. Those trends were reversed by low DM pectin, depending on the cell type, but not by high DM pectin [61].

Li et al. (2023) administered chows enriched with long-chain 57% DM citrus pectin for mice with *S. typhimurium*-induced colitis. The animals exhibited a reversal of colon shortening and improved disease activity index (DAI) scores. TJ proteins, including occludin, ZO-1, and ZO-2, downregulated by colitis, had their expression restored after pectin treatment. Additionally, the upregulation of TLR2 and NF-κB was reversed in the pectin-treated group [62]. It is important to note that all the studies mentioned above used pectins with high levels of Gal*p*A residues rather than neutral sugars like those found in rhamnogalacturonan (RG) fragments. Hyun et al. (2023) conducted an in vitro and in silico analysis of pectin heteropolysaccharides and discovered that pectin galactans are strongly affixed to the leucine-rich segment of TLR4. Interestingly, these fragments selectively bind to TLR4 and not TLR2, which activates macrophages toward an inflammatory profile [63]. Structure–function relationship studies of these carbohydrates are essential to understanding these variations in the effects of different pectic fragments. This approach will enable more refined preclinical and clinical testing in the near future.

A lesser-known prebiotic fiber, konjac glucomannan (KGM), was also shown to directly impact the immune system of mice beyond microbiota regulation. Once again, TJ proteins, such as occluding and ZO-2, in the gut were upregulated after KGM treatment, and the overall colonic barrier was enhanced following the pretreatment of KGM and injection of *S. typhimurium* (to induce colitis). The proposed mechanism was via TLR2, not via TLR4, resulting in the improved upregulation of TNF- α and CCL8, as well as the downregulation of IFNβ. Of course, some of the effects observed were also correlated but not limited to microbiota regulation [64].

4. Conclusions and Future Research

Overall, prebiotic consumption has been widely studied as a potential strategy for reducing inflammation by modulating the intestinal microbiota, with research often focusing on the development and progression of colorectal cancer (CRC) through in vitro and in vivo studies, as well as clinical trials. Although the definition of prebiotics varies across the literature [8,9], it generally includes dietary fibers, fructo- and galactooligosaccharides, specific conjugated lipids, and emerging candidates like polyphenols [38,46,48]. Clinical trials in humans have shown that various prebiotic and symbiotic treatments frequently modulate inflammatory markers such as IL-6, IL-10, and TNF-α. Such use aims to improve gut barrier function and structure, immune tolerance, and reasonable immune response buildup and is a promising alternative for co-adjuvant therapy. New studies have shown that some prebiotics can directly interact with cells from the immune system and the gut epithelial barrier, making the biological effects of prebiotics go beyond microbiota modulation. Nevertheless, some interindividual differences are limitations of a significant part of human clinical studies; therefore, more work focusing on avoiding and lowering such limitations is necessary for the progression of prebiotic usage protocol establishment. Hopefully, in the near future, it will be possible to further explore how particular prebiotics directly modulate immune molecules, enhancing our understanding of the impact this may have on specific immune-related disorders.

Future research on prebiotics should aim to provide more definitive evidence of both microbiota-dependent and direct immune-modulating effects, as these interactions offer promising therapeutic potential. The ability of prebiotics to support gut health and reduce inflammation has been well documented, primarily through microbiota modulation. However, emerging studies have revealed that prebiotics may also interact directly with immune cells and the epithelial barrier, suggesting a more complex biological role beyond microbiota influence alone. To build a stronger foundation for clinical applications, future studies should focus on understanding the structure–function relationship of diverse prebiotics and identifying which molecular features are the most effective for specific immune pathways. Standardizing methodologies and considering individual variability will also be crucial, as these factors often introduce limitations in clinical trials. By addressing these areas, future research can establish more explicit protocols for using prebiotics to target immune-related disorders, moving towards their potential role as adjunct therapies in managing inflammation and the associated diseases. Also worth mentioning are the applications of emerging prebiotics such as polyphenols and resistant starch, as well as the technologies such as the nanoparticle carriage system, especially in human studies, which should also contribute to further advances.

Overall, prebiotics are promising for the enhancement of health and resilience, especially by supporting gut integrity and immune function. This paper highlights evidence that prebiotics can assist in restoring the gut barrier and reducing inflammation, particularly following disturbances such as antibiotic use or other gut injuries. Prebiotics provide a valuable approach to maintaining health and preventing disease in vulnerable individuals by replenishing beneficial microbes and directly strengthening epithelial and immune cell functions.

Author Contributions: L.d.F.P.: Conceptualization, Methodology, Investigation, Writing—Original Draft Preparation; P.d.V.: Writing—Review and Editing; J.P.F.: Conceptualization, Writing—Review and Editing, Funding Acquisition. All authors have read and agreed to the published version of the manuscript.

Funding: The authors acknowledge the National Council for Scientific and Technological Development (CNPq) for the J.P.F productivity scholarship (CNPq Proc. #307842/2022-3) and the São Paulo Research Foundation (FAPESP) for L.d.F.P scholarship (#2020/08063-5 and #2022/12253-0). The study was financially supported by grants #2013/07914-8 and #2022/12834-2 from FAPESP.

Data Availability Statement: Data will be made available upon request.

Conflicts of Interest: The authors declare no conflicts of interest.

References

1. Roy, S.; Dhaneshwar, S. Role of Prebiotics, Probiotics, and Synbiotics in Management of Inflammatory Bowel Disease: Current Perspectives. *World J. Gastroenterol.* **2023**, *29*, 2078–2100. [CrossRef] [PubMed]
2. Castillo Andrade, A.I.; García Chávez, E.; Rivera Bautista, C.; Oros Ovalle, C.; Ruiz Cabrera, M.A.; Grajales Lagunes, A. Influence of Prebiotic Activity of Agave Salmiana Fructans on Mucus Production and Morphology Changes in Colonic Epithelium Cell of Healthy Wistar Rats. *Front. Plant Sci.* **2021**, *12*, 717460. [CrossRef]
3. Wu, R.Y.; Abdullah, M.; Määttänen, P.; Pilar, A.V.C.; Scruten, E.; Johnson-Henry, K.C.; Napper, S.; O'Brien, C.; Jones, N.L.; Sherman, P.M. Protein Kinase Cσ Signaling Is Required for Dietary Prebiotic-Induced Strengthening of Intestinal Epithelial Barrier Function. *Sci. Rep.* **2017**, *7*, 40820. [CrossRef]
4. Chang, S.C.; Chiang, H.H.; Liu, C.Y.; Li, Y.J.; Lu, C.L.; Lee, Y.P.; Huang, C.J.; Lai, C.L. Intestinal Mucosal Barrier Improvement with Prebiotics: Histological Evaluation of Longish Glucomannan Hydrolysates-Induced Innate T Lymphocyte Activities in Mice. *Nutrients* **2022**, *14*, 2220. [CrossRef]
5. Martel, J.; Chang, S.H.; Ko, Y.F.; Hwang, T.L.; Young, J.D.; Ojcius, D.M. Gut Barrier Disruption and Chronic Disease. *Trends Endocrinol. Metab.* **2022**, *33*, 247–265. [CrossRef] [PubMed]
6. Bevilacqua, A.; Campaniello, D.; Speranza, B.; Racioppo, A.; Sinigaglia, M.; Corbo, M.R. An Update on Prebiotics and on Their Health Effects. *Foods* **2024**, *13*, 446. [CrossRef] [PubMed]
7. Patra, S.; Sahu, N.; Saxena, S.; Pradhan, B.; Nayak, S.K.; Roychowdhury, A. Effects of Probiotics at the Interface of Metabolism and Immunity to Prevent Colorectal Cancer-Associated Gut Inflammation: A Systematic Network and Meta-Analysis With Molecular Docking Studies. *Front. Microbiol.* **2022**, *13*, 878297. [CrossRef]
8. Gibson, G.R.; Hutkins, R.; Sanders, M.E.; Prescott, S.L.; Reimer, R.A.; Salminen, S.J.; Scott, K.; Stanton, C.; Swanson, K.S.; Cani, P.D.; et al. Expert Consensus Document: The International Scientific Association for Probiotics and Prebiotics (ISAPP) Consensus Statement on the Definition and Scope of Prebiotics. *Nat. Rev. Gastroenterol. Hepatol.* **2017**, *14*, 491–502. [CrossRef]
9. Guarner, F.; Sanders, M.E.; Szajewska, H.; Cohen, H.; Eliakim, R.; Herrera-Deguise, C.; Karakan, T.; Merenstein, D.; Piscoya, A.; Ramakrishna, B.; et al. World Gastroenterology Organisation Global Guidelines: Probiotics and Prebiotics. *J. Clin. Gastroenterol.* **2024**, *58*, 533–553. [CrossRef]
10. Davani-Davari, D.; Negahdaripour, M.; Karimzadeh, I.; Seifan, M.; Mohkam, M.; Masoumi, S.J.; Berenjian, A.; Ghasemi, Y. Prebiotics: Definition, Types, Sources, Mechanisms, and Clinical Applications. *Foods* **2019**, *8*, 92. [CrossRef]
11. Andersen-Civil, A.I.S.; Arora, P.; Williams, A.R. Regulation of Enteric Infection and Immunity by Dietary Proanthocyanidins. *Front. Immunol.* **2021**, *12*, 637603. [CrossRef] [PubMed]
12. Ji, J.; Jin, W.; Liu, S.J.; Jiao, Z.; Li, X. Probiotics, Prebiotics, and Postbiotics in Health and Disease. *MedComm* **2023**, *4*, e420. [CrossRef]
13. Di Tommaso, N.; Gasbarrini, A.; Ponziani, F.R. Intestinal Barrier in Human Health and Disease. *Int. J. Environ. Res. Public Health* **2021**, *18*, 12836. [CrossRef]
14. Portincasa, P.; Bonfrate, L.; Khalil, M.; De Angelis, M.; Calabrese, F.M.; D'amato, M.; Wang, D.Q.H.; Di Ciaula, A. Intestinal Barrier and Permeability in Health, Obesity and NAFLD. *Biomedicines* **2022**, *10*, 83. [CrossRef]
15. Horrocks, V.; King, O.G.; Yip, A.Y.G.; Marques, I.M.; McDonald, J.A.K. Role of the Gut Microbiota in Nutrient Competition and Protection against Intestinal Pathogen Colonization. *Microbiology* **2023**, *169*, 001377. [CrossRef]
16. Suriano, F.; Nyström, E.E.L.; Sergi, D.; Gustafsson, J.K. Diet, Microbiota, and the Mucus Layer: The Guardians of Our Health. *Front. Immunol.* **2022**, *13*, 953196. [CrossRef] [PubMed]
17. Barbara, G.; Barbaro, M.R.; Fuschi, D.; Palombo, M.; Falangone, F.; Cremon, C.; Marasco, G.; Stanghellini, V. Inflammatory and Microbiota-Related Regulation of the Intestinal Epithelial Barrier. *Front. Nutr.* **2021**, *8*, 718356. [CrossRef] [PubMed]
18. Compare, D.; Sgamato, C.; Rocco, A.; Coccoli, P.; Ambrosio, C.; Nardone, G. The Leaky Gut and Human Diseases: "Can'T Fill the Cup If You Don'T Plug the Holes First". *Dig. Dis.* **2024**, *42*, 548–566. [CrossRef]
19. Jensen, S.K.; Pærregaard, S.I.; Brandum, E.P.; Jørgensen, A.S.; Hjortø, G.M.; Jensen, B.A.H. Rewiring Host-Microbe Interactions and Barrier Function during Gastrointestinal Inflammation. *Gastroenterol. Rep.* **2022**, *10*, goac008. [CrossRef]
20. Cosovanu, C.; Neumann, C. The Many Functions of Foxp3+ Regulatory T Cells in the Intestine. *Front. Immunol.* **2020**, *11*, 600973. [CrossRef]
21. Stockinger, B.; Diaz, O.E.; Wincent, E. The Influence of AHR on Immune and Tissue Biology. *EMBO Mol. Med.* **2024**, *16*, 2290–2298. [CrossRef] [PubMed]
22. Sung, H.; Ferlay, J.; Siegel, R.L.; Laversanne, M.; Soerjomataram, I.; Jemal, A.; Bray, F. Global Cancer Statistics 2020: GLOBOCAN Estimates of Incidence and Mortality Worldwide for 36 Cancers in 185 Countries. *CA. Cancer J. Clin.* **2021**, *71*, 209–249. [CrossRef] [PubMed]
23. Zhang, F.; Qiao, S. Research Progress on the Relationship Between Inflammation and Colorectal Cancer. *Ann. Gastroenterol. Surg.* **2022**, *6*, 204–211. [CrossRef] [PubMed]
24. Schmitt, M.; Greten, F.R. The Inflammatory Pathogenesis of Colorectal Cancer. *Nat. Rev. Immunol.* **2021**, *21*, 653–667. [CrossRef] [PubMed]

25. Suarez-Lopez, L.; Kong, Y.W.; Sriram, G.; Patterson, J.C.; Rosenberg, S.; Morandell, S.; Haigis, K.M.; Yaffe, M.B. MAPKAP Kinase-2 Drives Expression of Angiogenic Factors by Tumor-Associated Macrophages in a Model of Inflammation-Induced Colon Cancer. *Front. Immunol.* **2021**, *11*, 607891. [CrossRef]
26. Hirano, T.; Hirayama, D.; Wagatsuma, K.; Yamakawa, T.; Yokoyama, Y.; Nakase, H. Immunological Mechanisms in Inflammationassociated Colon Carcinogenesis. *Int. J. Mol. Sci.* **2020**, *21*, 3062. [CrossRef]
27. Sheikh, A.; Taube, J.; Greathouse, K.L. Contribution of the Microbiota and Their Secretory Products to Inflammation and Colorectal Cancer Pathogenesis: The Role of Toll-like Receptors. *Carcinogenesis* **2021**, *42*, 1133–1142. [CrossRef]
28. Lichtenstern, C.R.; Ngu, R.K.; Shalapour, S.; Karin, M. Immunotherapy, Inflammation and Colorectal Cancer. *Cells* **2020**, *9*, 618. [CrossRef]
29. Borowczak, J.; Szczerbowski, K.; Maniewski, M.; Kowalewski, A.; Janiczek-Polewska, M.; Szylberg, A.; Marszałek, A.; Szylberg, Ł. The Role of Inflammatory Cytokines in the Pathogenesis of Colorectal Carcinoma—Recent Findings and Review. *Biomedicines* **2022**, *10*, 1670. [CrossRef]
30. Kim, J.; Lee, H.K. Potential Role of the Gut Microbiome In Colorectal Cancer Progression. *Front. Immunol.* **2022**, *12*, 807648. [CrossRef]
31. Sánchez-Alcoholado, L.; Ramos-Molina, B.; Otero, A.; Laborda-Illanes, A.; Ordóñez, R.; Medina, J.A.; Gómez-Millán, J.; Queipo-Ortuño, M.I. The Role of the Gut Microbiome in Colorectal Cancer Development and Therapy Response. *Cancers* **2020**, *12*, 1406. [CrossRef] [PubMed]
32. Song, S.H.; Ghosh, T.; You, D.G.; Joo, H.; Lee, J.; Lee, J.; Kim, C.H.; Jeon, J.; Shin, S.; Park, J.H. Functionally Masked Antibody to Uncouple Immune-Related Toxicities in Checkpoint Blockade Cancer Therapy. *ACS Nano* **2023**, *17*, 10065–10077. [CrossRef]
33. Omstead, A.N.; Paskewicz, M.; Gorbunova, A.; Zheng, P.; Salvitti, M.S.; Mansoor, R.; Reed, P.; Ballengee, S.; Wagner, P.L.; Jobe, B.A.; et al. CSF-1R Inhibitor, Pexidartinib, Sensitizes Esophageal Adenocarcinoma to PD-1 Immune Checkpoint Blockade in a Rat Model. *Carcinogenesis* **2022**, *43*, 842–850. [CrossRef] [PubMed]
34. Lu, H.-M.; Fu, Y.-J.; Liu, N.; Xia, W.-Y.; Chen, H.-Y.; Liu, M.-Y.; Li, L.-F.; Gao, J.-X. A Novel Tumor-Specific Broad-Spectral Monoclonal Antibody to PL2L60 Is Highly Effective for the Treatment of Various Types of Cancers from Human and Mouse. *Am. J. Cancer Res.* **2022**, *12*, 265–279. [PubMed]
35. Chen, J.S.; Hsieh, Y.C.; Chou, C.H.; Wu, Y.H.; Yang, M.H.; Chu, S.H.; Chao, Y.S.; Chen, C.N. Chidamide plus Tyrosine Kinase Inhibitor Remodel the Tumor Immune Microenvironment and Reduce Tumor Progression When Combined with Immune Checkpoint Inhibitor in Naïve and Anti-PD-1 Resistant CT26-Bearing Mice. *Int. J. Mol. Sci.* **2022**, *23*, 10677. [CrossRef] [PubMed]
36. Hajiabadi, S.; Alidadi, S.; Montakhab Farahi, Z.; Ghahramani Seno, M.M.; Farzin, H.; Haghparast, A. Immunotherapy with STING and TLR9 Agonists Promotes Synergistic Therapeutic Efficacy with Suppressed Cancer-Associated Fibroblasts in Colon Carcinoma. *Front. Immunol.* **2023**, *14*, 1258691. [CrossRef]
37. Kang, X.; Liu, C.; Ding, Y.; Ni, Y.; Ji, F.; Lau, H.C.H.; Jiang, L.; Sung, J.J.Y.; Wong, S.H.; Yu, J. Roseburia Intestinalis Generated Butyrate Boosts Anti-PD-1 Efficacy in Colorectal Cancer by Activating Cytotoxic CD8 + T Cells. *Gut* **2023**, *72*, 2112–2122. [CrossRef]
38. Molinari, R.; Merendino, N.; Costantini, L. Polyphenols as Modulators of Pre-Established Gut Microbiota Dysbiosis: State-of-the-Art. *BioFactors* **2022**, *48*, 255–273. [CrossRef]
39. Macis, D.; Briata, I.M.; D'Ecclesiis, O.; Johansson, H.; Aristarco, V.; Buttiron Webber, T.; Oppezzi, M.; Gandini, S.; Bonanni, B.; DeCensi, A. Inflammatory and Metabolic Biomarker Assessment in a Randomized Presurgical Trial of Curcumin and Anthocyanin Supplements in Patients with Colorectal Adenomas. *Nutrients* **2023**, *15*, 3894. [CrossRef]
40. Moorthy, M.; Sundralingam, U.; Palanisamy, U.D. Polyphenols as Prebiotics in the Management of High-Fat Diet-Induced Obesity: A Systematic Review of Animal Studies. *Foods* **2021**, *10*, 299. [CrossRef]
41. Bush, J.R.; Baisley, J.; Harding, S.V.; Alfa, M.J. Consumption of Solnul™ Resistant Potato Starch Produces a Prebiotic Effect in a Randomized, Placebo-Controlled Clinical Trial. *Nutrients* **2023**, *15*, 1582. [CrossRef] [PubMed]
42. Rengadu, D.; Gerrano, A.S.; Mellem, J.J. Prebiotic Effect of Resistant Starch from Vigna Unguiculata (L.) Walp. (Cowpea) Using an in Vitro Simulated Digestion Model. *Int. J. Food Sci. Technol.* **2020**, *55*, 332–339. [CrossRef]
43. Das, M.; Rajan, N.; Biswas, P.; Banerjee, R. Dual Enzyme Treatment Strategy for Enhancing Resistant Starch Content of Green Banana Flour and in Vitro Evaluation of Prebiotic Effect. *LWT* **2022**, *160*, 113267. [CrossRef]
44. Wang, M.; Chen, X.; Zhou, L.; Li, Y.; Yang, J.; Ji, N.; Xiong, L.; Sun, Q. Prebiotic Effects of Resistant Starch Nanoparticles on Growth and Proliferation of the Probiotic Lactiplantibacillus Plantarum Subsp. Plantarum. *LWT* **2022**, *154*, 112572. [CrossRef]
45. Buhaș, M.C.; Candrea, R.; Gavrilaș, L.I.; Miere, D.; Tătaru, A.; Boca, A.; Cătinean, A. Transforming Psoriasis Care: Probiotics and Prebiotics as Novel Therapeutic Approaches. *Int. J. Mol. Sci.* **2023**, *24*, 11225. [CrossRef]
46. Neyrinck, A.M.; Rodriguez, J.; Taminiau, B.; Amadieu, C.; Herpin, F.; Allaert, F.A.; Cani, P.D.; Daube, G.; Bindels, L.B.; Delzenne, N.M. Improvement of Gastrointestinal Discomfort and Inflammatory Status by a Synbiotic in Middle-Aged Adults: A Double-Blind Randomized Placebo-Controlled Trial. *Sci. Rep.* **2021**, *11*, 1–12. [CrossRef]
47. Azuma, N.; Mawatari, T.; Saito, Y.; Tsukamoto, M.; Sampei, M.; Iwama, Y. Effect of Continuous Ingestion of Bifidobacteria and Dietary Fiber on Improvement in Cognitive Function: A Randomized, Double-Blind, Placebo-Controlled Trial. *Nutrients* **2023**, *15*, 4175. [CrossRef] [PubMed]

48. De Giani, A.; Sandionigi, A.; Zampolli, J.; Michelotti, A.; Tursi, F.; Labra, M.; Di Gennaro, P. Effects of Inulin-Based Prebiotics Alone or in Combination with Probiotics on Human Gut Microbiota and Markers of Immune System: A Randomized, Double-Blind, Placebo-Controlled Study in Healthy Subjects. *Microorganisms* **2022**, *10*, 1256. [CrossRef]
49. Moludi, J.; Khedmatgozar, H.; Nachvak, S.M.; Abdollahzad, H.; Moradinazar, M.; Sadeghpour Tabaei, A. The Effects of Co-Administration of Probiotics and Prebiotics on Chronic Inflammation, and Depression Symptoms in Patients with Coronary Artery Diseases: A Randomized Clinical Trial. *Nutr. Neurosci.* **2022**, *25*, 1659–1668. [CrossRef]
50. Visuthranukul, C.; Kwanbunbumpen, T.; Chongpison, Y.; Chamni, S.; Panichsillaphakit, E.; Uaariyapanichkul, J.; Maholarnkij, S.; Chomtho, S. The Impact of Dietary Fiber as a Prebiotic on Inflammation in Children with Obesity. *Foods* **2022**, *11*, 2856. [CrossRef]
51. Pan, Y.; Yang, Y.; Wu, J.; Zhou, H.; Yang, C. Efficacy of Probiotics, Prebiotics, and Synbiotics on Liver Enzymes, Lipid Profiles, and Inflammation in Patients with Non- Alcoholic Fatty Liver Disease: A Systematic Review and Meta-Analysis of Randomized Controlled Trials. *BMC Gastroenterol.* **2024**, *24*, 283. [CrossRef] [PubMed]
52. Qi, X.; Li, Y.; Fang, C.; Jia, Y.; Chen, M.; Chen, X.; Jia, J. The Associations between Dietary Fibers Intake and Systemic Immune and Inflammatory Biomarkers, a Multi-Cycle Study of NHANES 2015–2020. *Front. Nutr.* **2023**, *10*, 1242115. [CrossRef] [PubMed]
53. Yoshida, M.; Funasaka, Y.; Saeki, H.; Yamamoto, M.; Kanda, N. Dietary Fiber Inulin Improves Murine Imiquimod-Induced Psoriasis-like Dermatitis. *Int. J. Mol. Sci.* **2023**, *24*, 14197. [CrossRef]
54. Wang, L.; Wang, Z.; Lan, Y.; Tuo, Y.; Ma, S.; Liu, X. Inulin Attenuates Blood-Brain Barrier Permeability and Alleviates Behavioral Disorders by Modulating the TLR4/MyD88/NF-KB Pathway in Mice with Chronic Stress. *J. Agric. Food Chem.* **2023**, *71*, 13325–13337. [CrossRef]
55. Abreu, C.D.D.; Caldas, B.V.; Ribeiro, G.H.M.; Aguilar, C.M.; Farias, L.C.; Guimarães, A.L.S.; de Paula, A.M.B.; Glória, M.B.A.; Santos, S.H.S. Inulin Prebiotic Dietary Supplementation Improves Metabolic Parameters by Reducing the Toll-like Receptor 4 Transmembrane Protein Gene and Interleukin 6 Expression in Adipose Tissue. *PharmaNutrition* **2022**, *22*, 100316. [CrossRef]
56. Akkerman, R.; Oerlemans, M.M.P.; Ferrari, M.; Fernández-Lainez, C.; de Haan, B.J.; Faas, M.M.; Walvoort, M.T.C.; de Vos, P. Exopolysaccharide β-(2,6)-Levan-Type Fructans Have a Molecular-Weight-Dependent Modulatory Effect on Toll-like Receptor Signalling. *Food Funct.* **2023**, *15*, 676–688. [CrossRef]
57. Fernández-Lainez, C.; aan de Stegge, M.; Silva-Lagos, L.A.; López-Velázquez, G.; de Vos, P. B(2 → 1)-B(2 → 6) Branched Graminan-Type Fructans and B(2 → 1) Linear Fructans Impact Mucus-Related and Endoplasmic Reticulum Stress-Related Genes in Goblet Cells and Attenuate Inflammatory Responses in a Fructan Dependent Fashion. *Food Funct.* **2023**, *14*, 1338–1348. [CrossRef] [PubMed]
58. Armstrong, H.K.; Bording-Jorgensen, M.; Santer, D.M.; Zhang, Z.; Valcheva, R.; Rieger, A.M.; Sung-Ho Kim, J.; Dijk, S.I.; Mahmood, R.; Ogungbola, O.; et al. Unfermented β-Fructan Fibers Fuel Inflammation in Select Inflammatory Bowel Disease Patients. *Gastroenterology* **2023**, *164*, 228–240. [CrossRef]
59. Silva Lagos, L.; Klostermann, C.E.; López-Velázquez, G.; Fernández-Lainez, C.; Leemhuis, H.; Oudhuis, A.A.C.M.L.; Buwalda, P.; Schols, H.A.; de Vos, P. Crystal Type, Chain Length and Polydispersity Impact the Resistant Starch Type 3 Immunomodulatory Capacity via Toll-like Receptors. *Carbohydr. Polym.* **2024**, *324*, 121490. [CrossRef]
60. Jermendi, É.; Fernández-Lainez, C.; Beukema, M.; López-Velázquez, G.; van den Berg, M.A.; de Vos, P.; Schols, H.A. TLR 2/1 Interaction of Pectin Depends on Its Chemical Structure and Conformation. *Carbohydr. Polym.* **2023**, *303*, 120444. [CrossRef]
61. Gasaly, N.; Tang, X.; Chen, X.; Bellalta, S.; Hermoso, M.A.; de Vos, P. Effects of Pectin's Degree of Methyl Esterification on TLR2-Mediated IL-8 Secretion and Tight Junction Gene Expression in Intestinal Epithelial Cells: Influence of Soluble TLR2. *Food Funct.* **2023**, *15*, 569–579. [CrossRef] [PubMed]
62. Li, J.; Bai, J.; Song, Z.; Ji, Y.; Chen, Z.; Yang, Y.; Wu, Z. Dietary Pectin Attenuates Salmonella Typhimurium-Induced Colitis by Modulating the TLR2-NF-KB Pathway and Intestinal Microbiota in Mice. *Food Chem. Toxicol.* **2023**, *182*, 114100. [CrossRef] [PubMed]
63. Hyun, G.H.; Cho, I.H.; Yang, Y.Y.; Jeong, D.H.; Kang, Y.P.; Kim, Y.S.; Lee, S.J.; Kwon, S.W. Mechanisms of Interactions in Pattern-Recognition of Common Glycostructures across Pectin-Derived Heteropolysaccharides by Toll-like Receptor 4. *Carbohydr. Polym.* **2023**, *314*, 120921. [CrossRef] [PubMed]
64. Li, J.; Liu, H.; Fu, H.; Yang, Y.; Wu, Z. An Isofibrous Diet with Fiber Konjac Glucomannan Ameliorates Salmonella Typhimurium-Induced Colonic Injury by Regulating TLR2-NF-KB Signaling and Intestinal Microbiota in Mice. *J. Agric. Food Chem.* **2024**, *72*, 13415–13430. [CrossRef]

nutrients

MDPI

Review

The Impact of Bioactive Molecules from Probiotics on Child Health: A Comprehensive Review

Linda P. Guamán [1,*], Saskya E. Carrera-Pacheco [1], Johana Zúñiga-Miranda [1], Enrique Teran [2], Cesar Erazo [2] and Carlos Barba-Ostria [2,3,*]

[1] Centro de Investigación Biomédica (CENBIO), Facultad de Ciencias de la Salud Eugenio Espejo, Universidad UTE, Quito 170527, Ecuador; saskya.carrera@ute.edu.ec (S.E.C.-P.); johana.zuniga@ute.edu.ec (J.Z.-M.)
[2] Colegio de Ciencias de la Salud, Universidad San Francisco de Quito USFQ, Quito 170901, Ecuador; eteran@usfq.edu.ec (E.T.); cerazo@usfq.edu.ec (C.E.)
[3] Instituto de Microbiología, Universidad San Francisco de Quito USFQ, Quito 170901, Ecuador
* Correspondence: author: linda.guaman@ute.edu.ec (L.P.G.); cbarbao@usfq.edu.ec (C.B.-O.)

Abstract: Background: This review investigates the impact of bioactive molecules produced by probiotics on child health, focusing on their roles in modulating gut microbiota, enhancing immune function, and supporting overall development. Key metabolites, including short-chain fatty acids (SCFAs), bacteriocins, exopolysaccharides (EPSs), vitamins, and gamma-aminobutyric acid (GABA), are highlighted for their ability to maintain gut health, regulate inflammation, and support neurodevelopment. **Objectives**: The aim of this review is to examine the mechanisms of action and clinical evidence supporting the use of probiotics and postbiotics in pediatric healthcare, with a focus on promoting optimal growth, development, and overall health in children. **Methods**: The review synthesizes findings from clinical studies that investigate the effects of probiotics and their metabolites on pediatric health. The focus is on specific probiotics and their ability to influence gut health, immune responses, and developmental outcomes. **Results**: Clinical studies demonstrate that specific probiotics and their metabolites can reduce gastrointestinal disorders, enhance immune responses, and decrease the incidence of allergies and respiratory infections in pediatric populations. Additionally, postbiotics—bioactive compounds from probiotic fermentation—offer promising benefits, such as improved gut barrier function, reduced inflammation, and enhanced nutrient absorption, while presenting fewer safety concerns compared to live probiotics. **Conclusions**: By examining the mechanisms of action and clinical evidence, this review underscores the potential of integrating probiotics and postbiotics into pediatric healthcare strategies to promote optimal growth, development, and overall health in children.

Keywords: probiotics; short-chain fatty acids; exopolysaccharide; vitamins; peptides; postbiotics

1. Introduction

Childhood is a critical period characterized by the rapid growth, development, and maturation of various physiological systems, including the gastrointestinal tract and the immune system. During this stage, maintaining a balanced gut microbiota is essential, as it plays a vital role in nutrient absorption, metabolism, immune development, and protection against pathogens [1,2]. Disruptions in the gut microbiome, such as those caused by antibiotics, poor nutrition, or infections, can lead to various health issues in children, including gastrointestinal disorders, allergies, and impaired immune responses [3]. Recent research has highlighted the potential of probiotics—live microorganisms that confer health benefits to the host—to modulate the gut microbiota and improve pediatric health outcomes [4,5].

A key aspect of probiotics' benefits lies in the bioactive metabolites they produce. These metabolites, including short-chain fatty acids (SCFAs), bacteriocins, exopolysaccharides (EPSs), vitamins, and neuroactive compounds like gamma-aminobutyric acid

(GABA), are crucial mediators of the beneficial effects associated with probiotics [6,7]. For instance, SCFAs, such as acetate, propionate, and butyrate, play significant roles in maintaining gut barrier integrity, modulating immune function, and providing energy to colonocytes, which are particularly important in infants and young children whose immune and digestive systems are still developing [8,9].

Probiotic metabolites also contribute to the regulation of inflammatory responses, which is crucial in preventing conditions such as necrotizing enterocolitis (NEC), a severe gastrointestinal disorder that predominantly affects preterm infants [10,11]. Metabolites like bacteriocins, antimicrobial peptides produced by probiotics, help in protecting against pathogenic bacteria by disrupting their cell membranes, thereby preventing infections such as antibiotic-associated diarrhea. In addition, exopolysaccharides (EPSs) produced by probiotic strains support biofilm formation, promoting long-term colonization of the gut and protection against pathogens [12–14].

Furthermore, these metabolites have a direct impact on nutrient absorption, a vital function during childhood, a period marked by increased nutritional demands for growth and development. For example, certain metabolites enhance the bioavailability of essential vitamins and minerals, which are critical for bone growth, cognitive development, and overall health [15]. Additionally, bioactive molecules like GABA, produced by certain probiotic strains, have been implicated in supporting neurodevelopment and emotional regulation through their interaction with the gut-brain axis [16,17].

Given the diverse and critical roles of these bioactive metabolites in promoting child health, there is growing interest in understanding their specific mechanisms of action and potential applications in pediatric care. This review aims to provide a comprehensive overview of the various metabolites produced by probiotics and their impact on child health. We will discuss the specific types of metabolites, such as SCFAs, bacteriocins, EPS, vitamins, and neuroactive compounds, and explore how these molecules contribute to gastrointestinal health, immune modulation, nutrient absorption, and neurodevelopment in children. By examining the evidence, this review seeks to highlight the potential of incorporating probiotic-derived metabolites into pediatric healthcare strategies to enhance the health and well-being of children.

2. Adaptive Role of Metabolites Produced by Probiotic Bacteria

From an evolutionary perspective, the ability of probiotic bacteria to produce metabolites enables them to adapt effectively to their host [18,19]. Bacteria that generate beneficial metabolites gain a survival advantage by outcompeting harmful microbes and fostering symbiotic relationships with their hosts. This evolutionary trait allows beneficial bacteria to thrive and proliferate within the gut environment, supporting host health and contributing to the stability of the microbial community [20,21].

For example, probiotic bacteria produce various metabolites and molecules as part of their complex survival strategies within the gut environment, ensuring their survival and competitiveness while significantly contributing to the host's health [22]. They compete for space and nutrients by producing metabolites such as SCFAs and bacteriocins, which inhibit the growth of pathogenic bacteria and other microbes [23]. SCFAs, like acetate, propionate, and butyrate lower the gut pH, creating a less hospitable for harmful pathogens and favoring the growth of beneficial bacteria, thus maintaining their niche within the gut microbiota [24]. Additionally, bacteriocins, which are ribosomally synthesized antimicrobial peptides, target and kill competing bacterial strains by permeabilizing their cell membranes, ensuring the dominance of probiotic strains in their ecological niche [25].

Furthermore, EPSs, which are high-molecular-weight polymers secreted by probiotic bacteria, play a crucial role in biofilm formation. Biofilms are structured communities of bacteria encased in a self-produced matrix that adheres to surfaces that protect the bacteria from environmental stresses, including desiccation, antibiotics, and the host immune system [14]. This stable environment enhances bacterial survival and promotes long-term colonization in the gut [26].

In addition, the production of metabolites like vitamins and SCFAs fosters a symbiotic relationship with the host [27]. For example, vitamins such as B12 and folate, synthesized by bacteria like *Lactobacillus reuteri*, are essential for the host's metabolic processes, including DNA synthesis and energy metabolism [28,29]. By providing these essential nutrients, probiotics contribute to the host's nutritional status and overall health, creating a mutually beneficial relationship [30].

Moreover, SCFAs produced by probiotic bacteria serve as an energy source for colonocytes and play a role in maintaining the integrity of the gut barrier, which is crucial for preventing the translocation of pathogens and toxins from the gut into the bloodstream, thereby protecting the host from infections and inflammation [31]. Additionally, SCFAs have been shown to modulate the host immune system, promoting anti-inflammatory responses and enhancing immune tolerance [32]. A summary of the metabolites and their effects are illustrated in Figure 1.

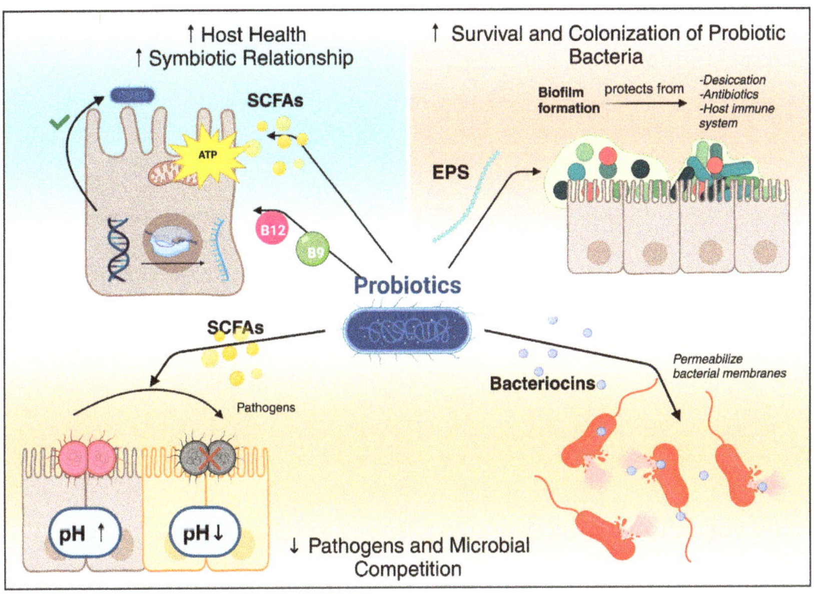

Figure 1. Role of metabolites in the adaptation of probiotic bacteria within the gut environment. This figure depicts how probiotic metabolites benefit both the bacteria and the host: 1. symbiotic interaction; 2. self-protection; and 3. pathogen inhibition. EPSs: exopolysaccharides; SCFAs: short-chain fatty acids; pH: potential of hydrogen; ATP: adenosine triphosphate.

In conclusion, probiotic bacteria produce these metabolites not out of altruism but as part of their survival strategies within the gut environment. The range of survival strategies employed by these bacteria is as diverse as the array of molecules they produce, which, while primarily serving their own persistence and competitiveness, ultimately have significant and beneficial effects on the host's health.

3. Specific Probiotic Strains for Children's Health

When it comes to enhancing children's health, selecting specific probiotic strains plays a crucial role in providing targeted benefits. Research has highlighted the efficacy of particular probiotic strains in promoting gut health, bolstering the immune system, and alleviating various health issues in children [33,34]. The gut microbiota of children is composed of numerous microorganisms, each potentially contributing to child health and development. However, this section will focus on those microorganisms with the strongest scientific evidence supporting their positive effects in children, as well as those with the most detailed descriptions of the metabolites they produce.

3.1. *Lactobacillus rhamnosus*

L. rhamnosus is a facultative anaerobic bacterium known for its extensive genomic diversity and metabolic flexibility. Its genome is highly adaptable, featuring a broad array of genes encoding transporters and enzymes that facilitate the utilization of various carbohydrates, including lactose, glucose, galactose, and fucose. This adaptability enables *L. rhamnosus* to thrive in different niches within the gastrointestinal tract [35,36].

The bacterium's ability to metabolize a wide range of carbohydrates is complemented by its production of key metabolites. In children, *L. rhamnosus* produces several important metabolites, including lactic acid, acetic acid, and short-chain fatty acids (SCFAs) such as butyrate. These metabolites are crucial for maintaining a low pH in the gut, which inhibits the growth of pathogenic bacteria and fosters a healthy gut environment [37].

Additionally, *L. rhamnosus* metabolizes fucose through unique pathways, resulting in the production of lactic acid, 1,2-propanediol, acetic acid, formic acid, and carbon dioxide. These metabolites are vital for energy production and biomass formation, which distinguish *L. rhamnosus* from other lactic acid bacteria [37].

L. rhamnosus also possesses several genes responsible for synthesizing antimicrobial peptides, including bacteriocins, which inhibit the growth of pathogenic bacteria [38]. Furthermore, it produces EPSs that are crucial for biofilm formation. These EPSs enhance the bacterium's ability to adhere to intestinal surfaces and resist environmental stressors, such as bile salts and low pH [39].

The presence of genes encoding antioxidant enzymes, such as superoxide dismutase (SOD) and glutathione reductase, further contributes to its protective capabilities. These enzymes help protect both the bacterium and host cells from oxidative stress [39,40].

L. rhamnosus has also been shown to modulate host immune responses by interacting with dendritic cells and promoting the production of anti-inflammatory cytokines, such as IL-10. This immunomodulatory effect underscores its value in maintaining gut health [35,36].

Moreover, *L. rhamnosus* produces inosine, a metabolite with antioxidant, anti-inflammatory, anti-infective, and neuroprotective properties. The production of inosine is notably upregulated in *L. rhamnosus* compared to other *Lactobacillus* species, suggesting a unique metabolic profile that enhances its probiotic and postbiotic activities [41].

3.2. *Bifidobacterium infantis*

B. longum subsp. *infantis* is particularly adapted to the infant gut environment, largely due to its unique genomic configuration that enables the metabolism of human milk oligosaccharides (HMOs). The genomics of *B. infantis* reveal specialized adaptations for thriving in the infant gut, including unique metabolic pathways, genetic stability, and competitive advantages through bacteriocin production [42].

The genome of *B. infantis* contains a vast array of genes encoding glycosyl hydrolases and transport proteins that specifically target and degrade HMOs, providing a competitive advantage in the infant gut [43,44].

Additionally, *B. infantis* produces key metabolites, such as acetate and lactate, which have been shown to strengthen the intestinal barrier by enhancing tight junction integrity [45]. The bacterium's genome also contains genes for producing various short-chain fatty acids (SCFAs), which serve as energy sources for colonocytes and play a role in maintaining gut homeostasis [46]. Although the genome of *B. infantis* does not directly contain genes for producing indole-3-lactic acid (ILA), its production of ILA results from its ability to metabolize tryptophan present in the growth medium, especially when grown on HMOs [47].

Furthermore, *B. infantis* stands out among *Bifidobacterium* species for its numerous bacteriocin gene clusters, including lanthipeptides and thiopeptides. These bacteriocins provide a competitive edge in colonizing the infant gut by inhibiting harmful microorganisms [48].

In addition to its gastrointestinal benefits, *B. infantis* produces inosine, a metabolite of *B. infantis*, has been shown to exert cardioprotective effects. It mitigates cardiac inflammation and cell death during ischemia/reperfusion injury by activating the adenosine A2A receptor, which reduces pro-inflammatory cytokines and supports ATP generation through the purine salvage pathway [49].

3.3. *Streptococcus thermophilus*

S. thermophilus, a widely recognized probiotic bacterium found in fermented dairy products, plays a pivotal role in influencing gut health and host metabolism through its diverse metabolic activities. One of its primary metabolites is lactate, produced via the glycolysis pathway [50]. Lactate is crucial for lowering the pH in the gut, which not only facilitates milk coagulation during fermentation but also modulates the colon epithelium. This modulation affects the expression of various transporters and proteins involved in cell cycle regulation, thereby positively impacting gut health and function [51].

In addition to lactate, *S. thermophilus* produces a range of amino acids, including leucine, isoleucine, proline, aspartic acid, and tryptophan. These amino acids are essential for the bacterium's growth and significantly contribute to the sensory properties and quality of fermented milk products. The strain also generates various fatty acids and other metabolites, such as 2-hydroxybutyric acid, D-glycerol-D-galactose-heptanol, and hydra starch, which influence the flavor profile and overall characteristics of the fermentation process [52].

Metabolic pathways in *S. thermophilus* involve the breakdown and utilization of several amino acids, including cysteine, methionine, glutamate, glutamine, arginine, aspartate, asparagine, and alanine. These amino acids are integral to the synthesis of glutathione, a potent antioxidant that helps the bacterium combat oxidative stress and environmental challenges [53]. Moreover, *S. thermophilus* affects gut metabolism by altering tryptophan metabolism, leading to reduced levels of indole derivatives and increased production of serotonin. This metabolic shift has broader implications for mood regulation and gut–brain interactions [54].

Furthermore, *S. thermophilus* produces a range of antibiotic-like compounds and bactericidal proteins, such as bacteriocins, which are instrumental in reducing uremia and inhibiting the growth of pathogenic microbes. These antimicrobial properties enhance the probiotic benefits of *S. thermophilus*, making it a valuable contributor to gut health and a potent preventive measure against infections caused by harmful bacteria [55]. Overall, the metabolic versatility and probiotic attributes of *S. thermophilus* underscore its significance in both industrial applications and health-related functions.

3.4. *Lactobacillus acidophilus*

Lactobacillus acidophilus is a probiotic microorganism known for producing a range of metabolites that contribute to gut health, including conjugated linoleic acid (CLA), exopolysaccharides (EPSs), and bacteriocins (BACs). The production of these metabolites is influenced by several factors, such as initial pH, temperature, incubation time, yeast extract concentration, and the availability of free linoleic acid [56].

Bacteriocins, which are antimicrobial peptides produced by *L. acidophilus*, play a key role in inhibiting the growth of pathogenic bacteria [57]. Unlike conventional antibiotics, these bacteriocins have a relatively narrow spectrum of activity and can be degraded by proteases in the gastrointestinal tract. This may limit their direct antimicrobial efficacy but also reduces their potential to disrupt beneficial microbiota [58]. Certain bacteriocin-producing strains, such as *L. acidophilus* JCM1132, have been shown to alter gut microbiota composition in healthy mice, reducing inflammatory responses and potentially preventing metabolic diseases. This finding underscores the probiotic potential of bacteriocin-producing strains in modulating gut health [59].

Another important metabolite produced by *L. acidophilus* is valeric acid, a short-chain fatty acid (SCFA) that has been demonstrated to suppress the development of non-alcoholic fatty liver disease-associated hepatocellular carcinoma (NAFLD-HCC) by inhibiting specific cellular pathways [60].

L. acidophilus also produces several antigenic proteins, such as GroEL (HSP60), enolase, and transcription factors EF-Ts and EF-Tu. These proteins are recognized by serum IgG antibodies in children, particularly those with autoimmune conditions like type 1 diabetes and celiac disease, suggesting a complex interaction between the immune system and commensal bacteria [61].

The strain *L. acidophilus* 5e2 synthesizes exopolysaccharides composed of glucose, galactose, and glucosamine, which may promote gut health by fostering beneficial microbial communities. Additionally, Lactobacillus species produce biosurfactants such as surlactin, which have the capacity to reduce surface tension and inhibit pathogen adhesion, thereby maintaining a balanced microbiota and providing protection against infections [62,63].

3.5. *Saccharomyces boulardii*

S. boulardii, a probiotic yeast, is well known for its therapeutic benefits, particularly in gastrointestinal health. Recent research has expanded our understanding of the metabolites produced by *S. boulardii*, highlighting its potential applications in gastrointestinal disorders, cancer treatment, and as a microbial cell factory.

Saccharomyces cerevisiae var. *boulardii* synthesizes a diverse range of bioactive metabolites, including polyphenolic compounds such as vanillic acid, cinnamic acid, and phenyl ethyl alcohol, as well as essential nutrients like vitamin B6. These compounds contribute to its antioxidant capacity and provide a foundation for its anti-carcinogenic, antibacterial, antiviral, and general health-promoting properties [64,65]. Additionally, *S. boulardii* produces molecules like erythromycin and amphetamine, further enhancing its therapeutic profile.

One of the notable metabolites produced by *S. boulardii*, particularly by the strain *S. boulardii*-B508, is the *Saccharomyces* anti-inflammatory factor (SAIF). This factor has been shown to reduce the burden of *Mycobacterium intracellulare* in human macrophages by inducing apoptosis in infected cells and inhibiting IL-8 expression through the suppression of NF-κB activation, a key regulator of the human inflammatory response [66]. *S. boulardii* also synthesizes a phosphatase capable of dephosphorylating endotoxins such as the lipopolysaccharide (LPS) of *Escherichia coli*, thereby reducing their cytotoxic effects. Furthermore, it produces a 54-kDa serine protease that decreases intestinal permeability and inhibits the secretion of water and electrolytes, enhancing its ability to protect against bacterial toxins and exert anti-inflammatory effects in the gastrointestinal tract [67,68].

Moreover, *S. boulardii* generates high levels of acetic acid at 37 °C, a characteristic linked to unique mutations in the SDH1 and WHI2 genes, which are not found in *S. cerevisiae*. These genetic traits enable the yeast to thrive in acidic environments, providing resistance to gastric conditions and supporting its efficacy as a probiotic [69,70]. Additionally, *S. boulardii* produces a serine protease that cleaves *Clostridioides difficile* toxin A, stimulates the production of antibodies against this toxin, and modulates inflammatory responses by promoting anti-inflammatory molecules like peroxisome proliferator-activated receptor-gamma (PPAR-γ) [71].

In summary, the array of bioactive compounds produced by *S. boulardii*, including organic acids, enzymes, polyphenols, and proteases, underlines its diverse probiotic activities and therapeutic potential, making it a valuable tool in managing various health conditions.

4. Bioactive Metabolites Produced by Probiotics: Mechanisms of Action and Their Role in Enhancing Pediatric Health

Probiotics, particularly *Lactobacillus* and *Bifidobacterium* species, are increasingly used in pediatric healthcare for their benefits in managing gastrointestinal disorders and enhancing immune function [72,73]. Research suggests that the benefits of probiotics in children are generally species-specific rather than strain-specific [4]. Probiotics have proven effective in treating conditions like necrotizing enterocolitis (NEC), antibiotic-associated diarrhea, and *Helicobacter pylori* infections [74–76]. Beyond gastrointestinal health, probiotics support immune modulation, nutrient absorption, and anti-inflammatory responses, potentially reducing the frequency of infections, allergies, and inflammation-related conditions [77]. Strains like *Lactobacillus rhamnosus* GG have shown promise in alleviating allergic symptoms and promoting overall child health [78]. Understanding the bioactive metabolites and mechanisms of action of probiotics is crucial for their effective integration into pediatric care.

4.1. Immunomodulatory Metabolites

Probiotic bacteria, particularly strains of *Bifidobacterium* and *Lactobacillus*, produce a variety of immunomodulatory metabolites that play crucial roles in maintaining gut homeostasis and modulating the host immune system. These metabolites, which include short-chain fatty acids (SCFAs), bacteriocins, indole derivatives, and vitamins, have been shown to suppress inflammation, enhance microbial diversity, and improve intestinal barrier function by altering intestinal permeability and strengthening intercellular junctions [79]. Additionally, *Lactobacillus* and *Bifidobacterium* species are well known for their production of SCFAs and vitamins, which enhance intestinal barrier integrity and promote anti-inflammatory responses [80]. Moreover, gut microbial-derived metabolites such as polyamines, choline-derived compounds, and secondary bile acids have been identified as immunoregulatory molecules that specifically affect adaptive immune responses, particularly T helper 17 and regulatory T cells, thereby influencing health and disease outcomes [81].

These strains produce immunomodulatory metabolites that positively impact the immune system in children by enhancing innate immunity, regulating pro-inflammatory cytokine expression, and preventing tissue damage from excessive inflammatory responses [82,83]. Specifically, *Lactobacillus rhamnosus* GG has been shown to induce beneficial Th1 immunomodulatory effects, thus helping to manage conditions like cow's milk allergy and atopic dermatitis by promoting IL-10 production [84].

Additionally, probiotics modulate the infant microbiota, induce immune mediator production, and influence cytokine production by intestinal cells, showcasing their ability to shape the immune response [82]. These effects are strain-specific and impact the immune system through various pathways, such as suppressing inflammation via the NF-κB pathway and enhancing phagocytic activity [85]. Furthermore, probiotics have been associated with reducing allergic reactions by downregulating Th2-related responses, inhibiting pro-inflammatory cytokine production, and modulating immune system components, ultimately promoting anti-inflammatory and immunomodulatory effects [86]. For example, lactic acid bacteria (LAB) are particularly effective in preventing allergic diseases like atopic eczema in infants by enhancing the body's capacity to produce immune-enhancing cytokines such as interferon-gamma (IFN-γ) [87]. Specific probiotic strains also significantly increase the production of intestinal immunoglobulin A (IgA), which enhances mucosal immunity and protects against gastrointestinal infections [88]. Moreover, probiotics can influence systemic immunity by promoting the activity of natural killer (NK) cells and the differentiation of T-helper cells, both crucial for fighting infections and maintaining immune balance [89].

Lactobacillus gasseri TCI515, a probiotic strain, exemplifies this role by enhancing the expression of innate immunity-regulating genes and inhibiting pro-inflammatory cytokine gene expression, thereby improving innate immunity and preventing tissue damage from excessive inflammatory responses [83] Similarly, LAB fermentation of herbal medicines generates metabolites like exopolysaccharides, SCFAs, and bacteriocins, which have immunomodulatory properties and interact with the immune system, potentially boosting the innate immune response in children [90].

Overall, probiotic bacteria play a crucial role in producing metabolites that interact with and modulate the immune system, offering significant benefits for children's health and well-being.

4.2. Anti-Inflammatory Metabolites

Probiotics have been studied for their potential to produce anti-inflammatory metabolites, such as short-chain fatty acids (SCFAs) like butyrate and acetate, which may modulate immune responses and reduce inflammation in children [91]. However, the evidence is mixed and often varies depending on the specific strains used, the health status of the children, and the biomarkers measured. For example, a recent study involving a combination of *Lactobacillus acidophilus* and *Bifidobacterium lactis* demonstrated a significant reduction in inflammatory markers, such as MPIF-1 and MIP-3α, in children, suggesting an anti-inflammatory effect through immune modulation [92]. Similarly, another study found that synbiotic supplementation in overweight children led to decreased levels of inflammatory markers like tumor necrosis factor-α and interleukin-6, although these effects were associated with weight reduction rather than a direct anti-inflammatory action [93,94].

In children with various diseases, probiotics have shown potential in reducing systemic inflammation, particularly in conditions such as allergies, autoimmune diseases, and severe illnesses [94]. However, a study investigating the long-term effects of *Lactobacillus paracasei* supplementation during weaning found no significant impact on metabolic and inflammatory profiles at school age, suggesting that early probiotic intervention may not have lasting anti-inflammatory benefits [95]. Thus, the effectiveness of these biotics appears to be influenced by the health status of the children, with greater benefits observed in those with specific conditions, indicating that they may be more effective for certain groups.

Given that the molecular mechanisms underlying the anti-inflammatory effects of specific probiotic molecules are not well understood, a recent in vitro study demonstrated that secreted metabolites from *Bifidobacterium infantis* and *Lactobacillus acidophilus* exert anti-inflammatory effects in immature human enterocytes by modulating genes involved in immune response, cell survival, and NF-κB signaling. These metabolites reduce IL-6 and IL-8 production, suggesting their potential to mitigate inflammation in conditions such as necrotizing enterocolitis in children [96].

Additionally, another in vitro study evaluating the anti-inflammatory effects of biomolecules in probiotics commonly used in children showed that intestinal bacterial metabolites produced by *Bifidobacterium animalis* subsp. *lactis* LKM512 can suppress TNF-alpha production in J774.1 cells stimulated by lipopolysaccharide (LPS), suggesting that consumption of yogurt containing this strain may help reduce inflammatory cytokines produced by macrophages [97].

In conclusion, while probiotics and their metabolites show promising potential as anti-inflammatory agents in children, further investigation is needed to fully elucidate the molecular mechanisms underlying these effects and to determine optimal conditions for their application. Such insights could guide the development of targeted probiotic therapies, particularly for conditions like necrotizing enterocolitis, allergies, and autoimmune diseases, ultimately improving therapeutic outcomes for pediatric populations. A summary of the role of probiotic metabolites in children's health is shown in Figure 2.

4.3. Nutrient Absorption-Enhancing Metabolites

Probiotics have been shown to enhance nutrient absorption in children by producing beneficial metabolites that improve gut health and nutrient uptake. These metabolites include vitamins, minerals, and short-chain fatty acids (SCFAs), which play crucial roles in maintaining a healthy gut microbiome and improving overall health outcomes in children [4].

A recent study demonstrated that probiotics significantly increased the blood levels of vitamins and minerals such as vitamin D, vitamin A, calcium, zinc, and iron in children over a 10-week period. This finding suggests that probiotics can enhance the absorption of these essential nutrients, potentially improving the nutritional status and immunity of children [98]. Probiotics have also been shown to regulate lipid metabolism, which is crucial for nutrient absorption. In overweight or obese children, probiotics helped reduce levels of LDL cholesterol and leptin while increasing HDL cholesterol and adiponectin, indicating improved lipid profiles and metabolic health [99,100].

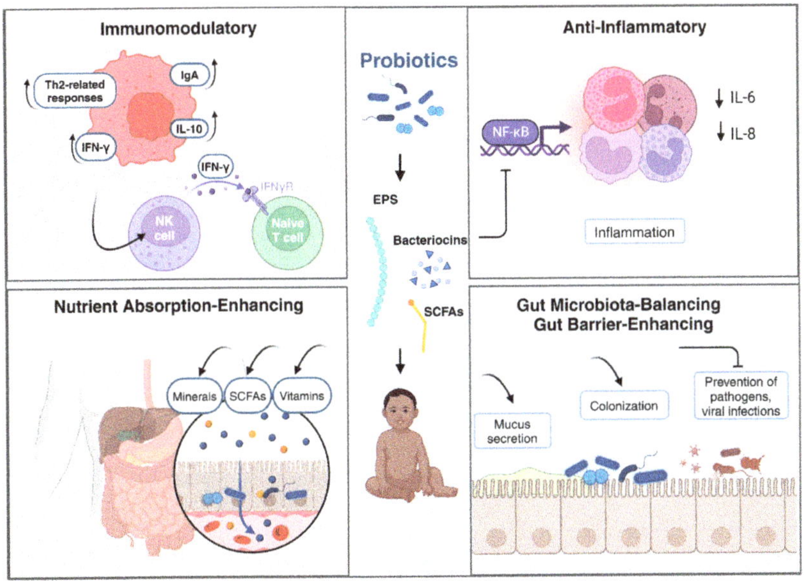

Figure 2. Role of probiotic metabolites in children's health. Short-chain fatty acids (SCFAs), exopolysaccharides (EPSs), Immunoglobulin A (IgA), interleukin-10 (IL-10), and interferon-gamma (IFN-γ) in immune cells. Nuclear factor kappa B (NF-κB) activation leading to reduced levels of interleukin 6 (IL 6) and interleukin-8 (IL-8).

The administration of prebiotic-enhanced lipid-based nutrient supplements (LNSs) in undernourished infants led to a significant increase in the production of SCFAs such as acetate, butyrate, and propionate. These SCFAs are known to enhance gut health and nutrient absorption by promoting a beneficial gut microbiota composition. Similarly, the use of multi-strain probiotics in obese children increased the abundance of beneficial bacteria like *Lactobacillus* spp. and *Bifidobacterium animalis*, which are associated with improved SCFA production and lipid metabolism [101]

A recent study assessed the effectiveness of probiotics in improving health outcomes for children with severe acute malnutrition (SAM) in the Democratic Republic of Congo. The findings suggested that probiotics contributed to better health metrics, including weight gain, shorter recovery times, and overall nutritional improvement. The study concluded that incorporating probiotics into rehabilitation protocols could enhance the recovery of malnourished children [102].

In vitro approaches have shown that a mixture of *Saccharomyces boulardii*, *Lactobacillus acidophilus*, *Lactobacillus rhamnosus*, and *Bifidobacterium breve*, in combination with the enzyme amylase, disrupts pathogenic gastrointestinal biofilms, improving nutrient absorption by enhancing permeability and increasing the penetration of proteins and vitamins through intestinal cell monolayers [103].

While the evidence supports the role of probiotics in enhancing nutrient absorption through the production of beneficial metabolites, the effects can vary based on the probiotic strains used, the duration of supplementation, and the health status of the children. Further studies are needed to optimize probiotic interventions for different pediatric populations and to fully understand the mechanisms behind these benefits.

4.4. Gut Microbiota-Balancing and Barrier-Enhancing Metabolites

A healthy microbiota involves maintaining an appropriate diversity and abundance of beneficial microorganisms that outnumber and outcompete potentially pathogenic or harmful ones [104]. When this delicate equilibrium is altered, the microbiota becomes imbalanced, a condition known as dysbiosis [105]. In children, dysbiosis is associated with alterations of gut function such as diarrhea [106], infant colic [107] or even autoimmune and atopic diseases such as asthma or rhinitis [108].

Balancing the gut microbiota is essential for optimal digestive health in children, and probiotic metabolites contribute significantly to this balance [109,110]. The bioactive molecules produced by probiotics help maintain a healthy gut environment by improving the balance of the gut microbiota [4]. By supporting a diverse and beneficial microbial community in the gut, these metabolites promote digestive health, prevent conditions such as diarrhea and colic, and contribute to overall well-being in children.

Short-chain fatty acids (SCFAs) are primary metabolites produced during the fermentation of dietary fibers by probiotic bacteria in the colon. These SCFAs, particularly butyrate, play a critical role in modulating the composition and activity of the gut microbiota [111]. Butyrate, for instance, is known for its ability to promote the growth of beneficial bacteria such as *Faecalibacterium prausnitzii*, a bacterium associated with anti-inflammatory properties [112]. At the same time, SCFAs help lower the pH of the gut environment, creating conditions unfavorable for pathogenic bacteria [113], thereby reducing the risk of gastrointestinal infections like diarrhea and colic in children.

In addition to promoting a healthy microbial balance, SCFAs have been found to enhance the production of mucin, a key component of the gut mucus layer that provides a protective barrier against pathogens [114]. By increasing mucin production, SCFAs help to fortify the gut lining, reducing the likelihood of pathogens adhering to and invading the epithelial cells of the intestine. Furthermore, butyrate serves as a primary energy source for colonocytes [115,116], the cells lining the colon, thereby maintaining the integrity and function of the intestinal barrier [117].

Tight junctions are critical components that regulate the permeability of the gut barrier. Probiotic-derived metabolites like butyrate have been demonstrated to upregulate the expression of tight junction proteins, including claudin and occluding [118], which are essential for maintaining the integrity of the epithelial barrier. By enhancing these tight junctions, probiotics help prevent "leaky gut," a condition characterized by increased intestinal permeability that allows toxins, microbes, and other harmful substances to enter the bloodstream, potentially triggering systemic inflammation and immune responses.

Additionally, certain metabolites produced by probiotics, such as lactate and hydrogen peroxide, have antimicrobial properties that contribute to maintaining a healthy gut barrier. Lactate, for example, can inhibit the growth of pathogenic bacteria by lowering the pH of the gut environment [119]. Meanwhile, hydrogen peroxide, produced by various microbial strains, has direct bactericidal effects against pathogens further protecting the integrity of the gut barrier [120].

5. Types and Mechanisms of Action of Probiotic Metabolites in Children

Probiotic metabolites, the bioactive compounds produced by beneficial microorganisms, play a crucial role in promoting and maintaining health in children. These metabolites, which include short-chain fatty acids (SCFAs), bacteriocins, exopolysaccharides (EPSs), and vitamins, among others, exert a variety of beneficial effects on the host. Understanding the types and mechanisms of action of these probiotic metabolites is essential for leveraging their full therapeutic potential. This section delves into the various types of probiotic metabolites and explores the specific mechanisms through which they exert their health-promoting effects in children. A summary of the main metabolites produced by probiotic bacteria in children is shown in Figure 3.

5.1. Vitamins

Probiotic bacteria play a vital role in synthesizing essential vitamins that significantly impact the overall health and development of children. These beneficial bacteria, particularly certain strains of lactic acid bacteria (LAB) and bifidobacteria, can produce various vitamins necessary for multiple physiological processes, such as growth, development, and immune function [121]. In this section, we explore the types of vitamins produced by probiotic bacteria, their impact on children's health, and the mechanisms through which these vitamins exert their beneficial effects.

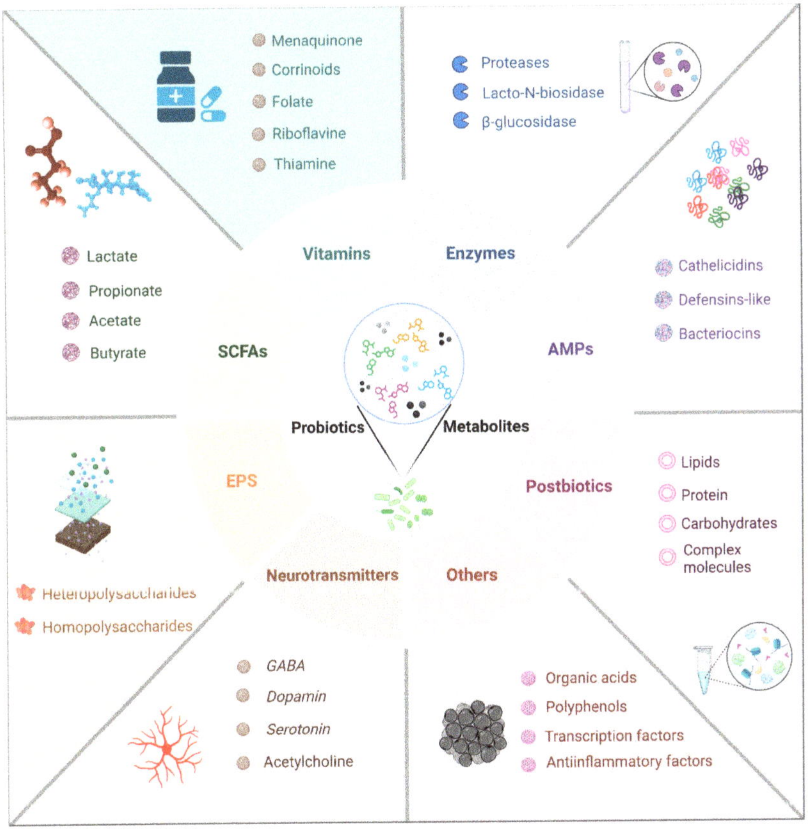

Figure 3. Major metabolites produced by probiotic bacteria in children. SCFAs: short-chain fatty acids; EPSs: exopolysaccharides; GABA: gamma-aminobutyric acid; AMPS: antimicrobial peptides.

5.1.1. Vitamin B Complex

Probiotic bacteria are known to synthesize several B vitamins, including B1 (thiamine), B2 (riboflavin), B3 (niacin), B5 (pantothenic acid), B6 (pyridoxine), B7 (biotin), B9 (folate), and B12 (cobalamin). These vitamins are critical for energy metabolism, DNA synthesis, red blood cell formation, and the proper functioning of the nervous system. Ensuring adequate levels of these vitamins is crucial for children's growth and development, as they support numerous metabolic pathways and physiological processes [15,121].

Research has shown that specific probiotic strains, such as those within the genera *Lactobacillus* and *Bifidobacterium*, are highly effective at producing B vitamins. For example, *Lactobacillus fermentum* has been identified as a robust producer of folate and vitamin B12, achieving production levels of up to 801.79 µg/mL for folate [122]. These strains can enhance the nutritional profile of various foods, such as fermented dairy products, making them an excellent source of B vitamins.

Oligosaccharides, which are prebiotic fibers that serve as a food source for probiotic bacteria, have been found to enhance the vitamin-producing capabilities of certain probiotic strains. Studies indicate that the presence of oligosaccharides can increase the synthesis of

B vitamins by enhancing the bacteria's properties, such as hydrophobicity, auto-aggregation, and biofilm formation [123].

Several probiotic strains, including *Lactobacillus reuteri*, *Lactobacillus acidophilus*, *Streptococcus thermophilus*, and *Lactobacillus rhamnosus* GG, are recognized for their ability to produce essential B-group vitamins, such as vitamin B12, riboflavin, folate, and thiamine, which are vital for numerous metabolic and physiological processes [124]. For example, *L. reuteri* strains produce corrinoids related to vitamin B12, while *L. acidophilus* strains increase riboflavin levels during fermentation, enhancing the nutritional value of foods like soymilk [125,126]. Additionally, *S. thermophilus* can boost the production of bioactive folate forms [127].

Leveraging these probiotic strains' vitamin-producing capabilities could help develop fortified, vitamin-rich functional foods, which are especially beneficial in promoting children's growth and development.

5.1.2. Vitamin K

Vitamin K, particularly in its K2 form (menaquinone), is another vital nutrient synthesized by probiotic bacteria [128]. This vitamin plays a crucial role in blood clotting, bone health, and cardiovascular function. The synthesis of vitamin K2 by gut bacteria is especially significant in children, who may have limited dietary intake of this nutrient due to selective eating habits or inadequate nutrition [129,130].

The probiotic bacterium *Bacillus clausii* has been demonstrated to produce vitamin K2, effectively correcting coagulation disorders in infants following antibiotic treatment by normalizing prothrombin levels [131]. Similarly, *Lactococcus lactis* is a known producer of vitamin K2, particularly in the context of fermented foods, where different strains produce varying amounts of the vitamin. Under specific cultivation conditions, these bacteria can enhance the delivery of vitamin K2 through extracellular vesicles [132].

Additionally, certain neonatal gut bacteria, such as *Enterobacter agglomerans*, *Serratia marcescens*, and *Enterococcus faecium*, have been identified as producers of menaquinones, a form of vitamin K2. This suggests that even in early life, the gut microbiota can contribute significantly to vitamin K production, which is essential for proper blood clotting processes [133]. However, antibiotic treatments in infants can disrupt the gut microbiota, leading to decreased vitamin K production. Studies have shown that administering probiotics like *B. clausii* can help restore normal prothrombin levels, highlighting the role of gut bacteria in vitamin K synthesis and coagulation [131].

There are far fewer bacteria known to produce vitamin K than those that synthesize B vitamins; however, as the gut microbiota of infants and children continues to be decoded, more microorganisms capable of synthesizing vitamin K are likely to be discovered, given its importance in pediatric health.

It is important to highlight that probiotics not only produce vitamins but also enhance their absorption and bioavailability. For instance, a study involving children aged 8–13 showed that probiotic supplementation significantly increased blood levels of vitamins D and A compared to a placebo group over a 10-week period, suggesting that probiotics may improve the absorption and serum concentrations of essential vitamins [98]. Another study reported improvements in gut health markers, such as increased bifidobacteria and reduced inflammatory markers, in children receiving prebiotics and vitamin supplements, demonstrating the potential of probiotics to enhance both gut health and vitamin status [134].

The ability of probiotics to synthesize and enhance the absorption of vitamins has broad implications for child health. By ensuring the production of essential nutrients, such as B vitamins and vitamin K, probiotics support various physiological processes, including energy metabolism, immune response, and bone mineralization. Furthermore, vitamins produced by commensal bacteria may influence immune responses, suggesting roles beyond basic nutrition, such as regulating gene expression and enhancing nutrient absorption [15,31]. A summary of the probiotic bacteria involved in the synthesis of vitamins is described in Table 1.

Table 1. Key vitamins produced by probiotic bacteria.

Probiotic Strain	Vitamin Produced	Mechanism
Lactobacillus fermentum [122]	Folate (B9), Vitamin B12 (Cobalamin)	Synthesis of folate and B12
Lactobacillus reuteri (DCM 20016, JCM1112, CRL1324, CRL1327) [31]	Corrinoids (related to Vitamin B12)	Production of corrinoids
Lactobacillus acidophilus (ATCC314, FTDC 8833) [121]	Riboflavin (B2)	Enhances riboflavin production
Streptococcus thermophilus (ABM5097) [15]	5-Methyltetrahydrofolate (5-MTHF) (Folate)	Increases folate production
Lactobacillus rhamnosus GG [31]	Folate (B9), Riboflavin (B2), Thiamine (B1)	Produces and releases folate and riboflavin efficiently; low production of intracellular thiamine
Bacillus clausii [131]	Vitamin K2 (Menaquinone)	Production of vitamin K2
Lactococcus lactis [132]	Vitamin K2 (Menaquinone)	Production of vitamin K2
Enterobacter agglomerans, Serratia marcescens, Enterococcus faecium [133]	Menaquinones (Vitamin K2)	Contributes to vitamin K production in the neonatal gut
Various strains (*Lactobacillus, Bifidobacterium*) [123]	B Vitamins (B1, B2, B3, B5, B6, B7, B9, B12), Vitamin K	Utilizes oligosaccharides to enhance hydrophobicity, auto-aggregation, and biofilm formation, thus improving B vitamin production
Lactobacillus gasseri (FTDC 8131) [135]	Riboflavin (B2)	Interacts with riboflavin; context suggests strain-dependent variability in production or consumption of the vitamin
Bifidobacterium strains (*B. longum, B. bifidum*) [31]	Thiamine (B1)	Low but significant production of intracellular thiamine without extracellular synthesis; does not produce folates or riboflavin
Children consuming probiotics [98]	Vitamin D, Vitamin A	Probiotics enhance absorption and serum concentrations of vitamins

5.2. Short-Chain Fatty Acids (SCFAs)

Short-chain fatty acids (SCFAs) such as acetate, propionate, and butyrate are key metabolites that play a vital role in supporting gut health in children. SCFAs are essential for maintaining a healthy digestive system [136]. One of their primary functions is to nourish the cells lining the intestines, known as colonocytes [115], which are crucial for forming a robust gut barrier that prevents harmful substances from entering the bloodstream [115]. Additionally, SCFAs are critical for the development, regulation, and maturation of the immune system [137,138]. Notably, butyrate, a specific SCFA, has demonstrated anti-inflammatory properties [136], helping to protect against conditions such as colic, gastrointestinal infections, and inflammatory bowel disease [139].

SCFAs also contribute to maintaining an optimal gut environment by lowering the pH [140,141], which supports the growth of beneficial bacteria while inhibiting gut pathogens [142]. Furthermore, these molecules enhance the gut barrier by promoting the production of mucins, substances that protect the gut lining and prevent the adherence and invasion of pathogens [143].

Certain probiotic strains, such as *Lactobacillus* and *Bifidobacterium*, are known to produce SCFAs in the gut [31,144–146]. These beneficial bacteria are commonly found in fermented foods like yogurt or in dietary supplements specifically formulated for children. When children consume foods rich in probiotics, these bacteria proliferate in the gut and produce SCFAs by fermenting dietary fibers present in fruits, vegetables, and whole grains [147,148].

5.3. Antimicrobial Peptides

Antimicrobial peptides (AMPs) are small molecules that serve as natural antibiotics produced by the body and certain probiotic bacteria [149,150]. AMPs represent a promising avenue for enhancing child health by modulating the gut microbiota and providing a natural defense against pathogens [150]. These bioactive molecules, which include bacteriocins and other peptide-based antimicrobials, have been shown to selectively inhibit harmful bacteria while promoting the growth of beneficial microbes [151]. Furthermore, unlike conventional antibiotics, AMPs have a lower propensity for developing resistance, making them a safer alternative for managing pediatric infections and promoting overall gut health [152–154].

Although there is currently limited direct evidence for the use of probiotic-derived antimicrobial peptides (AMPs) specifically in pediatric health, several lines of research suggest their potential as novel therapeutic agents for children. Extensive studies have demonstrated the broader benefits of probiotic-derived AMPs for human health, particularly their antimicrobial, immunomodulatory, and microbiota-regulating properties [151,155–157]. Additionally, AMPs from other sources have proven effective in managing pediatric infections, and clinical trials have highlighted cationic antimicrobial peptides as promising alternatives for treating infections in neonates and children, especially in cases involving antibiotic-resistant pathogens [158]. Further research into the use of probiotic-derived AMPs could open up new pathways for developing innovative, safe, and effective treatments to manage infections and enhance health outcomes in children.

5.4. Enzymes

The enzymes produced by probiotics hold significant potential in improving children's health by supporting essential digestive functions and enhancing nutrient absorption [159]. These beneficial microorganisms produce enzymes that break down complex nutrients, making them more accessible for a child's developing digestive system. As children's gut microbiota continues to evolve, the enzymatic activity provided by probiotics plays a pivotal role in promoting a balanced gut environment, which is crucial for immune function, growth, and overall well-being [160].

One of these enzymes derived from probiotics is β-glucosidase, an enzyme important for breaking down complex carbohydrates into simpler sugars, which can enhance the probiotic's effectiveness in the gut [161]. *Bifidobacterium* species, which are early colonizers of the infant gut, produce β-galactosidase to metabolize milk-based diets. This enzyme cleaves the glycosidic bond in lactose through hydrolysis, producing the monosaccharides glucose and galactose, which are essential for energy production and growth in infants [162]. In addition to hydrolysis, β-galactosidase can transfer galactosyl units to other sugar molecules through transgalactosylation. This activity forms galactooligosaccharides (GOSs), which are beneficial prebiotics that promote the growth of healthy gut microbiota [162–164]. In infants with nutritional disorders such as celiac disease and cystic fibrosis, β-galactosidase activity can be affected. For example, lactase activity is significantly reduced in celiac disease, while hetero-β-galactosidase activity remains relatively stable [165].

Similarly, lacto-N-biosidase (LNBase) plays a crucial role in the digestion of human milk oligosaccharides (HMOs). This enzyme, primarily found in *Bifidobacterium* spp., facilitates the breakdown of complex sugars into simpler forms that can be utilized by the infant's gut microbiota [160]. For instance, LNBase from *Bifidobacterium bifidum* (LnbB) is essential for the degradation of HMOs, specifically lacto-N-tetraose, into lacto-N-biose I and lactose, which is vital for the early life microbiota in infants [166].

LNBase operates via a substrate-assisted catalytic mechanism, with a unique metabolic pathway specific to lacto-N-biose I, a major core structure in HMOs [167]. Its activity is modulated by specific amino acids, such as His263, which plays a critical role in the catalytic process by altering the pKa of the acid/base residue [166]. The stability of LNBase during digestion and its ability to modulate gut microbiota composition, increasing the abundance of beneficial bacteria like *B. bifidum*, underscores its potential in alleviating infant food allergies and promoting overall gut health [160].

The enzymatic activities of probiotics contribute significantly to infant and child health by enhancing digestion, nutrient absorption, and gut microbiota composition. These findings highlight the importance of continued research into probiotic-derived enzymes as valuable tools for improving pediatric health outcomes.

5.5. Exopolysaccharides (EPSs)

EPSs are complex carbohydrate polymers secreted by probiotic bacteria such as *Lactobacillus*, *Bifidobacterium*, *Streptococcus*, *Weissella*, during metabolic processes [168]. Structurally, exopolysaccharides consist of repeating units of glucose, galactose, mannose and rhamnose, which can form homo or heteropolysaccharides, often linked by glycosidic bonds [169]. These polysaccharides can either be covalently anchored to the cell surface, forming capsular polysaccharides [170], or be secreted into the extracellular environment, leading to the formation of a mucilaginous layer. This latter form plays a critical role in the development of bacterial biofilms [170,171]. These types of polysaccharides can vary widely in composition, branching, and molecular weight, contributing to their diverse functional properties and activities [172].

One of the most notable is their bifidogenic activity, which refers to their ability to selectively stimulate the growth of specific members of the infant gut microbiota, specifically of *Bifidobacterium* species [173]. In infants, *Bifidobacterium* typically accounts for about 90% of intestinal bacteria [174].

The bifidogenic effects of EPSs arise from their complex carbohydrate structures, which are resistant to digestion in the upper gastrointestinal tract. Upon reaching the colon, these EPSs serve as prebiotics. Lv et al. (2024) isolated and purified EPSs from *Bifidobacterium animalis* subsp. Lactis SF (SF-EPS) from the feces sample of a healthy infant), and their probiotic potential was evaluated in vitro. SF-EPS regulated the gut microbiota by increasing the relative abundances of *Faecalibacterium*, *Anaerostipes*, and *Bifidobacterium*, while reducing the abundance of *Enterobacter* and *Klebsiella*. Furthermore, SF-EPS enhanced the production of SCFAs by intestinal microorganisms. These findings suggest that SF-EPS may serve as a potential prebiotic for use in functional foods [14].

Exopolysaccharides also exhibit potent biological activities, including antioxidant properties, free radical scavenging, and the reduction of oxidative stress [14,175]. Tarique et al. (2024) showed the antioxidant potential of EPSs from *Enterococcus faecium* and *S. thermophilus*. This potential could be due to the different sugars and their arrangements which can affect the ability of EPSs to interact with and neutralize free radicals [26].

EPSs also have immunoregulatory properties. EPSs from *Bifidobacterium longum* subsp. infantis E4 demonstrated significant immunomodulatory and anti-inflammatory effects in vitro. These EPSs enhanced macrophage activity and reduced inflammatory markers, indicating potential benefits for immune health in infants [176]. Additionally, some studies have found that EPSs from probiotics have other properties, such as antitumor activity [177–179], antibacterial activity [180], antiviral protection [181], and lipid regulation potential [182,183]. EPSs also enhance adherence and subsequent colonization of microflora on host cells [14].

In summary, their combination of different biological activities makes EPSs promising candidates for functional foods and therapeutic agents aimed at improving children's health.

5.6. Neurotransmitters

5.6.1. Gamma-Aminobutyric Acid (GABA)

GABA is a crucial neurotransmitter for children, playing a key role in brain development, emotional regulation, and cognitive functions like learning and memory. It helps maintain a balance between excitatory and inhibitory signals in the brain, which is important for reducing anxiety, promoting restful sleep, and managing stress [184–186]. Adequate

GABA levels support healthy neural circuit formation, contribute to emotional stability, and help children with behavioral control, making GABA essential for overall mental and physical well-being [187,188]. Interestingly, these levels have been found to increase with age [189], and this difference in GABA levels seems to be related to how fast children can learn. Experimental results indicate children show more flexible GABA-related inhibitory processing compared to adults, allowing for quicker adaptation to stabilize learning [186].

While *Lactobacillaceae* species are recognized as primary producers of GABA, Bifidobacterium species have been identified as the most efficient GABA producers [187]. Notably, *gad* genes responsible for GABA synthesis are also found in other probiotic strains [190,191]. The production of GABA by these microorganisms has suggested a link between the gut microbiota and neurological health [191]. GABA is produced by the enzyme glutamate decarboxylase (GAD), which requires pyridoxal-5′-phosphate (PLP) and works through the irreversible α-decarboxylation of l-glutamate, consuming one cytoplasmic proton [192].

Bifidobacterium species such as *Bifidobacterium adolescentis*, *Bifidobacterium dentium*, and *Bifidobacterium longum* have demonstrated the capacity to synthesize GABA through the decarboxylation of glutamate by GAD enzymes. This process not only helps in maintaining the gut's acid–base balance but also contributes to the pool of bioactive GABA within the host [190,193–196].

GABA-producing probiotic strains significantly impact the gut–brain axis, a communication network between the gastrointestinal tract and the central nervous system, involving the gut microbiota, immune system, enteric nervous system (ENS), and central nervous system [197]. GABA can modulate ENS activity—affecting gut motility, secretion, and blood flow [198]—and may influence the central nervous system via the vagus nerve, impacting neuropsychiatric conditions [199,200]. Increased GABA levels in the gut have been linked to reduced stress and anxiety-like behaviors [185], with specific bacteria like Bacteroides and GABA-producing Bifidobacterium strains playing roles in mental health and reducing systemic inflammation associated with mood disorders and neurodegenerative diseases [191,200].

Previous studies have demonstrated that the gut microbiota diversity in children with autism spectrum disorder (ASD) undergoes significant changes, with alterations in *Bifidobacterium* being linked to the severity of ASD [201]. Infants at higher risk for ASD have a decreased abundance of *Bifidobacterium* and an increased abundance of *Clostridium* and *Klebsiella* compared to those at lower risk. Additionally, fecal GABA levels were lower in infants with a higher likelihood of ASD, with GABA levels showing a positive correlation with Bifidobacterium [202].

Analyses on children with attention deficit hyperactivity disorder (ADHD) have found lower levels of GABA and reduced presence of lactic acid bacteria. These bacteria are known to be involved in the production of GABA, suggesting that the reduction of lactic acid bacteria in the gut of infants with ADHD could be associated with lower GABA levels [203,204].

Furthermore, clinical trials have suggested that probiotics containing GABA-producing *Bifidobacterium* or *Lactobacillaceae* species can be effective in treating gastrointestinal issues and enhancing overall mental health in children [205–207].

5.6.2. Other Neurotransmitters Produced by Probiotics

In addition to GABA, probiotic microorganisms are gaining recognition for their capacity to produce various neurotransmitters, which can significantly influence host health via the gut–brain axis. A well-documented example is serotonin (5-HT), a neurotransmitter crucial for regulating mood, appetite, and sleep. Certain probiotic strains, such as *Enterococcus* and *Streptococcus*, have been shown to synthesize serotonin, suggesting their potential impact on emotional and psychological well-being [208].

Dopamine, another essential neurotransmitter associated with reward and motivation, is produced by probiotic strains like *Bacillus* and *Lactobacillus* species [209]. The production of dopamine by these probiotics could affect neurological functions and behaviors, highlighting a possible route through which to influence the host's nervous system. Similarly,

Lactobacillus plantarum is known to produce acetylcholine, which plays a critical role in learning, memory, and muscle activation [210]. The presence of acetylcholine in the gut underscores the complex communication pathways of the gut–brain axis.

Probiotics like *Escherichia coli* and *Bacillus subtilis* have also demonstrated the ability to synthesize norepinephrine, a neurotransmitter involved in alertness and the body's "fight or flight" response [211]. This ability suggests that these microorganisms could help modulate stress and mood responses. Additionally, *Lactobacillus reuteri* is known to produce histamine, which is critical for immune responses and gut motility [212]. The production of glutamate, an excitatory neurotransmitter important for synaptic plasticity, has been observed in strains of *Lactobacillus* and *Bifidobacterium* [213].

Overall, the capacity of probiotics to synthesize these neurotransmitters supports their potential role in managing neuropsychiatric and gastrointestinal disorders by modulating the gut–brain axis, providing a promising avenue for therapeutic intervention.

5.7. Bioactive Postbiotic Fractions

Postbiotics, defined by the International Scientific Association for Probiotics and Prebiotics (ISAPP) as "inanimate microorganisms and/or their components that confer health benefits to the host", include microbial cells, metabolites, and fermentation byproducts [214]. Postbiotics are key mediators of microbiota–host interactions. Unlike probiotics, which are live microorganisms, postbiotics do not contain live bacteria but consist of beneficial byproducts released during the microorganisms' life cycle [215,216]. Postbiotics can be classified based on their composition, including lipids (e.g., butyrate, propionate), proteins (e.g., lactocepin), carbohydrates (e.g., polysaccharides, teichoic acids), vitamins, organic acids (e.g., lactic acid), and complex molecules (e.g., lipoteichoic acids) [215–217]. Additionally, they can be categorized by their physiological functions, such as immunomodulatory, anti-inflammatory, hypocholesterolemic, anti-obesity, antihypertensive, anti proliferative, and antioxidant effects (Figure 4) [218,219]. These compounds are widely present in foods such as yogurt, kefir, and pickled vegetables, and can also be intentionally applied in functional foods such as infant formulas [215,216,220].

The significance of postbiotics lies in their potential health benefits and greater stability compared to probiotics. Because postbiotics lack live bacteria, they are more stable and have a longer shelf life, making them easier to store and use in therapeutic and nutritional applications [221]. The unique structure of postbiotics can exert a range of beneficial effects on the host through diverse cellular and molecular mechanisms. Postbiotics play a crucial role in children's health by supporting gut balance, boosting immunity, and promoting digestive health. They help reduce inflammation, enhance nutrient absorption, and protect against infections, contributing to overall well-being in growing children [215,222]. Additionally, the use of postbiotics is associated with fewer safety concerns than probiotics, as there is no risk of infection from live bacteria. This makes them particularly suitable for vulnerable populations, including infants, the elderly, and immunocompromised individuals [215,216,223].

Postbiotics exhibit significant immunomodulatory and anti-inflammatory effects, playing a critical role in enhancing both innate and adaptive immune responses. One of the key mechanisms by which postbiotics exert immunomodulatory effects is through the activation of immune receptors such as Toll-like receptors (TLRs). For instance, heat-inactivated *Lactobacillus casei* has been shown to enhance macrophage-mediated innate immunity by increasing the transcription of TLRs (TLR2, TLR3, TLR4, and TLR9) and stimulating pro-inflammatory cytokines, which strengthen the body's defense against infections [224]. Moreover, postbiotics derived from *Lactobacillus gasseri* TMC0356 have demonstrated a more potent effect on immune activity than their probiotic counterparts, inducing higher levels of interleukin-12 (IL-12) in macrophages [225].

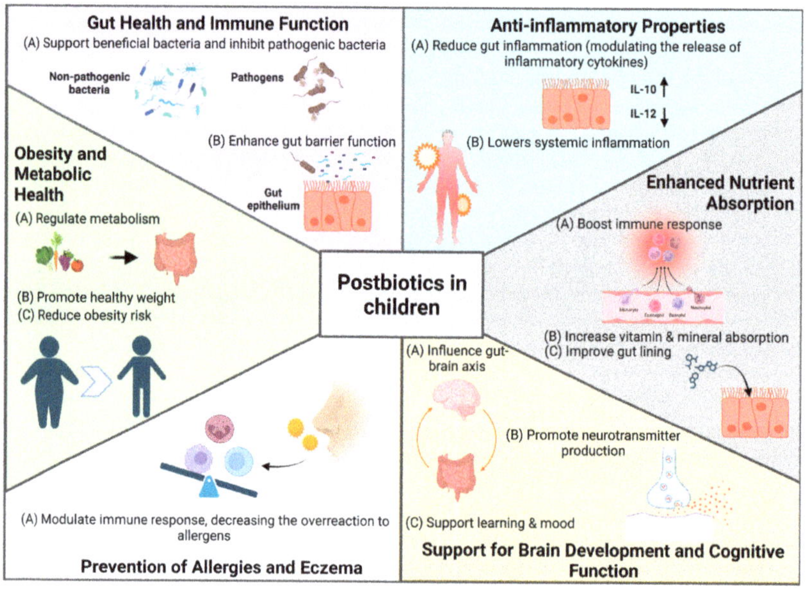

Figure 4. Postbiotics and their potential positive effects in children.

The anti-inflammatory properties of postbiotics are equally noteworthy. By regulating cytokine production, postbiotics can help reduce inflammatory responses. For example, supernatants of *Faecalibacterium prausnitzii* have been shown to alleviate colitis in mice by increasing the production of the anti-inflammatory cytokine IL-10 while reducing the pro-inflammatory cytokine IL-12, likely through the inhibition of NF-κB activation [226]. Additionally, *Lactobacillus paracasei* B21060 postbiotics have demonstrated protective effects against inflammation caused by *Salmonella* in human colon tissues [227]. Another example of the immunomodulatory action of postbiotics involves *Lactobacillus reuteri* 17938, which promotes the production of the anti-inflammatory cytokine IL-10 in dendritic cells, leading to an enhanced regulatory T-cell response [228].

Recent research has increasingly focused on the role of fermented infant formulas and their postbiotic components in enhancing infant health, particularly in reducing the severity of gastrointestinal and allergic conditions. A study by Béghin et al. (2021) explored the effects of a fermented infant formula (FF) with *Bifidobacterium breve* C50 and *Streptococcus thermophilus* O65 combined with prebiotic oligosaccharides on gut microbiota composition and immune function in healthy term infants. The findings indicated that this combination led to a gut microbiota composition and metabolic activity more similar to that of breastfed infants, with a significant increase in secretory IgA (SIgA) levels, highlighting its potential to enhance early immune defense [229]. In contrast, a study by Thibault et al. (2004) focused on the impact of an FF with the same probiotic bacteria on the incidence and severity of acute diarrhea in healthy infants aged 4 to 6 months [230].

In summary, postbiotics offer a range of immunomodulatory and anti-inflammatory benefits, making them promising therapeutic agents. Their ability to regulate immune responses, reduce inflammation, and support overall immune health highlights their potential in managing diverse health [215,231,232].

6. Clinical Applications and Health Implications

A comprehensive analysis published by Dronkers et al. in 2020 on 1341 studies retrieved from the ClinicalTrials.gov database (using the search term "probiotics") showed that 56% were conducted in the USA or Europe; around 100 studies have been registered annually since 2010, and this number has been increasing in recent years [233]. The vast majority of these studies were interventional (95.6%), but almost half were in healthy participants (43.8%), and only 31.8% of those studies were in children (from birth to 17 years old). Of the 852 studies that could be analyzed, *Lactobacillus rhamnosus* GG (LGG) was the probiotic strain most frequently registered (146 studies), followed by *Bifidobacterium animalis* ssp. *lactis* BB12 with 55 studies, while VSL#3, a consortium of three different *Bifidobacteria*, four *Lactobacillus*, and one *Streptococcus thermophilus* strains, was the most registered multispecies preparation, featuring in 74 studies.

Over the last five years, ClinicalTrials.gov has listed 730 studies, of which 167 were in children and 157 were interventional studies in children. Two studies were withdrawn (one in colic and other in peanut allergy); five were prematurely terminated (two in respiratory infections, one in atopic dermatitis, one in surgical procedures and one in healthy children), and forty-eight were completed, addressing a range of conditions including allergic diseases, respiratory infections, cystic fibrosis, obesity, ADHD, gastrointestinal diseases, and autism spectrum disorder (ASD). This indicates that 84 studies (53.5%) are registered but still ongoing. Despite this focus, the volume of published clinical trials in children has been decreasing according to the Pubmed.gov database, from 69 publications in 2020 and 2021 to 55 articles in 2022 and just 46 in the last year.

In recent years, the role of probiotics in managing ASD has been studied, as the potential benefits of probiotics have been suggested by explorations of the gut–brain axis, thus offering hope for improved management strategies [234]. However, a meta-analysis published in the Journal of Medical Microbiology in 2022 showed no significant benefit of probiotics for ASD treatment [235]. This study highlighted that the lack of standardization in trials—such as variations in strains, dosages, and protocols—complicates our ability to conduct comprehensive meta-analyses with robust findings.

Current Practice Guidelines

Recommendations for the clinical application of probiotics still show discrepancies between existing guidelines, particularly those for acute gastroenteritis in children, as summarized in Table 2.

American Academy of Pediatrics (AAP): Probiotics such as *L. rhamnosus* GG and *S. boulardii* are recommended for reducing the duration of acute gastroenteritis in children. However, routine use in healthy children is not broadly advised. Probiotics may help prevent antibiotic-associated diarrhea and necrotizing enterocolitis (NEC) in preterm infants, though caution is advised in immunocompromised patients [236].

European Society for Pediatric Gastroenterology, Hepatology and Nutrition (ESPGHAN): Probiotics for conditions like acute gastroenteritis and prevention of NEC in preterm infants are supported, but strain-specific efficacy and proper quality control are emphasized [237].

The British Society of Pediatric Gastroenterology, Hepatology and Nutrition (BSPGHAN) aligns with ESPGHAN, emphasizing selective use of probiotics for gastrointestinal issues and the importance of strain-specific evidence in children [236].

For instance, the American Gastroenterological Association (AGA) is against the use of probiotics [236,238]. These inconsistencies arise from several factors, including variability in literature review processes—such as incomplete searches or differing inclusion criteria—which can lead to conflicting recommendations. Additionally, some guidelines may place more weight on single randomized controlled trials (RCTs), while others incorporate a broader evidence base [239]. Differences in study populations, including socioeconomic levels, medical issues, and geographical features, also influence these recommendations. Potential conflicts of interest (COI) bias from industry sponsorship may further impact guideline development. Lastly, the evolving nature of research means that guidelines may not always reflect the most current evidence. These factors highlight the need for more standardized approaches and regular updates to clinical guidelines to support better decision making and foster ongoing research [239].

Table 2. Current Guidelines on Probiotics for most common conditions (modified from [240]).

Disorder	Probiotic Strain	Recommended Dose	Evidence Level **
Acute gastroenteritis [reduced the risk of diarrhea lasting ≥48 h; reduced the mean duration of diarrhea [241]]	Probiotics as a general group	N/A	1
	L. rhamnosus GG [242]	$\geq 10^{10}$ cfu/day, for 5–7 days	1
	S. boulardii * [243]	250–750 mg/day, for 5–7 days	1
	L. reuteri DSM 17938 [244]	1×10^8 to 4×10^8 cfu/day, for 5 days	1
	L. rhamnosus 19070-2 & *L. reuteri* DSM 12246 [245]	2×10^{10} cfu for each strain/day, for 5 days	1
	B. lactis B94 [246]	5×10^{10} cfu once daily, for 5 days	3
	L. paracasei B21060 [247]	2.5×10^9 cfu, twice daily, for 5 days	3
	L. rhamnosus strains 573L/1; 573L/2; 573L/3 [248]	1.2×10^{10} cfu, twice daily, for 5 days	3
	L. delbrueckii var. *bulgaricus*, *L. acidophilus*, *S. thermophilus*, *B. bifidum* (LMG-P17550, LMG-P 17549, LMG-P 17503, LMG-F 17500) [249]	10^9 cfu, 10^9 cfu, 10^9 cfu, 5×10^8 cfu/dose, for 5 days	3
	B. lactis Bi-07, *L. rhamnosus* HN001, and *L. acidophilus* NCFM [250]	Then, 1×10^{10} cfu once a day, for the duration of diarrhea plus 7 days	3
Prevention of AAD (reduced risk of AAD [251])	Probiotics as a general group	N/A	1
	S. boulardii * [252]	≥5 billion cfu per day, for the duration of antibiotic treatment	1
	L. rhamnosus GG [253]	≥5 billion cfu per day, for the duration of antibiotic treatment	1
	Multispecies probiotic (*Bifidobacterium bifidum W23*, *B. lactis W51*, *Lactobacillus acidophilus W37*, *Lactobacillus acidophilus W55*, *Lacticaseibacillus paracasei W20*, *Lactoplantibacillus plantarum W62*, *Lacticaseibacillus rhamnosus W71*, and *Ligilactobacillus salivarius W24*) [254]	10 billion cfu per day, for the duration of antibiotic treatment and for 7 days after	3
	L. rhamnosus (strains E/N, Oxy, and Pen) [255]	2×10^{10} cfu, twice daily, for the duration of antibiotic treatment	3
Prevention of *C. difficile* diarrhea	*S. boulardii* * [252]	250–500 mg	1
Prevention of nosocomial diarrhea	*L. rhamnosus GG* [256,257]	At least 10^9 cfu/day, for the duration of the hospital stay	1
Prevention of necrotizing enterocolitis [258–260]	Systematic reviews and meta-analyses (>10,000 neonates) of RCTs		1
	L. rhamnosus GG [261]	From 1×10^9 to 6×10^9 cfu	1
	B. infantis BB-02, *B. lactis* BB-12, and *S. thermophilus* TH-4 [261]	3.0 to 3.5×10^8 cfu (of each strain)	1
	B. animalis subsp. *lactis* Bb-12 or B94 [261]	5×10^9 cfu	3
	L. reuteri ATCC 55730 * or DSM 17938 * this strain is no longer available. [261,262]	1×10^8 cfu (various regimens)	1
	B. longum subsp. *infantis* ATCC 15697 + *L. acidophilus* ATCC 4356 [262,263]	125 mg/kg/dose twice daily with breast milk until discharge	3
	B. longum subsp. *longum* 35624 + *L. rhamnosus* GG [263]	5×10^8 cfu and 5×10^8 cfu, respectively	3

Table 2. *Cont.*

Disorder	Probiotic Strain	Recommended Dose	Evidence Level **
Helicobacter pylori infection [264–268]	Probiotics as a general group *S. boulardii* * [269,270]	 500 mg	1 1
Infantile colic [271–276]	Probiotics as a general group *L. reuteri* DSM 17938 [277,278] *B. lactis* Bb12 [279,280] *L. rhamnosus* 19070-2 and *L. reuteri* 12246 [281] *L. paracasei* DSM 24733, *L. plantarum* DSM 24730, *L. acidophilus* DSM 24735, *L. delbrueckii* subsp. *bulgaricus* DSM 24734, *B. longum* DSM 24736, *B. breve* DSM 24732, and *B. infantis* DSM 24737, and *S. thermophilus* DSM 24731 [282]	N/A 10^8 cfu/day for at least 21 days 1×10^9 cfu/day, for 21–28 days 250×10^6 cfu, respectively, for 28 days 5 billion cfu, for 21 days	1 1 2 3 3
Infantile colic prevention	*L. reuteri* DSM 17938 [283]	10^8 cfu/day, to newborns each day for 90 days	1
Functional abdominal pain/IBS	*L. reuteri* DSM 17938 [284,285] *L. rhamnosus GG* [284,286]	10^8 cfu to 2×10^8 cfu/day 10^9 cfu to 3×10^9 cfu twice daily	1 1
Ulcerative colitis [287]	Probiotics as a group A mixture of 8 strains (*L. paracasei* DSM 24733, *L. plantarum* DSM 24730, *L. acidophilus* DSM 24735, *L. delbrueckii* subsp. *bulgaricus* DSM 24734, *B. longum* DSM 24736, *B. infantis* DSM 24737, *B. breve* DSM 24732, and *S. thermophilus* DSM 2[illegible]7), as adjuvant therapy or in those intolerant to 5-ASA [288]	N/A Daily dosages: 4–6 y (17–23 kg): 450 billion; 7–9 y (24–33 kg): 900 billion; 11–14 y (34–53 kg): 1350 billion; 15–17 y (54–66 kg): 1800 billion.	1 3
Pouchitis	A mixture of 8 strains (*L. paracasei* DSM 24733, *L. plantarum* DSM 24730, *L. acidophilus* DSM 24735, *L. delbrueckii* subsp. *bulgaricus* DSM 24734, *B. longum* DSM 24736, *B. infantis* DSM 24737, *B. breve* DSM 24732, and *S. thermophilus* DSM 2[illegible]7) [289,290]	Daily dosages: 4–6 y (17–23 kg): 450 billion; 7–9 y (24–33 kg): 900 billion; 11–14 y (34–53 kg): 1350 billion; 15–17 y (54–66 kg): 1800 billion.	3

* Most studies with the strain *S. boulardii* CNCM I-745; ** (1) systematic review or meta-analysis of randomized control trials; (2) randomized control trials; (3) quasi-experimental studies; (4) non-experimental studies.

7. Challenges of Using Bioactive Molecules from Probiotics for Pediatric Diseases

The use of probiotics in children, particularly in the most vulnerable—such as preterm infants, immunocompromised, or those with underlying health conditions—is still limited by safety concerns. Although many studies confirm the safety of probiotics, there have been isolated cases of sepsis or gastrointestinal mucormycosis linked to contaminated probiotic products, highlighting the need for stringent quality control measures during manufacturing [291]. Furthermore, the reported risk of cross-colonization with probiotic strains in neonatal intensive care units (NICUs) suggested that despite probiotics offered potential benefits, they must be administered cautiously, supported by comprehensive safety assessments [291]. The Agency for Healthcare Research and Quality reviewed 622 studies on probiotic safety and found that while there is no evidence of increased risk from randomized controlled trials, the literature lacks systematic reporting on adverse events, making it difficult to assess rare risks with confidence [292]. Therefore, it is crucial to implement standardized protocols for safety outcomes in clinical trials, ensuring comprehensive and transparent data collection. Enhanced guidelines would enable a more accurate assessment of the benefits/risks associated with probiotic use and support informed decision making in both research and clinical practice.

Translating promising findings from preclinical and clinical studies into clinical guidelines is an imminent challenge. Research conducted in diverse geographic regions may not be generalizable due to variations in genetics, diet, sanitation, and endemic enteropathogens [293]. Systematic reviews have thus far failed to recommend specific strains due to inconsistency in results and lack of standardized protocols regarding species, dosage, and administration duration [233]. Variations in study designs, including differences in probiotic strains, dosages, and treatment durations, have led to inconsistent results and have hindered the development of clear guidelines. Moreover, the absence of standardized protocols and regulatory frameworks complicates the approval and adoption of probiotics for pediatric use. To overcome these obstacles, more well-designed randomized controlled trials (RCTs) and meta-analyses are needed to establish evidence-based guidelines for probiotic use in children.

Clinical trials in pediatric populations present additional ethical challenges and special regulatory compliance. Moreover, differences in diet, genetics, and environmental factors across different populations limit the generalization of the research findings. Regulatory agencies should develop frameworks that promote innovation while ensuring safety and efficacy, facilitating global standardization in probiotic research and applications.

A significant limitation in many studies is the underexplored interaction between different probiotic species, which may work synergistically. Most preclinical and clinical investigations focus on isolated strains, neglecting the complex interactions within the entire gut microbiota. The gut microbiome is a unique ecosystem for each individual, harboring diverse microorganisms. Interactions among multiple probiotic species can enhance therapeutic efficacy through synergistic mechanisms [294,295].

Current research primarily emphasizes individual probiotics, yet the potential for multi-strain synergism warrants further investigation. Combining strains with complementary roles, such as one that strengthens gut barrier integrity and another that modulates immune responses, could improve treatment strategies for complex pediatric conditions like inflammatory bowel disease (IBD) and allergies [296–298]. However, studying these interactions requires long-term research to navigate the microbiome's complexities.

Furthermore, probiotics do not function in isolation; their introduction can alter the native microbial community, yielding both positive and negative effects. Understanding these dynamics is crucial for developing effective, personalized probiotic therapies. Regulatory agencies should promote frameworks that foster innovation while ensuring safety and efficacy in probiotic research.

8. Future Perspectives and Opportunities

Successful integration of probiotic-derived bioactive molecules in the pediatric population is also an opportunity to enhance treatment options and potentially improve outcomes. In addressing the previously discussed challenges, we can make substantial strides toward safer, more effective, and widely accepted probiotic-based therapies for pediatric populations.

8.1. Fostering of Clinical Research

The current situation of clinical trials with probiotics in the pediatric population reveals a growing interest in the field, while the decrease in the number of publications might indicate stronger review criteria for the available data. Future allocation of resources, multidisciplinary teams, and active collaboration are essential to strengthen clinical research oriented to answer relevant and important questions.

8.2. Evidence-Based Clinical Guidelines

The current data available and their variability offer an opportunity to foster broader and global collaborations among healthcare providers, their institutions, research-oriented organizations, academia, and industry. Therefore, creating stronger evidence for the use of probiotics in children should be a must. Development of updated guidelines (based on evidence) that consider regional differences in diet, genetics, and environmental factors will further facilitate personalized treatments.

8.3. Strengthening Ethical Frameworks for Pediatric Research

Pediatric research presents unique opportunities, despite its ethical challenges, which can be addressed by developing innovative frameworks that balance patient protection with medical progress. Regulatory agencies can set international ethical standards for probiotic trials, including informed consent and independent ethical reviews. Transparent data sharing, and international collaboration can build trust and accelerate the transition from research to clinical practice.

8.4. Enhancing Professional Education and Interdisciplinary Collaboration

Promoting literacy in basic sciences (e.g., biochemistry and microbiology) among healthcare professionals through interdisciplinary workshops, seminars, and updated curricula fosters innovation and informed decision making. Integrating probiotic science into medical education and offering continuous professional development may lead to the discovery of new probiotic strains or bioactive compounds with unique benefits for children, thus enhancing clinical practice and research.

Author Contributions: Conceptualization, L.P.G. and C.B.-O.; writing—original draft preparation, L.P.G., C.B.-O., S.E.C.-P., J.Z.-M., E.T. and C.E.; writing—review and editing, L.P.G., C.B.-O., S.E.C.-P., J.Z.-M., E.T. and C.E.; supervision, L.P.G. and C.B.-O.; project administration, L.P.G. and C.B.-O. All authors have read and agreed to the published version of the manuscript.

Funding: This work was supported by the Universidad San Francisco de Quito through funding ID: 17884: "Development of a simple and efficient system for prototyping metabolic pathways in the probiotic yeast *S. boulardii*".

Informed Consent Statement: Not applicable.

Conflicts of Interest: The authors declare no conflicts of interest.

References

1. Buonocore, G. Microbiota and gut immunity in infants and young children. *Global Pediatr.* **2024**, *9*, 100202. [CrossRef]
2. Skvortsova, O.V.; Migacheva, N.B.; Kaganova, T.I.; Lyamin, A.V.; Antipov, V.A. Role of gut microbiota and selected metabolic products in the development of childhood obesity. *BCCM* **2024**, *17*, 81–87. [CrossRef]
3. Hoskinson, C.; Medeleanu, M.V.; Reyna, M.E.; Dai, D.L.Y.; Chowdhury, B.; Moraes, T.J.; Mandhane, P.J.; Simons, E.; Kozyrskyj, A.L.; Azad, M.B.; et al. Antibiotics taken within the first year of life are linked to infant gut microbiome disruption and elevated atopic dermatitis risk. *J. Allergy Clin. Immunol.* **2024**, *154*, 131–142. [CrossRef]

4. Sharma, S.; Singh, N. Probiotics for Children. In *Probiotics: A Comprehensive Guide to Enhance Health and Mitigate Disease*; CRC Press: Boca Raton, FL, USA, 2024; pp. 95–128. ISBN 9781003452249.
5. Bompalli, L.K. The Next-Gen Innovative Therapeutic Potential of Probiotics: Insights into Gut Microbiota Modulation and Health Promotion. *Int. J. Res. Anal. Rev.* **2024**. [CrossRef]
6. Eastwood, J.; van Hemert, S.; Poveda, C.; Elmore, S.; Williams, C.; Lamport, D.; Walton, G. The effect of probiotic bacteria on composition and metabolite production of faecal microbiota using in vitro batch cultures. *Nutrients* **2023**, *15*, 2563. [CrossRef]
7. Jang, H.J.; Lee, N.-K.; Paik, H.-D. A narrative review on the advance of probiotics to metabiotics. *J. Microbiol. Biotechnol.* **2024**, *34*, 487–494. [CrossRef]
8. Alsharairi, N.A. Therapeutic Potential of Gut Microbiota and Its Metabolite Short-Chain Fatty Acids in Neonatal Necrotizing Enterocolitis. *Life* **2023**, *13*, 561. [CrossRef]
9. Golpour, F.; Abbasi-Alaei, M.; Babaei, F.; Mirzababaei, M.; Parvardeh, S.; Mohammadi, G.; Nassiri-Asl, M. Short chain fatty acids, a possible treatment option for autoimmune diseases. *Biomed. Pharmacother.* **2023**, *163*, 114763. [CrossRef]
10. Calvo, L.N.; Greenberg, R.G.; Gray, K.D. Safety and Effectiveness of Probiotics in Preterm Infants with Necrotizing Enterocolitis. *Neoreviews* **2024**, *25*, e193–e206. [CrossRef]
11. Sharif, S.; Meader, N.; Oddie, S.J.; Rojas-Reyes, M.X.; McGuire, W. Probiotics to prevent necrotising enterocolitis in very preterm or very low birth weight infants. *Cochrane Database Syst. Rev.* **2023**, *7*, CD005496. [CrossRef]
12. Bartram, E.; Asai, M.; Gabant, P.; Wigneshweraraj, S. Enhancing the antibacterial function of probiotic *Escherichia coli* Nissle: When less is more. *Appl. Environ. Microbiol.* **2023**, *89*, e0097523. [CrossRef] [PubMed]
13. Thuy, T.T.D.; Lu, H.-F.; Bregente, C.J.B.; Huang, F.-C.A.; Tu, P.-C.; Kao, C.-Y. Characterization of the broad-spectrum antibacterial activity of bacteriocin-like inhibitory substance-producing probiotics isolated from fermented foods. *BMC Microbiol.* **2024**, *24*, 85. [CrossRef]
14. Lv, H.; Teng, Q.; Chen, J.; Peng, L.; Ren, Z.; Ma, L.; Yang, W.; Yu, B.; Wu, Z.; Wan, C. Probiotic potential of a novel exopolysaccharide produced by *Bifidobacterium animalis* subsp. Lactis SF. *LWT* **2024**, *193*, 115764. [CrossRef]
15. LeBlanc, J.G.; Laiño, J.E.; del Valle, M.J.; de Giori, G.S.; Sesma, F.; Taranto, M.P. B-Group Vitamins Production by Probiotic Lactic Acid Bacteria. In *Biotechnology of Lactic Acid Bacteria: Novel Applications*; Mozzi, F., Raya, R.R., Vignolo, G.M., Eds.; John Wiley & Sons, Ltd.: Chichester, UK, 2015; pp. 279–296. ISBN 9781118868386.
16. Chowdhury, S.R. Psychobiotics and gut–Brain axis—A new paradigm for improving mental health. In *Probiotics: A Comprehensive Guide to Enhance Health and Mitigate Disease*; CRC Press: Boca Raton, FL, USA, 2024; pp. 146–158. ISBN 9781003452249.
17. Lozano, J.; Fabius, S.; Fernández-Ciganda, S.; Urbanavicius, J.; Piccini, C.; Scorza, C.; Zunino, P. Beneficial effect of GABA-producing Lactiplantibacillus strain LPB145 isolated from cheese starters evaluated in anxiety- and depression-like behaviours in rats. *Benef. Microbes* **2024**, *15*, 465–479. [CrossRef] [PubMed]
18. Ambat, J.; Antonio, J.; Bala, A.; Suntornsaratoon, P.; Flores, J.; Li, W.V.; Jones, A. Metabolomic and transcriptomic correlative analyses in germ free mice link probiotic-associated metabolites to host intestinal fatty acid metabolism and b-oxidation. *Physiology* **2024**, *39*, 1503. [CrossRef]
19. Indira, M.; Venkateswarulu, T.C.; Abraham Peele, K.; Nazneen Bobby, M.; Krupanidhi, S. Bioactive molecules of probiotic bacteria and their mechanism of action: A review. *3 Biotech* **2019**, *9*, 306. [CrossRef]
20. Shealy, N.G.; Yoo, W.; Byndloss, M.X. Colonization resistance: Metabolic warfare as a strategy against pathogenic Enterobacteriaceae. *Curr. Opin. Microbiol.* **2021**, *64*, 82–90. [CrossRef]
21. Roy D'Souza, S.; Singh, S.; Ravi, L. Secondary metabolites produced from symbiotic microbes. In *Microbial Symbionts*; Elsevier: Amsterdam, The Netherlands, 2023; pp. 803–830. ISBN 9780323993340.
22. Kumar, M.K.; Morya, S. Probiotics: Role of bioactives in foods and dietary supplements. *Pharma Innov.* **2022**, *11*, 2592–2597. [CrossRef]
23. Sharma, D.; Rashid, G.; Sharma, L. Probiotics. In *Probiotics: A Comprehensive Guide to Enhance Health and Mitigate Disease*; CRC Press: Boca Raton, FL, USA, 2024; pp. 325–343. ISBN 9781003452249.
24. Baldassare, M.A.; Bhattacharjee, D.; Coles, J.D.; Nelson, S.; McCollum, C.A.; Seekatz, A.M. Butyrate enhances Clostridioides difficile sporulation in vitro. *J. Bacteriol.* **2023**, *205*, e0013823. [CrossRef]
25. Arbulu, S.; Kjos, M. Revisiting the multifaceted roles of bacteriocins: The multifaceted roles of bacteriocins. *Microb. Ecol.* **2024**, *87*, 41. [CrossRef]
26. Tarique, M.; Ali, A.H.; Kizhakkayil, J.; Liu, S.-Q.; Oz, F.; Dertli, E.; Kamal-Eldin, A.; Ayyash, M. Exopolysaccharides from Enterococcus faecium and Streptococcus thermophilus: Bioactivities, gut microbiome effects, and fermented milk rheology. *Food Chem. X* **2024**, *21*, 101073. [CrossRef]
27. Shenderov, B.A.; Sinitsa, A.V.; Zakharchenko, M.M.; Lang, C. Metabolic relationship between the host and its gut microbiota. In *Metabiotics: Present State, Challenges and Perspectives*; Springer International Publishing: Cham, Switzerland, 2020; pp. 15–21. ISBN 978-3-030-34166-4.
28. Oleskin, A.V.; Shenderov, B.A. Neuromodulatory effects and targets of the SCFAs and gasotransmitters produced by the human symbiotic microbiota. *Microb. Ecol. Health Dis.* **2016**, *27*, 30971. [CrossRef] [PubMed]
29. Engevik, M.A.; Morra, C.N.; Röth, D.; Engevik, K.; Spinler, J.K.; Devaraj, S.; Crawford, S.E.; Estes, M.K.; Kalkum, M.; Versalovic, J. Microbial metabolic capacity for intestinal folate production and modulation of host folate receptors. *Front. Microbiol.* **2019**, *10*, 2305. [CrossRef] [PubMed]

30. Chávarri, M.; Diez-Gutiérrez, L.; Marañón, I.; del Carmen Villarán, M.; Barrón, L.J.R. The role of probiotics in nutritional health: Probiotics as nutribiotics. In *Probiotics in the Prevention and Management of Human Diseases*; Elsevier: Amsterdam, The Netherlands, 2022; pp. 397–415. ISBN 9780128237335.
31. LeBlanc, J.G.; Chain, F.; Martín, R.; Bermúdez-Humarán, L.G.; Courau, S.; Langella, P. Beneficial effects on host energy metabolism of short-chain fatty acids and vitamins produced by commensal and probiotic bacteria. *Microb. Cell Fact.* **2017**, *16*, 79. [CrossRef]
32. Duan, H.; Wang, L.; Huangfu, M.; Li, H. The impact of microbiota-derived short-chain fatty acids on macrophage activities in disease: Mechanisms and therapeutic potentials. *Biomed. Pharmacother.* **2023**, *165*, 115276. [CrossRef]
33. Abavisani, M.; Ebadpour, N.; Khoshrou, A.; Sahebkar, A. Boosting vaccine effectiveness: The groundbreaking role of probiotics. *J. Agric. Food Res.* **2024**, *16*, 101189. [CrossRef]
34. Bradford, G.; Asgari, B.; Smit, B.; Hatje, E.; Kuballa, A.; Katouli, M. The Efficacy of Selected Probiotic Strains and Their Combination to Inhibit the Interaction of Adherent-Invasive *Escherichia coli* (AIEC) with a Co-Culture of Caco-2:HT29-MTX Cells. *Microorganisms* **2024**, *12*, 502. [CrossRef]
35. You, L.; Lv, R.; Jin, H.; Ma, T.; Zhao, Z.; Kwok, L.-Y.; Sun, Z. A large-scale comparative genomics study reveals niche-driven and within-sample intra-species functional diversification in Lacticaseibacillus rhamnosus. *Food Res. Int.* **2023**, *173*, 113446. [CrossRef]
36. Ceapa, C.; Davids, M.; Ritari, J.; Lambert, J.; Wels, M.; Douillard, F.P.; Smokvina, T.; de Vos, W.M.; Knol, J.; Kleerebezem, M. The Variable Regions of Lactobacillus rhamnosus Genomes Reveal the Dynamic Evolution of Metabolic and Host-Adaptation Repertoires. *Genome Biol. Evol.* **2016**, *8*, 1889–1905. [CrossRef]
37. Cheong, Y.E.; Kim, J.; Jin, Y.-S.; Kim, K.H. Elucidation of the fucose metabolism of probiotic Lactobacillus rhamnosus GG by metabolomic and flux balance analyses. *J. Biotechnol.* **2022**, *360*, 110–116. [CrossRef]
38. Zhang, J.; Li, K.; Bu, X.; Cheng, S.; Duan, Z. Characterization of the anti-pathogenic, genomic and phenotypic properties of a Lacticaseibacillus rhamnosus VHProbi M14 isolate. *PLoS ONE* **2023**, *18*, e0285480. [CrossRef]
39. Talbi, C.; Elmarrakchy, S.; Youssfi, M.; Bouzroud, S.; Belfquih, M.; Sifou, A.; Bouhaddou, N.; Badaoui, B.; Balahbib, A.; Bouyahya, A.; et al. Bacterial exopolysaccharides: From production to functional features. *Prog. Micobes Mol. Biol.* **2023**, *6*. [CrossRef]
40. Zheng, J.; Ahmad, A.A.; Yang, Y.; Liang, Z.; Shen, W.; Feng, M.; Shen, J.; Lan, X.; Ding, X. Lactobacillus rhamnosus CY12 Enhances Intestinal Barrier Function by Regulating Tight Junction Protein Expression, Oxidative Stress, and Inflammation Response in Lipopolysaccharide-Induced Caco-2 Cells. *Int. J. Mol. Sci.* **2022**, *23*, 11162. [CrossRef] [PubMed]
41. Spaggiari, L.; Pedretti, N.; Ricchi, F.; Pinetti, D.; Campisciano, G.; De Seta, F.; Comar, M.; Kenno, S.; Ardizzoni, A.; Pericolini, E. An Untargeted Metabolomic Analysis of *Lactobacillus* (L.) *rhamnosus*, *L. acidophilus*, *L. plantarum* and *L. reuteri* Reveals an Upregulated Production of Inosine from *L. rhamnosus*. *Preprints* **2024**. [CrossRef]
42. Tso, L.; Bonham, K.S.; Fishbein, A.; Rowland, S.; Klepac-Ceraj, V. Targeted High-Resolution Taxonomic Identification of Bifidobacterium longum subsp. infantis Using Human Milk Oligosaccharide Metabolizing Genes. *Nutrients* **2021**, *13*, 2833. [CrossRef]
43. Román, L.; Melis-Arcos, F.; Pröschle, T.; Saa, P.A.; Garrido, D. Genome-scale metabolic modeling of the human milk oligosaccharide utilization by Bifidobacterium longum subsp. infantis. *mSystems* **2024**, *9*, e0071523. [CrossRef]
44. Wong, C.B.; Huang, H.; Ning, Y.; Xiao, J. Probiotics in the New Era of Human Milk Oligosaccharides (HMOs): HMO Utilization and Beneficial Effects of Bifidobacterium longum subsp. infantis M-63 on Infant Health. *Microorganisms* **2024**, *12*, 1014. [CrossRef]
45. Reens, A.L.; Cosetta, C.M.; Saur, R.; Trofimuk, O.; Brooker, S.L.; Lee, M.L.; Sun, A.K.; McKenzie, G.J.; Button, J.E. Tunable control of B. infantis abundance and gut metabolites by co-administration of human milk oligosaccharides. *Gut Microbes* **2024**, *16*, 2304160. [CrossRef]
46. Sela, D.A.; Chapman, J.; Adeuya, A.; Kim, J.H.; Chen, F.; Whitehead, T.R.; Lapidus, A.; Rokhsar, D.S.; Lebrilla, C.B.; German, J.B.; et al. The genome sequence of Bifidobacterium longum subsp. infantis reveals adaptations for milk utilization within the infant microbiome. *Proc. Natl. Acad. Sci. USA* **2008**, *105*, 18964–18969. [CrossRef]
47. Ehrlich, A.M.; Pacheco, A.R.; Henrick, B.M.; Taft, D.; Xu, G.; Huda, M.N.; Mishchuk, D.; Goodson, M.L.; Slupsky, C.; Barile, D.; et al. Indole-3-lactic acid associated with Bifidobacterium-dominated microbiota significantly decreases inflammation in intestinal epithelial cells. *BMC Microbiol.* **2020**, *20*, 357. [CrossRef]
48. Yu, D.; Pei, Z.; Chen, Y.; Wang, H.; Xiao, Y.; Zhang, H.; Chen, W.; Lu, W. Bifidobacterium longum subsp. infantis as widespread bacteriocin gene clusters carrier stands out among the Bifidobacterium. *Appl. Environ. Microbiol.* **2023**, *89*, e0097923. [CrossRef] [PubMed]
49. Zhang, H.; Wang, J.; Shen, J.; Chen, S.; Yuan, H.; Zhang, X.; Liu, X.; Yu, Y.; Li, X.; Gao, Z.; et al. Prophylactic supplementation with Bifidobacterium infantis or its metabolite inosine attenuates cardiac ischemia/reperfusion injury. *iMeta* **2024**, *3*, e220. [CrossRef] [PubMed]
50. Kapse, N.; Pisu, V.; Dhakephalkar, T.; Margale, P.; Shetty, D.; Wagh, S.; Dagar, S.; Dhakephalkar, P.K. Unveiling the Probiotic Potential of Streptococcus thermophilus MCC0200: Insights from In Vitro Studies Corroborated with Genome Analysis. *Microorganisms* **2024**, *12*, 347. [CrossRef] [PubMed]
51. Rau, M.H.; Gaspar, P.; Jensen, M.L.; Geppel, A.; Neves, A.R.; Zeidan, A.A. Genome-Scale Metabolic Modeling Combined with Transcriptome Profiling Provides Mechanistic Understanding of Streptococcus thermophilus CH8 Metabolism. *Appl. Environ. Microbiol.* **2022**, *88*, e0078022. [CrossRef]
52. Li, Y.; Wang, Y.; Li, B.; Hou, B.; Hung, W.; He, J.; Jiang, Y.; Zhang, Y.; Man, C. Streptococcus thermophilus JM905-Strain Carbon Source Utilization and Its Fermented Milk Metabolic Profile at Different Fermentation Stages. *Foods* **2023**, *12*, 3690. [CrossRef]

53. Qiao, Y.; Liu, G.; Leng, C.; Zhang, Y.; Lv, X.; Chen, H.; Sun, J.; Feng, Z. Metabolic profiles of cysteine, methionine, glutamate, glutamine, arginine, aspartate, asparagine, alanine and glutathione in Streptococcus thermophilus during pH-controlled batch fermentations. *Sci. Rep.* **2018**, *8*, 12441. [CrossRef]
54. Yu, P.; Jiang, Y.; Pan, Y.; Guo, M.; Yang, B.; Liu, X.; Zhao, J.; Zhang, H.; Chen, W. Altered microbiome and metabolome features provide clues in understanding strain-specific regulation of Streptococcus thermophilus in the host. *Preprints* **2022**. [CrossRef]
55. Vitetta, L.; Llewellyn, H.; Oldfield, D. Gut Dysbiosis and the Intestinal Microbiome: Streptococcus thermophilus a Key Probiotic for Reducing Uremia. *Microorganisms* **2019**, *7*, 228. [CrossRef]
56. Amiri, S. Co-production of parabiotic metabolites by Lactobacillus acidophilus LA5 and Bifidobacterium animalis subsp. lactis BB12 in dairy effluents. *Coll. Res. Libr.* **2021**, *4*, 66–76. [CrossRef]
57. Rani, P.; Tiwari, S.K. Health benefits of bacteriocins produced by probiotic lactic acid bacteria. In *Microbial Biomolecules*; Elsevier: Amsterdam, The Netherlands, 2023; pp. 97–111. ISBN 9780323994767.
58. Deliorman Orhan, D. Bacteriocins produced by probiotic microorganisms. In *Advances in Probiotics*; Elsevier: Amsterdam, The Netherlands, 2021; pp. 277–291. ISBN 9780128229095.
59. Wang, G.; Yu, Y.; Garcia-Gutierrez, E.; Jin, X.; He, Y.; Wang, L.; Tian, P.; Liu, Z.; Zhao, J.; Zhang, H.; et al. Lactobacillus acidophilus JCM 1132 Strain and Its Mutant with Different Bacteriocin-Producing Behaviour Have Various in Situ Effects on the Gut Microbiota of Healthy Mice. *Microorganisms* **2019**, *8*, 49. [CrossRef]
60. Lau, H.C.-H.; Zhang, X.; Ji, F.; Lin, Y.; Liang, W.; Li, Q.; Chen, D.; Fong, W.; Kang, X.; Liu, W.; et al. Lactobacillus acidophilus suppresses non-alcoholic fatty liver disease-associated hepatocellular carcinoma through producing valeric acid. *EBioMedicine* **2024**, *100*, 104952. [CrossRef]
61. Prangli, A.-L.; Utt, M.; Talja, I.; Sepp, E.; Mikelsaar, M.; Rajasalu, T.; Uibo, O.; Tillmann, V.; Uibo, R. Antigenic proteins of Lactobacillus acidophilus that are recognised by serum IgG antibodies in children with type 1 diabetes and coeliac disease. *Pediatr. Allergy Immunol.* **2010**, *21*, e772–e779. [CrossRef] [PubMed]
62. Brzozowski, B.; Bednarski, W.; Dziuba, B. Functional properties of *Lactobacillus acidophilus* metabolites. *J. Sci. Food Agric.* **2009**, *89*, 2467–2476. [CrossRef]
63. Reid, G.; Heinemann, C.; Velraeds, M.; van der Mei, H.C.; Busscher, H.J. [31] Biosurfactants produced by Lactobacillus. In *Biofilms*; Methods in Enzymology; Elsevier: Amsterdam, The Netherlands, 1999; Volume 310, pp. 426–433. ISBN 9780121822118.
64. Datta, S.; Timson, D.J.; Annapure, U.S. Antioxidant properties and global metabolite screening of the probiotic yeast Saccharomyces cerevisiae var. boulardii. *J. Sci. Food Agric.* **2017**, *97*, 3039–3049. [CrossRef] [PubMed]
65. Ansari, F.; Alian Samakkhah, S.; Bahadori, A.; Jafari, S.M.; Ziaee, M.; Khodayari, M.T.; Pourjafar, H. Health-promoting properties of Saccharomyces cerevisiae var. boulardii as a probiotic; characteristics, isolation, and applications in dairy products. *Crit. Rev. Food Sci. Nutr.* **2023**, *63*, 457–485. [CrossRef] [PubMed]
66. Bai, A.; Weaver, M.; Bao, F.; Chan, E.D.; Bai, X. *Saccharomyces boulardii* Produces a Factor That Inhibits *Mycobacterium intracellulare* Burden in Human Macrophages. *AIM* **2016**, *6*, 965–974. [CrossRef]
67. Buts, J.-P.; Dekeyser, N.; Stilmant, C.; Delem, E.; Smets, F.; Sokal, E. Saccharomyces boulardii produces in rat small intestine a novel protein phosphatase that inhibits *Escherichia coli* endotoxin by dephosphorylation. *Pediatr. Res.* **2006**, *60*, 24–29. [CrossRef]
68. Vandenplas, Y.; Brunser, O.; Szajewska, H. Saccharomyces boulardii in childhood. *Eur. J. Pediatr.* **2009**, *168*, 253–265. [CrossRef]
69. Pais, P.; Almeida, V.; Yılmaz, M.; Teixeira, M.C. Saccharomyces boulardii: What Makes It Tick as Successful Probiotic? *J. Fungi* **2020**, *6*, 78. [CrossRef]
70. Offei, B.; Vandecruys, P.; De Graeve, S.; Foulquié-Moreno, M.R.; Thevelein, J.M. Unique genetic basis of the distinct antibiotic potency of high acetic acid production in the probiotic yeast Saccharomyces cerevisiae var. boulardii. *Genome Res.* **2019**, *29*, 1478–1494. [CrossRef]
71. Im, E.; Pothoulakis, C. Recent advances in Saccharomyces boulardii research. *Gastroenterol. Clin. Biol.* **2010**, *34* (Suppl. S1), S62–S70. [CrossRef] [PubMed]
72. Siddiqui, A.; Haider, R.; Aaqil, S.I.; Vohra, L.I.; Qamar, K.; Jawed, A.; Fatima, N.; Adnan, A.; Parikh, V.; Ochani, S.; et al. Probiotic formulations and gastro-intestinal diseases in the paediatric population: A narrative review. *Ann. Med. Surg.* **2024**, *86*, 2836–2847. [CrossRef] [PubMed]
73. Schneider, R.; Sant'Anna, A. Using probiotics in paediatric populations. *Paediatr. Child Health* **2022**, *27*, 482–491. [CrossRef] [PubMed]
74. Thomas, J.P.; Raine, T.; Reddy, S.; Belteki, G. Probiotics for the prevention of necrotising enterocolitis in very low-birth-weight infants: A meta-analysis and systematic review. *Acta Paediatr.* **2017**, *106*, 1729–1741. [CrossRef] [PubMed]
75. Zhang, Y.; Xu, Q.; Zhang, F.; Sun, C. Probiotics for Preventing Necrotizing Enterocolitis: A Meta-Analysis with Trial Sequential Analysis. *J. Clin. Pharm. Ther.* **2023**, *2023*, 1–15. [CrossRef]
76. Isaeva, G.; Isaeva, R. Probiotics in the treatment of Helicobacter pylori infection: Reality and perspective. *Minerva Gastroenterol.* **2022**, *68*, 277–288. [CrossRef]
77. Jain, M.; Agarwal, A.; Singh, K.; Jarouliya, U. Probiotics and immune response in disease prevention and treatment. In *Immune-Boosting Nutraceuticals for Better Human Health: Novel Applications*; Apple Academic Press: New York, NY, USA, 2023; pp. 99–129. ISBN 9781003371069.
78. Thoda, C.; Touraki, M. Immunomodulatory properties of probiotics and their derived bioactive compounds. *Appl. Sci.* **2023**, *13*, 4726. [CrossRef]

79. dos Santos Freitas, A.; da Silva Fernandes, L.J.; Coelho-Rocha, N.D.; de Jesus, L.C.L.; de Rezende Rodovalho, V.; da Silva, T.F.; de Oliveira Carvalho, R.D.; Azevedo, V. Immunomodulatory and antiinflammatory mechanisms of probiotics. In *Probiotics*; Elsevier: Amsterdam, The Netherlands, 2022; pp. 321–341. ISBN 9780323851701.
80. Tao, T.; Zhang, L.; Yu, T.; Ma, J.; Lu, S.; Ren, J.; Li, X.; Guo, X. Exopolysaccharide production by Lactobacillus plantarum T10 is responsible for the probiotic activity in enhancing intestinal barrier function in vitro and in vivo. *Food Funct.* **2024**, *15*, 3583–3599. [CrossRef]
81. Chia, T.-Y.; Zolp, A.; Miska, J. Polyamine immunometabolism: Central regulators of inflammation, cancer and autoimmunity. *Cells* **2022**, *11*, 896. [CrossRef]
82. Filidou, E.; Kandilogiannakis, L.; Shrewsbury, A.; Kolios, G.; Kotzampassi, K. Probiotics: Shaping the gut immunological responses. *World J. Gastroenterol.* **2024**, *30*, 2096–2108. [CrossRef]
83. Verhoeven, D. Immunometabolism and innate immunity in the context of immunological maturation and respiratory pathogens in young children. *J. Leukoc. Biol.* **2019**, *106*, 301–308. [CrossRef] [PubMed]
84. Fong, F.L.Y.; Shah, N.P.; Kirjavainen, P.; El-Nezami, H. Mechanism of Action of Probiotic Bacteria on Intestinal and Systemic Immunities and Antigen-Presenting Cells. *Int. Rev. Immunol.* **2016**, *35*, 179–188. [CrossRef] [PubMed]
85. Yousefi, B.; Eslami, M.; Ghasemian, A.; Kokhaei, P.; Salek Farrokhi, A.; Darabi, N. Probiotics importance and their immunomodulatory properties. *J. Cell. Physiol.* **2019**, *234*, 8008–8018. [CrossRef] [PubMed]
86. Kleniewska, P.; Pawliczak, R. Can probiotics be used in the prevention and treatment of bronchial asthma? *Pharmacol. Rep.* **2024**, *76*, 740–753. [CrossRef] [PubMed]
87. Nova, E.; Wärnberg, J.; Gómez-Martínez, S.; Díaz, L.E.; Romeo, J.; Marcos, A. Immunomodulatory effects of probiotics in different stages of life. *Br. J. Nutr.* **2007**, *98* (Suppl. 1), S90–S95. [CrossRef]
88. Fukushima, Y.; Kawata, Y.; Hara, H.; Terada, A.; Mitsuoka, T. Effect of a probiotic formula on intestinal immunoglobulin A production in healthy children. *Int. J. Food Microbiol.* **1998**, *42*, 39–44. [CrossRef]
89. Fong, F.L.Y.; El-Nezami, H.; Mykkänen, O.; Kirjavainen, P.V. The effects of single strains and mixtures of probiotic bacteria on immune profile in liver, spleen, and peripheral blood. *Front. Nutr.* **2022**, *9*, 773298. [CrossRef]
90. Zhu, H.; Guo, L.; Yu, D.; Du, X. New insights into immunomodulatory properties of lactic acid bacteria fermented herbal medicines. *Front. Microbiol.* **2022**, *13*, 1073922. [CrossRef]
91. Licciardi, P.V.; Wong, S.-S.; Tang, M.L.; Karagiannis, T.C. Epigenome targeting by probiotic metabolites. *Gut Pathog.* **2010**, *2*, 24. [CrossRef]
92. DeMuri, G.P.; Lehtoranta, L.M.; Eickhoff, J.C.; Lehtinen, M.J.; Wald, E.R. Ex vivo peripheral blood mononuclear cell response to R848 in children after supplementation with the probiotic Lactobacillus acidophilus NCFM/Bifidobacterium lactis Bi-07. *Benef. Microbes* **2021**, *12*, 85–93. [CrossRef]
93. Duan, Y.; Wang, L.; Ma, Y.; Ning, L.; Zhang, X. A meta-analysis of the therapeutic effect of probiotic intervention in obese or overweight adolescents. *Front. Endocrinol.* **2024**, *15*, 1335810. [CrossRef] [PubMed]
94. Atazadegan, M.A.; Heidari-Beni, M.; Entezari, M.H.; Sharifianjazi, F.; Kelishadi, R. Effects of synbiotic supplementation on anthropometric indices and body composition in overweight or obese children and adolescents: A randomized, double-blind, placebo-controlled clinical trial. *World J. Pediatr.* **2023**, *19*, 356–365. [CrossRef] [PubMed]
95. Karlsson Videhult, F.; Andersson, Y.; Öhlund, I.; Stenlund, H.; Hernell, O.; West, C.E. Impact of probiotics during weaning on the metabolic and inflammatory profile: Follow-up at school age. *Int. J. Food Sci. Nutr.* **2015**, *66*, 686–691. [CrossRef] [PubMed]
96. Guo, S.; Guo, Y.; Ergun, A.; Lu, L.; Walker, W.A.; Ganguli, K. Secreted Metabolites of Bifidobacterium infantis and Lactobacillus acidophilus Protect Immature Human Enterocytes from IL-1β-Induced Inflammation: A Transcription Profiling Analysis. *PLoS ONE* **2015**, *10*, e0124549. [CrossRef] [PubMed]
97. Matsumoto, M.; Benno, Y. Anti-inflammatory metabolite production in the gut from the consumption of probiotic yogurt containing Bifidobacterium animalis subsp. lactis LKM512. *Biosci. Biotechnol. Biochem.* **2006**, *70*, 1287–1292. [CrossRef]
98. Ballini, A.; Gnoni, A.; De Vito, D.; Dipalma, G.; Cantore, S.; Gargiulo Isacco, C.; Saini, R.; Santacroce, L.; Topi, S.; Scarano, A.; et al. Effect of probiotics on the occurrence of nutrition absorption capacities in healthy children: A randomized double-blinded placebo-controlled pilot study. *Eur. Rev. Med. Pharmacol. Sci.* **2019**, *23*, 8645–8657. [CrossRef]
99. Chen, A.-C.; Fang, T.-J.; Ho, H.-H.; Chen, J.-F.; Kuo, Y.-W.; Huang, Y.-Y.; Tsai, S.-Y.; Wu, S.-F.; Lin, H.-C.; Yeh, Y.-T. A multi-strain probiotic blend reshaped obesity-related gut dysbiosis and improved lipid metabolism in obese children. *Front. Nutr.* **2022**, *9*, 922993. [CrossRef]
100. Li, Y.; Liu, T.; Qin, L.; Wu, L. Effects of probiotic administration on overweight or obese children: A meta-analysis and systematic review. *J. Transl. Med.* **2023**, *21*, 525. [CrossRef]
101. Toe, L.C.; Kerckhof, F.-M.; De Bodt, J.; Morel, F.B.; Ouedraogo, J.-B.; Kolsteren, P.; Van de Wiele, T. A prebiotic-enhanced lipid-based nutrient supplement (LNSp) increases Bifidobacterium relative abundance and enhances short-chain fatty acid production in simulated colonic microbiota from undernourished infants. *FEMS Microbiol. Ecol.* **2020**, *96*, fiaa105. [CrossRef]
102. Kambale, R.M.; Ntagazibwa, J.N.; Kasengi, J.B.; Zigashane, A.B.; Francisca, I.N.; Mashukano, B.N.; Amani Ngaboyeka, G.; Bahizire, E.; Zech, F.; Bindels, L.B.; et al. Probiotics for children with uncomplicated severe acute malnutrition (PruSAM study): A randomized controlled trial in the Democratic Republic of Congo. *Am. J. Clin. Nutr.* **2023**, *117*, 976–984. [CrossRef]
103. Ghannoum, M.; Ghannoum, A.; Hager, C.; Retuerto, M.; Isham, N.; McCormick, T.S. The probiotic BIOHM improves nutrient absorption by disrupting gastrointestinal biofilms. *J. Prob. Health* **2019**, *7*. [CrossRef]

104. Fan, Y.; Pedersen, O. Gut microbiota in human metabolic health and disease. *Nat. Rev. Microbiol.* **2021**, *19*, 55–71. [CrossRef] [PubMed]
105. Lozupone, C.A.; Stombaugh, J.I.; Gordon, J.I.; Jansson, J.K.; Knight, R. Diversity, stability and resilience of the human gut microbiota. *Nature* **2012**, *489*, 220–230. [CrossRef] [PubMed]
106. Bandsma, R.H.J.; Sadiq, K.; Bhutta, Z.A. Persistent diarrhoea: Current knowledge and novel concepts. *Paediatr. Int. Child Health* **2019**, *39*, 41–47. [CrossRef] [PubMed]
107. Campbell, D.I.; McPhail, G.; Lunn, P.G.; Elia, M.; Jeffries, D.J. Intestinal inflammation measured by fecal neopterin in Gambian children with enteropathy: Association with growth failure, Giardia lamblia, and intestinal permeability. *J. Pediatr. Gastroenterol. Nutr.* **2004**, *39*, 153–157. [CrossRef]
108. Sjögren, Y.M.; Jenmalm, M.C.; Böttcher, M.F.; Björkstén, B.; Sverremark-Ekström, E. Altered early infant gut microbiota in children developing allergy up to 5 years of age. *Clin. Exp. Allergy* **2009**, *39*, 518–526. [CrossRef]
109. Bhattacharjee, R.; Pathak, K.; Das, J.; Bordoloi, S.; Pathak, M.P.; Barbhuiya, P.A.; Saikia, R. Therapeutic potential of gut microbiota in child health. *Anti-Infect. Agents* **2023**, *21*, 69–77. [CrossRef]
110. Navarro-Tapia, E.; Sebastiani, G.; Sailer, S.; Toledano, L.A.; Serra-Delgado, M.; García-Algar, Ó.; Andreu-Fernández, V. Probiotic supplementation during the perinatal and infant period: Effects on gut dysbiosis and disease. *Nutrients* **2020**, *12*, 2243. [CrossRef]
111. Maiuolo, J.; Bulotta, R.M.; Ruga, S.; Nucera, S.; Macrì, R.; Scarano, F.; Oppedisano, F.; Carresi, C.; Gliozzi, M.; Musolino, V.; et al. The Postbiotic Properties of Butyrate in the Modulation of the Gut Microbiota: The Potential of Its Combination with Polyphenols and Dietary Fibers. *Int. J. Mol. Sci.* **2024**, *25*, 6971. [CrossRef]
112. Dogra, S.K.; Dardinier, A.; Mainardi, F.; Siegwald, L.; Bartova, S.; le Roy, C.; Chou, C.J. Application of computational data modeling to a large-scale population cohort assists the discovery of specific nutrients that influence beneficial human gut bacteria *Faecalibacterium prausnitzii*. *Nutrients* **2023**, 15. [CrossRef]
113. Kadry, A.A.; El-Antrawy, M.A.; El-Ganiny, A.M. Impact of short chain fatty acids (SCFAs) on antimicrobial activity of new β-lactam/β-lactamase inhibitor combinations and on virulence of *Escherichia coli* isolates. *J. Antibiot.* **2023**, *76*, 225–235. [CrossRef] [PubMed]
114. Liang, L.; Liu, L.; Zhou, W.; Yang, C.; Mai, G.; Li, H.; Chen, Y. Gut microbiota-derived butyrate regulates gut mucus barrier repair by activating the macrophage/WNT/ERK signaling pathway. *Clin. Sci.* **2022**, *136*, 291–307. [CrossRef] [PubMed]
115. Ornelas, A.; Welch, N.; Countess, J.A.; Zhou, L.; Wang, R.X.; Dowdell, A.S.; Colgan, S.P. Mimicry of microbially-derived butyrate reveals templates for potent intestinal epithelial HIF stabilizers. *Gut Microbes* **2023**, *15*, 2267706. [CrossRef] [PubMed]
116. Recharla, N.; Geesala, R.; Shi, X.-Z. Gut microbial metabolite butyrate and its therapeutic role in inflammatory bowel disease: A literature review. *Nutrients* **2023**, *15*, 2275. [CrossRef]
117. Conder, E.; Shay, H.C.; Vekaria, H.; Erinkitola, I.; Bhogoju, S.; Goretsky, T.; Sullivan, P.; Barrett, T.; Kapur, N. Butyrate-induced mitochondrial function improves barrier function in inflammatory bowel disease (ibd). *Gastroenterology* **2023**, *164*, S91. [CrossRef]
118. Novella Nicese, M.; van den Berg, B.; Sol, W.; van Raalte, D.; Nieuwdorp, M.; Rotmans, J. MO618: Influence of butyrate and acetate on behavior and metabolism of human glomerular microvascular endothelial cells. *Nephrol. Dial. Transplant.* **2022**, *37*, gfac076-011. [CrossRef]
119. Wang, S.P.; Rubio, L.A.; Duncan, S.H.; Donachie, G.E.; Holtrop, G.; Lo, G.; Farquharson, F.M.; Wagner, J.; Parkhill, J.; Louis, P.; et al. Pivotal Roles for pH, Lactate, and Lactate-Utilizing Bacteria in the Stability of a Human Colonic Microbial Ecosystem. *mSystems* **2020**, *5*, 10-1128. [CrossRef]
120. Zhang, C.; Zhang, S.; Liu, W.; Guo, T.; Gu, R.; Kong, J. Potential Application and Bactericidal Mechanism of Lactic Acid-Hydrogen Peroxide Consortium. *Appl. Biochem. Biotechnol.* **2019**, *189*, 822–833. [CrossRef]
121. Gu, Q.; Li, P. Biosynthesis of vitamins by probiotic bacteria. In *Probiotics and Prebiotics in Human Nutrition and Health*; Rao, V., Rao, L.G., Eds.; InTech: Houston, TX, USA, 2016; ISBN 978-953-51-2475-7.
122. Odumosu, B.T.; Bamidele, T.A.; Ofem, D.W.; Agbozo, F.; Olasehinde, G.I. Screening, isolation and biotechnological potentials of foodborne Lactobacillus fermentum strains MT903311 and MT903312. *Heliyon* **2023**, *9*, e14959. [CrossRef]
123. Yin, J.; Peng, X.; Yang, A.; Lin, M.; Ji, K.; Dai, X.; Huang, J.; Li, L.; Feng, L. Impact of oligosaccharides on probiotic properties and B vitamins production: A comprehensive assessment of probiotic strains. *Int. J. Food Sci. Technol.* **2024**, *59*, 6044–6064. [CrossRef]
124. Uddin, M.; Uddin, S.; Deb, B.; Hussain, T.; Rafi, S.; Islam, A. The role of lactobacillus reuteri probiotic for preventing functional gastrointestinal disorders in toddlerhood. *Int. J. Gastroenterol.* **2024**, *8*, 14–21. [CrossRef]
125. Torres, A.C.; Vannini, V.; Font, G.; Saavedra, L.; Taranto, M.P. Novel pathway for corrinoid compounds production in lactobacillus. *Front. Microbiol.* **2018**, *9*, 2256. [CrossRef] [PubMed]
126. Langa, S.; Peirotén, Á.; Rodríguez, S.; Calzada, J.; Prieto-Paredes, R.; Curiel, J.A.; Landete, J.M. Riboflavin bio-enrichment of soy beverage by selected roseoflavin-resistant and engineered lactic acid bacteria. *Int. J. Food Microbiol.* **2024**, *411*, 110547. [CrossRef] [PubMed]
127. Meucci, A.; Rossetti, L.; Zago, M.; Monti, L.; Giraffa, G.; Carminati, D.; Tidona, F. Folates biosynthesis by Streptococcus thermophilus during growth in milk. *Food Microbiol.* **2018**, *69*, 116–122. [CrossRef]
128. Smajdor, J.; Jedlińska, K.; Porada, R.; Górska-Ratusznik, A.; Policht, A.; Śróttek, M.; Więcek, G.; Baś, B.; Strus, M. The impact of gut bacteria producing long chain homologs of vitamin K2 on colorectal carcinogenesis. *Cancer Cell Int.* **2023**, *23*, 268. [CrossRef]

129. Parama Iswara, R.; Yunus, K.; Pati, N. Vitamin K Deficiency. In *Causes and Management of Nutritional Deficiency Disorders*; Nayak, A., Misra, S., Eds.; Advances in Medical Diagnosis, Treatment, and Care; IGI Global: Hershey, PA, USA, 2024; pp. 214–218. ISBN 9798369329474.
130. Aljafary, M.A.; Alshwyeh, H.; Alahmadi, N.; Shehzad, A.; Tombuloglu, H.; Gaymalov, Z.; Homieda, A.; Al-Suhaimi, E. Physiological and cellular functions of vitamin K on cardiovascular function. In *Vitamin K-Recent Topics on the Biology and Chemistry*; Kagechika, H., Shirakawa, H., Eds.; Biochemistry; IntechOpen: London, UK, 2022; Volume 27, ISBN 978-1-83969-391-5.
131. Ivanko, O.G.; Solianyk, O.V. Antibiotic-associated disorders of prothrombin synthesis and their probiotic correction with B. Clausii in breastfeed infants. *Zaporozhye Med. J.* **2018**. [CrossRef]
132. Liu, Y. Good Things Come in Small Packages: Delivery of Vitamin K2 to Human Cells by Extracellular Vesicles from Lactococcus Lactis. Ph.D. Dissertation, Wageningen University, Wageningen, The Netherlands, 2022.
133. Cooke, G.; Behan, J.; Costello, M. Newly identified vitamin K-producing bacteria isolated from the neonatal faecal flora. *Microb. Ecol. Health Dis.* **2006**, *18*, 133–138. [CrossRef]
134. Kaushal, D.; Kalsi, G. Dietary intervention of prebiotics and vitamins on gut health of children. *Nutr. Food Sci.* **2022**, *52*, 1045–1054. [CrossRef]
135. Kim, J.-Y.; Choi, E.-J.; Lee, J.-H.; Yoo, M.-S.; Heo, K.; Shim, J.-J.; Lee, J.-L. Probiotic Potential of a Novel Vitamin B2-Overproducing Lactobacillus plantarum Strain, HY7715, Isolated from Kimchi. *Appl. Sci.* **2021**, *11*, 5765. [CrossRef]
136. Salvi, P.S.; Cowles, R.A. Butyrate and the intestinal epithelium: Modulation of proliferation and inflammation in homeostasis and disease. *Cells* **2021**, *10*, 1775. [CrossRef]
137. Round, J.L.; Mazmanian, S.K. Inducible Foxp3+ regulatory T-cell development by a commensal bacterium of the intestinal microbiota. *Proc. Natl. Acad. Sci. USA* **2010**, *107*, 12204–12209. [CrossRef] [PubMed]
138. Nilsen, M.; Nygaard, U.C.; Brodin, P.; Carlsen, K.C.L.; Fredheim, C.; Haugen, G.; Hedlin, G.; Jonassen, C.M.; Jonsmoen, U.L.A.; Lakshmikanth, T.; et al. Gut bacteria at 6 months of age are associated with immune cell status in 1-year-old children. *Scand. J. Immunol.* **2024**, *99*, e13346. [CrossRef] [PubMed]
139. Treem, W.R.; Ahsan, N.; Shoup, M.; Hyams, J.S. Fecal short-chain fatty acids in children with inflammatory bowel disease. *J. Pediatr. Gastroenterol. Nutr.* **1994**, *18*, 159–164. [CrossRef]
140. Lopetuso, L.R.; Scaldaferri, F.; Petito, V.; Gasbarrini, A. Commensal Clostridia: Leading players in the maintenance of gut homeostasis. *Gut Pathog.* **2013**, *5*, 23. [CrossRef] [PubMed]
141. Hossain, M.S.; Das, S.; Gazi, M.A.; Alam, M.A.; Haque, N.M.S.; Mahfuz, M.; Ahmed, T.; Damman, C.J. Association of faecal pH with childhood stunting: Results from a cross-sectional study. *BMJ Paediatr. Open* **2019**, *3*, e000549. [CrossRef]
142. van Limpt, C.; Crienen, A.; Vriesema, A.; Knol, J. 134 effect of colonic short chain fatty acids, lactate and PH on the growth of common gut pathogens. *Pediatr. Res.* **2004**, *56*, 487. [CrossRef]
143. Hatayama, H.; Iwashita, J.; Kuwajima, A.; Abe, T. The short chain fatty acid, butyrate, stimulates MUC2 mucin production in the human colon cancer cell line, LS174T. *Biochem. Biophys. Res. Commun.* **2007**, *356*, 599–603. [CrossRef]
144. Chen, Y. The SCFAs Production of Syntrophic Culture of L. johnsonii SZ-YL and A. Muciniphila in Different Macrobutrients. *Highlights Sci. Eng. Technol.* **2023**, *30*, 24–33. [CrossRef]
145. Ferrari, S.; Galla, R.; Mulè, S.; Rosso, G.; Brovero, A.; Macchi, V.; Ruga, S.; Uberti, F. The Role of Bifidobacterium bifidum novaBBF7, Bifidobacterium longum novaBLG2 and Lactobacillus paracasei TJB8 to Improve Mechanisms Linked to Neuronal Cells Protection against Oxidative Condition in a Gut-Brain Axis Model. *Int. J. Mol. Sci.* **2023**, *24*, 12281. [CrossRef]
146. Cheng, Y.; Liu, J.; Ling, Z. Short-chain fatty acids-producing probiotics: A novel source of psychobiotics. *Crit. Rev. Food Sci. Nutr.* **2022**, *62*, 7929–7959. [CrossRef]
147. Vital, M.; Karch, A.; Pieper, D.H. Colonic Butyrate-Producing Communities in Humans: An Overview Using Omics Data. *mSystems* **2017**, *2*, 10-1128. [CrossRef] [PubMed]
148. Hojsak, I.; Benninga, M.A.; Hauser, B.; Kansu, A.; Kelly, V.B.; Stephen, A.M.; Morais Lopez, A.; Slavin, J.; Tuohy, K. Benefits of dietary fibre for children in health and disease. *Arch. Dis. Child.* **2022**, *107*, 973–979. [CrossRef]
149. Cardoso, M.H.; Meneguetti, B.T.; Oliveira-Júnior, N.G.; Macedo, M.L.R.; Franco, O.L. Antimicrobial peptide production in response to gut microbiota imbalance. *Peptides* **2022**, *157*, 170865. [CrossRef] [PubMed]
150. Hassan, M.; Kjos, M.; Nes, I.F.; Diep, D.B.; Lotfipour, F. Natural antimicrobial peptides from bacteria: Characteristics and potential applications to fight against antibiotic resistance. *J. Appl. Microbiol.* **2012**, *113*, 723–736. [CrossRef] [PubMed]
151. Huang, F.; Teng, K.; Liu, Y.; Cao, Y.; Wang, T.; Ma, C.; Zhang, J.; Zhong, J. Bacteriocins: Potential for human health. *Oxid. Med. Cell. Longev.* **2021**, *2021*, 5518825. [CrossRef] [PubMed]
152. Spohn, R.; Daruka, L.; Lázár, V.; Martins, A.; Vidovics, F.; Grézal, G.; Méhi, O.; Kintses, B.; Számel, M.; Jangir, P.K.; et al. Integrated evolutionary analysis reveals antimicrobial peptides with limited resistance. *Nat. Commun.* **2019**, *10*, 4538. [CrossRef]
153. Andersson, D.I.; Hughes, D.; Kubicek-Sutherland, J.Z. Mechanisms and consequences of bacterial resistance to antimicrobial peptides. *Drug Resist. Updat.* **2016**, *26*, 43–57. [CrossRef]
154. Yeaman, M.R.; Yount, N.Y. Mechanisms of antimicrobial peptide action and resistance. *Pharmacol. Rev.* **2003**, *55*, 27–55. [CrossRef]
155. Mandal, S.M.; Pati, B.R.; Chakraborty, R.; Franco, O.L. New insights into the bioactivity of peptides from probiotics. *Front. Biosci* **2016**, *8*, 450–459. [CrossRef]
156. Heilbronner, S.; Krismer, B.; Brötz-Oesterhelt, H.; Peschel, A. The microbiome-shaping roles of bacteriocins. *Nat. Rev. Microbiol.* **2021**, *19*, 726–739. [CrossRef]

157. Shin, J.M.; Gwak, J.W.; Kamarajan, P.; Fenno, J.C.; Rickard, A.H.; Kapila, Y.L. Biomedical applications of nisin. *J. Appl. Microbiol.* **2016**, *120*, 1449–1465. [CrossRef] [PubMed]
158. Ashby, M.; Petkova, A.; Hilpert, K. Cationic antimicrobial peptides as potential new therapeutic agents in neonates and children: A review. *Curr. Opin. Infect. Dis.* **2014**, *27*, 258–267. [CrossRef] [PubMed]
159. Arun, K.B.; Madhavan, A.; Emmanual, S.; Sindhu, R.; Binod, P.; Pandey, A. Enzymes in probiotics and genetically modified foods. In *Value-Addition in Food Products and Processing Through Enzyme Technology*; Elsevier: Amsterdam, The Netherlands, 2022; pp. 13–23. ISBN 9780323899291.
160. Xie, W.; Li, W.; Gao, F.; Li, T. In vitro digestion and fermentation of lacto- *N* -biose, a core building block of human milk oligosaccharides. *Int. J. Food Sci. Technol.* **2023**, *58*, 4511–4518. [CrossRef]
161. Florindo, R.N.; Souza, V.P.; Manzine, L.R.; Camilo, C.M.; Marana, S.R.; Polikarpov, I.; Nascimento, A.S. Structural and biochemical characterization of a GH3 β-glucosidase from the probiotic bacteria Bifidobacterium adolescentis. *Biochimie* **2018**, *148*, 107–115. [CrossRef] [PubMed]
162. Ambrogi, V.; Bottacini, F.; O'Callaghan, J.; Casey, E.; van Breen, J.; Schoemaker, B.; Cao, L.; Kuipers, B.; O'Connell Motherway, M.; Schoterman, M.; et al. Infant-Associated Bifidobacterial β-Galactosidases and Their Ability to Synthesize Galacto-Oligosaccharides. *Front. Microbiol.* **2021**, *12*, 662959. [CrossRef]
163. Xin, Y.; Guo, T.; Zhang, Y.; Wu, J.; Kong, J. A new β-galactosidase extracted from the infant feces with high hydrolytic and transgalactosylation activity. *Appl. Microbiol. Biotechnol.* **2019**, *103*, 8439–8448. [CrossRef]
164. Enzymatic Production of Galactooligosaccharides by β-Galactosidase from Bifidobacterium longum BCRC 15708. Available online: https://pubs.acs.org/doi/10.1021/jf063126+ (accessed on 5 September 2024).
165. Antonowicz, I.; Shwachman, H.; Sotoo, I. Beta-galactosidase and beta-glucuronidase activities in intestinal mucosa of infants and children. *Pediatrics* **1971**, *47*, 737–744. [CrossRef]
166. Cuxart, I.; Coines, J.; Esquivias, O.; Faijes, M.; Planas, A.; Biarnés, X.; Rovira, C. Enzymatic Hydrolysis of Human Milk Oligosaccharides. The Molecular Mechanism of Bifidobacterium Bifidum Lacto-N-biosidase. *ACS Catal.* **2022**, *12*, 4737–4743. [CrossRef]
167. Ito, T.; Katayama, T.; Hattie, M.; Sakurama, H.; Wada, J.; Suzuki, R.; Ashida, H.; Wakagi, T.; Yamamoto, K.; Stubbs, K.A.; et al. Crystal structures of a glycoside hydrolase family 20 lacto-N-biosidase from Bifidobacterium bifidum. *J. Biol. Chem.* **2013**, *288*, 11795–11806. [CrossRef]
168. Ma, T.; Shen, X.; Shi, X.; Sakandar, H.A.; Quan, K.; Li, Y.; Jin, H.; Kwok, L.-Y.; Zhang, H.; Sun, Z. Targeting gut microbiota and metabolism as the major probiotic mechanism-An evidence-based review. *Trends Food Sci. Technol.* **2023**, *138*, 178–198. [CrossRef]
169. Kaur, N.; Dey, P. Bacterial exopolysaccharides as emerging bioactive macromolecules: From fundamentals to applications. *Res. Microbiol.* **2023**, *174*, 104024. [CrossRef] [PubMed]
170. Alessandri, G.; van Sinderen, D.; Ventura, M. The genus bifidobacterium: From genomics to functionality of an important component of the mammalian gut microbiota running title: Bifidobacterial adaptation to and interaction with the host. *Comput. Struct. Biotechnol. J.* **2021**, *19*, 1472–1487. [CrossRef] [PubMed]
171. Xie, Y.; Ye, Z.; Wan, X.; Deng, H.; Sun, W.; He, X.; Chen, K. Screening of exopolysaccharide-producing Enterobacter aerogenes NJ1023 and its cadaverine biosynthesis promotion. *Front. Microbiol.* **2023**, *14*, 1200123. [CrossRef] [PubMed]
172. Salimi, F.; Farrokh, P. Recent advances in the biological activities of microbial exopolysaccharides. *World J. Microbiol. Biotechnol.* **2023**, *39*, 213. [CrossRef] [PubMed]
173. Milani, C.; Duranti, S.; Bottacini, F.; Casey, E.; Turroni, F.; Mahony, J.; Belzer, C.; Delgado Palacio, S.; Arboleya Montes, S.; Mancabelli, L.; et al. The first microbial colonizers of the human gut: Composition, activities, and health implications of the infant gut microbiota. *Microbiol. Mol. Biol. Rev.* **2017**, *81*, 10-1128. [CrossRef]
174. Sadeghi, M.; Haghshenas, B.; Nami, Y. Bifidobacterium exopolysaccharides: New insights into engineering strategies, physicochemical functions, and immunomodulatory effects on host health. *Front. Microbiol.* **2024**, *15*, 1396308. [CrossRef]
175. Caggianiello, G.; Kleerebezem, M.; Spano, G. Exopolysaccharides produced by lactic acid bacteria: From health-promoting benefits to stress tolerance mechanisms. *Appl. Microbiol. Biotechnol.* **2016**, *100*, 3877–3886. [CrossRef]
176. Yue, Y.; Wang, Y.; Yang, R.; Liu, D.; Cheng, Y.; Li, S.; Sun, L.; Li, B.; Huo, G. Exopolysaccharides of Bifidobacterium longum subsp. infantis E4 on the immune and anti-inflammatory effects in vitro. *J. Funct. Foods* **2023**, *107*, 105699. [CrossRef]
177. Wu, J.; Zhang, Y.; Ye, L.; Wang, C. The anti-cancer effects and mechanisms of lactic acid bacteria exopolysaccharides in vitro: A review. *Carbohydr. Polym.* **2021**, *253*, 117308. [CrossRef]
178. Ren, Z.; Chen, S.; Lv, H.; Peng, L.; Yang, W.; Chen, J.; Wu, Z.; Wan, C. Effect of Bifidobacterium animalis subsp. lactis SF on enhancing the tumor suppression of irinotecan by regulating the intestinal flora. *Pharmacol. Res.* **2022**, *184*, 106406. [CrossRef]
179. Wang, L.; Wang, Y.; Li, Q.; Tian, K.; Xu, L.; Liu, G.; Guo, C. Exopolysaccharide, Isolated From a Novel Strain Bifidobacterium breve lw01 Possess an Anticancer Effect on Head and Neck Cancer-Genetic and Biochemical Evidences. *Front. Microbiol.* **2019**, *10*, 1044. [CrossRef] [PubMed]
180. Angelin, J.; Kavitha, M. Exopolysaccharides from probiotic bacteria and their health potential. *Int. J. Biol. Macromol.* **2020**, *162*, 853–865. [CrossRef] [PubMed]
181. Kim, K.; Lee, G.; Thanh, H.D.; Kim, J.-H.; Konkit, M.; Yoon, S.; Park, M.; Yang, S.; Park, E.; Kim, W. Exopolysaccharide from Lactobacillus plantarum LRCC5310 offers protection against rotavirus-induced diarrhea and regulates inflammatory response. *J. Dairy Sci.* **2018**, *101*, 5702–5712. [CrossRef] [PubMed]

182. Bhat, B.; Bajaj, B.K. Hypocholesterolemic potential and bioactivity spectrum of an exopolysaccharide from a probiotic isolate Lactobacillus paracasei M7. *Bioact. Carbohydr. Diet. Fibre* **2019**, *19*, 100191. [CrossRef]
183. Zhang, Z.; Zhou, Z.; Li, Y.; Zhou, L.; Ding, Q.; Xu, L. Isolated exopolysaccharides from Lactobacillus rhamnosus GG alleviated adipogenesis mediated by TLR2 in mice. *Sci. Rep.* **2016**, *6*, 36083. [CrossRef]
184. Porges, E.C.; Woods, A.J.; Edden, R.A.E.; Puts, N.A.J.; Harris, A.D.; Chen, H.; Garcia, A.M.; Seider, T.R.; Lamb, D.G.; Williamson, J.B.; et al. Frontal Gamma-Aminobutyric Acid Concentrations Are Associated With Cognitive Performance in Older Adults. *Biol. Psychiatry Cogn. Neurosci. Neuroimaging* **2017**, *2*, 38–44. [CrossRef]
185. Bravo, J.A.; Forsythe, P.; Chew, M.V.; Escaravage, E.; Savignac, H.M.; Dinan, T.G.; Bienenstock, J.; Cryan, J.F. Ingestion of Lactobacillus strain regulates emotional behavior and central GABA receptor expression in a mouse via the vagus nerve. *Proc. Natl. Acad. Sci. USA* **2011**, *108*, 16050–16055. [CrossRef]
186. Frank, S.M.; Becker, M.; Qi, A.; Geiger, P.; Frank, U.I.; Rosedahl, L.A.; Malloni, W.M.; Sasaki, Y.; Greenlee, M.W.; Watanabe, T. Efficient learning in children with rapid GABA boosting during and after training. *Curr. Biol.* **2022**, *32*, 5022–5030.e7. [CrossRef]
187. Yunes, R.A.; Poluektova, E.U.; Dyachkova, M.S.; Klimina, K.M.; Kovtun, A.S.; Averina, O.V.; Orlova, V.S.; Danilenko, V.N. GABA production and structure of gadB/gadC genes in Lactobacillus and Bifidobacterium strains from human microbiota. *Anaerobe* **2016**, *42*, 197–204. [CrossRef]
188. Allen, M.J.; Sabir, S.; Sharma, S. GABA Receptor. In *StatPearls*; StatPearls Publishing: Treasure Island, FL, USA, 2024.
189. Saleh, M.G.; Papantoni, A.; Mikkelsen, M.; Hui, S.C.N.; Oeltzschner, G.; Puts, N.A.; Edden, R.A.E.; Carnell, S. Effect of age on GABA+ and glutathione in a pediatric sample. *AJNR Am. J. Neuroradiol.* **2020**, *41*, 1099–1104. [CrossRef]
190. Monteagudo-Mera, A.; Fanti, V.; Rodriguez-Sobstel, C.; Gibson, G.; Wijeyesekera, A.; Karatzas, K.-A.; Chakrabarti, B. Gamma aminobutyric acid production by commercially available probiotic strains. *J. Appl. Microbiol.* **2023**, *134*, lxac066. [CrossRef] [PubMed]
191. Braga, J.D.; Thongngam, M.; Kumrungsee, T. Gamma-aminobutyric acid as a potential postbiotic mediator in the gut-brain axis. *npj Sci. Food* **2024**, *8*, 16. [CrossRef] [PubMed]
192. Ueno, H. Enzymatic and structural aspects on glutamate decarboxylase. *J. Mol. Catal. B Enzym.* **2000**, *10*, 67–79. [CrossRef]
193. Altaib, H.; Kozakai, T.; Badr, Y.; Nakao, H.; El-Nouby, M.A.M.; Yanase, E.; Nomura, I.; Suzuki, T. Cell factory for γ-aminobutyric acid (GABA) production using Bifidobacterium adolescentis. *Microb. Cell Fact.* **2022**, *21*, 33. [CrossRef] [PubMed]
194. Tamés, H.; Sabater, C.; Margolles, A.; Ruiz, L.; Ruas-Madiedo, P. Production of GABA in milk fermented by Bifidobacterium adolescentis strains selected on the bases of their technological and gastrointestinal performance. *Food Res. Int.* **2023**, *171*, 113009. [CrossRef]
195. Pokusaeva, K.; Johnson, C.; Luk, B.; Uribe, G.; Fu, Y.; Oezguen, N.; Matsunami, R.K.; Lugo, M.; Major, A.; Mori-Akiyama, Y.; et al. GABA-producing Bifidobacterium dentium modulates visceral sensitivity in the intestine. *Neurogastroenterol. Motil.* **2017**, *29*, e12904. [CrossRef]
196. Park, K.-B.; Ji, G.-E.; Park, M.-S.; Oh, S.-H. Expression of rice glutamate decarboxylase in Bifidobacterium longum enhances gamma-aminobutyric acid production. *Biotechnol. Lett.* **2005**, *27*, 1681–1684. [CrossRef]
197. Cocean, A.-M.; Vodnar, D.C. Exploring the gut-brain Axis: Potential therapeutic impact of Psychobiotics on mental health. *Prog. Neuropsychopharmacol. Biol. Psychiatry* **2024**, *134*, 111073. [CrossRef]
198. Annahazi, A.; Schemann, M. The enteric nervous system: "A little brain in the gut". *Neuroforum* **2020**, *26*, 31–42. [CrossRef]
199. Bonaz, B.; Sinniger, V.; Pellissier, S. The Vagus Nerve in the Neuro-Immune Axis: Implications in the Pathology of the Gastrointestinal Tract. *Front. Immunol.* **2017**, *8*, 1452. [CrossRef]
200. Barrio, C.; Arias-Sánchez, S.; Martín-Monzón, I. The gut microbiota-brain axis, psychobiotics and its influence on brain and behaviour: A systematic review. *Psychoneuroendocrinology* **2022**, *137*, 105640. [CrossRef] [PubMed]
201. Ding, X.; Xu, Y.; Zhang, X.; Zhang, L.; Duan, G.; Song, C.; Li, Z.; Yang, Y.; Wang, Y.; Wang, X.; et al. Gut microbiota changes in patients with autism spectrum disorders. *J. Psychiatr. Res.* **2020**, *129*, 149–159. [CrossRef] [PubMed]
202. Zuffa, S.; Schimmel, P.; Gonzalez-Santana, A.; Belzer, C.; Knol, J.; Bölte, S.; Falck-Ytter, T.; Forssberg, H.; Swann, J.; Diaz Heijtz, R. Early-life differences in the gut microbiota composition and functionality of infants at elevated likelihood of developing autism spectrum disorder. *Transl. Psychiatry* **2023**, *13*, 257. [CrossRef] [PubMed]
203. Cassidy-Bushrow, A.E.; Sitarik, A.R.; Johnson, C.C.; Johnson-Hooper, T.M.; Kassem, Z.; Levin, A.M.; Lynch, S.V.; Ownby, D.R.; Phillips, J.M.; Yong, G.J.M.; et al. Early-life gut microbiota and attention deficit hyperactivity disorder in preadolescents. *Pediatr. Res.* **2023**, *93*, 2051–2060. [CrossRef] [PubMed]
204. Pärtty, A.; Kalliomäki, M.; Wacklin, P.; Salminen, S.; Isolauri, E. A possible link between early probiotic intervention and the risk of neuropsychiatric disorders later in childhood: A randomized trial. *Pediatr. Res.* **2015**, *77*, 823–828. [CrossRef]
205. Braga, T.D.; da Silva, G.A.P.; de Lira, P.I.C.; de Carvalho Lima, M. Efficacy of Bifidobacterium breve and Lactobacillus casei oral supplementation on necrotizing enterocolitis in very-low-birth-weight preterm infants: A double-blind, randomized, controlled trial. *Am. J. Clin. Nutr.* **2011**, *93*, 81–86. [CrossRef]
206. Dilli, D.; Aydin, B.; Fettah, N.D.; Özyazıcı, E.; Beken, S.; Zenciroğlu, A.; Okumuş, N.; Özyurt, B.M.; İpek, M.Ş.; Akdağ, A.; et al. The propre-save study: Effects of probiotics and prebiotics alone or combined on necrotizing enterocolitis in very low birth weight infants. *J. Pediatr.* **2015**, *166*, 545–551.e1. [CrossRef]
207. Murphy, K.; Ross, R.P.; Ryan, C.A.; Dempsey, E.M.; Stanton, C. Probiotics, prebiotics, and synbiotics for the prevention of necrotizing enterocolitis. *Front. Nutr.* **2021**, *8*, 667188. [CrossRef]

208. O'Mahony, S.M.; Clarke, G.; Borre, Y.E.; Dinan, T.G.; Cryan, J.F. Serotonin, tryptophan metabolism and the brain-gut-microbiome axis. *Behav. Brain Res.* **2015**, *277*, 32–48. [CrossRef]
209. Wall, R.; Cryan, J.F.; Ross, R.P.; Fitzgerald, G.F.; Dinan, T.G.; Stanton, C. Bacterial neuroactive compounds produced by psychobiotics. *Adv. Exp. Med. Biol.* **2014**, *817*, 221–239. [CrossRef]
210. Yano, J.M.; Yu, K.; Donaldson, G.P.; Shastri, G.G.; Ann, P.; Ma, L.; Nagler, C.R.; Ismagilov, R.F.; Mazmanian, S.K.; Hsiao, E.Y. Indigenous bacteria from the gut microbiota regulate host serotonin biosynthesis. *Cell* **2015**, *161*, 264–276. [CrossRef] [PubMed]
211. Asano, Y.; Hiramoto, T.; Nishino, R.; Aiba, Y.; Kimura, T.; Yoshihara, K.; Koga, Y.; Sudo, N. Critical role of gut microbiota in the production of biologically active, free catecholamines in the gut lumen of mice. *Am. J. Physiol. Gastrointest. Liver Physiol.* **2012**, *303*, G1288–G1295. [CrossRef] [PubMed]
212. Thomas, C.M.; Versalovic, J. Probiotics-host communication: Modulation of signaling pathways in the intestine. *Gut Microbes* **2010**, *1*, 148–163. [CrossRef] [PubMed]
213. Barrett, E.; Ross, R.P.; O'Toole, P.W.; Fitzgerald, G.F.; Stanton, C. γ-Aminobutyric acid production by culturable bacteria from the human intestine. *J. Appl. Microbiol.* **2012**, *113*, 411–417. [CrossRef]
214. Salminen, S.; Collado, M.C.; Endo, A.; Hill, C.; Lebeer, S.; Quigley, E.M.M.; Sanders, M.E.; Shamir, R.; Swann, J.R.; Szajewska, H.; et al. The International Scientific Association of Probiotics and Prebiotics (ISAPP) consensus statement on the definition and scope of postbiotics. *Nat. Rev. Gastroenterol. Hepatol.* **2021**, *18*, 649–667. [CrossRef]
215. Ozma, M.A.; Abbasi, A.; Akrami, S.; Lahouty, M.; Shahbazi, N.; Ganbarov, K.; Pagliano, P.; Sabahi, S.; Köse, Ş.; Yousefi, M.; et al. Postbiotics as the key mediators of the gut microbiota-host interactions. *Infez. Med.* **2022**, *30*, 180–193. [CrossRef]
216. Aguilar-Toalá, J.E.; Garcia-Varela, R.; Garcia, H.S.; Mata-Haro, V.; González-Córdova, A.F.; Vallejo-Cordoba, B.; Hernández-Mendoza, A. Postbiotics: An evolving term within the functional foods field. *Trends Food Sci. Technol.* **2018**, *75*, 105–114. [CrossRef]
217. Konstantinov, S.R.; Kuipers, E.J.; Peppelenbosch, M.P. Functional genomic analyses of the gut microbiota for CRC screening. *Nat. Rev. Gastroenterol. Hepatol.* **2013**, *10*, 741–745. [CrossRef]
218. Sawada, H.; Furushiro, M.; Hirai, K.; Motoike, M.; Watanabe, T.; Yokokura, T. Purification and characterization of an antihypertensive compound from Lactobacillus casei. *Agric. Biol. Chem.* **1990**, *54*, 3211–3219. [CrossRef]
219. Nakamura, F.; Ishida, Y.; Sawada, D.; Ashida, N.; Sugawara, T.; Sakai, M.; Goto, T.; Kawada, T.; Fujiwara, S. Fragmented Lactic Acid Bacterial Cells Activate Peroxisome Proliferator-Activated Receptors and Ameliorate Dyslipidemia in Obese Mice. *J. Agric. Food Chem.* **2016**, *64*, 2549–2559. [CrossRef]
220. Chaluvadi, S.; Hotchkiss, A.T.; Yam, K.L. Gut Microbiota. In *Probiotics, Prebiotics, and Synbiotics*; Elsevier: Amsterdam, The Netherlands, 2016; pp. 515–523. ISBN 9780128021897.
221. Venema, K. Beneficial microbes. Foreword. *Benef. Microbes* **2012**, *3*, 1–2. [CrossRef] [PubMed]
222. Ozma, M.A.; Khodadadi, E.; Rezaee, M.A.; Asgharzadeh, M.; Aghazadeh, M.; Zeinalzadeh, E.; Ganbarov, K.; Kafil, H.S. Bacterial proteomics and its application in pathogenesis studies. *Curr. Pharm. Biotechnol.* **2022**, *23*, 1245–1256. [CrossRef] [PubMed]
223. Shigwedha, N.; Sichel, L.; Jia, L.; Zhang, L. Probiotical cell fragments (pcfs) as "novel nutraceutical ingredients". *J. Biosci. Med.* **2014**,
2, 43–55. [CrossRef]
224. Wang, Y.; Xie, J.; Wang, N.; Li, Y.; Sun, X.; Zhang, Y.; Zhang, H. Lactobacillus casei Zhang modulate cytokine and toll-like receptor expression and beneficially regulate poly I:C-induced immune responses in RAW264.7 macrophages. *Microbiol. Immunol.* **2013**, *57*, 54–62. [CrossRef] [PubMed]
225. Miyazawa, K.; He, F.; Kawase, M.; Kubota, A.; Yoda, K.; Hiramatsu, M. Enhancement of immunoregulatory effects of Lactobacillus gasseri TMC0356 by heat treatment and culture medium. *Lett. Appl. Microbiol.* **2011**, *53*, 210–216. [CrossRef]
226. Sokol, H.; Pigneur, B.; Watterlot, L.; Lakhdari, O.; Bermúdez-Humarán, L.G.; Gratadoux, J.-J.; Blugeon, S.; Bridonneau, C.; Furet, J.-P.; Corthier, G.; et al. Faecalibacterium prausnitzii is an anti-inflammatory commensal bacterium identified by gut microbiota analysis of Crohn disease patients. *Proc. Natl. Acad. Sci. USA* **2008**, *105*, 16731–16736. [CrossRef]
227. Tsilingiri, K.; Barbosa, T.; Penna, G.; Caprioli, F.; Sonzogni, A.; Viale, G.; Rescigno, M. Probiotic and postbiotic activity in health and disease: Comparison on a novel polarised ex-vivo organ culture model. *Gut* **2012**, *61*, 1007–1015. [CrossRef]
228. Haileselassie, Y.; Navis, M.; Vu, N.; Qazi, K.R.; Rethi, B.; Sverremark-Ekström, E. Postbiotic Modulation of Retinoic Acid Imprinted Mucosal-like Dendritic Cells by Probiotic Lactobacillus reuteri 17938 In Vitro. *Front. Immunol.* **2016**, *7*, 96. [CrossRef]
229. Béghin, L.; Tims, S.; Roelofs, M.; Rougé, C.; Oozeer, R.; Rakza, T.; Chirico, G.; Roeselers, G.; Knol, J.; Rozé, J.C.; et al. Fermented infant formula (with Bifidobacterium breve C50 and Streptococcus thermophilus O65) with prebiotic oligosaccharides is safe and modulates the gut microbiota towards a microbiota closer to that of breastfed infants. *Clin. Nutr.* **2021**, *40*, 778–787. [CrossRef]
230. Thibault, H.; Aubert-Jacquin, C.; Goulet, O. Effects of long-term consumption of a fermented infant formula (with Bifidobacterium breve c50 and Streptococcus thermophilus 065) on acute diarrhea in healthy infants. *J. Pediatr. Gastroenterol. Nutr.* **2004**, *39*, 147–152. [CrossRef]
231. Lukovac, S.; Belzer, C.; Pellis, L.; Keijser, B.J.; de Vos, W.M.; Montijn, R.C.; Roeselers, G. Differential modulation by Akkermansia muciniphila and Faecalibacterium prausnitzii of host peripheral lipid metabolism and histone acetylation in mouse gut organoids. *mBio* **2014**, *5*, 10-1128. [CrossRef] [PubMed]

232. Aguilar-Toalá, J.E.; Santiago-López, L.; Peres, C.M.; Peres, C.; Garcia, H.S.; Vallejo-Cordoba, B.; González-Córdova, A.F.; Hernández-Mendoza, A. Assessment of multifunctional activity of bioactive peptides derived from fermented milk by specific Lactobacillus plantarum strains. *J. Dairy Sci.* **2017**, *100*, 65–75. [CrossRef] [PubMed]
233. Dronkers, T.M.G.; Ouwehand, A.C.; Rijkers, G.T. Global analysis of clinical trials with probiotics. *Heliyon* **2020**, *6*, e04467. [CrossRef] [PubMed]
234. Critchfield, J.W.; van Hemert, S.; Ash, M.; Mulder, L.; Ashwood, P. The potential role of probiotics in the management of childhood autism spectrum disorders. *Gastroenterol. Res. Pract.* **2011**, *2011*, 161358. [CrossRef] [PubMed]
235. Song, W.; Zhang, M.; Teng, L.; Wang, Y.; Zhu, L. Prebiotics and probiotics for autism spectrum disorder: A systematic review and meta-analysis of controlled clinical trials. *J. Med. Microbiol.* **2022**, *71*, 001510. [CrossRef]
236. Depoorter, L.; Vandenplas, Y. Probiotics in pediatrics. A review and practical guide. *Nutrients* **2021**, *13*, 2176. [CrossRef]
237. van den Akker, C.H.P.; van Goudoever, J.B.; Shamir, R.; Domellöf, M.; Embleton, N.D.; Hojsak, I.; Lapillonne, A.; Mihatsch, W.A.; Berni Canani, R.; Bronsky, J.; et al. Probiotics and preterm infants: A position paper by the European Society for Paediatric Gastroenterology Hepatology and Nutrition Committee on nutrition and the European Society for Paediatric Gastroenterology Hepatology and Nutrition Working Group for probiotics and prebiotics. *J. Pediatr. Gastroenterol. Nutr.* **2020**, *70*, 664–680. [CrossRef]
238. Su, G.L.; Ko, C.W.; Bercik, P.; Falck-Ytter, Y.; Sultan, S.; Weizman, A.V.; Morgan, R.L. AGA clinical practice guidelines on the role of probiotics in the management of gastrointestinal disorders. *Gastroenterology* **2020**, *159*, 697–705. [CrossRef]
239. Weizman, Z.; Vandenplas, Y. Why do clinical guidelines for probiotics differ? *Benef. Microbes* **2024**, *15*, 411–415. [CrossRef]
240. Guarner, F.; Sanders, M.E.; Szajewska, H.; Cohen, H.; Eliakim, R.; Herrera-deGuise, C.; Karakan, T.; Merenstein, D.; Piscoya, A.; Ramakrishna, B.; et al. World gastroenterology organisation global guidelines: Probiotics and prebiotics. *J. Clin. Gastroenterol.* **2024**, *58*, 533–553. [CrossRef]
241. Collinson, S.; Deans, A.; Padua-Zamora, A.; Gregorio, G.V.; Li, C.; Dans, L.F.; Allen, S.J. Probiotics for treating acute infectious diarrhoea. *Cochrane Database Syst. Rev.* **2020**, *12*, CD003048. [CrossRef] [PubMed]
242. Szajewska, H.; Berni Canani, R.; Domellöf, M.; Guarino, A.; Hojsak, I.; Indrio, F.; Lo Vecchio, A.; Mihatsch, W.A.; Mosca, A.; Orel, R.; et al. ESPGHAN Special Interest Group on Gut Microbiota and Modifications Probiotics for the management of pediatric gastrointestinal disorders: Position paper of the ESPGHAN special interest group on gut microbiota and modifications. *J. Pediatr. Gastroenterol. Nutr.* **2023**, *76*, 232–247. [CrossRef] [PubMed]
243. Szajewska, H.; Kołodziej, M.; Zalewski, B.M. Systematic review with meta-analysis: Saccharomyces boulardii for treating acute gastroenteritis in children-a 2020 update. *Aliment. Pharmacol. Ther.* **2020**, *51*, 678–688. [CrossRef] [PubMed]
244. Patro-Gołąb, B.; Szajewska, H. Systematic Review with Meta-Analysis: Lactobacillus reuteri DSM 17938 for Treating Acute Gastroenteritis in Children. An Update. *Nutrients* **2019**, *11*, 2762. [CrossRef] [PubMed]
245. Rosenfeldt, V.; Michaelsen, K.F.; Jakobsen, M.; Larsen, C.N.; Møller, P.L.; Tvede, M.; Weyrehter, H.; Valerius, N.H.; Paerregaard, A. Effect of probiotic Lactobacillus strains on acute diarrhea in a cohort of nonhospitalized children attending day-care centers. *Pediatr. Infect. Dis. J.* **2002**, *21*, 417–419. [CrossRef]
246. İşlek, A.; Sayar, E.; Yılmaz, A.; Baysan, B.Ö.; Mutlu, D.; Artan, R. The role of Bifidobacterium lactis B94 plus inulin in the treatment of acute infectious diarrhea in children. *Turk. J. Gastroenterol.* **2014**, *25*, 628–633. [CrossRef]
247. Passariello, A.; Terrin, G.; Cecere, G.; Micillo, M.; De Marco, G.; Di Costanzo, M.; Cosenza, L.; Leone, L.; Nocerino, R.; Canani, R.B. Randomised clinical trial: Efficacy of a new synbiotic formulation containing Lactobacillus paracasei B21060 plus arabinogalactan and xilooligosaccharides in children with acute diarrhoea. *Aliment. Pharmacol. Ther.* **2012**, *35*, 782–788. [CrossRef]
248. Szymański, H.; Pejcz, J.; Jawień, M.; Chmielarczyk, A.; Strus, M.; Heczko, P.B. Treatment of acute infectious diarrhoea in infants and children with a mixture of three Lactobacillus rhamnosus strains—A randomized, double-blind, placebo-controlled trial. *Aliment. Pharmacol. Ther.* **2006**, *23*, 247–253. [CrossRef]
249. Canani, R.B.; Cirillo, P.; Terrin, G.; Cesarano, L.; Spagnuolo, M.I.; De Vincenzo, A.; Albano, F.; Passariello, A.; De Marco, G.; Manguso, F.; et al. Probiotics for treatment of acute diarrhoea in children: Randomised clinical trial of five different preparations. *BMJ* **2007**, *335*, 340. [CrossRef]
250. Chen, K.; Xin, J.; Zhang, G.; Xie, H.; Luo, L.; Yuan, S.; Bu, Y.; Yang, X.; Ge, Y.; Liu, C. A combination of three probiotic strains for treatment of acute diarrhoea in hospitalised children: An open label, randomised controlled trial. *Benef. Microbes* **2020**, *11*, 339–346. [CrossRef]
251. Guo, Q.; Goldenberg, J.Z.; Humphrey, C.; El Dib, R.; Johnston, B.C. Probiotics for the prevention of pediatric antibiotic-associated diarrhea. *Cochrane Database Syst. Rev.* **2019**, *4*, CD004827. [CrossRef] [PubMed]
252. Szajewska, H.; Canani, R.B.; Guarino, A.; Hojsak, I.; Indrio, F.; Kolacek, S.; Orel, R.; Shamir, R.; Vandenplas, Y.; van Goudoever, J.B.; et al. ESPGHAN Working Group for ProbioticsPrebiotics Probiotics for the Prevention of Antibiotic-Associated Diarrhea in Children. *J. Pediatr. Gastroenterol. Nutr.* **2016**, *62*, 495–506. [CrossRef] [PubMed]
253. Szajewska, H.; Kołodziej, M. Systematic review with meta-analysis: Lactobacillus rhamnosus GG in the prevention of antibiotic-associated diarrhoea in children and adults. *Aliment. Pharmacol. Ther.* **2015**, *42*, 1149–1157. [CrossRef] [PubMed]
254. Lukasik, J.; Dierikx, T.; Besseling-van der Vaart, I.; de Meij, T.; Szajewska, H.; Multispecies Probiotic in AAD Study Group. Multispecies Probiotic for the Prevention of Antibiotic-Associated Diarrhea in Children: A Randomized Clinical Trial. *JAMA Pediatr.* **2022**, *176*, 860–866. [CrossRef] [PubMed]

255. Ruszczyński, M.; Radzikowski, A.; Szajewska, H. Clinical trial: Effectiveness of Lactobacillus rhamnosus (strains E/N, Oxy and Pen) in the prevention of antibiotic-associated diarrhoea in children. *Aliment. Pharmacol. Ther.* **2008**, *28*, 154–161. [CrossRef] [PubMed]
256. Szajewska, H.; Wanke, M.; Patro, B. Meta-analysis: The effects of Lactobacillus rhamnosus GG supplementation for the prevention of healthcare-associated diarrhoea in children. *Aliment. Pharmacol. Ther.* **2011**, *34*, 1079–1087. [CrossRef]
257. Hojsak, I.; Szajewska, H.; Canani, R.B.; Guarino, A.; Indrio, F.; Kolacek, S.; Orel, R.; Shamir, R.; Vandenplas, Y.; van Goudoever, J.B.; et al. ESPGHAN Working Group for Probiotics/Prebiotics Probiotics for the prevention of nosocomial diarrhea in children. *J. Pediatr. Gastroenterol. Nutr.* **2018**, *66*, 3–9. [CrossRef]
258. Beghetti, I.; Panizza, D.; Lenzi, J.; Gori, D.; Martini, S.; Corvaglia, L.; Aceti, A. Probiotics for preventing necrotizing enterocolitis in preterm Infants: A network meta-analysis. *Nutrients* **2021**, *13*, 192. [CrossRef]
259. Chi, C.; Li, C.; Buys, N.; Wang, W.; Yin, C.; Sun, J. Effects of probiotics in preterm infants: A network meta-analysis. *Pediatrics* **2021**, *147*, e20200706. [CrossRef]
260. Gao, X.; Wang, Y.; Shi, L.; Feng, W.; Yi, K. Effect and Safety of Saccharomyces boulardii for Neonatal Necrotizing Enterocolitis in Pre-term Infants: A Systematic Review and Meta-Analysis. *J. Trop. Pediatr.* **2021**, *67*, fmaa022. [CrossRef]
261. van den Akker, C.H.P.; van Goudoever, J.B.; Szajewska, H.; Embleton, N.D.; Hojsak, I.; Reid, D.; Shamir, R. ESPGHAN Working Group for Probiotics, Prebiotics & Committee on Nutrition Probiotics for Preterm Infants: A Strain-Specific Systematic Review and Network Meta-analysis. *J. Pediatr. Gastroenterol. Nutr.* **2018**, *67*, 103–122. [CrossRef] [PubMed]
262. Athalye-Jape, G.; Rao, S.; Patole, S. Lactobacillus reuteri DSM 17938 as a Probiotic for Preterm Neonates: A Strain-Specific Systematic Review. *JPEN J. Parenter. Enteral Nutr.* **2016**, *40*, 783–794. [CrossRef] [PubMed]
263. Lin, H.-C.; Su, B.-H.; Chen, A.-C.; Lin, T.-W.; Tsai, C.-H.; Yeh, T.-F.; Oh, W. Oral probiotics reduce the incidence and severity of necrotizing enterocolitis in very low birth weight infants. *Pediatrics* **2005**, *115*, 1–4. [CrossRef] [PubMed]
264. Feng, J.-R.; Wang, F.; Qiu, X.; McFarland, L.V.; Chen, P.-F.; Zhou, R.; Liu, J.; Zhao, Q.; Li, J. Efficacy and safety of probiotic-supplemented triple therapy for eradication of Helicobacter pylori in children: A systematic review and network meta-analysis. *Eur. J. Clin. Pharmacol.* **2017**, *73*, 1199–1208. [CrossRef] [PubMed]
265. Wen, J.; Peng, P.; Chen, P.; Zeng, L.; Pan, Q.; Wei, W.; He, J. Probiotics in 14-day triple therapy for Asian pediatric patients with Helicobacter pylori infection: A network meta-analysis. *Oncotarget* **2017**, *8*, 96409–96418. [CrossRef]
266. Zhou, B.-G.; Chen, L.-X.; Li, B.; Wan, L.-Y.; Ai, Y.-W. Saccharomyces boulardii as an adjuvant therapy for Helicobacter pylori eradication: A systematic review and meta-analysis with trial sequential analysis. *Helicobacter* **2019**, *24*, e12651. [CrossRef]
267. Szajewska, H.; Horvath, A.; Kołodziej, M. Systematic review with meta-analysis: Saccharomyces boulardii supplementation and eradication of Helicobacter pylori infection. *Aliment. Pharmacol. Ther.* **2015**, *41*, 1237–1245. [CrossRef]
268. Fang, H.-R.; Zhang, G.-Q.; Cheng, J.-Y.; Li, Z.-Y. Efficacy of Lactobacillus-supplemented triple therapy for Helicobacter pylori infection in children: A meta-analysis of randomized controlled trials. *Eur. J. Pediatr.* **2019**, *178*, 7–16. [CrossRef]
269. Hurduc, V.; Plesca, D.; Dragomir, D.; Sajin, M.; Vandenplas, Y. A randomized, open trial evaluating the effect of Saccharomyces boulardii on the eradication rate of Helicobacter pylori infection in children. *Acta Paediatr.* **2009**, *98*, 127–131. [CrossRef]
270. Bin, Z.; Ya-Zheng, X.; Zhao-Hui, D.; Bo, C.; Li-Rong, J.; Vandenplas, Y. The Efficacy of Saccharomyces boulardii CNCM I-745 in Addition to Standard Helicobacter pylori Eradication Treatment in Children. *Pediatr. Gastroenterol. Hepatol. Nutr.* **2015**, *18*, 17–22. [CrossRef]
271. Sung, V.; D'Amico, F.; Cabana, M.D.; Chau, K.; Koren, G.; Savino, F.; Szajewska, H.; Deshpande, G.; Dupont, C.; Indrio, F.; et al. Lactobacillus reuteri to Treat Infant Colic: A Meta-analysis. *Pediatrics* **2018**, *141*, e20171811. [CrossRef] [PubMed]
272. Skonieczna-Żydecka, K.; Janda, K.; Kaczmarczyk, M.; Marlicz, W.; Łoniewski, I.; Łoniewska, B. The Effect of Probiotics on Symptoms, Gut Microbiota and Inflammatory Markers in Infantile Colic: A Systematic Review, Meta-Analysis and Meta-Regression of Randomized Controlled Trials. *J. Clin. Med.* **2020**, *9*, 999. [CrossRef] [PubMed]
273. Ong, T.G.; Gordon, M.; Banks, S.S.; Thomas, M.R.; Akobeng, A.K. Probiotics to prevent infantile colic. *Cochrane Database Syst. Rev.* **2019**, *3*, CD012473. [CrossRef] [PubMed]
274. Dryl, R.; Szajewska, H. Probiotics for management of infantile colic: A systematic review of randomized controlled trials. *Arch. Med. Sci.* **2018**, *14*, 1137–1143. [CrossRef]
275. Gutiérrez-Castrellón, P.; Indrio, F.; Bolio-Galvis, A.; Jiménez-Gutiérrez, C.; Jimenez-Escobar, I.; López-Velázquez, G. Efficacy of Lactobacillus reuteri DSM 17938 for infantile colic: Systematic review with network meta-analysis. *Medicine* **2017**, *96*, e9375. [CrossRef]
276. Schreck Bird, A.; Gregory, P.J.; Jalloh, M.A.; Risoldi Cochrane, Z.; Hein, D.J. Probiotics for the treatment of infantile colic: A systematic review. *J. Pharm. Pract.* **2017**, *30*, 366–374. [CrossRef]
277. Harb, T.; Matsuyama, M.; David, M.; Hill, R.J. Infant Colic-What works: A Systematic Review of Interventions for Breast-fed Infants. *J. Pediatr. Gastroenterol. Nutr.* **2016**, *62*, 668–686. [CrossRef]
278. Szajewska, H.; Gyrczuk, E.; Horvath, A. Lactobacillus reuteri DSM 17938 for the management of infantile colic in breastfed infants: A randomized, double-blind, placebo-controlled trial. *J. Pediatr.* **2013**, *162*, 257–262. [CrossRef]
279. Nocerino, R.; De Filippis, F.; Cecere, G.; Marino, A.; Micillo, M.; Di Scala, C.; de Caro, C.; Calignano, A.; Bruno, C.; Paparo, L.; et al. The therapeutic efficacy of Bifidobacterium animalis subsp. lactis BB-12® in infant colic: A randomised, double blind, placebo-controlled trial. *Aliment. Pharmacol. Ther.* **2020**, *51*, 110–120. [CrossRef]

280. Chen, K.; Zhang, G.; Xie, H.; You, L.; Li, H.; Zhang, Y.; Du, C.; Xu, S.; Melsaether, C.; Yuan, S. Efficacy of Bifidobacterium animalis subsp. lactis, BB-12® on infant colic-a randomised, double-blinded, placebo-controlled study. *Benef. Microbes* **2021**, *12*, 531–540. [CrossRef]
281. Gerasimov, S.; Gantzel, J.; Dementieva, N.; Schevchenko, O.; Tsitsura, O.; Guta, N.; Bobyk, V.; Kaprus, V. Role of Lactobacillus rhamnosus (FloraActive™) 19070-2 and Lactobacillus reuteri (FloraActive™) 12246 in Infant Colic: A Randomized Dietary Study. *Nutrients* **2018**, *10*, 1975. [CrossRef] [PubMed]
282. Baldassarre, M.E.; Di Mauro, A.; Tafuri, S.; Rizzo, V.; Gallone, M.S.; Mastromarino, P.; Capobianco, D.; Laghi, L.; Zhu, C.; Capozza, M.; et al. Effectiveness and Safety of a Probiotic-Mixture for the Treatment of Infantile Colic: A Double-Blind, Randomized, Placebo-Controlled Clinical Trial with Fecal Real-Time PCR and NMR-Based Metabolomics Analysis. *Nutrients* **2018**, *10*, 195. [CrossRef] [PubMed]
283. Indrio, F.; Di Mauro, A.; Riezzo, G.; Civardi, E.; Intini, C.; Corvaglia, L.; Ballardini, E.; Bisceglia, M.; Cinquetti, M.; Brazzoduro, E.; et al. Prophylactic use of a probiotic in the prevention of colic, regurgitation, and functional constipation: A randomized clinical trial. *JAMA Pediatr.* **2014**, *168*, 228–233. [CrossRef] [PubMed]
284. Trivić, I.; Niseteo, T.; Jadrešin, O.; Hojsak, I. Use of probiotics in the treatment of functional abdominal pain in children-systematic review and meta-analysis. *Eur. J. Pediatr.* **2021**, *180*, 339–351. [CrossRef] [PubMed]
285. Weizman, Z.; Abu-Abed, J.; Binsztok, M. Lactobacillus reuteri DSM 17938 for the Management of Functional Abdominal Pain in Childhood: A Randomized, Double-Blind, Placebo-Controlled Trial. *J. Pediatr.* **2016**, *174*, 160–164.e1. [CrossRef]
286. Gawrońska, A.; Dziechciarz, P.; Horvath, A.; Szajewska, H. A randomized double-blind placebo-controlled trial of Lactobacillus GG for abdominal pain disorders in children. *Aliment. Pharmacol. Ther.* **2007**, *25*, 177–184. [CrossRef]
287. Kaur, L.; Gordon, M.; Baines, P.A.; Iheozor-Ejiofor, Z.; Sinopoulou, V.; Akobeng, A.K. Probiotics for induction of remission in ulcerative colitis. *Cochrane Database Syst. Rev.* **2020**, *3*, CD005573. [CrossRef]
288. Miele, E.; Pascarella, F.; Giannetti, E.; Quaglietta, L.; Baldassano, R.N.; Staiano, A. Effect of a probiotic preparation (VSL#3) on induction and maintenance of remission in children with ulcerative colitis. *Am. J. Gastroenterol.* **2009**, *104*, 437–443. [CrossRef]
289. Mimura, T.; Rizzello, F.; Helwig, U.; Poggioli, G.; Schreiber, S.; Talbot, I.C.; Nicholls, R.J.; Gionchetti, P.; Campieri, M.; Kamm, M.A. Once daily high dose probiotic therapy (VSL#3) for maintaining remission in recurrent or refractory pouchitis. *Gut* **2004**, *53*, 108–114.
290. Gionchetti, P.; Rizzello, F.; Venturi, A.; Brigidi, P.; Matteuzzi, D.; Bazzocchi, G.; Poggioli, G.; Miglioli, M.; Campieri, M. Oral bacteriotherapy as maintenance treatment in patients with chronic pouchitis: A double-blind, placebo-controlled trial. *Gastroenterology* **2000**, *119*, 305–309. [CrossRef]
291. Blanchetière, A.; Dolladille, C.; Goyer, I.; Join-Lambert, O.; Fazilleau, L. State of the Art of Probiotic Use in Neonatal Intensive Care Units in French-Speaking European Countries. *Children* **2023**, *10*, 1889. [CrossRef] [PubMed]
292. Hempel, S.; Newberry, S.; Ruelaz, A.; Wang, Z.; Miles, J.N.V.; Suttorp, M.J.; Johnsen, B.; Shanman, R.; Slusser, W.; Fu, N.; et al. Safety of probiotics used to reduce risk and prevent or treat disease. *Evid. Rep. Technol. Assess* **2011**, *200*, 1–645.
293. Zhao, F.; Bai, X.; Zhang, J.; Kwok, L.-Y.; Shen, L.; Jin, H.; Sun, T.; Sun, Z.; Zhang, H. Gut Bifidobacterium responses to probiotic Lactobacillus casei Zhang administration vary between subjects from different geographic regions. *Appl. Microbiol. Biotechnol.* **2022**, *106*, 2665–2675. [CrossRef] [PubMed]
294. Lee, S.-H.; Cho, S.-Y.; Yoon, Y.; Park, C.; Sohn, J.; Jeong, J.-J.; Jeon, B.-N.; Jang, M.; An, C.; Lee, S.; et al. Bifidobacterium bifidum strains synergize with immune checkpoint inhibitors to reduce tumour burden in mice. *Nat. Microbiol.* **2021**, *6*, 277–288. [CrossRef] [PubMed]
295. Bomba, A.; Jonecová, Z.; Koščová, J.; Nemcová, R.; Gancarčíková, S.; Mudroňová, D.; Sciranková, L.; Buleca, V.; Lazar, G.; Pošivák, J.; et al. The improvement of probiotics efficacy by synergistically acting components of natural origin: A review. *Biologia* **2006**, *61*, 729–734. [CrossRef]
296. Gasaly, N.; de Vos, P.; Hermoso, M.A. Impact of bacterial metabolites on gut barrier function and host immunity: A focus on bacterial metabolism and its relevance for intestinal inflammation. *Front. Immunol.* **2021**, *12*, 658354. [CrossRef]
297. Weber, T.K.; Polanco, I. Gastrointestinal microbiota and some children diseases: A review. *Gastroenterol. Res. Pract.* **2012**, *2012*, 676585. [CrossRef]
298. Weichert, S.; Schroten, H.; Adam, R. The role of prebiotics and probiotics in prevention and treatment of childhood infectious diseases. *Pediatr. Infect. Dis. J.* **2012**, *31*, 859–862. [CrossRef]

nutrients

MDPI

Review

Probiotics in Infancy and Childhood for Food Allergy Prevention and Treatment

Margherita Di Costanzo [1,2,*], Adriana Vella [3], Claudia Infantino [3], Riccardo Morini [3], Simone Bruni [3], Susanna Esposito [3] and Giacomo Biasucci [1,2]

[1] Pediatrics and Neonatology Unit, Guglielmo da Saliceto Hospital, 29121 Piacenza, Italy; giacomo.biasucci@unipr.it
[2] Department of Medicine and Surgery, University of Parma, 43126 Parma, Italy
[3] Pediatric Clinic, Department of Medicine and Surgery, University of Parma, 43126 Parma, Italy; adrivella.95@gmail.com (A.V.); infantinoclaudia@gmail.com (C.I.); riccardomorini95@gmail.com (R.M.); simone.bruni@unipr.it (S.B.); susannamariaroberta.esposito@unipr.it (S.E.)
* Correspondence: margherita.dicostanzo@unipr.it; Tel.: +39-0523-303370

Abstract: Food allergy represents a failure of oral tolerance mechanisms to dietary antigens. Over the past few years, food allergies have become a growing public health problem worldwide. Gut microbiota is believed to have a significant impact on oral tolerance to food antigens and in initiation and maintenance of food allergies. Therefore, probiotics have also been proposed in this field as a possible strategy for modulating both the gut microbiota and the immune system. In recent years, results from preclinical and clinical studies suggest a promising role for probiotics in food allergy prevention and treatment. However, future studies are needed to better understand the mechanisms of action of probiotics in food allergies and to design comparable study protocols using specific probiotic strains, defined doses and exposure times, and longer follow-up periods.

Keywords: oral tolerance; gut microbiota; gut dysbiosis; cow's milk allergy; egg allergy; shellfish allergy; food specific immunotherapy

1. Introduction

Food allergy represents a major health issue in Western countries due to its increasing prevalence in the last several decades, reaching rates of 8% in children and 3% in adults [1]. In the last few years, an increase in the severity of food-induced allergic reactions, such as anaphylaxis, has been reported in children [2]. Allergens responsible for allergic reactions are usually different in children and adults; indeed, peanut (2.2%), milk (1.9%), shellfish (1.3%), and tree nuts (1.2%) are the most common allergens in children, whereas shellfish (2.9%), milk (1.9%), peanut (1.8%), tree nuts (1.2%), and finfish (0.9%) are the most common ones in adults [3]. Furthermore, different allergens generally result in food allergies with a different clinical course. Thus, childhood food allergies to cow's milk, egg, wheat, or soy typically resolve during childhood, while food allergies to peanuts, tree nuts, fish, and shellfish are usually persistent in adulthood. However, a changing pattern in food allergies has been observed in the last thirty years, with an increased risk of persistence up to later ages [3,4]. Beside this, it is well known that individuals with food allergies are at a higher risk of developing other allergic conditions later in life as part of the atopic march. These conditions may include allergic rhinitis, conjunctivitis, and asthma [5]. Recent epidemiological changes have placed a significant burden on patients, their caregivers, and healthcare systems. Therefore, it is very important to develop effective strategies for the management of food allergies, starting at a young age. The elimination diet is currently the first-line treatment for all children with food allergies. There is increasing evidence that the gut microbiota plays a crucial role in overall health. It is likely that gut dysbiosis, which is an imbalance in gut microbiota composition and functions, anticipates the development of food allergies [6]. Thus, modulation of gut microbiota has become a potential tool for prevention and treatment of food allergies. Hence, the use of probiotics has been claimed as one of the possible strategies to modulate gut microbiota composition and functions.

This paper aims to review the most recent and relevant preclinical and clinical studies on the use of probiotics in the management of food allergies, analyzing the real possibility of an effective strategy for their prevention and treatment in the future.

2. Gut Dysbiosis and Food Allergies

Based on the hypothesis that gut dysbiosis may play a role in the development of food allergies, many studies have been carried out to assess whether gut microbiota composition and functions could be associated with the development, persistence, or resolution of food allergies, as well as which biological mechanisms could be involved.

Several observational studies have shown that gut dysbiosis plays a role in the development of food allergies [6–15]. The currently available studies are quite diverse, and no specific bacterial group has been definitively linked to the onset or clinical course of food allergy [6–15]. However, the results of observational studies suggest that gut dysbiosis may precede the onset of food allergies. Additionally, research indicates that gut microbiota during early life, particularly in the first 6 months of life, play a crucial role not only in the development but also in the persistence of food allergies until adulthood [10]. A recent study analyzed the fecal microbiome and metabolome of food allergy concordant or discordant twin pairs, suggesting a potential role of gut dysbiosis in food allergies beyond infancy and into adulthood. Data analysis revealed a diverse gut microbiota and metabolites in twins with food allergies compared to healthy twins, even within the same twin pairs, both in infancy and adulthood. These findings suggest that the gut microbiota may have a protective role against food allergies, even in adulthood [16].

Further studies suggested that *Akkermansia muciniphila* and *Faecalibacterium prausnitzii* contribute to allergic diseases and that their colonization in the gut microbiota is modified in atopic patients compared to healthy controls [17–19]. However, current studies in pediatric patients are preliminary and only focused on atopic dermatitis and allergic asthma [17–19].

In a cross-sectional observational pilot study, Fieten et al. analyzed the fecal microbiome of children with atopic dermatitis with or without a concomitant food allergy and found that *F. prausnitzii* and *A. muciniphila* discriminate between the presence and absence of food allergy in children with atopic dermatitis (p = 0.001). The fecal microbiome of children with atopic dermatitis and food allergies harbored relatively less *F. prausnitzii* and *A. muciniphila* than that of children with atopic dermatitis without food allergies [20].

De Filippis et al. identified specific microbial signatures in the gut microbiome of allergic children affected by food or respiratory allergies, such as a higher abundance of *Ruminococcus gnavus* and *F. prausnitzii* and a depletion of *Bifidobacterium longum, Bacteroides dorei, B. vulgatus,* and fiber-degrading taxa [21]. The authors hypothesized that the increased abundance of *F. prausnitzii* reported in allergic subjects in this and previous studies was probably linked to an increase in *F. prausnitzii* clade A, previously associated with the Westernized lifestyle [22]. Interestingly, Song et al. found an increase in *F. prausnitzii* strain L2-6 (belonging to clade A) in atopic dermatitis, suggesting a role of this *F. prausnitzii* clade in allergy development [23].

Observational studies in humans, however, provide no evidence about a causal relationship between gut dysbiosis and the development of food allergies and do not elucidate the mechanisms involved. Animal models show that antibiotic-treated mice exhibit a predisposition to allergy development, while germ-free mice do not develop oral tolerance but maintain a Th2 immune response to oral administration of food antigens [24,25]. This condition may only be reversed by early gut microbiota remodulation. These data support the pivotal role of gut microbiota in establishing oral tolerance to dietary antigens early in life. Indeed, germ-free mice, colonized with feces from healthy donors, are protected from developing cow's milk allergy (CMA) upon sensitization and challenge to cow's milk proteins. In contrast, germ-free mice colonized with feces from infants with CMA exhibit severe allergic responses, including anaphylaxis [26].

There are multiple mechanisms by which the gut microbiota may influence food allergy predisposition. Murine models of food allergy have shown several effects of the gut microbiota, including modulation of the Th2 immune response, regulation of the development of mucosal immunity and oral tolerance, regulation of basophil populations, and promotion of gut barrier function through reduced gut permeability and increased mucus production [27].

Also, even metabolites resulting from the gut microbiota functions have an emerging role in food allergies. Butyrate is a short-chain fatty acid produced by the fermentation of dietary fiber in the colon. It has a strong immunoregulatory effect, which is expressed through both immune and non-immune mechanisms of action. Indeed, butyrate can improve the integrity of the intestinal epithelial barrier by increasing the thickness of the mucus layer and the expression of tight junctions. Alongside these effects, it has several direct and indirect effects on immune cells that contribute to the induction and maintenance of oral tolerance [28].

As for butyrate production, a Canadian longitudinal study showed that infants who develop allergic sensitization do not differ from those who already have gut dysbiosis with reduced butyrate production at 3 months of age [29].

On the other hand, the protective effects of breastfeeding on food allergies may be partly explained by the human milk butyrate content, which has been demonstrated to modulate several tolerogenic mechanisms. Human milk butyrate could at least partly explain the breastfeeding protective effect towards food allergies. This effect has been tested in animal models *in vivo* and in cellular models *in vitro*. Butyrate can regulate gut barrier function, promote the activation of regulatory T cells (Tregs), and modulate the Th1/Th2 response in favor of a tolerogenic Th1 immune response [30].

3. Probiotic-Induced Gut Microbiota Modulation for Food Allergy Prevention and Treatment

3.1. Probiotics and Their Mechanisms of Action

According to the widely recognized FAO/WHO definition [31], revised in a consensus statement by the International Scientific Association for Probiotics and Prebiotics [32], probiotics are defined as "live microorganisms that, when administered in adequate amounts, confer a health benefit on the host". To meet this definition, probiotics must be present in a reasonable amount within the product. It has been suggested that at least 1×10^9 colony-forming units (CFU) are required to ensure gut colonization and exert measurable beneficial effects [33]. Lower amounts can be used in cases where robust scientific evidence supports the specific strain's colonization ability. Probiotics are available on the market in many different forms, including medicinal products, medical devices, dietary supplements, and foodstuffs. Globally, the probiotics market has steadily grown in the last few years.

This is true for all probiotic formulations available and at any age, both in childhood and adulthood [34]. Considering this increase in marketing, it is worthy of being emphasized that not all probiotics are the same and/or provide the same beneficial effects. Growing evidence shows that the efficacy of probiotics is strain- and disease-specific [35]. Therefore, health professionals should take these two aspects into account when recommending probiotics.

Different strains of probiotics have distinct mechanisms of action. This may explain why some probiotics, unlike others, are effective against specific diseases and conditions [35]. Current preclinical and clinical data regarding the possible role for probiotics in the prevention and treatment of allergies, specifically food allergies, are encouraging but not yet sufficient to strongly recommend the use of probiotics in these conditions. In the following sections, we will discuss the main preclinical and clinical studies that have evaluated the role of probiotics and their mechanisms of action in the prevention and treatment of pediatric food allergies.

3.2. Data from Animal Models of Food Allergy

Mice are the most widely used animal models to study the pathophysiological mechanisms underlying IgE-mediated food allergies, and the possible preventive and therapeutic strategies for cow's milk, egg, and shellfish allergies. Herein, we report the most recent studies, limited to the last five years, focused on the role of specific probiotic strains in mouse models of food allergies.

3.2.1. Cow's Milk Allergy

In a recent study, *Lactobacillus* (L.) *plantarum* HM-22 has been administered by gavage to α-lactalbumin-induced allergic mice for five weeks to investigate its possible effect on gut inflammation and microbiota. The study demonstrated that *L. plantarum* HM-22 induced a significant increase in serum levels of tolerogenic cytokines, including IL-10, IFN-γ, and TGF-β, and a significant decrease in serum total IgE and IL-4 levels in mice with α-lactalbumin-induced allergy. The colonic tissue crypt structure of α-lactalbumin-induced allergic mice was initially altered, resulting in reduced goblet cells and increased inflammatory corpuscles, but *L. plantarum* HM-22 administration was found to attenuate these effects. Furthermore, *L. plantarum* HM-22 significantly increased the expression of occludin and claudin-1 in the colon of α-lactalbumin-induced allergic mice, thereby reducing gut permeability. In addition, *L. plantarum* HM-22 enhanced gut microbiota colonization in α-lactalbumin-induced allergic mice [36].

Lactobacillus acidophilus KLDS 1.0738 was found to alleviate β-lactoglobulin-induced allergic inflammation in a mouse model of cow's milk allergy (CMA) [37]. Furthermore, Li et al. investigated its molecular regulation mechanism in β-lactoglobulin-induced macrophages, treated with viable or non-viable strains of *L. acidophilus* KLDS 1.0738 and Toll-like receptor 4 (TLR4) inhibitors or miR-146a inhibitors. The results showed that treatment with *L. acidophilus* KLDS 1.0738 may suppress the TLR4/NF-κB signaling pathway by modulating miR-146a expression, thereby reducing the overexpression of downstream inflammatory factors [38].

Fu et al. assessed the impact of three *Lactobacillus* strains on the immune system, gut barrier, and gut microbiota in β-lactoglobulin-induced allergic mouse model. Oral administration of *L. plantarum* ZDY2013 and *L. rhamnosus* GG (LGG) suppressed the allergic response by reducing serum total IgE levels, attenuating anaphylaxis symptoms, and inducing Th1 immune cells or Tregs differentiation to inhibit the Th2 immune response. In addition, *L. plantarum* ZDY2013 and LGG improved gut barrier function through tight junction regulation, and *L. plantarum* ZDY2013 and *L. plantarum* WLPL04 regulated gut dysbiosis in allergic mice [39].

3.2.2. Egg Allergy

In a recent study, the use of *Bifidobacterium longum* subsp. longum 51A (BL51A) was evaluated in a mouse model of ovalbumin (OVA) food allergy. BL51A was orally administered and resulted in a reduction of OVA-specific serum IgE levels, gut permeability, proximal jejunal damage, eosinophil and neutrophil recruitment, and levels of eotaxin-1, CXCL1/KC, IL-4, IL-5, IL-6, IL-13, and TNF. In addition, this treatment increased IL-10 levels [40]. Recently, similar results were obtained for *Akkermansia muciniphila* BAA-835 in a mouse model of OVA food allergy [41]. However, a recent study found that the impact of *A. muciniphila* is context dependent and can be detrimental to food allergies when the microbiota is deprived of dietary fiber. To investigate the causal role of *A. muciniphila* in modulating food allergies, the authors used germ-free mice colonized with a fully characterized 14-member synthetic human gut microbiota, in which *A. muciniphila* can be included or excluded. The study found that the presence of *A. muciniphila* in the microbiota, combined with fiber deprivation, led to stronger anti-commensal IgE coating and innate type 2 immune responses. This worsened food allergy symptoms in animal models of OVA and peanut allergy [42].

Duan et al. investigated the effects of oral administration of *L. plantarum* JC7 using a mouse model of OVA sensitization. The authors showed that *L. plantarum* JC7 significantly alleviated allergic manifestations; it also reduced plasma histamine levels, OVA-specific serum IgE levels, and shifted Th1/Th2 and Treg/Th17 imbalances. This was achieved by promoting the secretion of IL-10 and IFN-γ tolerogenic cytokines, meantime, inhibiting secretion of those involved in the allergic response, such as IL-4 and Th17. The observed effects may be attributed to the activation of the NF-κB signaling pathway. In addition, OVA-sensitized group showed gut dysbiosis that was restored by *L. plantarum* JC7 oral administration. Specifically, oral administration of *L. plantarum* JC7 increased the richness, diversity, and uniformity of cecum microbiota, which was characterized by a higher abundance of Bacteroidetes and reduced Firmicutes colonization [43].

In another study by Miranda et al., the probiotic effect of *Saccharomyces cerevisiae* UFMG A-905 was evaluated in an OVA food allergy model. The authors also evaluated if *Saccharomyces cerevisiae* UFMG A-905 might be effective after inactivation. The study found that oral administration of only viable probiotics significantly reduced tissue damage and myeloperoxidase activity, as well as IL-17 levels. However, this study did not find any significant changes in the serum OVA-specific IgE and IgG levels. This suggests that the observed effects in the evaluated murine model were local rather than systemic [44].

3.2.3. Shellfish Allergy

Fu et al. demonstrated different effects of oral administration of five distinct strains of lactic acid bacteria in alleviating gut allergic inflammation and symptoms related to food-induced anaphylaxis in a mouse model of food allergy to shrimp tropomyosin, a major shrimp allergen. The most effective strain in reducing allergies was *Bacillus coagulans* 09.712, which significantly improved epithelial barrier function and increased lymphocyte proliferation. *Bacillus coagulans* 09.712 induces CD4+Foxp3+Tregs production, which suppresses the pro-inflammatory Th17 response in this allergic mouse model. Also, *Bacillus coagulans* 09.712 administration suppresses mTOR activation, resulting in up-regulation of FOXP3 and down-regulation of GATA-3, which, in turn, facilitates the control of tropomyosin-induced pro-inflammatory Th2 and Th17 immune responses [45].

Oral administration of *L. casei* Zhang probiotic strain reduced allergy symptoms and gut epithelial damage in a mouse model of tropomyosin-induced food allergy. In addition, administration of *L. casei* Zhang changed development and function of dendritic cells (DCs), T cells, and B cells, resulting in a tropomyosin-specific antibody isotypes shift towards a more tolerogenic pattern through the activation of the NF-κB signaling pathway [46]. Moreover, in a previous study, the same authors demonstrated that even *Bifidobacterium (B.) infantis* can alleviate shrimp tropomyosin-induced food allergy in mice by tolerogenic DCs-dependent Treg induction, by and a favorable gut microbiota modulation [47] (Table 1).

In conclusion, though the results of probiotics use in animal models of food allergies look promising, they cannot be immediately translated to humans due to many genetic and environmental factors that may influence food allergies onset and course. Nonetheless, well-designed animal models may be useful for future studies to better understand the mechanisms that underlie specific probiotic strains effects in food allergies.

Table 1. Data from animal models of food allergies.

<table>
<tr><th>Food</th><th>Study</th><th>Probiotic</th><th>Findings</th></tr>
<tr><td rowspan="3">Cow's milk</td><td>Jiang et al. [36]</td><td>L. plantarum HM-22</td><td>Increased serum levels of IL-10, IFN-γ, and TGF-β;
Reduced serum levels of total IgE and IL-4;
Reduced gut permeability (increased expression of occludin and claudin-1 in the colon).</td></tr>
<tr><td>Ni et al.
Li et al.
[37,38]</td><td>L. acidophilus KLDS 1.0738</td><td>Suppression of the TLR4/NF-kB signaling pathway</td></tr>
<tr><td>Fu et al. [39]</td><td>L. plantarum ZDY2013 and L. rhamnosus GG (LGG)</td><td>Reduced serum levels of total IgE;
Promoted Th1 differentiation, inhibiting Th2 responses;
Improved gut barrier function.</td></tr>
<tr><td rowspan="5">Egg</td><td>Santos et al. [40]</td><td>B. longum 51A</td><td rowspan="2">Reduced serum levels of total IgE, gut permeability, proximal jejunal damage, eosinophil and neutrophil recruitment, and levels of eotaxin-1, CXCL1/KC, IL-4, IL-5, IL-6, IL-13, and TNF;
Increased serum levels of IL-10.</td></tr>
<tr><td>Miranda et al. [41]</td><td>A. muciniphila BAA-835</td></tr>
<tr><td>Parrish et al. [42]</td><td>A. muciniphila</td><td>The presence of A. muciniphila in the microbiota, combined with fiber deprivation, led to stronger anti-commensal IgE coating and innate type 2 immune responses. This worsened food allergy symptoms in animal models of OVA and peanut allergy.</td></tr>
<tr><td>Duan et al. [43]</td><td>L. plantarum JC7</td><td>Reduced plasma histamine levels, OVA-specific IgE serum levels, shift in Th1/Th2 immune response, and Treg/Th17 imbalance.</td></tr>
<tr><td>Miranda et al. [44]</td><td>Saccharomyces cerevisiae UFMG A-905</td><td>Reduced tissue damage, myeloperoxidase activity levels, and IL-17 serum levels.</td></tr>
<tr><td rowspan="3">Shellfish</td><td>Fu et al. [45]</td><td>Bacillus coagulans 09.712</td><td>Improved gut barrier function;
Suppression of the pro-inflammatory Th17 response.</td></tr>
<tr><td>Fu et al. [46]</td><td>L. casei Zhang</td><td>Attenuated allergy symptoms and gut epithelial damage;
Favoring a tolerogenic pattern through the activation of the NF-κB signaling pathway.</td></tr>
<tr><td>Fu et al. [47]</td><td>B. infantis</td><td>Attenuated allergy symptoms;
Induction of Tregs.</td></tr>
</table>

3.3. Data from Human Studies Related to Food Allergy Prevention

Clinical studies that have evaluated the role of probiotics in the prevention of food allergies are based on three different approaches:

- administration of probiotics only to the mother during pregnancy and breastfeeding;
- administration of probiotics to mother and infant in the perinatal period;
- administration of probiotics only to infants after delivery.

3.3.1. Administration of Probiotics Only to the Mother during Pregnancy and Breastfeeding

Boyle et al. conducted a randomized controlled trial to investigate the effects of prenatal treatment with LGG on pregnant mothers from 36 weeks of gestation until delivery. The study found that this treatment did not reduce the risk of eczema and food sensitization to eggs, peanuts, and cow's milk in infants at high risk of developing allergic diseases, based on a one-year follow-up [48].

In a double-blind, randomized trial, pregnant women were given either probiotic-supplemented milk or placebo from 36 weeks of gestation until three months postpartum while breastfeeding. The probiotic milk contained LGG, *L. acidophilus* La-5, and *B. animalis* subsp. *lactis* Bb-12. At two years of age, their children underwent assessments for atopic sensitization, atopic dermatitis, asthma, and allergic rhino-conjunctivitis. The authors concluded that administering probiotics to non-selected mothers reduced atopic dermatitis overall incidence, but did not have any impact on atopic sensitization. The tested trophoallergens included cow's milk, hen egg white, cod, hazelnut, and peanut [49].

Rautava et al. conducted a double-blind, placebo-controlled study to determine if probiotic supplementation during pregnancy and breastfeeding in mothers with allergic diseases and atopic sensitization could reduce the risk of eczema development in infants, with a two-year follow-up. Mothers were randomized to receive: *L. rhamnosus* LPR and *B. longum* BL999; *L. paracasei* ST11 and *B. longum* BL999; or placebo, starting two months before delivery throughout the first two months after delivery, while breastfeeding. Infants of mothers who received *L. rhamnosus* LPR and *B. longum* BL999 or *L. paracasei* ST11 and *B. longum* BL999 had a significantly lower risk of developing eczema during the first two years of life compared to placebo. In contrast, the two probiotic mixtures did not affect the risk of infants' atopic sensitization. The tested trophoallergens included cow's milk, hen egg white, wheat and rice flour, cod, soybean, potato, carrot, and banana [50].

3.3.2. Administration of Probiotics to Mother and Infant in the Perinatal Period

One of the first large sample size studies was a double-blind, placebo-controlled trial conducted in mothers with infants at high risk of allergy. Pregnant women were randomized to receive a probiotic mixture consisting of two lactobacilli, bifidobacteria, and propionibacteria (a capsule containing freeze-dried LGG, *L. rhamnosus* LC705, *B. breve* Bb99, and *Propionibacterium freudenreichii* ssp. *shermanii* JS) or placebo during the last month of pregnancy, and their infants were to receive it from birth until age 6 months. Infants also received a prebiotic galactooligosaccharide or placebo. At five years, the cumulative incidence of allergic diseases (eczema, food allergy, allergic rhinitis, and asthma) and IgE sensitization did not differ between the two study groups. However, there were fewer IgE-associated allergic diseases in cesarean section delivered children who received probiotics. The authors concluded that probiotic supplementation during the last month of pregnancy, and during the first six months of infants' life, is not effective in reducing the incidence of allergic diseases at five years of age [51].

Indeed, in a previous double-blind, randomized, placebo-controlled study, a single LGG probiotic strain was administered prenatally to mothers with infants at high risk allergy and postnatally to their infants for six months. The authors found that LGG was effective in preventing early atopic disease in high-risk children throughout a two-year follow-up period [52].

3.3.3. Administration of Probiotics Only to Infants after Delivery

Postnatal administration of a probiotic mixture consisting of *B. infantis*, *B. lactis*, and *Streptococcus thermophilus* did not affect the incidence of allergic manifestations or atopic sensitization during the first two years of life in very preterm newborns [53]. In this study, food allergy was defined according to a parental report of a physician-diagnosed allergy to cow's milk, soy, egg, wheat, or peanut. Skin prick tests were performed only for egg white, cow's milk, and peanut. Additionally, this study did not evaluate the strain- and disease-specific probiotic effect.

In a multicenter, randomized, double-blind, controlled study, a non-hydrolyzed fermented infant formula containing heat-killed *B. breve* C50 and *Streptococcus thermophilus* 065 (HKBBST) was administered to infants at high risk of atopy during their first year of life. The use of HKBBST milk did not affect the proportion of CMA but reduced the proportion of positive skin prick tests to cow's milk and the occurrence of allergy-like events in the first two years of life [54] (Table 2).

Table 2. Data from human studies related to food allergy prevention.

Study	Probiotic	Findings
Boyle et al. [48]	LGG to a pregnant mother	No reduction in the risk of eczema and food sensitization to eggs, peanuts, and cow's milk in offspring
Dotterud et al. [49]	LGG, *L. acidophilus* La-5, and *B. animalis* subsp. lactis Bb-12 to a pregnant and lactating mother	Lower cumulative incidence of atopic dermatitis and no effect on atopic sensitization in offspring
Rautava et al. [50]	*L. rhamnosus* LPR and *B. longum* BL999 or *L. paracasei* ST11 and *B. longum* BL999 to pregnant and lactating mothers	Lower risk of eczema and no effect on atopic sensitization in offspring
Kuitunen et al. [51]	Probiotic mixture (lactobacilli, bifidobacteria, and propionibacteria) for pregnant mothers and their infants after birth	No difference in the cumulative incidence of allergic diseases and IgE sensitization at 5 years of life in offspring
Kalliomäki et al. [52]	LGG to the pregnant mother and their infants after birth	Lower incidence of early atopic diseases in high-risk children
Plummer et al. [53]	*B. infantis, B. lactis,* and *Streptococcus thermophilus* in very preterm newborns	No effect on the incidence of allergic diseases or atopic sensitization during the first 2 years of life
Morisset et al. [54]	Not hydrolyzed fermented formula containing heat-killed *B. breve* C50 and *Streptococcus thermophilus* 065 to infants at high risk of atopy	No effect on the incidence of CMA, a lower proportion of positive skin prick tests in cow's milk, or a lower occurrence of allergy-like events in the first 2 years of life

In 2015, the World Allergy Organization (WAO) issued guidelines on probiotics for the prevention of allergic diseases. The WAO guideline panel suggests supplementation with probiotics in pregnant women at high risk of having an allergic child, in women who breastfeed infants at high risk of developing allergies, and in infants at high risk of developing allergies. The authors specified that all recommendations are conditional and supported by very low-quality evidence [55].

In 2016, Zhang et al. conducted a PRISMA-compliant systematic review and meta-analysis of randomized controlled trials on probiotics for the prevention of atopy and food hypersensitivity in early childhood. The results indicated that administering probiotics prenatally and postnatally could reduce the risk of atopy and food hypersensitivity in young children [56].

To date, many studies have evaluated the role of probiotics in preventing food sensitization without assessing their effects on confirmed food allergy prevention. Therefore, further studies are needed to determine the effectiveness of probiotics as a global prevention strategy for food allergies. Future studies should also assess the optimal probiotic strains, dosing, and duration of therapy and should be designed with a long-term follow-up period.

3.4. Data from Human Studies Related to Food Allergy Treatment

Most of the available clinical studies on the use of probiotics as a possible therapeutic strategy for pediatric food allergies focus on IgE-mediated CMA, which is the earliest and most common food allergy in pediatrics [57]. It is usually resolved at school age, but the natural history of food allergies has changed in recent years, and persistent forms of food allergy in adulthood are increasingly common [47]. Currently, CMA therapy is based on cow's milk proteins elimination diet replaced by the use of special alternative formulas in non-breastfed infants [58]. Special formulas mostly used for the management of CMA are: extensively hydrolyzed whey formula (eHWF), extensively hydrolyzed casein formula (eHCF), soy formula (SF), hydrolyzed rice formula (HRF), and amino acid-based formula (AAF) [59]. These hypoallergenic formulas resolve allergic symptoms by lacking IgE-binding epitopes [60]. However, besides ameliorating allergic symptoms, it should be crucial to find strategies to promote oral tolerance in patients with food allergies.

Berni Canani et al. demonstrated that in children with IgE-mediated CMA, LGG-supplemented eHCF resulted in higher rates of oral tolerance compared to eHCF without LGG and other hypoallergenic formulas used in CMA treatment [61].

These findings were consistent with those of a one-year follow-up study conducted in the United States, which showed better outcomes using eHCF plus LGG compared to eHCF alone or AAF, as first-line CMA dietary management in infants [62].

The use of eHCF plus LGG for the treatment of IgE-mediated CMA in children is associated with a higher rate of oral tolerance acquisition and a lower incidence of atopic manifestations compared to the use of eHCF alone, or other special formulas for CMA treatment (e.g., HRF, SF, eHWF, AAF), even after a 36-months follow-up [63,64]. These results align with those of a retrospective study performed in the United Kingdom based on a large cohort of formula-fed CMA infants extracted from the Health Improvement Network database, which indicated that eHCF plus LGG is not only more effective than eHWF in managing CMA symptoms, but it also has greater potential to prevent the occurrence of other atopic manifestations in these patients [65].

Basturk et al. conducted a randomized, double-blind, placebo-controlled trial in CMA infants who received oral LGG for 4 weeks. The mothers of all breastfed patients were put on a milk-free diet, and all formula-fed patients were offered eHF. The probiotic group showed statistically significant improvement in symptoms such as bloody stools, diarrhea, restiveness, and abdominal distension, as well as improvement in mucous stools and vomiting, compared to the placebo group. In contrast, a statistically significant improvement in abdominal pain, constipation, and dermatitis was not observed. Although the probiotic group had higher complete recovery rates than the placebo group, the difference was not statistically significant [66].

All the clinical studies presented so far have evaluated the role of LGG alone, or in addition to formula, in the management of infants and children with IgE-mediated CMA. However, there is also evidence regarding the role of some *Bifidobacteria* strains in the treatment of CMA.

In a randomized, double-blind, placebo-controlled trial, Jing et al. demonstrated that *B. bifidum* TMC3115 supplementation reduced allergic symptoms, improved anti-inflammatory responses, reduced serum IgE levels, increased serum IgG2 levels, and improved gut microbiota in infants with CMA [67].

Strisciuglio et al. investigated the effect of *Bifidobacteria* on the phenotype and activation status of peripheral basophils and lymphocytes in children with CMA. The treatment with *Bifidobacteria* resulted in a decrease in circulating naive and activated CD4+ T cells, as well as degranulating basophils. The authors concluded that *Bifidobacteria* may have beneficial effects on in modulating oral tolerance in children with CMA [68] (Table 3).

Table 3. Data from human studies related to cow's milk allergy treatment.

Study	Probiotic	Findings
Berni Canani et al. [61]	eHCF + LGG	eHCF + LGG induced a higher oral tolerance rate than eHCF alone or other special formulas in children with CMA
Guest et al. [62]	eHCF + LGG	eHCF + LGG induced a higher tolerance rate than eHCF alone or AAF in children with CMA
Berni Canani et al. [63]	eHCF + LGG	eHCF + LGG induced a higher oral tolerance rate and a lower incidence of atopic manifestations than eHCF alone in children with CMA with a follow-up of 36 months
Nocerino et al. [64]	eHCF + LGG	eHCF + LGG induced a higher oral tolerance rate and a lower incidence of atopic manifestations than other special formulas in children with CMA with a follow-up of 36 months
Guest et al. [65]	eHCF + LGG	eHCF + LGG is more effective than eHWF in both managing symptoms of CMA and preventing the occurrence of other atopic manifestations in children with CMA
Basturk et al. [66]	Milk-free diet + LGG	Milk-free diet + LGG improved symptoms such as bloody stools, diarrhea, restiveness, abdominal distension, mucous stools and vomiting in infants with CMA
Jing et al. [67]	Milk-free diet + *B. bifidum* TMC3115	Reduced allergic scores, improved anti-inflammatory responses, reduced serum IgE levels, increased serum IgG2 levels and improved gut microbiota in infants with CMA
Strisciuglio et al. [68]	Milk-free diet + Bifidobacteria	Decreased circulating naive and activated CD4+ T cells, as well as degranulating basophils, in infants with CMA

In 2019, Qamer et al. evaluated the use of probiotics for CMA in the first systematic review of randomized controlled trials. The authors concluded that there is limited, low-quality evidence indicating that probiotic supplementation may be associated with earlier acquisition of oral tolerance to cow's milk proteins in children with CMA. However, the authors specified that large, well-designed trials are necessary to confirm these findings [69].

In 2022, the Global Allergy and Asthma European Network (GA^{2}LEN) made no recommendation for or against any probiotics in managing food allergies, whether used as a supplement or added to infant formulas. GA^{2}LEN suggested addressing high-quality prospective trials on infants and young children with documented food allergies [70].

3.5. Probiotics in Food-Specific Immunotherapy

Oral immunotherapy is one of the possible allergen-specific therapeutic strategies proposed for the management of food allergies. The primary goal of oral immunotherapy is to induce desensitization to the allergen, but it is often burdened by allergic reactions. Probiotics have been evaluated in combination with oral immunotherapy to enhance their effectiveness or mitigate their adverse effects.

In 2015, Tang and colleagues published the first double-blind, placebo-controlled, randomized trial of a combined therapy with a probiotic, *L. rhamnosus* CGMCC 1.3724, and peanut oral immunotherapy in children with peanut allergies. They found that probiotic and peanut oral immunotherapy were highly effective, with seven children achieving possible sustained unresponsiveness if nine were treated [71].

The same study group later demonstrated that combined therapy with *L. rhamnosus* CGMCC 1.3724 and peanut oral immunotherapy provided long-lasting clinical benefit compared to placebo, with two-thirds of treated participants symptom-free after peanut ingestion 4 years after completing treatment [72].

Moreover, the authors described another study protocol of a multicentre, randomized, controlled trial evaluating the effectiveness of probiotic *L. rhamnosus* CGMCC 1.3724 and peanut oral immunotherapy in inducing desensitization or tolerance in children with peanut allergy compared with oral immunotherapy alone and with placebo [73].

In a multicenter, randomized, phase 2b trial, another probiotic, *L. rhamnosus* ATCC 53103, plus peanut oral immunotherapy, was compared to peanut oral immunotherapy plus placebo in children aged 1–10 years with a confirmed diagnosis of peanut allergy through oral challenge. The authors concluded that both treatments were able to induce desensitization, and the addition of the probiotic did not improve treatment efficacy but might offer a safety benefit [74].

Based on the observations in peanut allergy, Loke et al. planned the first double-blind, placebo-controlled, randomized trial to examine the effectiveness of probiotic and egg oral immunotherapy in inducing desensitization or sustained unresponsiveness in children with egg allergy compared to placebo [75].

4. Conclusions, Limitations and Future Perspectives

This narrative review outlines the current preclinical and clinical studies on the use of probiotics in the management of food allergies in infancy and childhood, exploring available data from animal models of cow's milk, egg, and shellfish allergies as well as data from human studies related to food allergy prevention and treatment. The available evidence is not conclusive, but it suggests that probiotics may have a role in preventing and treating food allergies in pediatrics. The lack of consistency is due to the wide range of probiotic strains used in studies based on different study protocols.

The most promising results available concern the use of specific probiotic strains as adjuvants in the management of children with IgE-mediated cow's milk allergies, as well as the use of specific probiotic strains in oral immunotherapy for children with IgE-mediated peanut allergies.

Nevertheless, current studies lay the groundwork for future well-designed studies to eventually identify specific probiotic strains that may be effective in the management of food allergies, the relative optimal dose to be administered, and the proper duration and timing of administration.

It is also critical that these studies do not confuse food allergy with allergic sensitization to food antigens, as this could be an additional confounding factor in interpretation. Indeed, the diagnosis of food allergies should be based on a positive oral food challenge.

Another future goal should be to clarify the molecular mechanisms by which probiotics interact with host cells and the gut microbiota and how these interactions affect the immune response to food antigens, including complex epigenetic mechanisms.

Author Contributions: Conceptualization, M.D.C.; writing—original draft preparation, M.D.C.; writing—review and editing, M.D.C., A.V., C.I., R.M., S.B., S.E. and G.B.; visualization, M.D.C., A.V., C.I., R.M., S.B., S.E. and G.B.; supervision, S.E. and G.B. All authors have read and agreed to the published version of the manuscript.

Funding: This research received no external funding.

Conflicts of Interest: The authors declare no conflict of interest.

References

1. Prescott, S.; Allen, K.J. Food allergy: Riding the second wave of the allergy epidemic. *Pediatr. Allergy Immunol.* **2011**, *22*, 155–160. [CrossRef] [PubMed]
2. Jeong, K.; Lee, S. The Natural course of IgE-mediated food allergy in children. *Clin. Exp. Pediatr.* **2023**, *66*, 504–511. [CrossRef] [PubMed]

3. Sicherer, S.H.; Warren, C.M.; Dant, C.; Gupta, R.S.; Nadeau, K.C. Food allergy from infancy through adulthood. *J. Allergy Clin. Immunol. Pract.* **2020**, *8*, 1854–1864. [CrossRef] [PubMed]
4. Wood, R.A.; Sicherer, S.H.; Vickery, B.P.; Jones, S.M.; Liu, A.H.; Fleischer, D.M.; Henning, A.K.; Mayer, L.; Burks, A.W.; Grishin, A.; et al. The natural history of milk allergy in an observational cohort. *J. Allergy Clin. Immunol.* **2013**, *131*, 805–812. [CrossRef] [PubMed]
5. Tsuge, M.; Ikeda, M.; Matsumoto, N.; Yorifuji, T.; Tsukahara, H. Current Insights into Atopic March. *Children* **2021**, *8*, 1067. [CrossRef]
6. Azad, M.B.; Konya, T.; Guttman, D.S.; Field, C.J.; Sears, M.R.; HayGlass, K.T.; Mandhane, P.J.; Turvey, S.E.; Subbarao, P.; Becker, A.B.; et al. Infant gut microbiota and food sensitization: Associations in the first year of life. *Clin. Exp. Allergy* **2015**, *45*, 632–643. [CrossRef]
7. Ling, Z.; Li, Z.; Liu, X.; Cheng, Y.; Luo, Y.; Tong, X.; Yuan, L.; Wang, Y.; Sun, J.; Li, L.; et al. Altered Fecal Microbiota Composition Associated with Food Allergy in Infants. *Appl. Environ. Microbiol.* **2014**, *80*, 2546–2554. [CrossRef] [PubMed]
8. Chen, C.; Chen, K.; Kong, M.; Chang, H.; Huang, J. Alterations in the gut microbiotas of children with food sensitization in early life. *Pediatr. Allergy Immunol.* **2016**, *27*, 254–262. [CrossRef] [PubMed]
9. Berni Canani, R.; Sangwan, N.; Stefka, A.; Nocerino, R.; Paparo, L.; Aitoro, R.; Calignano, A.; Khan, A.A.; Gilbert, J.A.; Nagler, C.R. Lactobacillus rhamnosus GG-supplemented formula expands butyrate-producing bacterial strains in food allergic infants. *ISME J.* **2016**, *10*, 742–750. [CrossRef]
10. Bunyavanich, S.; Shen, N.; Grishin, A.; Wood, R.; Burks, W.; Dawson, P.; Jones, S.M.; Leung, D.Y.; Sampson, H.; Sicherer, S.; et al. Early-life gut microbiome composition and milk allergy resolution. *J. Allergy Clin. Immunol.* **2016**, *138*, 1122–1130. [CrossRef]
11. Inoue, R.; Sawai, T.; Sawai, C.; Nakatani, M.; Romero-Pérez, G.A.; Ozeki, M.; Nonomura, K.; Tsukahara, T. A pre-liminary study of gut dysbiosis in children with food allergy. *Biosci. Biotechnol. Biochem.* **2017**, *81*, 2396–2399. [CrossRef]
12. Kourosh, A.; Luna, R.A.; Balderas, M.; Nance, C.; Anagnostou, A.; Devaraj, S.; Davis, C.M. Fecal microbiome signatures are different in food-allergic children compared to siblings and healthy children. *Pediatr. Allergy Immunol.* **2018**, *29*, 545–554. [CrossRef] [PubMed]
13. Fazlollahi, M.; Chun, Y.; Grishin, A.; Wood, R.A.; Burks, A.W.; Dawson, P.; Jones, S.M.; Leung, D.Y.M.; Sampson, H.A.; Sicherer, S.H.; et al. Early-life gut microbiome and egg allergy. *Allergy Eur. J. Allergy Clin. Immunol.* **2018**, *73*, 1515–1524. [CrossRef] [PubMed]
14. Dong, P.; Feng, J.-J.; Yan, D.-Y.; Lyu, Y.-J.; Xu, X. Early-life gut microbiome and cow's milk allergy—A prospective case-control 6-month follow-up study. *Saudi J. Biol. Sci.* **2018**, *25*, 875–880. [CrossRef] [PubMed]
15. Savage, J.H.; Lee-Sarwar, K.A.; Sordillo, J.; Bunyavanich, S.; Zhou, Y.; O'Connor, G.; Sandel, M.; Bacharier, L.B.; Zeiger, R.; Sodergren, E.; et al. A prospective microbiome-wide association study of food sensitization and food allergy in early childhood. *Allergy Eur. J. Allergy Clin. Immunol.* **2018**, *73*, 145–152. [CrossRef] [PubMed]
16. Bao, R.; Hesser, L.A.; He, Z.; Zhou, X.; Nadeau, K.C.; Nagler, C.R. Fecal microbiome and metabolome differ in healthy and food-allergic twins. *J. Clin. Investig.* **2021**, *131*, e141935. [CrossRef]
17. Demirci, M.; Tokman, H.; Uysal, H.; Demiryas, S.; Karakullukcu, A.; Saribas, S.; Cokugras, H.; Kocazeybek, B. Reduced Akkermansia muciniphila and Faecalibacterium prausnitzii levels in the gut microbiota of children with allergic asthma. *Allergol. Immunopathol.* **2019**, *47*, 365–371. [CrossRef] [PubMed]
18. Koga, Y.; Tokunaga, S.; Nagano, J.; Sato, F.; Konishi, K.; Tochio, T.; Murakami, Y.; Masumoto, N.; Tezuka, J.-I.; Sudo, N.; et al. Age-associated effect of kestose on Faecalibacterium prausnitzii and symptoms in the atopic dermatitis infants. *Pediatr. Res.* **2016**, *80*, 844–851. [CrossRef]
19. Candela, M.; Rampelli, S.; Turroni, S.; Severgnini, M.; Consolandi, C.; De Bellis, G.; Masetti, R.; Ricci, G.; Pession, A.; Brigidi, P. Unbalance of intestinal microbiota in atopic children. *BMC Microbiol.* **2012**, *12*, 95. [CrossRef] [PubMed]
20. Fieten, K.B.; Totté, J.E.E.; Levin, E.; Reyman, M.; Meijer, Y.; Knulst, A.; Schuren, F.; Pasmans, S.G.M.A. Fecal Micro-biome and Food Allergy in Pediatric Atopic Dermatitis: A Cross-Sectional Pilot Study. *Int. Arch. Allergy Immunol.* **2018**, *175*, 77–84. [CrossRef]
21. De Filippis, F.; Paparo, L.; Nocerino, R.; Della Gatta, G.; Carucci, L.; Russo, R.; Pasolli, E.; Ercolini, D.; Berni Canani, R. Specific gut microbiome signatures and the associated pro-inflamatory functions are linked to pediatric allergy and acquisition of immune tolerance. *Nat. Commun.* **2021**, *12*, 5958. [CrossRef] [PubMed]
22. De Filippis, F.; Pasolli, E.; Ercolini, D. Newly Explored Faecalibacterium Diversity Is Connected to Age, Lifestyle, Geography, and Disease. *Curr. Biol.* **2020**, *30*, 4932–4943. [CrossRef] [PubMed]
23. Song, H.; Yoo, Y.; Hwang, J.; Na, Y.-C.; Kim, H.S. Faecalibacterium prausnitzii subspecies–level dysbiosis in the human gut microbiome underlying atopic dermatitis. *J. Allergy Clin. Immunol.* **2016**, *137*, 852–860. [CrossRef] [PubMed]
24. Russell, S.L.; Gold, M.J.; Hartmann, M.; Willing, B.P.; Thorson, L.; Wlodarska, M.; Gill, N.; Blanchet, M.-R.; Mohn, W.W.; McNagny, K.M.; et al. Early life antibiotic-driven changes in microbiota enhance susceptibility to allergic asthma. *EMBO Rep.* **2012**, *13*, 440–447. [CrossRef] [PubMed]
25. Rodriguez, B.; Prioult, G.; Bibiloni, R.; Nicolis, I.; Mercenier, A.; Butel, M.-J.; Waligora-Dupriet, A.-J. Germ-free status and altered caecal subdominant microbiota are associated with a high susceptibility to cow's milk allergy in mice. *FEMS Microbiol. Ecol.* **2011**, *76*, 133–144. [CrossRef] [PubMed]

26. Noval Rivas, M.; Burton, O.T.; Wise, P.; Zhang, Y.-Q.; Hobson, S.A.; Lloret, M.G.; Chehoud, C.; Kuczynski, J.; DeSantis, T.; Warrington, J.; et al. A microbiota signature associated with experimental food allergy promotes allergic sensitization and anaphylaxis. *J. Allergy Clin. Immunol.* **2013**, *131*, 201–212. [CrossRef] [PubMed]
27. Zhao, W.; Ho, H.-E.; Bunyavanich, S. The gut microbiome in food allergy. *Ann. Allergy Asthma Immunol.* **2019**, *122*, 276–282. [CrossRef]
28. Di Costanzo, M.; De Paulis, N.; Biasucci, G. Butyrate: A Link between Early Life Nutrition and Gut Microbiome in the Development of Food Allergy. *Life* **2021**, *11*, 384. [CrossRef]
29. Cait, A.; Cardenas, E.; Dimitriu, P.A.; Amenyogbe, N.; Dai, D.; Cait, J.; Sbihi, H.; Stiemsma, L.; Subbarao, P.; Mandhane, P.J.; et al. Reduced genetic potential for butyrate fermentation in the gut microbiome of infants who develop allergic sensitization. *J. Allergy Clin. Immunol.* **2019**, *144*, 1638–1647.e3. [CrossRef]
30. Paparo, L.; Nocerino, R.; Ciaglia, E.; Di Scala, C.; De Caro, C.; Russo, R.; Trinchese, G.; Aitoro, R.; Amoroso, A.; Bruno, C.; et al. Butyrate as a bioactive human milk protective component against food allergy. *Allergy Eur. J. Allergy Clin. Immunol.* **2021**, *76*, 1398–1415. [CrossRef]
31. FAO/WHO. *Probiotics in Food: Health and Nutritional Properties and Guidelines for Evaluation*; Report; FAO Food and Nutrition Paper 85; Food and Agriculture Organization of the United Nations; World Health Organization: Rome, Italy, 2006.
32. Hill, C.; Guarner, F.; Reid, G.; Gibson, G.R.; Merenstein, D.J.; Pot, B.; Morelli, L.; Canani, R.B.; Flint, H.J.; Salminen, S.; et al. Expert consensus document: The International Scientific Association for Probiotics and Prebiotics consensus statement on the scope and appropriate use of the term probiotic. *Nat. Rev. Gastroenterol. Hepatol.* **2014**, *11*, 506–514. [CrossRef] [PubMed]
33. Bertazzoni, E.; Donelli, G.; Midtvedt, T.; Nicoli, J.; Sanz, Y. Probiotics and clinical effects: Is the number what counts? *J. Chemother.* **2013**, *25*, 193–212. [CrossRef] [PubMed]
34. Research and Markets—Market Research Reports—Welcome. Available online: https://www.researchandmarkets.com/?gad_source=1&gclid=EAIaIQobChMInpedpKTYggMV5JKDBx1npA8BEAAYASAAEgJXu_D_BwE (accessed on 22 November 2023).
35. McFarland, L.V.; Evans, C.T.; Goldstein, E.J.C. Strain-specificity and disease-specificity of probiotic efficacy: A systematic review and meta-analysis. *Front. Med.* **2018**, *5*, 124. [CrossRef]
36. Jiang, S.; Hou, Y.; Meng, L.; Pu, X.; Zhu, X.; Tuo, Y.; Qian, F.; Mu, G. Effect of *Lactiplantibacillus plantarum* HM-22 on immunoregulation and intestinal microbiota in α-lactalbumin-induced allergic mice. *Food Funct.* **2021**, *12*, 8887–8898. [CrossRef]
37. Ni, W.W.; Zhang, Q.M.; Zhang, X.; Li, Y.; Yu, S.S.; Wu, H.Y.; Chen, Z.; Li, A.L.; Du, P.; Li, C. Modulation effect of Lactobacillus acidophilus KLDS 1.0738 on gut microbiota and TLR4 expression in β-lactoglobulin-induced allergic mice model. *Allergol. Immunopathol.* **2020**, *48*, 149–157. [CrossRef]
38. Li, A.; Yang, J.; Zhang, C.; Chi, H.; Zhang, C.; Li, T.; Zhang, J.; Du, P. Lactobacillus acidophilus KLDS 1.0738 inhibits TLR4/NF-KB inflammatory pathway in β-lactoglobulin-induced macrophages via modulating MiR-146a. *J. Food Biochem.* **2021**, *45*, e13662. [CrossRef]
39. Fu, G.; Zhao, K.; Chen, H.; Wang, Y.; Nie, L.; Wei, H.; Wan, C. Effect of 3 Lactobacilli on immunoregulation and intestinal microbiota in a β-lactoglobulin–induced allergic mouse model. *J. Dairy Sci.* **2019**, *102*, 1943–1958. [CrossRef] [PubMed]
40. Santos, S.S.; Miranda, V.C.; Trindade, L.M.; Cardoso, V.N.; Reis, D.C.; Cassali, G.D.; Nicoli, J.R.; Cara, D.C.; Martins, F.S. Bifidobacterium longum subsp. longum 51A Attenuates Signs of Inflammation in a Murine Model of Food Allergy. *Probiotics Antimicrob. Proteins* **2021**, *15*, 63–73. [CrossRef]
41. Miranda, V.C.; Souza, R.O.; Quintanilha, M.F.; Gallotti, B.; Assis, H.C.; Faria, A.M.C.; Nicoli, J.R.; Cara, D.C.; Martins, F.S. A Next-Generation Bacteria (*Akkermansia muciniphila* BAA-835) Presents Probiotic Potential Against Ovalbumin-Induced Food Allergy in Mice. *Probiotics Antimicrob. Proteins*, 2023; 1–15, *Epub ahead of print*. [CrossRef]
42. Parrish, A.; Boudaud, M.; Grant, E.T.; Willieme, S.; Neumann, M.; Wolter, M.; Craig, S.Z.; De Sciscio, A.; Cosma, A.; Hunewald, O.; et al. *Akkermansia muciniphila* exacerbates food allergy in fibre-deprived mice. *Nat. Microbiol.* **2023**, *8*, 1863–1879. [CrossRef]
43. Duan, C.; Ma, L.; Yu, J.; Sun, Y.; Liu, L.; Ma, F.; Li, X.; Li, D. Oral Administration of *Lactobacillus plantarum* JC7 alleviates OVA-induced murine food allergy through immunoregulation and restoring disordered intestinal microbiota. *Eur. J. Nutr.* **2023**, *62*, 685–698. [CrossRef]
44. Miranda, V.C.; Santos, S.S.; Assis, H.C.; Faria, A.M.C.; Quintanilha, M.F.; Morão, R.P.; Nicoli, J.R.; Cara, D.C.; Martins, F.S. Effect of *Saccharomyces cerevisiae* UFMG A-905 in a murine model of food allergy. *Benef. Microbes* **2020**, *11*, 255–268. [CrossRef] [PubMed]
45. Fu, L.; Peng, J.; Zhao, S.; Zhang, Y.; Su, X.; Wang, Y. Lactic acid bacteria-specific induction of CD4+Foxp3+T cells ameliorates shrimp tropomyosin-induced allergic response in mice via suppression of mTOR signaling. *Sci. Rep.* **2017**, *7*, 1987. [CrossRef]
46. Fu, L.; Xie, M.; Wang, C.; Qian, Y.; Huang, J.; Sun, Z.; Zhang, H.; Wang, Y. Lactobacillus Ccsei Zhang alleviates shrimp tropomyosin-induced food allergy by switching antibody isotypes through the NF-KB-dependent immune tolerance. *Mol. Nutr. Food Res.* **2020**, *64*, e1900496. [CrossRef] [PubMed]
47. Fu, L.; Song, J.; Wang, C.; Fu, S.; Wang, Y. Bifidobacterium infantis Potentially Alleviates Shrimp Tropomyosin-Induced Allergy by Tolerogenic Dendritic Cell-Dependent Induction of Regulatory T Cells and Alterations in Gut Microbiota. *Front. Immunol.* **2017**, *8*, 1536. [CrossRef] [PubMed]
48. Boyle, R.J.; Ismail, I.H.; Kivivuori, S.; Licciardi, P.V.; Robins-Browne, R.M.; Mah, L.J.; Axelrad, C.; Moore, S.; Donath, S.; Carlin, J.B.; et al. Lactobacillus GG treatment during pregnancy for the prevention of eczema: A randomized controlled trial. *Allergy Eur. J. Allergy Clin. Immunol.* **2011**, *66*, 509–516. [CrossRef]

49. Dotterud, C.K.; Storrø, O.; Johnsen, R.; Øien, T. Probiotics in pregnant women to prevent allergic disease: A random-ized, double-blind trial. *Br. J. Dermatol.* **2010**, *163*, 616–623. [CrossRef]
50. Rautava, S.; Kainonen, E.; Salminen, S.; Isolauri, E. Maternal probiotic supplementation during pregnancy and breast-feeding reduces the risk of eczema in the infant. *J. Allergy Clin. Immunol.* **2012**, *130*, 1355–1360. [CrossRef] [PubMed]
51. Kuitunen, M.; Kukkonen, K.; Juntunen-Backman, K.; Korpela, R.; Poussa, T.; Tuure, T.; Haahtela, T.; Savilahti, E. Probiotics prevent IgE-associated allergy until age 5 years in cesarean-delivered children but not in the total cohort. *J. Allergy Clin. Immunol.* **2009**, *123*, 335–341. [CrossRef]
52. Kalliomäki, M.; Salminen, S.; Arvilommi, H.; Kero, P.; Koskinen, P.; Isolauri, E. Probiotics in primary prevention of atopic disease: A randomised placebo-controlled trial. *Lancet* **2001**, *357*, 1076–1079. [CrossRef]
53. Plummer, E.L.; Lozinsky, A.C.; Tobin, J.M.; Uebergang, J.B.; Axelrad, C.; Garland, S.M.; Jacobs, S.E.; Tang, M.L.K.; The ProPrems Study Group. Postnatal probiotics and allergic disease in very preterm infants: Sub-study to the *ProPrems* randomized trial. *Allergy Eur. J. Allergy Clin. Immunol.* **2020**, *75*, 127–136. [CrossRef] [PubMed]
54. Morisset, M.; Aubert-Jacquin, C.; Soulaines, P.; Moneret-Vautrin, D.-A.; Dupont, C. A non-hydrolyzed, fermented milk formula reduces digestive and respiratory events in infants at high risk of allergy. *Eur. J. Clin. Nutr.* **2010**, *65*, 175–183. [CrossRef]
55. Fiocchi, A.; Pawankar, R.; Cuello-Garcia, C.; Ahn, K.; Al-Hammadi, S.; Agarwal, A.; Beyer, K.; Burks, W.; Canonica, G.W.; Ebisawa, M.; et al. World Allergy Organization-McMaster University Guidelines for Allergic Disease Prevention (GLAD-P): Probiotics. *World Allergy Organ. J.* **2015**, *8*, 4. [CrossRef] [PubMed]
56. Zhang, G.Q.; Hu, H.J.; Liu, C.Y.; Zhang, Q.; Shakya, S.; Li, Z.Y. Probiotics for Prevention of Atopy and Food Hypersensitivity in Early Childhood: A PRISMA-Compliant Systematic Review and Meta-Analysis of Randomized Controlled Trials. *Medicine* **2016**, *95*, e2562. [CrossRef] [PubMed]
57. Flom, J.D.; Sicherer, S.H. Epidemiology of Cow's Milk Allergy. *Nutrients* **2019**, *11*, 1051. [CrossRef] [PubMed]
58. Vandenplas, Y.; Brough, H.A.; Fiocchi, A.; Miqdady, M.; Munasir, Z.; Salvatore, S.; Thapar, N.; Venter, C.; Vieira, M.C.; Meyer, R. Current Guidelines and Future Strategies for the Management of Cow's Milk Allergy. *J. Asthma Allergy* **2021**, *14*, 1243–1256. [CrossRef]
59. Fiocchi, A.; Schunemann, H.; Ansotegui, I.; Assa'ad, A.; Bahna, S.; Canani, R.B.; Bozzola, M.; Dahdah, L.; Dupont, C.; Ebisawa, M.; et al. The global impact of the DRACMA guidelines cow's milk allergy clinical practice. *World Allergy Organ. J.* **2018**, *11*, 2. [CrossRef]
60. Linhart, B.; Freidl, R.; Elisyutina, O.; Khaitov, M.; Karaulov, A.; Valenta, R. Molecular approaches for diagnosis, therapy and prevention of cow's milk allergy. *Nutrients* **2019**, *11*, 1492. [CrossRef]
61. Berni Canani, R.; Nocerino, R.; Terrin, G.; Frediani, T.; Lucarelli, S.; Cosenza, L.; Passariello, A.; Leone, L.; Granata, V.; Di Costanzo, M.; et al. Formula Selection for Management of Children with Cow's Milk Allergy Influences the Rate of Acquisition of Tolerance: A Prospective Multicenter Study. *J. Pediatr.* **2013**, *163*, 771–777.e1. [CrossRef]
62. Guest, J.F.; Kobayashi, R.H.; Mehta, V.; Neidich, G. Cost-effectiveness of using an extensively hydrolyzed casein formula containing *Lactobacillus rhamnosus* GG in managing infants with cow's milk allergy in the US. *Curr. Med. Res. Opin.* **2017**, *34*, 1539–1548. [CrossRef]
63. Berni Canani, R.; Di Costanzo, M.; Bedogni, G.; Amoroso, A.; Cosenza, L.; Di Scala, C.; Granata, V.; Nocerino, R. Extensively hydrolyzed casein formula containing *Lactobacillus rhamnosus* GG reduces the occurrence of other allergic manifestations in children with cow's milk allergy: 3-year randomized controlled trial. *J. Allergy Clin. Immunol.* **2017**, *139*, 1906–1913.e4. [CrossRef]
64. Nocerino, R.; Bedogni, G.; Carucci, L.; Cosenza, L.; Cozzolino, T.; Paparo, L.; Palazzo, S.; Riva, L.; Verduci, E.; Canani, R.B. The Impact of Formula Choice for the Management of Pediatric Cow's Milk Allergy on the Occurrence of Other Allergic Manifestations: The Atopic March Cohort Study. *J. Pediatr.* **2021**, *232*, 183–191.e3. [CrossRef] [PubMed]
65. Guest, J.F.; Fuller, G.W. Effectiveness of using an extensively hydrolyzed casein formula supplemented with Lactobacillus rhamnosus GG compared with an extensively hydrolysed whey formula in managing cow's milk protein allergic infants. *J. Comp. Eff. Res.* **2019**, *8*, 1317–1326. [CrossRef] [PubMed]
66. Basturk, A.; Isik, I.; Atalay, A.; Yılmaz, A. Investigation of the Efficacy of Lactobacillus rhamnosus GG in Infants with Cow's Milk Protein Allergy: A Randomised Double-Blind Placebo-Controlled Trial. *Probiotics Antimicrob. Proteins* **2019**, *12*, 138–143. [CrossRef] [PubMed]
67. Jing, W.; Liu, Q.; Wang, W. *Bifidobacterium bifidum*TMC3115 ameliorates milk protein allergy in by affecting gut microbiota: A randomized double-blind control trial. *J. Food Biochem.* **2020**, *44*, e13489, Erratum in *J. Food Biochem.* **2021**, *45*, e13591. [CrossRef] [PubMed]
68. Strisciuglio, C.; Vitale, A.; Perna, F.; Garziano, F.; Dolce, P.; Vitale, S.; Micillo, T.; Oglio, F.; del Giudice, M.M.; Matarese, G.; et al. Bifidobacteria modulate immune response in pediatric patients with cow's milk protein allergy. *Pediatr. Res.* **2023**, *94*, 1111–1118. [CrossRef]
69. Qamer, S.; Deshmukh, M.; Patole, S. Probiotics for cow's milk protein allergy: A systematic review of randomized controlled trials. *Eur. J. Pediatr.* **2019**, *178*, 1139–1149. [CrossRef] [PubMed]
70. Muraro, A.; de Silva, D.; Halken, S.; Worm, M.; Khaleva, E.; Arasi, S.; Dunn-Galvin, A.; Nwaru, B.I.; De Jong, N.W.; Del Río, P.R.; et al. Managing food allergy: GA2LEN guideline 2022. *World Allergy Organ. J.* **2022**, *15*, 100687. [CrossRef]
71. Tang, M.L.; Ponsonby, A.L.; Orsini, F.; Tey, D.; Robinson, M.; Su, E.L.; Licciardi, P.; Burks, W.; Donath, S. Administration of a probiotic with peanut oral immunotherapy: A randomized trial. *J. Allergy Clin. Immunol.* **2015**, *135*, 737–744.e8. [CrossRef]

72. Hsiao, K.C.; Ponsonby, A.L.; Axelrad, C.; Pitkin, S.; Tang, M.L.K.; PPOIT Study Team. Long-term clinical and immunological effects of probiotic and peanut oral immunotherapy after treatment cessation: 4-year follow-up of a randomised, double-blind, placebo-controlled trial. *Lancet Child Adolesc. Health* **2017**, *1*, 97–105; Erratum in *Lancet Child Adolesc. Health* **2017**, *1*, e2. [CrossRef]
73. Lozinsky, A.C.; Loke, P.; Orsini, F.; O'sullivan, M.; Prescott, S.L.; Gold, M.S.; Quinn, P.; Dunn Galvin, A.; Tang, M.L. Study protocol of a multicentre, randomised, controlled trial evaluating the effectiveness of probiotic and peanut oral immunotherapy (PPOIT) in inducing desensitisation or tolerance in children with peanut allergy compared with oral immunotherapy (OIT) alone and with placebo (the PPOIT-003 study). *BMJ Open* **2020**, *10*, e035871. [CrossRef]
74. Loke, P.; Orsini, F.; Lozinsky, A.C.; Gold, M.; O'Sullivan, M.D.; Quinn, P.; Lloyd, M.; Ashley, S.E.; Pitkin, S.; Axelrad, C.; et al. Probiotic peanut oral immunotherapy versus oral immunotherapy and placebo in children with peanut allergy in Australia (PPOIT-003): A multicentre, randomised, phase 2b trial. *Lancet Child Adolesc. Health* **2022**, *6*, 171–184. [CrossRef] [PubMed]
75. Loke, P.; Lozinsky, A.C.; Orsini, F.; Wong, L.S.-Y.; Leung, A.S.-Y.; Tham, E.H.; Lopata, A.L.; Shek, L.P.-C.; Tang, M.L. Study protocol of a phase 2, dual-centre, randomised, controlled trial evaluating the effectiveness of probiotic and egg oral immunotherapy at inducing desensitisation or sustained unresponsiveness (remission) in participants with egg allergy compared with placebo (Probiotic Egg Allergen Oral Immunotherapy for Treatment of Egg Allergy: PEAT study). *BMJ Open* **2021**, *11*, e044331. [CrossRef] [PubMed]

MDPI AG
Grosspeteranlage 5
4052 Basel
Switzerland
Tel.: +41 61 683 77 34

Nutrients Editorial Office
E-mail: nutrients@mdpi.com
www.mdpi.com/journal/nutrients

www.ingramcontent.com/pod-product-compliance
Lightning Source LLC
Chambersburg PA
CBHW061838210926
51850CB00006B/916
* 9 7 8 3 7 2 5 8 6 0 9 7 5 *